changing the way the world learns

To get extra value from this book for no additional cost, go to:

http://www.thomson.com/wadsworth.html

thomson.com is the World Wide Web site for Wadsworth/ITP and is your direct source to dozens of on-line resources. *thomson.com* helps you find out about supplements, experiment with demonstration software, search for a job, and send e-mail to many of our authors. You can even preview new publications and exciting new technologies.

thomson.com: *It's where you'll find us in the future.*

Constructions of Deviance

Social Power, Context, and Interaction

Second Edition

PATRICIA A. ADLER
University of Colorado

PETER ADLER
University of Denver

Wadsworth Publishing Company

I⊤P® An International Thomson Publishing Company

Belmont, CA • Albany, NY • Bonn • Boston • Cincinnati • Detroit • Johannesburg
London • Madrid • Melbourne • Mexico City • New York • Paris
San Francisco • Singapore • Tokyo • Toronto • Washington

Sociology Editor: Eve Howard
Editorial Assistant: Deirdre McGill
Marketing Manager: Mike Dew
Project Editor: Jennie Redwitz
Print Buyer: Karen Hunt
Permissions Editor: Jeanne Bosschart
Copy Editor: Donald Pharr
Cover Design: Malcolm Tarlofsky
Compositor: Thompson Type
Printer: Malloy Lithographing, Inc.

Printed in the United States of America
  2 3 4 5 6 7 8 9 10

For more information, contact Wadsworth Publishing Company, 10 Davis Drive,
Belmont, CA 94002, or electronically at http://www.thomson.com/wadsworth.html

International Thomson Publishing Europe
Berkshire House 168-173
High Holborn
London, WC1V 7AA, England

International Thomson Editores
Campos Eliseos 385, Piso 7
Col. Polanco
11560 México D.F. México

Thomas Nelson Australia
102 Dodds Street
South Melbourne 3205
Victoria, Australia

International Thomson Publishing Asia
221 Henderson Road
#05-10 Henderson Building
Singapore 0315

Nelson Canada
1120 Birchmount Road
Scarborough, Ontario
Canada M1K 5G4

International Thomson Publishing Japan
Hirakawacho Kyowa Building, 3F
2-2-1 Hirakawacho
Chiyoda-ku, Tokyo 102, Japan

International Thomson Publishing GmbH
Königswinterer Strasse 418
53227 Bonn, Germany

International Thomson Publishing
Southern Africa
Building 18, Constantia Park
240 Old Pretoria Road
Halfway House, 1685 South Africa

Library of Congress Cataloging-in-Publication Data

Constructions of deviance: social power, context, and interaction /
    [edited by] Patricia A. Adler, Peter Adler.—2nd ed.
          p.   cm.
    Includes bibliographical references.
    ISBN 0-534-50454-X (pbk.)
    1. Deviant behavior. 2. Social interaction.   I. Adler, Patricia A.
II. Adler, Peter.
HM291.C642   1996   96-15865
302.59 42—dc20

To Diane and Dana
Who remind us that the zest of life lies near the margins

and

To Chuck
Who brings out the deviance in everyone

Contents

Preface

Recently, British sociologist Colin Sumner announced the death of the sociology of deviance. In a scathing critique of the contemporary state of affairs, he wrote that "the terrain now resembles the [battle of] Somme of 1918. It is barren, fruitless, full of empty trenches and craters, littered with unexploded mines and eerily silent. No one fights for hegemony over a dangerous graveyard. It is now time to drop arms and show respect for the dead" (Sumner 1994: ix). Although his metaphor carries vivid imagery, we beg to differ with his pronouncement. True, the sociology of deviance had its efflorescence in the early 1970s, with the full burgeoning of labeling theory and its proponents. Largely as an outgrowth of the "second Chicago School," authors from the postwar period "had a remarkable impact on the study of crime and deviant behavior" (Galliher 1995: 168). Ranging in scope, theoretical persuasion, and methodological approach, the contributions of Erving Goffman, Howard Becker, James Short, Lloyd Ohlin, Albert Reiss, Kai Erikson, Joseph Gusfield, and Fred Davis, to name just a few, all made a major impact on the sociology of deviant behavior. However, the contemporary scene is not devoid of its own controversies, debates, dilemmas, and empirical donations. New theories are being developed or honed (the movements toward constructionism and feminism serve as just two examples), and new research is being produced. As the shifting sands of morality in American culture continue to transform our cultural milieu, deviance and its changing definitions are at the fore. As Gusfield (1967) showed us nearly thirty years ago, with these social changes, activities that were considered nondeviant a generation ago have taken on deviant characteristics (the near prohibition on cigarettes in public settings is a case in point), while activities that were unacceptable are now widely acknowledged as commonplace, albeit still stigmatized (the fad of tattooing and piercing, especially among teenagers, perhaps being the most remarkable). With this type of backdrop, it would be impossible to think of the field of deviant behavior as dead. It is our hope that, in this second edition of *Constructions of Deviance,* we have been able to highlight some of the recent changes in the field. We want to show students the liveliness of the debates, the graphic images that sociologists have painted, and the wide array of activities that fall under the aegis of deviance.

NEW TO THIS EDITION

You will find twenty new readings in this edition. In selecting new empirical articles, we have tried to address concerns that are of interest to collegians in the 1990s. Thus, the book now includes more work on rape, cigarette smoking, domestic violence, homophobia, delinquency, blood donors, obesity, women athletes, and sexual asphyxia, as well as the topics that worked well in the first edition. While this second edition is a showcase for the latest studies in the 1990s, in some ways it also represents a "back to the future." At the recommendation of some of our reviewers, we have included more of the classic statements in the field. In so doing, we have tried to stay as close as possible to our original intent: to provide a text that builds on our own intellectual backgrounds in symbolic interactionism and ethnographic research. Toward that end, we now include some of the earlier theoretical statements on deviance so that students can see the roots from which these empirical studies derive. We have also added new, original text to every section of the book. We have dramatically added theoretical material to the General Introduction at the insistence of reviewers: now, rather than a brief snippet on the history of deviance theory, we have a full-blown chapter that includes a chronology of deviance in sociology. In addition, each section has a much longer introduction, and a better preview of the articles so that readers know why these articles were selected.

Other changes we have made include subdividing the book to make the text and the articles more accessible. We broke up the longest part, "Organizing Deviants and Deviance" into three new parts: "Relations Among Deviants,"

"Deviant Acts," and "Deviant Careers." There is a general Part Introduction at the beginning of each of these new parts. Our desire is to present a text that fully represents the constructionist perspective, in its divergences, commonalities, and parallels to other theories. We still strongly favor the ethnographic method to illustrate these points because we believe it best suited to penetrate the inner worlds of deviants. As we have shown elsewhere (Adler and Adler 1995), ethnography is still dominated by deviance research.

Thus, instead of the obituary Sumner (1994) would deliver, we prefer to offer this collection as a testimonial to the continuing vibrancy of deviance research. This is both a time to sit back and admire what we have created as well as to take the opportunity to undertake new ventures, scale greater heights, and, in our maturity, provide greater wisdom. Our sense is that the battle is still being waged, with increasing ferocity and interest. This book has many competitors in the marketplace, and we welcome these too, as students continue to be interested in the compelling arena of deviant behavior.

ACKNOWLEDGMENTS

In the scant three years that have passed since we put together the first edition of this book, many people have provided critical feedback that has helped us in fashioning this second edition. With twenty new selections, the book has been almost completely remodeled; half of the papers here are fresh examples. It is not that we thought the first edition off base, but we have evaluated our first foray and reacted to our readers' wishes to make the book increasingly more relevant, intellectually sound, and interesting to read. First and foremost are the many students who have provided the bulk of feedback. Especially in Patti's class, "Deviance in U.S. Society" at the University of Colorado, where more than 400 students cram in each semester, we have been able to get a sense of the wide swath of opinions that are generated by contemporary collegians. These students remind us of the diversity of sentiments, moral and immoral, normative and deviant, radical and conservative, that exist. Second, our friends in the discipline continue to suggest studies, supply encouragement, and lend support for our endeavors. Whether during a quick conversation in the hallway of a convention hotel, an e-mail message, a lengthy letter, or a harangue over the telephone, they remind us that we should keep the edge and continue to search for the latest examples to hold their students interested. Colleagues from around the country responded to our survey to tell us how we could improve the book. Some of these are Theresa Chandler, Thad Coreno, Phil Davis, Lauren Dundes, David Friedrichs, Ruth Horowitz, Eric Ling,

Robert Lynxwiler, Michael Olson, Joel Powell, R. Penn Reeve, Carol Rambo Ronai, and Lise Vogel. We were particularly fortunate to get a set of constructive comments from the reviewers who Wadsworth commissioned as well. These include Tom Cook, Wayne State College; William M. Hall, Alfred University; John D. Hewlitt, Northern Arizona University; Ruth Horowitz, University of Delaware; Phyllis G. Kitzerow, Westminster College; and Michael R. Nusbaumer, Indiana University—Purdue University at Fort Wayne. The stalwart staff at Wadsworth has provided unending succor during the process of revision. We are fortunate to be working with such diligent professionals as Jeanne Bosschart, Eve Howard, Deirdre McGill, Andrew Ogus, Jennie Redwitz, and Susan Shook. And while you can't always tell a book by its cover, we would like to take this opportunity to thank Malcolm Tarlofsky, artist of the cover of the first edition, who added to the book's success by producing, in our opinion, the finest cover in the field and has designed another outstanding cover for this edition as well. One of the pleasures of editing this book has been sharing it with our friends. As we respectfully dedicated the first edition to Diane and Dana, to that list we now add our dear confederate, Chuck Gallmeier, who happily has turned his life around. Finally, our children, now teenagers, remind us that deviance is often a youthful endeavor and that we need to be respectful of their desires to test the boundaries. To all of the readers of the first edition, thanks for the support; to the new readers of this second edition, welcome to the journey.

About the Editors

Patricia A. Adler (Ph.D., University of California, San Diego) is Associate Professor of Sociology at the University of Colorado. She has written and taught in the areas of deviance, drugs in society, social theory, and the sociology of children. A second edition of her book *Wheeling and Dealing* (Columbia University Press), a study of upper-level drug traffickers, was released in 1993.

Peter Adler (Ph.D., University of California, San Diego) is Professor of Sociology at the University of Denver, where he served as Chair from 1987 to 1993. His research interests include social psychology, qualitative methods, and the sociology of sport and leisure.

Together the Adlers edited the *Journal of Contemporary Ethnography* and were the founding editors of *Sociological Studies of Child Development*. Their most recent book, *Backboards and Blackboards,* based on a five-year participant-observation study of college athletes, was published by Columbia University Press in 1991. Currently, they are studying the culture of elementary school-children. Their book based on this research, *Peer Power,* is forthcoming from Rutgers University Press.

About the Contributors

Elijah Anderson is currently the Charles and William L. Day Professor of the Social Sciences and Professor of Sociology at the University of Pennsylvania. An expert on the sociology of black America, he is the author of *A Place on the Corner* and *Streetwise,* which was honored with the Robert E. Park Award of the American Sociological Association. He is interested in the social psychology of organizations, field methods of social research, social interaction, and social organization.

Tom Barker is Dean of the College of Criminal Justice at Jacksonville State University, Alabama.

Deborah R. Baskin is Professor and Chairperson of the Department of Criminal Justice at California State University–Los Angeles. She is the author of numerous articles on women's involvement in violent street crime and drug dealing. She has also published in the area of forensic mental health and community mediation. She is coauthor with Ira Sommers on two forthcoming books related to varying aspects of the community context of women's involvement in street life.

Howard S. Becker is Professor of Sociology at the University of Washington. He is the author and editor of numerous books, including *Outsiders, The Other Side, Art Worlds, Writing for Social Scientists, Doing Things Together,* and *Symbolic Interactionism and Cultural Studies.*

Elaine Blinde is an Associate Professor of Physical Education and Sociology at Southern Illinois University at Carbondale. Her research involves the sociological analysis of sport, with a particular focus on gender issues. Recent work and publications relate to various dimensions of the sport experience of women athletes, including empowerment, exploitation, homophobia, and sport retirement. Current research explores the sport experience of individuals with disabilities.

David Carter is a Professor of Criminal Justice at Michigan State University who has conducted extensive research on police behavior and misconduct. He is the author of numerous articles and books, including his most recent, *The Police and the Community,* 4th edition (Macmillan/Prentice-Hall). He currently serves as Director of the National Center for Community Policing at MSU as well as Director of the Overseas Study Program in England.

William J. Chambliss is Professor of Sociology at George Washington University. He is the author of *Law, Order, and Power* with Robert Seidman, *Making Law* with Marjorie Zatz, and *On the Take: From Petty Criminals to Presidents.* He is the past president of the American Society of Criminology (1989–1990) and of the Society for the Study of Social Problems (1992–1993). He has received the American Sociological Association Lifetime Achievement Award and the Academy of Criminal Justice Sciences Outstanding Achievement Award.

Douglas Degher is Professor of Sociology at Northern Arizona University. His primary interests are in deviance theory and the sociology of sport. His most recent published work involves research on status assumption and identity change.

Emile Durkheim (1858–1917) was a French sociologist who is generally considered to be the father of sociology. His major works are *Suicide, The Rules of the Sociological Method, The Division of Labor in Society,* and *The Elementary Forms of the Religious Life.*

Carolyn Ellis received her Ph.D. from SUNY at Stony Brook and is currently Professor of Communication and Sociology and Director of the Institute for Interpretive Human Studies at the University of South Florida. Her most recent book is *Final Negotiations: A Story of Love, Loss, and Chronic Illness* (Temple University Press).

Jeffrey Fagan is a Professor in the School of Public Health at Columbia University. He has published extensively on street gangs, drugs, juvenile delinquency and juvenile justice, and domestic violence. His current research examines the situational contexts and processual dynamics of violent events involving adolescents.

Kathryn J. Fox received her Ph.D. from the University of California, Berkeley, and is currently an Assistant Professor in the Department of Sociology at the University of Vermont. Her areas of specialization include deviant behavior, qualitative methods, social problems theory, and the sociology of organizations.

Lisa Frohmann received her Ph.D. in sociology from UCLA and is currently an Assistant Professor of Criminology at the University of Illinois, Chicago. She coauthored *The Guide to Writing Sociology Papers* with the Sociology Writing Group at UCLA and is now working on a book about prosecutorial decision making in sexual assault cases.

Naomi Gerstel, Professor of Sociology at the University of Massachusetts, Amherst, wrote, with Harriet Gross, *Commuter Marriage* and *Families and Work,* and has contributed articles to *Social Forces, Gender & Society, Social Problems,* and *Journal of Marriage and the Family.*

Neil Gilbert is Chernin Professor of Social Welfare at the University of California, Berkeley. His most recent book is *Restoring Social Equity* (Yale University Press).

Roy Godson is Associate Professor in the Department of Government at Georgetown University and director of the National Strategy Information Center. He has served as a consultant to the president's Foreign Intelligence Advisory Board, the National Security Council, and related agencies of the U.S. government. He has written, coauthored, or edited over sixteen books on intelligence and national security.

Erving Goffman was the Benjamin Franklin Professor of Sociology and Anthropology at the University of Pennsylvania when he died in 1983.

Nancy J. Herman earned her Ph.D. from McMaster University and is Professor of Sociology at Central Michigan University. She has published a number of articles on the sociology of mental illness, ex-psychiatric patients, and symbolic interactionism. She is the editor of *Deviance: A Symbolic Interactionist Approach* and *Symbolic Interaction: An Introduction to Social Psychology.*

Columbus B. Hopper is Professor of Sociology at the University of Mississippi. Professor Hopper's writing has been in the areas of prison, the role of pharmacists in hospitals, outlaw motorcycle gangs, and popular culture. He was president of the Alabama-Mississippi Sociological Association.

Gerald Hughes is a Professor of Sociology at Northern Arizona University. His primary interests are deviance and human services program planning and evaluation.

Robert A. Hummer received his Ph.D. from Florida State University and is currently Assistant Professor of Sociology at Louisiana State University and senior research scientist at the Louisiana Population Data Center. His research focuses on adult and infant mortality.

Laud Humphreys was a Professor of Sociology at Pomona College, California.

David A. Karp received his Ph.D. from New York University in 1971. He is currently Professor and Departmental Chairperson of Sociology at Boston College. Professor Karp has coauthored several books in the areas of the sociology of everyday life, aging, and urban social psychology. His most recent book is *Speaking of Sadness: Depression, Disconnection, and the Meanings of Illness.* Along with research on family dynamics during the year that high school students apply to college, Professor Karp has become increasingly interested in the social dynamics associated with gaining or losing respect.

Robert J. Kelly is Broeklundian Professor of Social Science and Professor of Criminology and Sociology in the Graduate School, City University of New York, Brooklyn College.

Martin J. Kretzmann is a Ph.D. candidate in the Department of Sociology at the University of Denver.

Shearon A. Lowery is affiliated with Florida International University, Miami.

Betsy Lucal received her Ph.D. from the Department of Sociology at Kent State University in 1996. Her dissertation examines college students' experiences of the interrelatedness of race, gender, and class. Other interests include approaches to teaching about inequalities, social theory, and family violence.

Gerald E. Markle received his Ph.D. from Florida State University. Currently, he is Professor of Sociology at Western Michigan University. His most recent book is *Meditations of a Holocaust Traveler* (State University of New York Press, 1995).

Joseph Marolla is currently Chairperson in the Department of Sociology and Anthropology at Virginia Commonwealth University. He is now doing research in the sociology of sport but also continues his interest in the social construction of deviance.

Patricia Yancey Martin is Professor of Sociology and Daisy Parker Flory Alumni Professor at Florida State University. Her scholarly interests are gender and organizations. She co-edited a book in 1995 (with Myra Marx Ferree, University of Connecticut), *Feminist Organizations: Harvest of the New Women's Movement,* and published a paper in 1994 on legal organizations' treatment of rape victims ("Accounting for the 'Second Assault': Legal Organizations' Framing of Rape Victims," in *Law & Social Inquiry,* with Marlene Powell). She is writing two books, one on rape processing in the community and one on gender and organizations (with David Collinson).

Donald L. McCabe is Associate Professor of Management at Rutgers University in Newark, New Jersey. He received his Ph.D. in Management from New York University in 1988 after a 22-year career in industry. His research interests currently center on business ethics, with particular emphasis on the moral education of future business leaders.

Penelope A. McLorg is a Ph.D. candidate in Anthropology at Southern Illinois University at Carbondale. She is specializing in biological anthropology, with interests in aging, women's health, epidemiology, and human adaptability. Her published works involve infant feeding practices, gender socialization in eating disorders, and bone anatomy and measurement.

Jody Miller is an Assistant Professor of Criminology and Criminal Justice at the University of Missouri at St. Louis. Her most recent work is a study of female gang involvement.

Johnny Moore, a former president of Satan's Dead, a Gulf Coast outlaw motorcycle club, is now employed in the Food Services Division of the University of Mississippi. He has earned a degree from the University of Mississippi and continues research on motorcycle clubs.

James Myers earned his Ph.D. from the University of California, Berkeley, and is a Professor of Anthropology at California State University, Chico. He is the coauthor (with Arthur Lehmann) of *Magic, Witchcraft, and Religion: An Anthropological Approach to the Supernatural.*

William J. Olson most recently was deputy assistant secretary of state in the Bureau of International Narcotics Matters. He was also director and served as acting deputy assistant secretary of defense for low intensity conflict. His published works include books and articles on U.S. strategic interests in the Persian Gulf, guerilla warfare, counterinsurgency, and the war on drugs. He is currently a senior fellow at the National Strategy Information Center.

Harriet Pollack has a Ph.D. in constitutional law from Columbia University. For 25 years she taught Civil Liberties and Criminal Justice courses at John Jay College (CUNY). She has written numerous articles and five books, mostly coauthored with Alexander B. Smith. Her most recent book is *Criminal Justice: An Overview* (West, 1991).

Robert Prus is a sociologist at the University of Waterloo. He is the author of *Road Hustler* (with C. R. D. Sharper), *Hookers, Rounders, and Desk Clerks* (with Styllianoss Irini), *Making Sales, Pursuing Customers,* and most recently *Symbolic Interactionism and Ethnographic Research*. He is currently doing research on consumer behavior and the "courtship" of corporate investors by economic developers.

Richard Quinney is Professor of Sociology at Northern Illinois University. His books in criminology include *The Social Reality of Crime, Critique of Legal Order, Class, State, and Crime, Criminology as Peacemaking* (with Harold E. Pepinsky), and *Criminal Behavior Systems* (with Marshall B. Clinard and John Wildeman). His autobiographical reflections are contained in *Journey to a Far Place*.

Craig Reinarman is currently Professor of Sociology at the University of California, Santa Cruz. He is the author of *American States of Mind* and coauthor of *Cocaine Changes*.

Carol Rambo Ronai received her Ph.D. in Sociology from the University of Florida and is an Assistant Professor at the University of Memphis. She is currently interviewing survivors of childhood sex abuse and assembling an edited collection with two other editors on the topic of gender discrimination.

Don Sabo is a Professor of Social Science at D'Youville College in Buffalo, New York. His recent books include *Sex, Power & Violence in Sport: Rethinking Masculinity* (1995) and *Men's Health & Illness: Gender, Power, and the Body* (1996). His current research and writing focus on men in prison, psychosocial aspects of health, and men's violence.

Martín Sánchez Jankowski is Associate Professor of Sociology and Chair of the Chicano/Latino Policy Project of the Institute for the Study of Social Change at the University of California at Berkeley. He has written *City Bound: Urban Life and Political Attitudes Among Chicano Youth* and *Islands in the Street: Gangs and American Urban Society.*

Clinton R. Sanders is Professor of Sociology at the University of Connecticut. He is the author of *Customizing the Body: The Art and Culture of Tattooing,* the editor of *Marginal Conventions: Popular Culture, Mass Media, and Social Deviance,* and co-editor of *Cultural Criminology.* He has most recently done fieldwork in a veterinary clinic and a guide dog training program. His book *Regarding Animals* (with Arnold Arluke) is forthcoming from Temple University Press.

Diana Scully is Director of Women's Studies and Professor of Sociology at Virginia Commonwealth University in Richmond, Virginia. She received her Ph.D. from the University of Illinois, Chicago. Her books include *Men Who Control Women's Health: The Miseducation of Obstetrician-Gynecologists* and *Understanding Sexual Violence: A Study of Convicted Rapists.*

C. R. D. Sharper is a pseudonym.

Alexander B. Smith, LL.B., Ph.D., practiced law for 13 years. After serving in the Counter Intelligence Corps of the U.S. Army during World War II, he became a New York State Parole Officer, and a supervisor in the Probation Department of the Kings County Supreme Court. After earning his doctorate at NYU, he became a professor, chair, and dean at John Jay College (CUNY). He has served as a consultant to a number of U.S. governmental commissions. He has coauthored 70 articles and 14 books, most of them with Harriet Pollack. His most recent book, coauthored with Harriet Pollack, is *Criminal Justice: An Overview* (West, 1991).

Ira Sommers heads the Women's Research Project through the Research Foundation, City University of New York. Over the past several years he has obtained numerous grants to study women and various types of street crime, the community adjustment of forensic parolees, community-based substance abuse prevention programs, and drug market and violent crime connections. With Deborah Baskin, he is writing two books integrating themes discovered in earlier studies.

Diane E. Taub is Associate Professor of Sociology at Southern Illinois University at Carbondale. Her research primarily involves the sociology of deviance, with a focus on the experiences of women. Recent publications concern eating disorders among adolescent females and the stigma associated with individuals with AIDS.

Ronald J. Troyer is currently Professor of Sociology and Dean of the College of Arts and Sciences at Drake University. He has written *Cigarettes: The Battle Over Smoking,* is the senior editor and author of five chapters of *Social Control in the People's Republic of China,* and has written journal articles on social problems and social movements.

Charles V. Wetli is affiliated with the Medical Examiner Office, Dade County, Florida, and the University of Miami School of Medicine.

CONSTRUCTIONS OF DEVIANCE

General Introduction

The topic of deviance has held an enduring fascination for students of sociology, gripping their interest for several reasons. Some people hold career plans that include law or law enforcement and want to expand their base of practical knowledge. Others feel a special affinity for the subject of deviance based on personal experience or inclination. A third group is drawn to deviance merely because it is different, offering the promise of excitement or the exotic. The sociological study of deviance can fulfill all these goals, taking us deep into the criminal underworld, inward to the familiar, and outward to the fascinating and bizarre. In the following pages we peer into the deviant realm, looking at both deviants and those who define them as such. In so doing, we look at a range of deviant behaviors, discuss why people engage in these, and analyze how the behaviors are sociologically organized. We begin in Part I by defining deviance in an effort to lay down the parameters of its scope.

STUDYING DEVIANCE

Reasonable theories and social policies pertaining to deviance must be based on a firm foundation of accurate knowledge. Social scientists have an array of different methodologies at their disposal, including survey research, experimental design, historical methods, and field research. All of these methods have obvious strengths and weaknesses, and all of them have been used by sociologists in studying deviance. However, it is our belief that due to the often secretive nature of deviant acts, it is improbable that methods that objectify or distance the researcher from the people being researched will accurately portray deviant worlds. Thus, the works in this book are tied together by the belief that researchers must study deviance as it naturally occurs in the real world. The most appropriate methodology to do this, participant observation or ethnography, advocates that sociologists should get as close as possible to the people they are studying in order to understand their worlds (Adler and Adler 1987). Despite the problems that arise from the clandestine and hidden nature of deviant acts, sociologists have devised techniques to penetrate secluded deviant worlds. These methodological ploys often come complete with perils, so it is wise for the person who is considering the study of deviance to be aware of these issues. Part II discusses the methods used by most of the authors in the subsequent chapters of the book.

CONSTRUCTING DEVIANCE

In Part III we delve further into the origin and definitions of deviant behavior. Many people have traditionally considered defining deviance a simple task, suggesting that a widespread agreement exists about what is deviant and what is not. This view corresponds to the **objectivist** position on deviance, the perspective that something obvious within an act, belief, or condition makes it different from the norm in everybody's eyes. People have backed up this assumption by pointing to seemingly universal taboos, such as bans against murder, incest, and lying. These might clearly seem deviant, but other behaviors such as getting an abortion, gambling, and consuming illicit drugs may not generate universal agreement as simply. Delving into the former cases, we also find that there are conditions under which these acts would be considered nondeviant. To kill somebody for personal gain, for vengeance, for freedom, or through negligence might be deviant (and even criminal), but when these acts are committed by the state (for executions, war, or covert intelligence), in self-defense, or on one's own property (in states that have "make my day" laws), they are considered not only nondeviant, but heroic. Similarly, anthropologists have found cultures that condone certain forms of incest to quiet or soothe infants (Henry 1964; Weatherford 1986). Often, too, people differentiate between normal lies and "white" lies, those designed to spare the feelings of someone other than the speaker or differentiated by situational circumstances, which are deemed to be acceptable. These definitional variations suggest that deviance may be lodged in the eyes of the beholder rather than in the act itself, and that the social context in which an act occurs frames and gives meaning to how it is interpreted. We should thus regard definitions of deviance as relative to their situation and surroundings. This relativistic view of deviance constitutes the **subjectivist** position and arises from **interactionist theory** or the **social constructionist perspective** on deviance.

Assuming shared agreement, the objectivist position on deviance guides researchers toward studying the structural conditions fostering deviance, leading them to generate causal correlations that might predict and ultimately modify or control deviance through deterrence and social policy. However, the strength of the objectivist position's simplistic approach is also its weakness, for general agreement on societal norms and values cannot always be found. Interpretations may be complex and varied, especially in a society as broad and diverse as the United States, because people belong to many divergent subcultural groups. Yet even if people shared broad agreement, definitions of deviance would not necessarily be objective. In contrast, the constructionist perspective lodges the definition and hence focus of attention on the audience that frames and reacts to people's actions. Deviance is not located in the act but in the societal reaction. As Becker (1963: 9) argued in advancing **labeling theory,** one of the main components of an interactionist approach to deviance is that

> . . . *social groups create deviance by making the rules whose infraction constitutes deviance,* and by applying these rules to particular people and labeling them as outsiders. From this point of view, deviance is *not* a quality of the act the person commits, but rather a consequence of the application by others of rules and sanctions to an "offender." *(italics in original)*

In labeling theory we see the overlap between the interactionist and power/conflict approaches to deviance. Deviance is defined through the social meanings collectively applied to people's behavior or conditions, and is rooted in interaction. Those who have the power to make and apply rules over others control the normative order. The politically, socially, and economically dominant enforce their definitions over the downtrodden and powerless. Deviance is thus a representation of unequal power in society.

THEORIES OF DEVIANCE

Deviance has long held an intrigue for scholars. Given its pervasive nature in society, its enigmatic conditions, and its generic appeal, even the earliest sociologists attempted to explain how and why deviance occurs. Especially because of people's inclination to conformity, the pressing question for scholars has dealt with *why* people engage in norm-violating behavior. Explanations for deviant behavior are as divergent as the acts they explain, ranging from acts of delinquency to professional theft, acts of integrity to a search for kicks, acts of desperation to those of bravado and daring. Below, we outline some of the major attempts at understanding deviant behavior.

Structural Explanations

The dominant theory in sociology for the first half of the twentieth century, structural functionalism, also commanded the greatest amount of sway in explaining deviant behavior. Emile Durkheim, the French sociologist generally considered to be the founding father of sociology, advanced the theory that society is a moral phenomenon, that at its root, the morals (norms, values, and laws) that individuals are taught constrain their behavior. Durkheim believed that youngsters are taught the "rights" and "wrongs" of society early in life, with most people conforming to these expectations throughout adulthood. These moral beliefs, in large measure, determine how people behave, what they want, and who they are. Durkheim posited that societies with high degrees of social integration (bonding, community involvement) would increase the conformity of their members. However, and this is what concerned Durkheim, in the modern French society in which he was living, more and more people were becoming distanced from each other, people were partially losing their sense of belonging to their communities, and the norms and expectations of their groups were becoming less clearly defined. He believed that this condition, which he referred to as *anomie,* was producing a concomitant amount of social disintegration, leading to greater degrees of deviance. Thus, for Durkheim, while norms still existed on the societal level, the lack of social integration created a situation in which they were no longer becoming as significant a part of each individual.

Despite his concerns about the increasing rates of deviance that society would produce, Durkheim also subscribed to the idea that deviance is functional for society. Curiously, despite its obvious deleterious effects, Durkheim felt that deviance also produces some positive benefits as well. At a time when people are worrying about the moral breakdown and social disintegration of society, deviance serves to remind us of the moral boundaries in society. Each time a deviant act is committed and publicly announced, society is united in indignation against the perpetrator. This serves to bring people together, rather than to tear them apart. At the same time, the society is reminded about what is "right" and "wrong," and, for those who conform, greater social integration ensues. These ideas were perhaps best illustrated by Yale sociologist Kai Erikson, who, in his 1966 book *Wayward Puritans,* demonstrated the role of deviance in defining morality and bringing people together. Erikson examined Puritan patterns of isolating and treating offenders. He believed that deviance serves as a means to promote a *contrast* with the rest of the community, thus giving members of the larger society more strength in their moral convictions. Erikson's analysis focused on the transformation of the seventeenth-century Bay Colony, as a group of revolutionaries tried to establish a new community in New England. These deviants, the revolutionaries, played an important role in the transformation of norms and values—their behavior elicited societal reaction, which served to *clearly define* the new community's norms and values. In addition, punishing some people for norm violations reminded others of the rewards for conformity.

It was Robert Merton, a mid-twentieth-century sociologist from Columbia University, who actually extended Durkheim's ideas and built them into a specific structural theory of deviant behavior. In a wide-sweeping and influential article, "Social Structure and Anomie" (1938), he claimed that contradictions are implicit in a stratified system in which the culture dictates success goals for all citizens, while institutional access is limited to just the middle and upper strata. In other words, despite the American dream of rags to riches, some people, most often lower-class individuals, are systematically excluded from the competition. Instead of merely going through the motions while knowing that their legitimate path to success (measured in American society by financial wealth) is blocked, some members of the lower class retaliate by choosing a deviant alternative. Merton believed that these people have accepted society's goals (to be comfortable, to get rich), but they have insufficient access to the approved means of attaining these goals (deferred gratification, education, hard work). The problem lies in the social structure, where even if people follow the approved means, there are "roadblocks" prohibiting them from rising through the stratification system. Deviant behavior thus occurs when socially sanctioned means are not available for the realization of highly desirable goals. The only way to achieve these goals is to "detour" around them, to bypass the approved means in order to get at the approved goals. For example, for young men raised in urban ghettos with poor housing facilities, dilapidated schools, and inconsistent family lives, their road to "success" is more likely to be through dealing drugs, pimping, or robbing than it is through the normative route of school and hard work. According to Merton, then, *anomie* results from the lack of access to culturally prescribed goals and the lack of availability of legitimate means for attaining those goals. Deviance (or, more specifically, crime) is the obvious alternative.

Cloward and Ohlin, in *Delinquency and Opportunity* (1960), thought that Merton was correct in directing us toward the notion that the lower class has less opportunity for achieving success in a legitimate manner, but they thought that Merton wrongly assumed that the lower classes, when confronted with the problem of differential opportunity, could automatically choose deviance and crime. Cloward and Ohlin posited that all lower-class people have some lack of opportunity for legitimate pursuits, but they do not have the same opportunity for participating in illegitimate practices. What Cloward and Ohlin believed was that deviant behavior depends on people's access to illegitimate opportunities. They found that three types of deviant opportunities are present: (1) criminal: similar to the type Merton described, these opportunities arise from access to deviant subcultures, though not all lower-class youth enjoy these avenues; (2) conflict: these opportunities attract people who have a propensity for violence and fighting; and (3) retreatist: these opportunities attract people who are not inclined toward illegitimate means or violent actions, but who want to withdraw from society, such as drug users. Thus, Cloward and Ohlin extended Merton's theory by specifying the existence of differential illegitimate opportunities available to members of the lower class. All these structural theories place the blame for the incidence of deviance on the structures of society, rather than on individuals and their problems.

Cultural Explanations

While Merton's theory had enormous impact on sociologists' thinking about deviance, other authors arose who felt that his was not an all-encompassing description. These theorists believed that deviance was a collective act, driven by and carried out by groups of people. Thorsten Sellin, in "Culture Conflict and Crime" (1938), suggested that people can follow the norms and values of their group and still produce behavior that becomes defined as deviant. Different groups have different cultural codes, and these disparities become apparent when people from one culture cross over into the territory of another culture, or when the laws of one cultural group are extended to apply to another. Sellin was particularly thinking about the deviance of immigrant ethnic or racial groups in the United States, but his theory applies equally well to the large number or diverse subcultural groups in our country. A pluralistic nation that was once thought of as the world's "melting pot," we have become, instead, a nation of many different groups, each with its own distinct subculture. Overarching all these particularistic subcultural norms and values is the dominant American culture, with its norms, values, and laws. People who follow the different norms and values of their subculture may thus find their behavior in conflict with the cultural codes of the dominant culture, and hence deviant.

Building on this idea, Albert Cohen, in *Delinquent Boys* (1955), showed that working-class adolescent males make up a subculture with a different value system from the dominant American culture. These boys, Cohen asserted, have the greatest degree of difficulty in achieving success, since the establishment's standards are so different from their own. Exposed to middle-class aspirations and judgments that they cannot reasonably fulfill, they develop a conflict (or strain) that leads them to experience "status frustration." What occurs from this frustration is the formation of a subculture that allows them to achieve status based on nonutilitarian, malicious, and negativistic behavior. These boys, in reaction to society's unfairness towards them, substitute norms that reverse those of the larger society. Cohen claimed that delinquent boys turned the society's norms "upside down," rejecting middle-class standards and adopting values in direct opposition to those of the majority.

Walter Miller, writing just after Cohen, further delineated the importance of subcultural values for the development of deviant behavior. He believed that the values of the lower-class culture produce deviance because they are "naturally" in discord with middle-class values. Young people who conform to the lower-class culture in which they were born almost automatically become deviant. The culture of the lower class, by which members attain status in the eyes of their peers, is characterized by several "focal concerns": trouble, toughness, autonomy, street-smartness, and excitement. When individuals follow the norms of their culture, they become deviant according to the predominantly middle-class societal norms and values.

The lasting impact of subcultural theories has been to suggest that conflicting values may exist in society. When one part of society can impose its definitions on other parts, the dominant group has the ability to label minority groups' behavior as deviant. Thus, any act can be considered deviant if it is so defined. These theories are suited to illustrating the motivations of people from minority or disadvantaged subcultures that are not well-aligned with the dominant culture.

Interactionist Explanations

While these previous theories shed insight into some types of deviance, there are interactional forces that inevitably intervene between the causes they propose and the way deviant behavior takes shape. When people confront the problems, pressures, excitements, and allures of the world, they most often do so in conjunction with their peer group. It is within peer groups that people make decisions about what they will do and how they will do it. Their core feelings about themselves develop and become rooted in such groups. People's actions and reactions are thus guided by the collective perceptions, interpretations, and actions of their peer group. Edwin Sutherland (1934) recognized this point when he proposed his **differential association** theory of deviance, which suggests that deviant behavior is learned from people's intimate friends. The more people associate with others who hold favorable definitions of deviant acts, the more likely they are to develop these same attitudes. The more their friends engage in deviant behavior, the more likely they are to follow suit. Sutherland suggested that people learn a variety of elements critical to deviance from their associates: the norms and values of the deviant subculture; the rationalizations for legitimizing deviant behavior; the techniques necessary to commit the deviant acts; and the status system of the subculture, by which members evaluate themselves and others. Thus, people do not decide, at a fixed point in time, to become deviant, but they move toward these attitudes and behavior as they shift their circle of associates from more normative friends to more deviant friends. Matza (1964) noted that this movement into deviant subcultures occurs through a process of "drift," as people gradually leave their old crowd and become enmeshed in a circle of deviant associates. Quitting deviance is a similarly gradual and difficult process, requiring the abandonment of the group of deviant friends and reintegration in conventional circles before normative behavior becomes thoroughly ingrained. The key emphasis of the interactionist approach to deviance, then, is on the importance of human peer interaction in understanding the root cause of human behavior.

Peer groups and the social contexts within which they form operate not only in the interactional setting but within social organizations and institutions as well. Pressures can arise that shape collective attitudes, propelling people into deviance and providing legitimations that neutralize the perceived consequences of the deviance. In Part IV we examine the political and organizational contexts that surround people and push them into deviance for the sake of the group. We then look at the way organizations, especially the type of bureaucracies that process norm violators, treat their clients and the effects this has on them.

The Labeling Perspective

Many people dabble to greater or lesser degrees in forms of deviance. Studies of juvenile delinquency suggest that rates of youthful participation are extremely widespread, nearly universal. How many people can claim to have reached adulthood without experimenting in illicit drinking, drug use, stealing, or vandalism? Yet do all these people consider themselves deviants? Most do not. Many people retire from deviance as they mature, avoiding developing the deviant identity altogether. Others go on to engage in what Becker (1963) has called "secret deviance," conducting their acts of norm violation without ever seriously encountering the deviant label. Yet others, many of them no more experienced in the ways of deviance than the youthful delinquent or the secret deviant, become identified and identify themselves as deviants. What causes this difference? One critical difference, interactionist theory suggests, lies in who gets caught. Getting caught sets off a chain reaction of events that leads to profound social and self-conceptual consequences. Tannenbaum (1938) has described how individuals are publicly identified as norm violators and branded with that tag. They may go through official or unofficial social sanctioning in which people identify and treat them as deviant. Returning to labeling theory, Becker (1963: 9) has noted that "the deviant is one to whom that label has successfully been applied; deviant behavior is be-havior that people so label." Deviance exists at the macro societal level of social norms and definitions through the collective attitudes we assign to certain acts and conditions. But it also comes into being at the micro everyday life level when the deviant label is applied to someone. This process, along with the consequences for people who become labeled, is the focus of Part V.

The thrust of labeling theory is thus twofold, focusing on diverse levels and forces. Schur (1979: 160) summarized its complexity:

> The twin emphases in such an approach are on *definition* and *process* at all the levels that are involved in the production of deviant situations and outcomes. Thus, the perspective is concerned not only with what happens to specific individuals when they are branded with deviantness ("labeling," in the nar-row sense) but also with the wider domains and processes of social definitions and collective rule-making that frequently lie behind such concrete applica-tions of negative labels. *(italics in original)*

THE SOCIAL ORGANIZATION OF DEVIANCE

We conclude this volume, in Parts VI, VII, and VIII, with a discussion of how deviants and deviance are socially organized. Earlier sections of the book have concentrated on, first, macro and then micro levels of addressing deviance. Here, we take a mid-level focus by looking at how deviants organize their social rela-tionships, activities, and careers, in conjunction with others. We begin with the study of deviant associations, examining various types of relationships among members of deviant scenes. These range from loose subcultures to more tightly bound gangs and finally to highly committed cartels. We then consider the orga-nization of deviant acts. Whereas some forms of deviance can be committed alone, others are more complex, requiring multiple persons performing different roles. Some involve cooperation between the participants, with people mutually ex-changing illicit goods or services. Others are characterized by conflict, with some parties to the act taking advantage of others, often against their will. Finally, we look at the contours of deviant careers, beginning with people's entry into the world of deviant behavior and associates, continuing with the way they fashion their involvement in deviance, and concluding with their often problematic, oc-casionally inconclusive, retirements from the compelling world of deviance.

PART I

Defining Deviance

In order to study the topic of deviance, we must first clarify what we mean by the term. What behaviors or conditions fall into this category, and what is the relation between deviance and other categories, such as criminal? When we speak of deviance, we refer to violations of social norms. Norms are behavioral codes or prescriptions that guide people into actions and self-presentations conforming to social acceptability. Norms need not be agreed upon by every member of the group doing the defining, but a clear or vocal majority must agree.

One of the founding sociologists, William Sumner (1906), conceptualized norms into three categories: **folkways, mores,** and **laws.** He defined folkways as simple everyday norms based on custom, tradition, or etiquette. Violations of folkway norms do not generate serious outrage but might cause people to think of the violator as odd. Common folkway norms include standards of dress, demeanor, physical closeness to or distance from others, and eating behavior. People who come to class dressed in bathing suits, who never seem to be paying attention when they are spoken to, who sit or stand too close to others, or who eat with their hands instead of silverware would be violating a folkway norm. We would not arrest them, nor would we impugn their moral character, but we might think that there was something wrong with them.

Mores are norms based on broad societal morals whose infraction would generate more serious social condemnation. Interracial marriage, illegitimate childbearing, and drug addiction all constitute more violations. Upholding these norms is seen as critical to the fabric of society, for their violation threatens the social order. Interracial marriage threatens racial purity and the stratification hierarchy based on race; illegitimate childbearing threatens the institution of marriage and the transference of money, status, and family responsibility from one generation to the next; and drug addiction represents the triumph of hedonism over rationality, threatening the responsible behavior necessary to hold society together and accomplish its necessary tasks. People who violate mores may be considered both wicked and potentially harmful to society.

Laws are the strongest norms because they are supported by codified social sanctions. People who violate them are subject to arrest and punishments ranging from fines to imprisonment. Many laws are directed toward behavior that used to be folkway or, especially, more violations but became encoded into laws. Others are regarded as necessary for maintaining social order. Although violating a law will bring the stigma associated with arrest, it will not necessarily brand the violator as deviant.

This discussion returns us to the question about the relationship between **deviance** and **crime.** Are they identical terms, is one a subset of the other, or are they overlapping categories? To answer this question, we must consider one facet of it at a time. First, do some things fall into both categories, crime and deviance? The overlap between these two is extensive, with crimes of violence, crimes of harm, and theft of personal property considered both deviant and illegal. Second, are there types of deviance that are not crimes? Actually, much deviance is non-criminal, such as obesity, stuttering, physical handicaps, racial intermarriage, and unwed pregnancy. Deviance is not a subset of crime, then. Finally, is there crime that is nondeviant? Although much crime is considered deviant, and derives from various lesser deviant categories, some criminal violations do not violate norms or bring moral censure. Examples of this include some white-collar crimes commonly regarded as merely aggressive business practices, such as income tax evasion and some forms of civil disobedience, in which people break laws to protest them. Thus, crime is not a subset of deviance. Crime and deviance, then, are overlapping categories with independent dimensions.

People can be labeled deviant as the result of their **behavior, beliefs,** or **condition.** It is the behavioral category that is the most familiar, with people coming to be regarded as deviant for their actions. These behaviors may be intentional or inadvertent, and they include such activities as violating dress or speech conventions, kinky sexual behavior, or murder. People can also be branded deviant for alternative attitudes or belief systems. These beliefs can fall into the religious or political category, with people who hold radical or unusual views of the supernatural (cult members, satanists, fundamentalists) or hold extreme political attitudes (far leftists or rightists, terrorists) considered deviant. Mental illness also falls into the deviant belief category, for people with deviant attitudes are often considered mentally ill and people with emotional deficiencies are considered deviant. People cast into the deviant realm for their acts or attitudes have an **achieved deviant status:** they have earned the deviant label through something they have done.

People regarded as deviant because of their condition often have an **ascribed deviant status:** it is something they acquire at birth. This would include having a deviant socioeconomic status, such as being poor; a deviant racial status, such as being a person of color (in a dominantly Caucasian society); a congenital physical handicap; or a height deviance (too tall or too short). Here, there is nothing that such people have done to become deviant and nothing they can do to repair their deviant status. Moreover, there is nothing necessarily inherent in these statuses that makes them deviant: they become deviant through the result of a socially defining process that gives unequal weight to powerful and dominant groups in society.

With all of these categories, we can see the differentiation between the deviant and the conventional. Our first selection, "The Normal and the Pathological," by the eminent French sociologist Emile Durkheim, discusses the inevitability of deviance in all societies. In an ironic twist, Durkheim argues that deviance is normal rather than pathological, serving a positive function in society. To achieve the maximum benefit, however, a society needs a manageable amount of deviance. When the numbers of people declared deviant by current moral standards rises or falls too much, society alters its moral criteria to maintain the level of deviance in the optimal range. At different times it may "define deviancy down," as Moynihan (1993) has suggested in looking at the way the bar defining normality has been lowered by normalizing the high levels of violence, divorce, and deinstitutionalized mental patients inhabiting the streets. Or it may "define deviancy up," as Krauthammer (1993) has suggested, raising the bar defining normality so that behavior formerly considered innocent, such as the way that parents discipline their children, the sex that people may have on dates, and the prejudice that people feel towards members of other groups, has been redefined as child abuse, date rape, and thought crime. What deviance does for society is to define the moral boundaries for everyone. Violation of norms serves to remind the masses what is acceptable and what is not; in Durkheim's words, it enforces the "collective conscience" of the group. Perhaps it is difficult to imagine that behaviors that disgust, revile, or even nauseate you are not the acts of immoral, sick, or evil people, but are normal and even beneficial parts of all societies. In our second selection, "Deviance and the Response of Others," Howard Becker, considered the founder of the labeling approach, posits that deviance lies in the eye of the beholder. There is nothing inherently deviant in any particular act until some group, usually one with power over others, defines the act as deviant. Taking the onus off the individual, Becker believes that we must look at the *process* by which people are labeled deviant, and understand that deviance is a consequence of others' reactions. This approach forces us to look, then, at how people are defined as deviant, why some acts are labeled and others ignored, and the circumstances that surround the commission of the act. Thus, deviance exists only when it is created by society. Finally, our third selection, "Deviance as Crime, Sin, and Poor Taste," by Alexander Smith and Harriet Pollack, builds on some of the variations in deviance discussed above. Smith and Pollack, drawing on Sumner's three categories of norms (folkways, mores, and laws), categorize different types of norm violations. They recognize Becker's notion about the designation of deviant labels, but they also admit that some violations, such as violent assault, are nearly universally rejected. The lesson to be learned here is that we can acknowledge that some behavior is heinous, but once we get past that initial layer many of the other acts are ones that we can accept, if we can see them as artifacts of religious, moral, or cultural ideals. Serving to broaden the base that Becker lays out, Smith and Pollack offer us a way to admit that, although a small classification of acts are inherently deviant, there is a wide array of other acts that are defined as such only relative to their context.

1

The Normal and the Pathological

EMILE DURKHEIM

Crime is present not only in the majority of societies of one particular species but in all societies of all types. There is no society that is not confronted with the problem of criminality. Its form changes; the acts thus characterized are not the same everywhere; but, everywhere and always, there have been men who have behaved in such a way as to draw upon themselves penal repression. If, in proportion as societies pass from the lower to the higher types, the rate of criminality, i.e., the relation between the yearly number of crimes and the population, tended to decline, it might be believed that crime, while still normal, is tending to lose this character of normality. But we have no reason to believe that such a regression is substantiated. Many facts would seem rather to indicate a movement in the opposite direction. From the beginning of the [nineteenth] century, statistics enable us to follow the course of criminality. It has everywhere increased. In France the increase is nearly 300 percent. There is, then, no phenomenon that presents more indisputably all the symptoms of normality, since it appears closely connected with the conditions of all collective life. To make of crime a form of social morbidity would be to admit that morbidity is not something accidental, but, on the contrary, that in certain cases it grows out of the fundamental constitution of the living organism; it would result in wiping out all distinction between the physiological and the pathological. No doubt it is possible that crime itself will have abnormal forms, as, for example, when its rate is unusually high. This excess is, indeed, undoubtedly morbid in nature. What is normal, simply, is the existence of criminality, provided that it attains and does not exceed, for each social type, a certain level, which it is perhaps not impossible to fix in conformity with the preceding rules.[1]

Here we are, then, in the presence of a conclusion in appearance quite paradoxical. Let us make no mistake. To classify crime among the phenomena of normal sociology is not to say merely that it is an inevitable, although regrettable phenomenon, due to the incorrigible wickedness of men; it is to

---

From Emile Durkheim, *The Rules of Sociological Method,* translated by S. A. Solovay and J. H. Mueller. Edited by George E. G. Catlin. Copyright © 1938 by George E. G. Catlin; copyright renewed 1966 by Sarah A. Solovay, John H. Mueller, George E. G. Catlin. Reprinted by permission of The Free Press, a division of Simon & Schuster.

crime is part of a healthy society

affirm that it is a factor in public health, an integral part of all healthy societies. This result is, at first glance, surprising enough to have puzzled even ourselves for a long time. Once this first surprise has been overcome, however, it is not difficult to find reasons explaining this normality and at the same time confirming it.

In the first place crime is normal because a society exempt from it is utterly impossible. Crime, we have shown elsewhere, consists of an act that offends certain very strong collective sentiments. In a society in which criminal acts are no longer committed, the sentiments they offend would have to be found without exception in all individual consciousnesses, and they must be found to exist with the same degree as sentiments contrary to them. Assuming that this condition could actually be realized, crime would not thereby disappear; it would only change its form, for the very cause which would thus dry up the sources of criminality would immediately open up new ones.

a society who crime is impossible

Indeed, for the collective sentiments which are protected by the penal law of a people at a specified moment of its history to take possession of the public conscience or for them to acquire a stronger hold where they have an insufficient grip, they must acquire an intensity greater than that which they had hitherto had. The community as a whole must experience them more vividly, for it can acquire from no other source the greater force necessary to control these individuals who formerly were the most refractory. For murderers to disappear, the horror of bloodshed must become greater in those social strata from which murderers are recruited; but, first it must become greater throughout the entire society. Moreover, the very absence of crime would directly contribute to produce this horror; because any sentiment seems much more respectable when it is always and uniformly respected.

One easily overlooks the consideration that these strong states of the common consciousness cannot be thus reinforced without reinforcing at the same time the more feeble states, whose violation previously gave birth to mere infraction of convention—since the weaker ones are only the prolongation, the attenuated form, of the stronger. Thus robbery and simple bad taste injure the same single altruistic sentiment, the respect for that which is another's. However, this same sentiment is less grievously offended by bad taste than by robbery; and since, in addition, the average consciousness has not sufficient intensity to react keenly to the bad taste, it is treated with greater tolerance. That is why the person guilty of bad taste is merely blamed, whereas the thief is punished. But, if this sentiment grows stronger, to the point of silencing in all consciousnesses the inclination which disposes man to steal, he will become more sensitive to the offenses which, until then, touched him but lightly. He will react against them, then, with more energy; they will be the object of greater opprobrium, which will transform certain of them from the simple moral faults that they were and give them the quality of crimes. For example, improper contracts, or contracts improperly executed, which only incur public blame or civil damages, will become offenses in law.

Imagine a society of saints, a perfect cloister of exemplary individuals. Crimes, properly so called, will there be unknown; but faults which appear venial to the layman will create there the same scandal that the ordinary offense does in ordinary consciousness. If, then, this society has the power to judge and punish, it will define these acts as criminal and will treat them as such. For the same reason, the perfect and upright man judges his smallest failings with a severity that the majority reserve for acts more truly in the nature of an offense. Formerly, acts of violence against persons were more frequent than they are today, because respect for individual dignity was less strong. As this has increased, these crimes have become more rare; and also, many acts violating this sentiment have been introduced into the penal law which were not included there in primitive times.[2]

In order to exhaust all the hypotheses logically possible, it will perhaps be asked why this unanimity does not extend to all collective sentiments without exception. Why should not even the most feeble sentiment gather enough energy to prevent all dissent? The moral consciousness of the society would be present in its entirety in all the individuals, with a vitality sufficient to prevent all acts offending it—the purely conventional faults as well as the crimes. But a uniformity so universal and absolute is utterly impossible; for the immediate physical milieu in which each one of us is placed, the hereditary antecedents, and the social influences vary from one individual to the next, and consequently diversify consciousnesses. It is impossible for all to be alike, if only because each one has his own organism and that these organisms occupy different areas in space. That is why, even among the lower peoples, where individual originality is very little developed, it nevertheless does exist.

Thus, since there cannot be a society in which the individuals do not differ more or less from the collective type, it is also inevitable that, among these divergences, there are some with a criminal character. What confers this character upon them is not the intrinsic quality of a given act but that definition which the collective conscience lends them. If the collective conscience is stronger, if it has enough authority practically to suppress these divergences, it will also be more sensitive, more exacting; and, reacting against the slightest deviations with the energy it otherwise displays only against more considerable infractions, it will attribute to them the same gravity as formerly to crimes. In other words, it will designate them as criminal.

Crime is, then, necessary; it is bound up with fundamental conditions of all social life, and by that very fact it is useful, because these conditions of which it is part are themselves indispensable to the normal evolution of morality and law.

Indeed, it is no longer possible today to dispute the fact that law and morality vary from one social type to the next, nor that they change within the same type if the conditions of life are modified. But, in order that these transformations may be possible, the collective sentiments at the basis of morality must not be hostile to change, and consequently must have but moderate energy. If they were too strong, they would no longer be plastic. Every pattern is an obstacle to new patterns, to the extent that the first pattern is inflexible. The better a structure is articulated, the more it offers a healthy resistance to all modification; and this is equally true of functional, as of anatomical, organization. If there were no crimes, this condition could not have been fulfilled;

for such a hypothesis presupposes that collective sentiments have arrived at a degree of intensity unexampled in history. Nothing is good indefinitely and to an unlimited extent. The authority which the moral conscience enjoys must not be excessive; otherwise no one would dare criticize it, and it would too easily congeal into an immutable form. To make progress, individual originality must be able to express itself. In order that the originality of the idealist whose dreams transcend his century may find expression, it is necessary that the originality of the criminal, who is below the level of his time, shall also be possible. One does not occur without the other.

Nor is this all. Aside from this indirect utility, it happens that crime itself plays a useful role in this evolution. Crime implies not only that the way remains open to necessary changes but that in certain cases it directly prepares these changes. Where crime exists, collective sentiments are sufficiently flexible to take on a new form, and crime sometimes helps to determine the form they will take. How many times, indeed, it is only an anticipation of future morality—a step toward what will be! According to Athenian law, Socrates was a criminal, and his condemnation was no more that just. However, his crime, namely, the independence of his thought, rendered a service not only to humanity but to his country. It served to prepare a new morality and faith which the Athenians needed, since the traditions by which they had lived until then were no longer in harmony with the current conditions of life. Nor is the case of Socrates unique; it is reproduced periodically in history. It would never have been possible to establish the freedom of thought we now enjoy if the regulations prohibiting it had not been violated before being solemnly abrogated. At that time, however, the violation was a crime, since it was an offense against sentiments still very keen in the average conscience. And yet this crime was useful as a prelude to reforms which daily became more necessary. Liberal philosophy had as its precursors the heretics of all kinds who were justly punished by secular authorities during the entire course of the Middle Ages and until the eve of modern times.

From this point of view the fundamental facts of criminality present themselves to us in an entirely new light. Contrary to current ideas, the criminal no longer seems a totally unsociable being, a sort of parasitic element, a strange and unassimilable body, introduced into the midst of society.[3] On the contrary, he plays a definite role in social life. Crime, for its part, must no longer be conceived as an evil that cannot be too much suppressed. There is no occasion for self-congratulation when the crime rate drops noticeably below the average level, for we may be certain that this apparent progress is associated with some social disorder. Thus, the number of assault cases never falls so low as in times of want.[4] With the drop in the crime rate, and as a reaction to it, comes a revision, or the need of a revision in the theory of punishment. If, indeed, crime is a disease, its punishment is its remedy and cannot be otherwise conceived; thus, all the discussions it arouses bear on the point of determining what the punishment must be in order to fulfill this role of remedy. If crime is not pathological at all, the object of punishment cannot be to cure it, and its true function must be sought elsewhere.

NOTES

1. From the fact that crime is a phenomenon of normal sociology, it does not follow that the criminal is an individual normally constituted from the biological and psychological points of view. The two questions are independent of each other. This independence will be better understood when we have shown, later on, the difference between psychological and sociological facts.

2. Calumny, insults, slander, fraud, etc.

3. We have ourselves committed the error of speaking thus of the criminal, because of a failure to apply our rule (*Division du travail social,* pp. 395–96).

4. Although crime is a fact of normal sociology, it does not follow that we must not abhor it. Pain itself has nothing desirable about it; the individual dislikes it as a society does crime, and yet it is a function of normal physiology. Not only is it necessarily derived from the very constitution of every living organism, but it plays a useful role in life, for which reason it cannot be replaced. It would, then, be a singular distortion of our thought to present it as an apology for crime. We would not even think of protesting against such an interpretation, did we not know to what strange accusations and misunderstandings one exposes oneself when one undertakes to study moral facts objectively and to speak of them in a different language from that of the layman.

Crime is necessary for Progress

2

Deviance and the
Response of Others

HOWARD S. BECKER

The interactionist perspective . . . defines deviance as the infraction of some agreed-upon rule. It then goes on to ask who breaks rules, and to search for the factors in their personalities and life situations that might account for the infractions. This assumes that those who have broken a rule constitute a homogeneous category, because they have committed the same deviant act.

Such an assumption seems to me to ignore the central fact about deviance: it is created by society. I do not mean this in the way it is ordinarily understood, in which the causes of deviance are located in the social situation of the deviant or in "social factors" which prompt his action. I mean, rather, that *social groups create deviance by making the rules whose infraction constitutes deviance,* and by applying those rules to particular people and labeling them as outsiders. From this point of view, deviance is *not* a quality of the act the person commits, but rather a consequence of the application by others of rules and sanctions to an "offender." The deviant is one to whom the label has successfully been applied; deviant behavior is behavior that people so label.[1]

Since deviance is, among other things, a consequence of the responses of others to a person's act, students of deviance cannot assume that they are dealing with a homogeneous category when they study people who have been labeled deviant. That is, they cannot assume that those people have actually committed a deviant act or broken some rule, because the process of labeling may not be infallible; some people may be labeled deviant who in fact have not broken a rule. Furthermore, they cannot assume that the category of those labeled deviant will contain all those who actually have broken a rule, for many offenders may escape apprehension and thus fail to be included in the population of "deviants" they study. Insofar as the category lacks homogeneity and fails to include all the cases that belong in it, one cannot reasonably expect to find common factors of personality or life situation that will account for the supposed deviance. What, then, do people who have been labeled deviant have in common? At the least, they share the label and the

From Howard S. Becker, *Outsiders: Studies in the Sociology of Deviance.* Copyright © 1963 by The Free Press. Reprinted by permission of The Free Press, a division of Simon & Schuster.

experience of being labeled as outsiders. I will begin my analysis with this basic similarity and view deviance as the product of a transaction that takes place between some social group and one who is viewed by that group as a rule-breaker. I will be less concerned with the personal and social characteristics of deviants than with the process by which they come to be thought of as outsiders and their reactions to that judgement. . . .

The point is that the response of other people has to be regarded as problematic. Just because one has committed an infraction of a rule does not mean that others will respond as though this had happened. (Conversely, just because one has not violated a rule does not mean that he may not be treated, in some circumstances, as though he had.)

The degree to which other people will respond to a given act as deviant varies greatly. Several kinds of variation seem worth noting. First of all, there is variation over time. A person believed to have committed a given "deviant" act may at one time be responded to much more leniently than he would be at some other time. The occurrence of "drives" against various kinds of deviance illustrates this clearly. At various times, enforcement officials may decide to make an all-out attack on some particular kind of deviance, such as gambling, drug addiction, or homosexuality. It is obviously much more dangerous to engage in one of these activities when a drive is on than at any other time. (In a very interesting study of crime news in Colorado newspapers, Davis found that the amount of crime reported in Colorado newspapers showed very little association with actual changes in the amount of crime taking place in Colorado. And, further, that people's estimate of how much increase there had been in crime in Colorado was associated with the increase in the amount of crime news but not with any increase in the amount of crime.)[2]

The degree to which an act will be treated as deviant depends also on who commits the act and who feels he has been harmed by it. Rules tend to be applied more to some persons than others. Studies of juvenile delinquency make the point clearly. Boys from middle-class areas do not get as far in the legal process when they are apprehended as do boys from slum areas. The middle-class boy is less likely, when picked up by the police, to be taken to the station; less likely when taken to the station to be booked; and it is extremely unlikely that he will be convicted and sentenced.[3] This variation occurs even though the original infraction of the rule is the same in the two cases. Similarly, the law is differentially applied to Negroes and whites. It is well known that a Negro believed to have attacked a white woman is much more likely to be punished than a white man who commits the same offense; it is only slightly less well known that a Negro who murders another Negro is much less likely to be punished than a white man who commits murder.[4] This, of course, is one of the main points of Sutherland's analysis of white-collar crime: crimes committed by corporations are almost always prosecuted as civil cases, but the same crime committed by an individual is ordinarily treated as a criminal offense.[5]

Some rules are enforced only when they result in certain consequences. The unmarried mother furnishes a clear example. Vincent[6] points out that illicit sexual relations seldom result in severe punishment or social censure for the offenders. If, however, a girl becomes pregnant as a result of such activities the reaction of others is likely to be severe. (The illicit pregnancy is also an interesting example of the differential enforcement of rules on different categories of people. Vincent notes that unmarried fathers escape the severe censure visited on the mother.)

Why repeat these commonplace observations? Because, taken together, they support the proposition that deviance is not a simple quality, present in some kinds of behavior and absent in others. Rather, it is the product of a process which involves responses of other people to the behavior. The same behavior may be an infraction of the rules at one time and not at another; may be an infraction when committed by one person, but not when committed by another; some rules are broken with impunity, others are not. In short, whether a given act is deviant or not depends in part on the nature of the act (that is, whether or not it violates some rule) and in part on what other people do about it.

Some people may object that this is merely a terminological quibble, that one can, after all, define terms any way he wants to and that if some people want to speak of rule-breaking behavior as deviant without reference to the reactions of others they are free to do so. This, of course, is true. Yet it might be worthwhile to refer to such behavior as *rule-breaking behavior* and reserve the term *deviant* for those labeled as deviant by some segment of society. I do not insist that this usage be followed. But it should be clear that insofar as a scientist uses "deviant" to refer to any rule-breaking behavior and takes as his subject of study only those who have been *labeled* deviant, he will be hampered by the disparities between the two categories.

If we take as the object of our attention behavior which comes to be labeled as deviant, we must recognize that we cannot know whether a given act will be categorized as deviant until the response of others has occurred. Deviance is not a quality that lies in behavior itself, but in the interaction between the person who commits an act and those who respond to it. . . .

In any case, being branded as deviant has important consequences for one's further social participation and self-image. The most important consequence is a drastic change in the individual's public identity. Committing the improper act and being publicly caught at it place him in a new status. He has been revealed as a different kind of person from the kind he was supposed to be. He is labeled a "fairy," "dope fiend," "nut" or "lunatic," and treated accordingly.

In analyzing the consequences of assuming a deviant identity let us make use of Hughes' distinction betwen master and auxiliary status traits.[7] Hughes notes that most statuses have one key trait which serves to distinguish those who belong from those who do not. Thus the doctor, whatever else he may be, is a person who has a certificate stating that he has fulfilled certain requirements and is licensed to practice medicine; this is the master trait. As Hughes points out, in our society a doctor is also informally expected to have a number of auxiliary traits: most people expect him to be upper middle-class, white, male, and Protestant. When he is not, there is a sense that he has in some way failed to fill the bill. Similarly, though skin color is the master status trait determining who is Negro and who is white, Negroes are informally expected to have certain status traits and not to have others; people are surprised and find it anomalous if a Negro turns out to be a doctor or a college professor. People often have the master status trait but lack some of the auxiliary, informally expected characteristics; for example, one may be a doctor but be a female or a Negro.

Hughes deals with this phenomenon in regard to statuses that are well thought of, desired, and desirable (noting that one may have the formal qualifications for entry into a status but be denied full entry because of lack of the proper auxiliary traits), but the same process occurs in the case of deviant statuses. Possession of one deviant trait may have a generalized symbolic value, so that people automatically assume that its bearer possesses other undesirable traits allegedly associated with it.

To be labeled a criminal one need only commit a single criminal offense, and this is all the term formally refers to. Yet the word carries a number of connotations specifying auxiliary traits characteristic of anyone bearing the label. A man who has been convicted of housebreaking and thereby labeled criminal is presumed to be a person likely to break into other houses; the police, in rounding up known offenders for investigation after a crime has been committed, operate on this premise. Further, he is considered likely to commit other kinds of crimes as well, because he has shown himself to be a person without "respect for the law." Thus, apprehension for one deviant act exposes a person to the likelihood that he will be regarded as deviant or undesirable in other respects.

There is one other element in Hughes' analysis we can borrow with profit: the distinction between master and subordinate statuses.[8] Some statuses, in our society as in others, override all other statuses and have a certain priority. Race is one of these. Membership in the Negro race, as socially defined, will override most other status considerations in most other situations; the fact that one is a physician or middle-class or female will not protect one from being treated as a Negro first and any of these other things second. The status of deviant (depending on the kind of deviance) is this kind of master status. One receives the status as a result of breaking a rule, and the identification proves to be more important than most others. One will be identified as a deviant first, before other identifications are made. . . .

NOTES

1. The most important earlier statements of this view can be found in Frank Tannenbaum, *Crime and the Community* (New York: Columbia University Press, 1938), and E. M. Lemert, *Social Pathology* (New York: McGraw-Hill Book Co., 1951). A recent article stating a position very similar to mine is John Kitsuse, "Societal

Reaction to Deviance: Problems of Theory and Method," *Social Problems,* 9 (Winter, 1962), 247–256.

2. F. James Davis, "Crime News in Colorado Newspapers," *American Journal of Sociology,* LVII (January, 1952), 325–330.

3. See Albert K. Cohen and James F. Short, Jr., "Juvenile Delinquency," p. 87 in Robert K. Merton and Robert A. Nisbet, eds., *Contemporary Social Problems.* New York: Harcourt, Brace and World, 1961.

4. See Harold Garfinkel, "Research Notes on Inter- and Intra-Racial Homicides," *Social Forces* 27 (May, 1949): 369–81.

5. Edwin Sutherland, "White Collar Criminality," *American Sociological Review* V (February, 1940): 1–12.

6. Clark Vincent, *Unmarried Mothers* (New York: The Free Press of Glencoe, 1961): 3–5.

7. Everett C. Hughes, "Dilemmas and Contradictions of Status," *American Journal of Sociology* L (March, 1945): 353–359.

8. *Ibid.*

3

Deviance as Crime, Sin, and Poor Taste

ALEXANDER B. SMITH
AND HARRIET POLLACK

Superficially, it is very easy to define deviance. A deviant person is one who does something we wouldn't do. In the words of Howard Becker, he is an outsider, one who is outside the consensus of what constitutes proper conduct. The problem is that from someone's point of view we are all outsiders in one respect or another. Discussions of deviance, therefore, really turn on searches for universals, for modes of conduct that all human societies consider unacceptable.

In the classroom, anthropology professors like to upset their students by pointing out that there are no such universally disapproved modes of conduct. Even a killing that we would consider murder is acceptable in some societies: infanticide was common in Sparta, as was deliberate starvation of old people by Eskimos. In actuality, however, assaultive acts against the person or someone else's property, such as murder, assault, rape, and robbery, are considered taboo in almost all human societies, and people who perform such acts are clearly deviant. These acts, however, constitute only a tiny fraction of all the modes of conduct that our own and other societies have from time to time labeled as wrong.

If today we were to ask a middle-class, middle-aged, white American what kinds of acts (outside of assaultive crime) he considered deviant, he might respond as follows:

- Being a homosexual; reading dirty books or seeing pornographic movies; going to prostitutes; engaging in sex outside of marriage; having illegitimate children (especially if the children wind up on welfare).

- Using drugs—not prescription drugs or over-the-counter items like Alka Seltzer or Geritol or Vitamin E—but heroin, LSD, and pep pills.

- Drinking too much; eating enough to make you fat; smoking cigarettes (maybe); smoking marijuana (positively).

- Not taking care of your obligations; being lazy or shiftless; losing money at gambling; swearing and using bad language publicly.

From "Deviance as a Method of Coping," Alexander B. Smith and Harriet Pollack, *Crime and Delinquency*, Vol. 22, No. 1, 1976. Reprinted by permission of Sage Publications, Inc.

If we accept this list as typical, it is as interesting for the conduct it omits as for that which it includes. Many acts which once were or now are attacked as highly immoral are not even mentioned: for example, contraception, abortion, and sexual and racial discrimination. Our Everyman also seems unconcerned about profiteering, sharp dealing, tax evasion, consumer fraud, and other kinds of white-collar crime. To be sure, if questioned specifically about these unmentioned acts, he would disapprove of all of them (except possibly for contraception), but the term "deviant conduct" would not bring them immediately to mind.

The reason for our Everyman's perceiving deviance selectively lies in our description of Everyman: middle-class, middle-aged, and white. From where he stands, some acts affect his world adversely, others have little effect, and some are simply irrelevant. He doesn't care especially about racial or sexual discrimination because he is neither black nor female. He believes in sexual regularity because he is a family man and his world is stabilized by the nuclear families of his friends and neighbors. Furthermore, illegitimacy (as he sees it) is a direct and undeserved burden on taxpayers like him because of its effect on the welfare rolls. On the other hand, contraception doesn't seem wrong to him since his middle-class status probably depends on his success in limiting the size of his family. Even abortion has much to be said for it, since anyone can get into trouble and anyway maybe abortion will keep some of those babies off welfare. He doesn't worry too much about tax evasion because he is not aware of the activities of large-scale tax evaders, such as giant corporations and wealthy individuals whose accountants and tax lawyers have created tax shelters for them; and small-scale tax evasion is probably a fairly common and socially acceptable activity in his milieu. Sharp dealings (such as exploitative landlord-tenant or seller-consumer transactions) are likewise a middle-class way of making a living; and in any case, most middle-class persons are able to cope with dishonest landlords or tradesmen. On the other hand, persons who take or sell drugs are enormously threatening, both because drug use frequently leads to assaultive or dangerous criminal conduct and because drug addicts threaten the stability of the social system by their aberrant attitudes toward work and other social obligations. In fact, if there is one thread that runs through the fabric of Everyman's scheme of desirable social conduct, it is the desire to maintain stability, to preserve the status quo. As a member of the middle class, he has made it, and he recognizes that life is as good for him as it is ever likely to be. He doesn't want to lose what he has. Change is threatening and makes him very uncomfortable.

The laundry list of unacceptable conduct varies with the age and status of the person compiling it. Inner city blacks, for example, might list racial discrimination first and not list gambling at all. Marijuana smoking might be quite acceptable to middle-class university students, but tax evasion, sharp dealing, and profiteering would be high on their list of forbidden conduct. In the Bible Belt of the Deep South, blasphemy, secularism, and atheism are still heinous offenses, yet relatively free use of firearms, moonshining, and blatant racial discrimination are regarded with considerable tolerance.

Obviously, deviance is to some extent in the eye of the beholder—but only to some extent. All classes and status groups reject violent assaultive crime.[1] They differ, however, in respect to other types of unacceptable conduct, some of which in our system are illegal, some of which are immoral, and some of which are merely displays of poor taste. In considering these widely varying perceptions of what constitutes deviant conduct, we must ask not who is right and who is wrong but what kinds of conduct society can tolerate and still exist as a viable society and what kinds it cannot accept. Part of the answer to this basic question must lie in one's perception of a desirable society. For purposes of this discussion we are assuming an ideal closely akin to the traditional Jeffersonian model: an open society predicated on a belief in equality of opportunity and equality before the law, with a reasonable level of material comfort and economic security for all. In such a society, what kinds of behavior are necessarily beyond the pale? In this connection, we propose to discuss three categories of conduct: crime, sin, and actions that are in poor taste.

DEVIANCE: CRIME

Clearly, heading the list are murder, rape, arson, assault, robbery, burglary, and larceny, acts which are totally unacceptable and which can be condoned, if at all, only under very special circumstances.[2] We label these acts *crimes,* meaning that their violation of the public order is so severe that they must be handled punitively and coercively by the police, courts, and prisons. Even those who commit them agree that this type of conduct is wrong. A housebreaker does not want his own house to be burglarized and, except in Robin Hood legends, robbers do not argue that what they do is legitimate. This type of conduct is taboo because if it is tolerated a viable society is not possible. The control of such conduct, indeed, is one of the central problems faced throughout history by philosophers who have attempted to construct model societies. Whatever their point of view and whatever type of Utopia they have created, they all agree at least that this type of act must be forbidden. While Hobbes and Locke, for example, differed radically in their perceptions of the fundamental nature of man and in their prescriptions for social control of human conduct, they agreed that the principal difficulty in human society is the governance of violent assault by one individual against another.

However, assaultive conduct is only one category of crime. So-called "white-collar crime," while nonviolent, is basically an attack on legitimate property arrangements in society. Acts such as tax fraud, stock manipulations, commercial bribery, misrepresentation in advertising and salesmanship, short weighting and misgrading of commodities, embezzlement, etc., are all methods of obtaining money or other property illegitimately. Since the function of an economic system is to prescribe how one may properly obtain property, white-collar criminals are subversive of accepted economic relationships. As such, like their more violent criminal counterparts, they are a threat to a viable society and it is reasonable that their acts be included in the penal codes.[3] Although the prescribed penalties for white-collar crimes may sometimes be as severe as those for burglary or larceny, these acts do not carry the stigma or the punishment of violent crimes.

Basically our law is ambivalent. Property crimes are crimes, but they are not really heinous if they are not violent or potentially violent. Far less ambivalence in regard to so-called "economic crimes" is exhibited in the Soviet Union, where some offenses of this type, such as currency manipulation, are punishable by death sentences whereas certain kinds of homicide are treated relatively leniently. This probably reflects the orientation of the legal system toward preservation of the Soviet economic and social order rather than, as in this country, protection of individual rights. From this point of view, the inconsistency of the American system, which punishes personal crimes more severely than property crimes, is understandable. Whatever our ambivalence, however, it is clear that nonviolent crimes of property must be handled punitively, at least to the extent necessary to maintain the legitimacy of both our property arrangements and our system of law. The latent admiration of Americans for Robber Baron types may never disappear from the culture. Nevertheless, if business dealings are to be conducted in an orderly way and if prohibitions on assaultive crimes are to be taken seriously, there must be reasonable enforcement of the law relating to white-collar offenses. As the public conscience grows in sensitivity, moreover, the criminal sanction will be extended to dealings which are now considered unsavory but not illegal. The basic push in the developing field of poverty law is to extend the criminal law to cover some actions of landlords against tenants and of merchants against customers that were not considered illegal before. For example, may a landlord be paid rent by his tenants if he has failed to provide the agreed-upon level of services? May a merchant misrepresent the quality of the merchandise he is selling and demand continued performance of a time-payment contract if the goods in question have already deteriorated? These practices are probably permissible at present. The trend, however, is toward making such actions illegal—probably an indication of our feeling that even nonassaultive crimes of property are a threat to the viability of our society.

Our penal law, thus, contains prohibitions against both assaultive crimes against persons and property and nonassaultive crimes against property. Assaultive crimes offend our notions of natural justice; nonassaultive property crimes undermine the economic arrangements that are basic to the stability of society. The penal code contains, however, strictures against a number of modes of conduct which are included because of a relatively parochial cultural determination that they are immoral: drinking, gambling, homosexuality, doing business on Sunday, prostitution, drug addiction, abortion, etc. While at the time these prohibitions were enacted, the particular legislative majority which enacted them doubtless felt they were preventing subversion of the legitimate social system, many societies quite similar to ours do, in fact, tolerate such prohibited conduct quite well or, in any case, handle it nonpunitively.

Many of these regulations are, moreover, both inconsistent and incomplete in their regulatory schemes. Prostitutes are punished but not their customers; heroin is forbidden but not amphetamines; it is permissible to bet on a race but not on a football game, etc.

DEVIANCE: SIN

Many of these modes of conduct were originally thought of as sin and were *religiously* prohibited. Our use of secular law to regulate them is a relic of the time when the authority of the state was used to enforce the rules of an established church. That era is past, but we can see our cultural heritage most clearly perhaps in the laws we inherited from the Puritan theocracy in New England. We have (or have had in the recent past) laws against blasphemy, obscenity, contraception, Sabbath breaking, extramarital sexual relations, lewdness, homosexuality, gambling, and drunkenness. We also have inherited a distrust of self-indulgence and hedonism: even a rich man is expected to be constructively, if not gainfully, employed.

This heritage reflects a culture in which religion once was dominant. As our culture has changed, as religion has waned in importance, as our economic system has developed, as scientific discoveries have occurred, and as improved communications and the development of the mass media have reduced both social and cultural isolation, our feelings about what constitutes sin have undergone a marked change. Some behavior once regarded as sinful has become virtually acceptable today—for example, blasphemy; some, like heroin use, is still taboo. About other forms of conduct such as gambling, drinking, homosexuality, and abortion, we have ambivalent feelings. Some of this conduct is still subject to criminal sanction; some is not. If we remove the religious component, the criterion for whether the conduct in question should be forbidden should rest on whether *there is any demonstrable, objectively measurable social harm resulting from it*. To determine this, we must separately consider and evaluate each mode of conduct. In a totally rational world we would expect to find a correlation between the prohibition of conduct and its objective harmfulness. But this is not a rational world and the correlation does not exist.

Of all the modes of conduct in this culturally determined category, drinking is probably the most harmful and also the most widely accepted. Alcohol is involved in at least half of all fatal automobile accidents, a majority of private airline crashes, thousands of industrial accidents, millions of lost man–days annually, etc. We have in this country approximately 9,000,000 alcoholics who are unable to support their families, do their jobs, or function normally in the community. Alcohol use is involved in 55 percent of the arrests made by police. From a medical point of view, furthermore, even moderate drinking puts a strain on the liver and complicates many maladies.

Yet alcohol consumption is widely accepted today in the United States, where nondrinkers constitute only a small minority of the population.

Historically, the temperance movement waxed and waned in strength for over a century before it culminated in the "noble experiment" of Prohibition in 1920. However, within a few years after enactment of the Eighteenth Amendment, it became apparent that Prohibition was a disaster and, since the repeal in 1933, the temperance movement appears to be all but moribund. Thus, drinking has been handled both coercively and noncoercively, and while our current noncoercive approach has fewer adverse effects in the form of enforcement difficulties and police corruption, alcohol abuse still presents a problem—a problem not reflected in public attitudes.

Even more permissive than our attitudes toward drinking are our feelings about cigarette smoking and overeating. The medical evidence against both smoking and obesity is overwhelming but to forbid them by law would be ludicrous, a civil liberty horror. Even attempts to regulate cigarette advertising have met with great resistance. Though smoking and overeating are seriously harmful, medically and sociologically speaking, and though there is considerable consensus that people should not smoke or get fat, Americans who do not smoke and who are not overweight probably constitute a minority.

In contrast to drinking, smoking, and overeating, there is no medical evidence that moderate use of marijuana is harmful and no medical evidence of physiological harm from reasonable heroin consumption. That many heroin or marijuana users exhibit undesirable psychological symptoms is undoubtedly true. It is not clear, however, whether these symptoms are a result of drug use or whether both drug use and behavioral dysfunction result from a prior existing pathological, psychological, or sociological condition. Most of the other adverse sociological effects of drug use, such as crime and prostitution, result from our present coercive handling of the drug problem rather than from drug use per se. Yet few modes of conduct are looked upon with more social disapproval than heroin use and only recently has a similar attitude toward marijuana been softening. Moreover, in certain respects our method of handling drug use has been precisely opposite from our handling of alcohol: alcohol, formerly handled punitively, is now handled nonpunitively; opiates and marijuana, formerly handled nonpunitively, are now handled punitively. Neither punitive handling nor extreme social disapproval has resulted in a decline (or even a stabilization) of the number of marijuana and heroin users in the United States. In 1967 there were in the United States about 100,000 addicts, of whom 50,000 were in New York City; five years later the estimates had precisely tripled: 300,000 addicts in the United States, with 150,000 in New York City.

In contrast to our attitude toward alcohol and drug use, which has fluctuated between acceptance and rejection, our attitude toward deviant sexual conduct has become consistently more permissive. During the eighteenth and nineteenth centuries in this country, man–woman relationships reflected a society that placed high value on premarital chastity and monogamy. Divorce was frowned upon, and premarital dalliance (except possibly for young men who were sowing their "wild oats") was strictly taboo. Prostitution, at least from the middle–class point of view, was considered degrading and abhorrent, and the fallen woman became a stock figure in literature. In the same period, homosexuality was considered so dreadful that there was no public discussion of the subject and, except for some very guarded indirect references, no literary mention of the problem. Today we are permissive in regard to premarital sex, we permit divorce, we have ambivalent attitudes toward prostitution, and we are slowly coming to a grudging acceptance of homosexual conduct. Some of these attitudinal changes have been reflected in changes in either the criminal law or its application; others have not. Nevertheless, few people would dispute the proposition that our attitudes toward sexual conduct have changed substantially even if the conduct in question has not.

To understand this phenomenon one must appreciate that the older rules for sexual conduct were drawn up in a society which had vastly different needs: until the twentieth century the need was for more population rather than less; venereal disease was an uncontrollable plague; and production of goods and services was directly dependent on the family in a way that no longer exists. Twentieth century advances in public health and medical knowledge have changed all this.

Medical knowledge and technology have turned the older rationale for monogamous units upside down. One hundred years ago, a couple might have to produce ten or twelve children to be certain that five or six would survive them; today the parents of two can reasonably expect to raise both to adulthood. Formerly children represented a source of income and social security for one's old age; today children are economic liabilities at least until they reach adulthood, and some thereafter.

In the face of these substantial changes, it is understandable that many of the older rules of sexual conduct are anachronistic. This is not to say that our commitment to monogamous union as the basis of family structure has diminished. Nor does it mean that actual sexual practices (as opposed to the accepted social standard for what those practices should be) have changed very much. What it means is that deviation from these sexual norms is accepted more readily and less fearfully than before. We are not so hysterically defensive about our rules of sexual conduct because we no longer regard deviations from them as subversive of the entire social order. We no longer need a strict sexual code to provide for population maintenance or growth, industrial or agricultural production, or prophylaxis against rampant venereal disease. We adhere to our family structure—and hence our sexual code—not so much to meet societal needs as to fulfill our own, the achievement of personal happiness and an optimal setting in which to raise children. Under these circumstances the desire of some individuals to find personal happiness through premarital sex, homosexuality, prostitution, etc., becomes less terrifying and is, if not acceptable, at least understandable.

Gambling, however, is a mode of conduct which probably has come closest of all to shedding the stigma of immorality inherited from the past. American attitudes toward gambling have always been ambivalent. Even in Puritan times we find mention of gaming and lotteries at the same time the churches were exhorting against such worldly pleasures. Gradually, however, our attitudes have softened, probably because of the general relaxation of the personal standards of behavior and possibly because of the possibilities of relief for the hardpressed taxpayer through state-sponsored lotteries. In any case, at the present time, not only does Nevada have legalized gambling and New York the OTB (a public corporation to conduct off-track betting) but increasingly the criminal justice system is refusing to use its resources to enforce antigambling laws. The police protest openly at the futility of picking up small-time gamblers who are doing no more than the OTB employees, and such gamblers as are prosecuted are handled by the courts perfunctorily and with minimal penalties. The change in public opinion, the negative attitudes of police and prosecutors toward gambling law enforcement, and general awareness that illegal gambling is a major source of income for organized crime have combined to hasten the repeal of many—perhaps all—gambling statutes. There is virtually no effective interest group in the United States that espouses the retention of gambling laws. Apparently, legislative repeal is retarded only by public apathy and the fear of criticism by zealots.

DEVIANCE: POOR TASTE

In contrast to acts which are crimes or sins are actions which are matters of taste and which, even when disapproved, are rarely regulated by law. Manners and style fall within this category. Pants on women were once an object of scandal; girls' bobbed hair in the 1920s was viewed as dubiously as boys' long hair in the 1960s. In Puritan New England it was a misdemeanor for a man and a woman to kiss in public even if they were married; we think nothing of more overt expressions of affection although we become increasingly offended as the conduct becomes more explicitly sexual. Adults smile benignly at little Boy Scouts and Girl Scouts in their uniforms, but glare at black-jacketed Hell's Angels and similarly dressed members of black and Puerto Rican youth gangs. Frenchmen may kiss each other heartily; American men may not. It is alright to wear a cross or a mezuzah, but a swastika arm-band, a hooded sheet, and a clenched fist salute are perceived with considerable hostility and, under certain circumstances, are forbidden by the authorities.

To the visitor from Mars, all of this can be very confusing. Why, for example, is it all right for an adult to appear in public wearing a skimpy bathing suit but not his underwear? To us, however, it is not confusing at all, although few people when pressed could rationalize all the idiosyncrasies of manners and style that go to make up taste. It is clear that to a great extent these modes of conduct are cultural accidents. Pants are no more ordained by nature for men than skirts are for women, and in some tribal societies men do wear skirts and women pants. There is nothing in the shape of a cross that necessarily suggests Christianity and nothing in the shape of the swastika that necessarily equals fascism. As a method of greeting, handshaking is neither more nor less rational than a kiss on the cheek or a deep curtsy. But while the conduct in question may be irrational, the inferences drawn from it may be highly rational. The wearing of the swastika by American fascists is a reliable indicator of a belief in racial inequality, a totalitarian system of government, etc. A man who appears in public in a woman's dress probably is sexually deviant. What we object to in these modes of conduct, therefore, is that they suggest or antici- pate other actions to which we take exception. They are in a sense symbolic conduct, symbolic of some type of overt action to which there is or may be a rational objection. Thus, the objection to the swastika is an objection to fas- cism; and the more we object to fascism as a mode of conduct, the more we will object to the swastika. Many modes of dress are objectionable because they appear to anticipate undesirable sexual conduct: slacks and bikinis on women, long hair and feminine looking clothing on men. Interpersonal con- duct—modes of greeting and communicating with other people—is evaluated by our interpretation of the hidden messages those modes send out. When at- tempts are made to change matters of manner and style, objection is frequently vigorous simply because such changes are viewed as a precedent to change in more serious forms of nonsymbolic conduct. Opposition fades away when the symbolic conduct loses its symbolism. In Victorian times a woman who showed her ankles freely was considered "fast," aggressively inviting promiscu- ous conduct. When enough women wore short skirts without the occurrence of the undesirable sexual conduct that had been anticipated, short skirts be- came acceptable. The first men wearing long hair in the current style were considered to be homosexually inclined. When the majority of adolescent youths and young men adopted the fashion, long hair as a symbol of homo- sexuality faded.

Thus the problem in regard to matters of taste is to recognize, first of all, that they are cultural accidents and may be intrinsically quite irrational. We must also recognize, however, that such conduct is symbolic conduct and may be the surface manifestation of far more meaningful attitudes and actions. In regulating such matters of taste, then, we must know when the surface con- duct is truly symbolic and when it has lost its symbolism. If the symbolism is extant and the conduct to which it refers is truly harmful, it is possible that even symbolic action may need to be regulated socially. . . .

CONCLUSION

To sum up, deviant conduct is ubiquitous in a society such as ours. While de- viance lies, to some extent, in the eye of the beholder, certain forms of it that are objectively and measurably harmful to the community or that violate ra- tional institutionalized expectations are always deviant. The roots of deviance lie in sociological and psychological pressures generated within the individual by social forces frequently beyond his control. Since, however, the very notion of a free society is based on the responsibility of each individual for his own conduct, the responsibility for the control of deviant conduct lies with both the individual and the community at large. . . . Physically coercive punishment must be used only as a last resort and for the protection of the community, for it has almost no rehabilitative effect and serves only to keep the offender away from the community. For this reason the criminal process should be reserved almost exclusively for all persons who must be restrained at all costs or who are so seriously disruptive of the peace and good order of the community (e.g., swindlers and embezzlers who commit nonviolent property crimes) that rehabilitative counseling should be carried on in a semicoercive setting such as probation. For all others, either we should attempt education and persuasion by appropriate therapists or, in regard to those whose conduct really harms no one but themselves, *we should let them alone,* recognizing that to some extent we are all deviants.

NOTES

1. An exception might be black revolutionaries such as George Jackson, who, while imprisoned in San Quentin for armed robbery, wrote extensively on the place of blacks in white society. Jackson felt that because "Amerika" was a "society above society" in which blacks were "captive," they were under no obligation to obey the laws. All crime, therefore, was an act of rebellion. Even Jackson concedes, however, that noneconomic crime—e.g., "the rape of a Black woman by a Black man"—is an expression of racial violence turned inward. It is "autodestructive" and hence presumably wrong even if understandable. Tad Szulc, "George Jackson Radicalizes the Brothers in Soledad and San Quentin," *New York Times Magazine,* Aug. 1, 1971, p. 10.

2. We are referring here, of course, to random acts by individuals or small groups such as gangs and are omitting discussion of governmentally organized and sponsored violence such as that practiced during the Hitler period in Germany, the Spanish Inquisition, or any war. Whether this kind of organized violence is ever justifiable depends on one's politics, religion, nationality, and time in history.

3. While many political theorists have attacked the American economic system and consequent property arrangements as illegitimate—as violations of "natural justice"—none has seriously suggested that the types of fraud usually encompassed by the term "white collar crime" are justified as an attempt to remedy economic inequity. The embezzlers and stock manipulators have not yet produced their George Jackson.

PART II

Studying Deviance

Accurate and reliable knowledge about deviance is critically important to many groups of people in society. First, policy makers are very concerned with deviant groups such as the homeless and transient, the chronically mentally ill, high school dropouts, criminal offenders, prostitutes, juvenile delinquents, gang members, runaways, and other members of disadvantaged and disenfranchised populations. These people pose social problems in society that lawmakers and social welfare agencies want to help alleviate. Second, sociologists and other researchers have an interest in deviance based on their goal of understanding human nature, human behavior, and human society. Deviants are a critical group to this enterprise because they reside near the margins of social definition: they help define the boundaries of what is considered acceptable and unacceptable by given groups of people.

Information about deviance in American society can come from one of three primary data sources. Government officials and employees of social service agencies routinely collect information about their clients as they process them. This information includes arrest data that are compiled by police and published by the FBI (the *Uniform Crime Reports*), census data on various shifting populations (such as the homeless), victim data from helping agencies (such as battered women's shelters), medical data from emergency rooms (such as DAWN, the Drug Abuse Warning Network) or from state public health agencies (such as the coroner or medical examiner offices), and prosecution data on cases that are tried in the courts. These statistical **precollected data** are then compiled by the various government organizations responsible for collecting them and made available to the public.

Another source of statistical data about deviance is **survey research.** Rather than relying on information that the government might collect, sociologists gather their own data through large-scale questionnaire surveys. Prominent ones include the National Youth Survey, a self-report questionnaire about delinquent behavior, the Annual High School Survey conducted by the National Institute on Drug Abuse on the drug use of high school seniors, and some of the Kinsey surveys about sexual behavior.

A third kind of information, richly descriptive and analytical rather than numerical, comes from sociologists who conduct **participant-observation field research** on deviance. Much like anthropologists who go out to live among native peoples, sociological fieldworkers live among members of deviant groups and become intimately familiar with their lives. This type of research yields information more deeply based on the subjects' own perspectives, detailing how they see the world, the allure of deviance for them, the problems that they encounter, the ways in which they resolve these, the significant individuals and groups in their lives, and their role among these others. Unlike the other sources of information, participant observation is generally a longitudinal method that entails years of involvement with subjects. Researchers must gain acceptance by group members, develop meaningful relationships with them, and learn about the deepest core of their thoughts and feelings.

There are many differences among these types of data and among the methods used to gather them. Although the quantitative data yield information about a broad spectrum of people, they may be fairly shallow and unreliable in nature. The *Uniform Crime Reports* fail to include a host of crimes for several reasons: crimes may be unrecognized by victims who do not notice their occurrence or lack the power to define them as deviant; crimes may be unreported by individuals who see no gain by calling police attention to their victimization or fear embarrassment, censure, or retaliation; or crimes may be unrecorded by police officers who use their broad discretion to handle problems informally. For these reasons, official statistics vastly underrepresent criminal activity.

Survey research is plagued by problems of inaccuracy and differential interpretation. It is problematic that people, especially deviants, will fill out a questionnaire and readily disclose information about the covert aspects of their lives. Questionnaire studies are also plagued by two particular problems of language and interpretation. First, subjects may not define their behavior the same way or use the same terms as researchers who are writing the questions (prostitutes' conceptions of a "date" may be different from those of survey researchers, and runaways may mean different things when they refer to their "home" than researchers intend). Researchers are then likely to misinterpret the nature and extent of behavior from the answers they receive. Second, surveys yield correlational connections between factors—that is, they tell us what trends occur together, such as divorce and deviance or violence in the media and violence in everyday life. We may see these factors occurring together and mistakenly think that one is causing the other. But survey research findings cannot tell us *why* or *how* people act; they can only tell us *what* people are doing, even if these trends range across a broad spread of the population.

Participant observation, in contrast, cannot reach as many people, but it yields deeper and more accurate information about research subjects, backed up by the researchers' own observations, to enhance its validity. Participant-observation researchers spend long amounts of time in the field becoming intimate with the people they study and learning how their subjects perceive, interpret, and act upon the complex and often contradictory nature of their social worlds. Survey research is thus more controlled, since researchers are not influenced by their relationships with research subjects in gathering data about them, whereas participant observation relies on its insight, attained through the researchers' close relationships with people, to get behind false fronts and find out what is really going on. Depth penetration is especially important when studying a topic such as deviance, where so much behavior is hidden due to its negative social stigma and illicit status. Also critically important is the ability of participant observation to study deviance as, Polsky (1967) urged, it occurs *in situ,* in its natural setting, not via the structural constraints of police reporting, or the interpretation and recollection of questionnaire research.

The empirical selections that fill the remainder of this book are primarily based on participant observation studies of deviance for two main reasons. First, as Becker (1973) has noted, participation observation is the method of the interactionist perspective, since it offers direct access to the way definitions and laws are socially constructed, to the way people's actions are influenced by their associates, and to the way people's identities are affected by the deviant labels cast on them. Second, these types of studies offer a deeper view of people's feelings, experiences, motivations, and social psychological states, giving a richer and more vivid portrayal of deviance than charts of numbers. To help readers understand why we have made this decision, we include three methodological selections. The first, "Was It Rape? An Examination of Sexual Abuse Statistics," by Neil Gilbert, discusses some of the differences in definition and interpretation that may occur with survey research. Gilbert critiques several prominent surveys on sexual abuse, arguing that their criteria for defining sexual abuse, for interpreting behavior, and for extrapolating from their findings are seriously flawed. You may want to compare the arguments from this selection with several later selections that deal with sexual abuse to see how you feel about these differences of interpretation and analysis.

Next, Robert Kelly's "Field Research Among Deviants" addresses some of the common issues that participant observers face in studying deviance. Drawing primarily on organized criminological research, Kelly outlines some of the practical and ethical issues that both plague and distinguish the fieldwork approach.

Finally, in "Researching Dealers and Smugglers," by Patricia Adler, we get a glimpse of what it is like to carry out participant-observation research with a deviant group. This natural history carefully explains the process used in field research, the relationships we formed with setting members, and the feelings that we, as researchers, experienced. We found how putting ourselves inside a deviant world can profoundly affect researchers as well as cause them serious potential dangers, but we also found that only from this vantage point can researchers fully comprehend the forces at play in deviant worlds. This article offers a greater understanding of some of the research roles, research concerns, and problems and issues that may arise in field research.

4

Was It Rape?

An Examination of Sexual Abuse Statistics

NEIL GILBERT

- At least one-quarter of young women will be sexually assaulted before they leave high school.
- Twenty-seven percent of female college students have been victims of rape or attempted rape.
- Almost half of all women will be victims of rape or attempted rape sometime in their lives.

These alarming statistics have informed many media reports about the prevalence of sexual assault against women in our society. By whatever measure, the problem is a serious one. But should we trust the statistics? A careful review of some of the research that has been done suggests that the problem of sexual assault has been magnified and the data misinterpreted, in part because the issue of sexual violence against women has become enmeshed in politics.

One should not underestimate the difficulty of research in this area. Not only is it often hard to get people to talk about their deeply traumatic experiences, but deciding what to include in a definition of the problem is particularly tricky. In determining the prevalence of child sexual abuse, for example, should a one-time encounter with an exhibitionist be lumped together with repeated forcible rape? A review of the surveys reveals more about the ambiguities of definition than the magnitude of the problem. Legal definitions of these crimes vary from state to state and thus provide no clear reference points for researchers. Beyond this, the subject of sexual violence against women is an emotional one and this makes impartial research very difficult. Finally, researchers must deal with wide discrepancies between reported crimes and the projections their surveys yield.

The estimated prevalence rates cited above, for example, are considerably higher than what one would infer from the official figures on child sexual abuse and rape. Of course, since many if not most incidents of child sexual abuse are never reported to authorities and rape is also underreported, official figures yield conservative estimates of the problem. Still, the size of the discrepancies should make us skeptical of the research findings cited above. In 1992, for

From "Was It Rape? An examination of Sexual Abuse Statistics," Neil Gilbert, *The American Enterprise*, Sept./Oct. 1994. Reprinted by permission.

example, an estimated 499,120 cases of child sexual abuse were reported to child protective services, of which approximately 40 percent were substantiated. An annual incidence rate of about 200,000 substantiated cases (or 3 in 1,000 children) is serious, but it is only a fraction of the prevalence rate of 25 percent plus cited on the previous page. Similarly, Bureau of Justice Statistics findings reveal that in 1990 approximately 130,000 women or about 1.2 women in 1,000 over 12 years of age were victims of rape or attempted rape. No trivial number, of course, but that annual figure translates into a 4 to 7 percent lifetime prevalence rate—not the nearly 50 percent rate cited previously. While the officially reported cases underestimate the full extent of sexual violence against women, the highly publicized prevalence rates overestimate it. What follows is an analysis of some of the research cited most often.

PROBLEMS OF DEFINITION

Child Sexual Abuse

The different views of the full range of offenses that constitute child sexual abuse are apparent in a review of 15 surveys conducted since 1976 that attempt to estimate the prevalence of this problem. According to these surveys, the proportion of females sexually molested as children ranges from 6 percent to 62 percent of the population (for males the figures range from 3 percent to 31 percent); half of the studies showed a female prevalence rate of 6 percent to 15 percent. Discrepancies among these studies are due in large measure to differences in the researchers' operational definitions of sexual abuse. Sexual abuse in the studies that yielded the highest results included everything from sexual propositions, exposure to an exhibitionist, and unwanted touches and kisses and fondling to sexual intercourse and other physical contact.

Two of the largest and most widely cited surveys of the prevalence of child sexual abuse illustrate how expansively the problem is defined by researchers. Diana Russell (professor emeritus of sociology at Mills College) surveyed 930 women in San Francisco in the late 1970s and reported that 54 percent of her respondents were victims of incestuous or extrafamilial sexual abuse at least once before the age of 18. There were, however, sampling problems with this survey. Although efforts were made to achieve a random sample of participants, the fact that researchers were able to complete interviews with less than 50 percent of the original sample (930 out of 2,000) makes generalizations to the population of the United States or even San Francisco extremely speculative. Another problem with the survey was the way sexual abuse was defined. The 54 percent prevalence rate reflects a definition of child sexual abuse under which children who receive unwanted hugs and kisses are classified as victims, as are others who are not touched at all (children who encounter exhibitionists). Using a slightly narrower definition, Russell also calculated a lower rate of 38 percent. This narrower measure eliminated cases that did not involve physical contact, but it did include unwanted sexual experiences ranging from a single incident of attempted petting to repeated rape.

Rape

The most highly publicized figures on rape do not hold up under close examination any better than the numbers just cited on child sexual abuse. For example, in her 1984 book, in addition to finding a child sexual abuse prevalence rate of 38 percent to 54 percent, Diana Russell also found that 44 percent of the women in her survey were victims of rape or attempted rape an average of twice in their lives. Beyond sampling bias there is also the question of how rates of rape and attempted rape were measured. Russell's estimates were derived from responses to 38 questions, of which only one of these questions asked respondents whether they had been a victim of rape or attempted rape any time in their lives. And to this question, only 22 percent of the sample, or one-half of women defined as victims by Russell, answered in the affirmative.

A considerable proportion of the cases counted as rape and attempted rape in this study are based on the researchers' interpretation of experiences described by respondents. Thus, in assessing these interpretations, one might bear in mind Russell's perceptions about what constitutes rape. She says, "If one were to see sexual behavior as a continuum with rape at one end and sex liberated from sex-role stereotyping at the other, much of what passes for normal heterosexual intercourse would be seen as close to rape."

Rape on Campus

Quoted in newspapers and journals, on television, and during the 1990 and 1993 Senate hearings on the Violence Against Women Act, the *Ms.* Magazine Campus Project on Sexual Assault directed by Mary Koss is the most widely cited study of rape on college campuses.

Based on a survey of 6,159 students at 32 colleges, the *Ms.* study reported that 15 percent of female college students have been victims of rape and 12 percent of attempted rape, an average of two times between the ages of 14 and 21. The vast majority of offenders were acquaintances, often dates. Using questions that Koss claimed represented a strict legal description of the crime, she also calculated that during a 12-month period 6.6 percent of college women were victims of rape and 10 percent attempted rape with more than one-half of all these victims assaulted twice. If the victimization reported in the *Ms.* study continued at this annual rate over four years, one would expect well over half of all college women to suffer an incident of rape or attempted rape during that period, and more than one-quarter of them to be victimized twice.

The *Ms.* project was funded by the National Institute of Mental Health and its findings published in several professional journals, which is some indication of the clout the project has. Nevertheless, there are compelling reasons to question the magnitude of rape and attempted rape conveyed by this study. To begin with, there is a notable discrepancy between Koss's definition of rape and the way most of the women she labeled as victims interpreted their experiences: almost three-quarters of the students whom Koss categorized as victims of rape did not think that they had been raped. Moreover, in support of the students' own account of their experiences, 42 percent of those identified as victims had sex again with the man whom Koss (but not the students) claimed had raped them; of those whom Koss categorized as victims of *attempted* rape, 35 percent later had sex with their purported offender.

As Koss reported in Judiciary Committee hearings in 1990, "Among college women who had an experience that met legal requirements for rape, only a quarter labeled their experience as rape. Another quarter thought their experience was some kind of crime, but not rape. The remaining half did not think their experience qualified as any type of crime."

A somewhat different account of how these same students labeled their experience was reported in 1988 by Koss and others:

a.   11 percent of the students said they "don't feel vicitimized,"

b.   49 percent labeled the experience "miscommunication,"

c.   14 percent labeled it, "crime, but not rape," and

d.   27 percent said it was "rape."

The fact that 42 percent of students classified as victims by Koss had sex again with the men who supposedly raped them was not mentioned in her testimony before the Judiciary Committee. In 1988 she originally reported: "Surprisingly, 42 percent of the women indicated that they had sex again with the offender on a later occasion, but it is not known if this was forced or voluntary; most relationships (87 percent) did eventually break up subsequent to the victimization."

But in response to questions raised elsewhere, Koss again offered different accounts of her data. In 1991, in a letter to the *Wall Street Journal,* Koss was no longer surprised by this finding and claimed to know that when the students had sex again with the offenders they were raped a second time and that the relationship broke up not "eventually" (as do most college relationships), but immediately after the second rape. In this revised version of her findings Koss explained, "Many victims reacted to the first rape with self-blame and thought that if they tried harder to be clear they could influence the man's behavior. Only after the second rape did they realize the problem was the man, not themselves. Afterwards, 87 percent of the women ended the relationship with the man who raped them." Koss went on to suggest that these students did not know they were raped because many of them were sexually inexperienced and "lacked familiarity with what consensual intercourse should be like."

One part of the problem in the definition of rape used in the *Ms.* study is the vagueness of a couple of the questions in the survey. Although the exact legal definition of rape varies by state, most definitions involve sexual penetration accomplished against a person's will by means of physical force, threat of bodily harm, or when the victim is incapable of giving consent; the latter condition usually includes cases in which the victim is mentally ill, developmentally disabled, or intentionally incapacitated through the administration of intoxicating or anesthetic substances. Rape and attempted rape were operationally defined in the *Ms.* study by five questions, three of which referred to the threat or use of "some degree of physical force." But the other two questions asked were, "Have you had a man attempt sexual intercourse (get on top of you, attempt to insert his penis) when you didn't want to by giving you alcohol or drugs?" and "Have you had sexual intercourse when you didn't want to because a man gave you alcohol or drugs?" But what does it mean to have sex when you don't want to "because" a man gives you alcohol or drugs? How does one attempt intercourse "by" giving alcohol or drugs? Positive responses to these survey questions do not indicate whether duress, intoxication, force, or the threat of force was present; whether the woman's judgment or control was substantially impaired; or whether the man purposefully got the woman drunk in order to prevent her from resisting his sexual advances. They could conceivably mean that a woman was trading sex for drugs or that a few drinks lowered the respondent's inhibitions and she consented to an act she later regretted. While the items could have been clearly worded to denote the legal standard of "intentional incapacitation of the victim," as the questions stand there is no way to detect whether an affirmative response corresponds to the legal definition of rape. In 1993, Koss acknowledged that the drug and alcohol questions were ambiguous. When these questions are removed, according to Koss's revised estimates, the prevalence rate of rape and attempted rape declines by one-third.

Ignoring Discrepancies

One of the reasons the *Ms.* study has gained so much credibility is that Koss claims her findings are corroborated by many other studies. This creates the appearance that the *Ms.* findings are independently verified by a cumulative block of scientific evidence. However, an examination of these other studies reveals the highly variable methodology and questionable definitions employed in them, along with the tremendous discrepancies among the findings. Koss and Sarah Cook (a doctoral candidate in community psychology at the University of Virginia) refer to one such study this way: "In a just released telephone survey of more than 4,000 women in a nationally representative sample, the rate of rape was reported to be 14 percent, although this rate excluded rapes of women unable to consent." This, of course, is close to the 15 percent rate of rape claimed by the *Ms.* study. What Koss fails to tell the reader, however, is that approximately 40 percent of the rapes reported in this study affected women between the ages of 14 and 24, resulting in about a 5 percent rape prevalence estimate for that age group. According to the *Ms.* study, 15 percent of college women are victims of rape between the ages of 14 to 21. Thus, for almost the same age group, the *Ms.* study found a rate of rape *three times higher* than that detected by the National Victim Center survey. Only by sweeping the relevant details under the carpet of "supporting literature" can the results of studies such as these be construed as independently confirming the *Ms.* findings.

In addition, advocates claim that data in this field all tend in the same direction. All the data do not tend in the same direction. If the work of Koss, Russell, and others suggests that prevalence rates of rape and attempted rape rise from considerably above 10 percent to as much as one out of every two women, there are other studies that point to rates well below 10 percent. These studies, of course, rarely make headlines. Linda George (professor of psychiatry and sociology at Duke University), Idee Winfield (assistant professor of sociology at Louisiana State University), and Dan Blazer (Gibbons Professor of Psychiatry at Duke University), for example, found a 5.9 percent lifetime prevalence rate of sexual assault in North Carolina. This finding was based on responses to the broad question, "Has someone ever pressured you against your will into forced contact with the sexual parts of your body or their body?" While regional differences may account for some of the disparity between the *Ms.* findings and those of the Duke University team, in 1990 the rate of rapes reported to the police in North Carolina was 83 percent of the national average. An even lower prevalence rate was detected by Stephanie Riger and Margaret Gordon (policy researchers at Northwestern University), who reported that among 1,620 respondents randomly selected in Chicago, San Francisco, and Philadelphia, only 2 percent had been raped or sexually assaulted in their lifetime.

The most startling disparity is between the *Ms.* study's finding on the annual incidence of rape and attempted rape and the number of these offenses actually reported to the authorities on college campuses. Using her survey questions, Koss found that 166 women in 1,000 were victims of rape and attempted rape in just one year on campuses across the country (each victimized an average of 1.5 times). In sharp contrast, the 1993 FBI figures show that in 1992 at about 500 major colleges and universities with an overall population of 5 million students, only 408 cases of rape and attempted rape were reported to the police, less than one incident of rape and attempted rape per campus. This number yields an annual rate of .16 in 1,000 for female students, which is 1,000 times smaller than Koss's finding. Although it is generally agreed that many rape victims do not report their ordeal, no one to my knowledge publicly claims that the problem is 1,000 times greater than the cases reported to the police—a rate at which almost every woman in the country would be raped at least once every year.

Demonizing Men/Infantilizing Women/
Trivializing Violence

By promoting their figures on rape and violence, researchers who are also advocates seek to raise public consciousness about this issue. Raising consciousness can sometimes be positive. But they also seek to alter public consciousness by molding perceptions of the basic nature of the problem and of what constitutes the common experience of heterosexual relations. Thus, because they distort the issue, the effects can be harmful.

If it is true that one-third of female children are sexually abused and almost half of all women will suffer an average of two incidents of rape or attempted rape at some time in their lives, one is ineluctably driven to conclude that most men are pedophiles or rapists. This view of men is repeatedly expressed by some prominent activists in the child sexual abuse prevention movement and the rape crisis movement. As Diana Russell puts it, "Efforts to explain rape as a psychopathological phenomenon are inappropriate. How could it be that all these rapes are being perpetrated by a tiny segment of the male population?" Her explanation of "the truth that must be faced is that this culture's notion of masculinity—particularly as it is applied to male sexuality—predisposes men to violence, to rape, to sexually harass, and to sexually abuse children." In a similar vein, Koss notes that her findings support the view that sexual violence against woman "rests squarely in the middle of what our culture defines as 'normal' interaction between men and women." University of Michigan law professor Catherine MacKinnon offers a vivid rendition of the theme that rape is a social disease afflicting most men. Writing in the *New York Times,* she advises that when men charged with the crime of rape come to trial the court should ask, "Did this member of a group sexually trained to woman-hating aggression commit this particular act of woman-hating sexual aggression?"

Given such perspectives, the sexual politics of advocacy research on violence against women demonizes men and defines the common experience in heterosexual relations as inherently violent and menacing. This is the message that is being delivered on college campuses, and a frightening atmosphere is the result. As social critic Louis Menand observes: "The assumption that sexual relations among students at a progressive liberal-arts college should be thought of as per se fraught with the potential for violence is now taken for granted by just about everyone." A similar message is being delivered to students at a very early age by sexual abuse prevention programs in the schools. Many of these programs make a point of teaching kids as young as three that male family members, particularly fathers, uncles, and grandfathers may sexually abuse children.

Taking Back the Night

A puzzling question remains: If advocacy research trivializing violence against woman and its estimates of rape on campus are so greatly exaggerated compared to the cases reported to police and counseling centers, why are thousands of female students across the country marching by candlelight to take back the night? Something is going on here. Those in the rape crisis movement may take this anguished behavior as confirmation of widespread violence against college women. But there are several alternative explanations that suggest a climate in which statistics are used and abused and heavily influenced by the social dynamics of fear, power, conflict, and sex.

Inflated Fear In light of the statistics on sexual violence being bandied about, the view that take-back-the-night protests confirm an epidemic of rape can, of course, be turned on its head. Rather than responding to real violence against women on campus, these demonstrations may reflect a level of fear inflated by the exaggerated reports of rape, which create a premonition of danger that is vastly out of proportion to the actual risks of campus life. According to Katie Roiphe, a significant number of students are walking around with the alarming belief that 50 percent of women are raped. "This hyperbole," she notes, "contains within it a state of perpetual fear."

Power of Victimization One might argue as well that demonstrations to take back the night are incited less by irrational fear than by the sense of righteousness and power that often accompanies the invocation of victim status. "Even the privileged," as political scientist Charles Sykes tells us, "have found that being oppressed has its advantage."

Increased Gender Conflict Over the last few decades more women have entered college and joined the paid labor force than ever before. In 1960, most women were married before their twenty-first birthday. Today at that age they are competing with men for graduate school slots in almost every field and for jobs throughout the economy. An unprecedented number of women have also become heads of single-parent families, to which absent fathers often contribute minimal support. These developments enlarge occasions for friction and resentment in economic and social relations between the sexes, particularly among the college age group. The anxieties and animosities expressed in rallies to take back the night, in part, may reflect a real increase in tensions between men and women. This explanation suggests that "sexual assault" has come to represent a broad and vaguely formulated category of offenses, which ranges from rape to the incivilities of heightened gender conflict.

Sexual Turmoil Explanations that refer to fear, power, and conflict downplay the extent to which sexual activities influence the highly emotional responses to the threat of rape on campus. Sexual relations between young men and women, never exactly serene, have become more troublesome and confused in recent times. This is because there is more of it to manage, sex has become more dangerous, and it is harder to say no. Since 1960, as the median age for premarital intercourse declined and the median age for first marriage rose, the period of premarital sexual activity has grown from an average of 1 to 7.5 years for women. The sexual revolution in the late 1960s discounted the moral strictures against premarital sex. Without the shield of morality, it became more socially awkward to avoid having sex in situations where one's feelings were ambiguous, or even in situations where sex was totally unwanted. (It is easier to assert morality and say no because premarital sex is wrong, than to say no for all the other reasons: I don't love [or even like] you enough; I am not ready yet.) Shortly after the time span for sexual activity began to increase and the shield of morality fell, moreover, sexual relations became more dangerous as the risks of contracting AIDS and other sexually transmitted diseases rose to alarming levels.

These explanations are not mutually exclusive. They describe a cultural climate of strained gender relations under which advocacy research has been relatively successful in furthering radical sexual politics. The statistics fuel a powerful crusade. Theirs is a politics based on the convictions that violence is the norm in heterosexual relations, that all women are oppressed and deserve special treatment by government, and that from an early age women need to be empowered by publicly supported training and prevention programs. The radical feminist agenda coincides with the interests of some sexual abuse prevention professionals. This agenda does not, however, serve the interests of poor women, minorities, teenage boys, and other groups most in need of social protection from violence.

Misdirected Policy

In the deliberations on federal policy regarding violence against women, advocacy research on rape was accepted almost at face value. In the opening statement during Senate hearings on the Violence Against Women Act in 1990, Chairman Joseph Biden explained:

> One out of every four college women will have been attacked by a rapist before they graduate, and one in seven will have been raped. Less than 5 percent of these women will report these rapes to the police. Rape remains the least reported of all major crimes. . . . Dr. Koss will tell us today the actual number of college women raped is more than 14 times the number reported by official governmental statistics. Indeed, while studies suggest that about 1,275 women were raped at America's three largest universities last year, only three rapes—only three—were reported to the police.

The panel was informed that unreported rapes of college women are more than 400 times the number of incidents reported to the police. In fact, as noted earlier, according to Koss's figures the unreported rate is about 1,000 times higher than the rate of rape and attempted rape reported on college campuses. And Koss's conclusions were never seriously challenged.

Due to the credibility attributed to advocacy research, under Title IV (Safe Campuses for Women), the Violence Against Women Act of 1993 proposes to appropriate $20 million for rape education and prevention programs to make college campuses safe for women. Extrapolating from the 408 reported cases of rape and attempted rape on 500 campuses in 1992 yields approximately 1,000 reported incidents among all colleges in the United States. The appropriation proposed under Title IV thus amounts to $20,000 per reported case.

Compared to the $20 million designated for college campuses, the Violence Against Women Act proposes to appropriate $65 million (approximately $650 per reported case) for prevention and education programs to serve the broader community. Under this arrangement, a disproportionate sum will be distributed for education and prevention programs on college campuses, where most victims, according to the *Ms.* study, do not know they have been raped. And comparatively meager resources will be invested in similar programs for low-income and minority communities. Whatever the value of the college programs, the cost is remarkably high compared to the funds allocated per reported case of victims outside of college campuses, among whom poor and minority women are vastly overrepresented, according to the Bureau of Justice Statistics. According to the 1993 FBI data, overall there were 109,062 reported cases of rape and attempted rape in 1992. Based on these rates, to make the rest of society as safe as college campuses would require an expenditure of $2 billion on educational and prevention programs.

It is difficult to criticize advocacy research without giving the impression that one cares less about the problems of victims than do those who magnify the size of problems such as rape and child abuse. Advocacy researchers who uncover a problem, measure it with reasonable accuracy, and bring it to public attention perform a valuable service by raising public consciousness. But the current trend in research on sexual violence against women is to inflate the problem and redefine it in line with the advocates' ideological preferences. The few impose their definition of social ills on the many. By creating sensational headlines, this type of advocacy research invites the formulation of social policies that are likely to be neither effective nor fair.

5

Field Research Among Deviants

A Consideration of Some Methodological Recommendations

ROBERT J. KELLY

Field work depends to a great extent on complicated and continuous concerns, the most central of which is how one is seen by others. It is a matter filled with risks and contradictions.

Constructing and managing fronts, manipulating the impression of self (or selves) in order to gain access, and maintain acceptance with deviant populations under study, have been the subject of much discussion among sociologists.[1] A key element of the debate in the literature focuses on questions of interviewer-observer effects, on issues of objectivity and how observer presence potentially contaminates the phenomenon under study. Despite these and other threats to scientific integrity, most field workers seem to be satisfied that their research accurately depicts the social reality selected for study and that their results are scientifically valid. In writing about his field study of black streetcorner men, Liebow (1967) remarked that "the people I was observing knew that I was observing them, yet they allowed me to participate in their activities and take part in their lives to a degree that continues to surprise me" (p. 253). Gans (1962), in writing about his experiences in an Italian-American lower class ghetto, was more cautious but no less optimistic. For him, the problem endangering the research was the likelihood of overidentification with his subjects, a problem that is especially troublesome when those being studied are deprived, victimized, or oppressed. Empathizing strongly with social underdogs detracts (at least theoretically) from the objectivity of the research, according to Gans. Nevertheless, these problems are not insoluble in his view.

With the study of career criminals and criminal groups, however, the methodological problems are exacerbated and become significant obstacles in the conduct of field work. Regarding organized crime, for example, Cressey (1967) wrote

> The secrecy of participants, the confidentiality of materials collected by investigative agencies, and filters or screens on the perceptive apparatus of informants and investigators pose serious methodological problems for the social scientist. (p. 101)

From "Field Research Among Deviants," Robert Kelly, Deviant Behavior V. 3, No. 3, Taylor & Francis, Inc., Washington, D.C., pp. 219–228, 1982. Reprinted with permission. All rights reserved.

And in *Theft of the Nation* (1969), he noted that the activities of organized criminals are not accessible to observation by the social scientist. Moreover, even gaining access to law enforcement groups that specialize in organized crime investigations requires, in his words, "connections."

For Cressey and many others, the vaunted secretive nature of criminal syndicate activity precludes the use of many, if not all, of the usual instruments for gathering data. If crime families and syndicates take pains to conceal their activities from law enforcement agents, then it is unlikely that they would disclose sensitive information about their structure to a social scientist or permit him to examine their inner workings. Thus, social scientists must work at several removes from their data.

To a great extent, most of the information available to law enforcement agencies and social scientists derives from two basic sources: the participants themselves and electronic eavesdropping devices (Salerno, 1978). The consequences these methodological hurdles present are ominous. For one, social scientists must rely on secondary and tertiary accounts of organized crime and therefore cannot independently evaluate their findings in accordance with scientific methods. Distortions invariably arise. Second, accounts of organized crime prepared by governmental agencies focus on the criminal act and criminal actor and tend to ignore the complicity of the upperworld in sustaining criminal syndicate activity (Chambliss, 1971; Ianni, 1977). Thus, a bias permeates the bulk of the writing on organized crime from which most social scientists draw their information.

In recent years, some social scientists, notably Ianni (1973, 1974), Chambliss (1971) and Klockars (1974) have attempted field work among organized criminals and have collected data using participant-observation techniques. Chambliss (1975) indicated that a wealth of data about syndicate activity is to be found on the streets, where, incidentally, the absence of sociologists is all too obvious. A review of these studies and others devoted to career criminals and other deviants reveals that much of their methodological inspiration obtains directly or indirectly from the ethnographic work and methodological suggestions of Polsky (1967).

Polsky's refreshing iconoclasm injected some vitality into the debate on research methods and directly challenged a widespread point of view among law enforcement officials that "those who know won't tell and those who tell don't know." Put another way, Polsky's views clash with the traditional belief that criminals, especially those affiliated with crime syndicates, are not only difficult to identify but are disinclined to provide information about themselves and their illegal activities.

The few advocates of open-air research among deviants, "street sociologists" as they are sometimes known, have been subjected to attack because it is claimed that their work does not lend itself to conventional standards of verification and replication. Research samples, it is argued, are not ordinarily unbiased. The most important criticism, however, is that the collection instrument itself, the field researcher, can scarcely be duplicated objectively. In short, the utilization of precisely defined and controlled techniques of observation cannot be employed in field research. In reply to this, Polsky (1977) argues, quite persuasively, that "precise controlled techniques of observation" are something of a fetish and as a guideline to research quite ambiguous. Even among non-criminal populations, field research, as many widely hailed accounts attest,[2] hardly follows exactly reproducible lines of data collection. The insistence that one strictly adhere to prefield protocols amounts to more of an imposition than an aid.

Another dimension of the participant-observer role frequently ignored in discussions of qualitative research concerns the changing role relationships among researchers and subjects. Ianni, Klockars, and others point out that whatever the initial focus of the research and the role relationships envisioned by the researcher, these are bound to change. According to Klockars (1976), "participant-observation research is never conducted within the bounds of a single role relationship." Informants and subjects are not merely information sources, but often guides, informal teachers, and occasionally friends. Ianni (1972) reported similar results: a bargain emerged between him and his principal informants in which he provided assistance and advice concerning a range of private and public matters about which he had knowledge in exchange for access to subjects.

The implications of these social relationships have important consequences for both the quantity and the quality of knowledge generated in the research setting. Because field research usually involves relatively prolonged periods of contact and interaction with subjects, the characteristics that these relationships take on can affect what is learned about them. In addition, and no less important, interacting with subjects in their natural settings may transform researchers themselves, socializing them to some degree (going native) to the extent that they begin to become, at last, somewhat different people in interests, tastes and purpose. Because the investigators are the data collection instruments, changing role relationships and attitudes occasioned by deeper immersion into the settings may dramatically affect knowledge production. In part, this is what Polsky (1967) is taken to mean when he says that studying the criminal in his natural setting

> (1) allows you to make observations about his lifestyle you ordinarily wouldn't make, (2) causes you to think of important questions about him that ordinarily you wouldn't think of, (3) causes him to think of relevant things to tell you that he otherwise wouldn't think of, and (4) causes him to make explanations of certain events to you that he ordinarily wouldn't make. (pp. 135–136)

Briefly, Polsky's suggestions for the conduct of field work among criminals in their natural settings would apply to any type of observation. In fact, participant observation among deviants is for Polsky merely a special case of general qualitative techniques. The questions Polsky addresses are these:

- How does one determine the culture and social structure of a group without creating complications for the research?

- How does one determine and then preserve the "natural setting" that criminals inhabit?

- To what extent can research interests be revealed without jeopardizing the study by giving the subjects an orientation by which they provide dramatic presentations of what the researcher wants?

As noted, the rationale justifying studies of illegal activities and occupations is that such occupations and lifestyles can and should be perceived as instances of general occupational theory.

In his essay, "Research Method, Morality, and Criminology," Polsky (1967) enumerates some rules of thumb to guide the field worker in deviant groups and insists that these procedural recipes apply in other research environments as well. Participant observation of criminals, as with participant observation of any others, involves getting thoroughly familiar with their frame of reference, language, and interests. This means close observation without becoming too obtrusive of criminals in their hangouts and play scenes.[3] Blending into the setting by discarding most of the usual research paraphernalia (tape recorders, questionnaires, etc.) may reduce potential information gathering but sharpen observational skills. The researcher who is faced with restrictions on presence and participation and who knows that opportunities to observe particular pieces of action may not occur again may evolve into a more skillful and astute observer.

Once entry has been accomplished, the development, maintenance, and extension of contacts, in short, the sustenance of the research situation, require the clear separation of the researcher from criminal activity and identification and a lack of pretense regarding the purposes of the investigation. It is important that subjects understand that the field worker is neither a law enforcement official nor an informant (in the pejorative sense in which the term is customarily understood in the under- and upperworlds).

Cultivating rapport triggers a snowball effect in which a network of informants is likely to emerge. In work among criminal syndicate members, I found that not until after it was established to everyone's satisfaction that I was not interested in names, places, and specific criminal acts was I introduced to others as someone around whom one could talk freely. In most cases, however, it was clear that unguarded conversations were exceptional despite assurances to the contrary. For the most part, even those with whom I thought I had created implicit trust were often reserved in conversations. On some occasions, though, as I later discovered, I was actually present during some illegal transaction involving disbursements of cash, distributions of hijacked goods (swag), and large drug (marijuana) deals without realizing what was taking place.

Exposure to crimes raises some delicate ethical issues for the field researcher. In a taped interview, Polsky (1977) indicated that he preferred not to witness criminal acts and informed his subjects accordingly. And nowhere does Ianni say that he observed criminal acts. A careful inspection of Polsky's and Ianni's statements about their field work experience shows that their field work, like that of most others, with the possible exceptions of Chambliss and

Klockars, was based largely on verbal accounts of criminal acts. Thus, neither Polsky nor Ianni could be cited for complicity in criminal acts.

As data, verbal accounts are conceptually distinguishable from data based on observation in the pure sense. But in practical terms, as Irwin (1972) points out, "the great bulk of data sought in field studies does appear in conversations" (p. 119). Moreover, participant observation need not necessarily entail actually participating with subjects in their activities in order to be meaningful. Second, the question naturally arises as to techniques and solutions available to determine the reliability of data derived mainly from interactions grounded in unstructured talk and conversations as distinct from organized interviews. Here it seems that a network of informants provides some foundation for cross-checking the dependability of information elicited from subjects. Also, extensive contacts with other sources, such as law enforcement agencies and others on the periphery of crime (such as journalists) can be of immense value in determining how well the data agree with verified facts.

An expanding informant network, immersion in the social surrounds of subjects, and exchanges of services and resources for investigator access and information are all extremely important devices and techniques that can be profitably exploited in field work. . . .

NOTES

1. On the front management, see Goffman (1962), Irwin (1972), Ianni (1973), and Polsky (1967). The literature on participant observation among nondeviant populations contains classic statements by Whyte (1943), Levi-Strauss (1968), and Liebow (1967).

2. See, on this issue, a collection of essays on field experiences (Hammond, 1964).

3. Whyte (1943) frittered away his time in Cornerville bars until he asked a social worker to introduce him to some slum dwellers. Simply "hanging out" is a necessary but not sufficient condition for the creation of informant networks.

REFERENCES

Chambliss, William. 1971. "Vice, corruption, bureaucracy and power." *Wisconsin Law Review* 4: 1150–1173.

———. 1975. "On the paucity of original research on organized crime: A footnote to Galliher and Cain." *American Sociologist* 10: 36–39.

Cressey, Donald. 1967. "Methodological problems in the study of organized crime as a social problem." *Annals of American Academy of Political and Social Science* 374: 101–112.

———. 1969. *Theft of the Nation: The Structure and Operations of Organized Crime in America*. New York: Harper & Row.

Gans, Herbert. 1962. *The Urban Villagers: Group and Class in the Life of Italian-Americans*. New York: Free Press.

Goffman, Erving. 1962. *Asylums: Essays on the Social Situation of Mental Patients and Other Inmates*. Chicago: Aldine.

Hammond, Phillip E., Ed. 1964. *Sociologists at Work: Essays on the Craft of Social Research*. New York: Basic Books.

Ianni, Francis. 1972. *A Family Business: Kinship and Social Control in Organized Crime*. New York: Russell Sage Foundation.

———. 1974. *Black Mafia: Ethnic Succession in Organized Crime*. New York: Simon and Schuster.

———. 1977. "Ideology and Field Research Theory on Organized Crime." Unpublished paper.

Irwin, John. 1972. "Participant observation of criminals." In Jack Douglas (ed.), *Research on Deviance*. New York: Random House.

Klockars, Carl. 1974. The Professional Fence. New York: Free Press.

———. 1976. Field ethics for the life history. Unpublished paper presented to Workshop on Ethnography of Drugs and Crime, Miami, Florida, May 5–7.

Levi-Strauss, Claude. 1968. *Triste Tropiques*. New York: Atheneum.

Liebow, Elliot. 1967. *Tally's Corner: A Study of Negro Streetcorner Men*. Boston: Little, Brown.

Polsky, Ned. 1967. *Hustlers, Beats and Others*. Chicago: Aldine.

———. 1977. Unpublished tape-recorded interview, Feb. 15.

Salerno, Ralph. 1978. "The structure of organized crime." Testimony before the House of Representatives Committee on Assassinations.

Whyte, William F. 1943. *Street Corner Society: The Social Structure of an Italian Slum*. Chicago: University of Chicago Press.

6

Researching Dealers and Smugglers

PATRICIA A. ADLER

I strongly believe that investigative field research (Douglas 1976), with emphasis on direct personal observation, interaction, and experience, is the only way to acquire accurate knowledge about deviant behavior. Investigative techniques are especially necessary for studying groups such as drug dealers and smugglers because the highly illegal nature of their occupation makes them secretive, deceitful, mistrustful, and paranoid. To insulate themselves from the straight world, they construct multiple false fronts, offer lies and misinformation, and withdraw into their group. In fact, detailed, scientific information about upper-level drug dealers and smugglers is lacking precisely because of the difficulty sociological researchers have had in penetrating into their midst. As a result, the only way I could possibly get close enough to these individuals to discover what they were doing and to understand their world from their perspectives (Blumer 1969) was to take a membership role in the setting. While my different values and goals precluded my becoming converted to complete membership in the subculture, and my fears presented my ever becoming "actively" involved in their trafficking activities, I was able to assume a "peripheral" membership role (Adler and Adler 1987). I became a member of the dealers' and smugglers' social world and participated in their daily activities on that basis. In this chapter, I discuss how I gained access to this group, established research relations with members, and how personally involved I became in their activities.

GETTING IN

When I moved to Southwest County [California] in the summer of 1974, I had no idea that I would soon be swept up in a subculture of vast drug trafficking and unending partying, mixed with occasional cloak-and-dagger subterfuge. I had moved to California with my husband, Peter, to attend graduate school in sociology. We rented a condominium townhouse near the beach and started taking classes in the fall. We had always felt that socializing exclusively

From Patricia A. Adler, *Wheeling and Dealing* (New York: Columbia University Press, 1985). © Columbia University Press. Reprinted by permission of the publisher.

with academicians left us nowhere to escape from our work, so we tried to meet people in the nearby community. One of the first friends we made was our closest neighbor, a fellow in his late twenties with a tall, hulking frame and gentle expression. Dave, as he introduced himself, was always dressed rather casually, if not sloppily, in T-shirts and jeans. He spent most of his time hanging out or walking on the beach with a variety of friends who visited his house, and taking care of his two young boys, who lived alternately with him and his estranged wife. He also went out of town a lot. We started spending much of our free time over at his house, talking, playing board games late into the night, and smoking marijuana together. We were glad to find someone from whom we could buy marijuana in this new place, since we did not know too many people. He also began treating us to a fairly regular supply of cocaine, which was a thrill because this was a drug we could rarely afford on our student budgets. We noticed right away, however, that there was something unusual about his use and knowledge of drugs: while he always had a plentiful supply and was fairly expert about marijuana and cocaine, when we tried to buy a small bag of marijuana from him he had little idea of the going price. This incongruity piqued our curiosity and raised suspicion. We wondered if he might be dealing in larger quantities. Keeping our suspicions to ourselves, we began observing Dave's activities a little more closely. Most of his friends were in their late twenties and early thirties and, judging by their lifestyles and automobiles, rather wealthy. They came and left his house at all hours, occasionally extending their parties through the night and the next day into the following night. Yet throughout this time we never saw Dave or any of his friends engage in any activity that resembled a legitimate job. In most places this might have evoked community suspicion, but few of the people we encountered in Southwest County seemed to hold traditionally structured jobs. Dave, in fact, had no visible means of financial support. When we asked him what he did for a living, he said something vague about being a real estate speculator, and we let it go at that. We never voiced our suspicions directly since he chose not to broach the subject with us.

We did discuss the subject with our mentor, Jack Douglas, however. He was excited by the prospect that we might be living among a group of big dealers, and urged us to follow our instincts and develop leads into the group. He knew that the local area was rife with drug trafficking, since he had begun a life history case study of two drug dealers with another graduate student several years previously. That earlier study was aborted when the graduate student quit school, but Jack still had many hours of taped interviews he had conducted with them, as well as an interview that he had done with an undergraduate student who had known the two dealers independently, to serve as a cross-check on their accounts. He therefore encouraged us to become friendlier with Dave and his friends. We decided that if anything did develop out of our observations of Dave, it might make a nice paper for a field methods class or independent study.

Our interests and background made us well suited to study drug dealing. First, we had already done research in the field of drugs. As undergraduates at

Washington University we had participated in a nationally funded project on urban heroin use (see Cummins et al. 1972). Our role in the study involved using fieldwork techniques to investigate the extent of heroin use and distribution in St. Louis. In talking with heroin users, dealers, and rehabilitation personnel, we acquired a base of knowledge about the drug world and the subculture of drug trafficking. Second, we had a generally open view toward soft drug use, considering moderate consumption of marijuana and cocaine to be generally nondeviant. This outlook was partially etched by our 1960s-formed attitudes, as we had first been introduced to drug use in an environment of communal friendship, sharing, and counterculture ideology. It also partially reflected the widespread acceptance accorded to marijuana and cocaine use in the surrounding local culture. Third, our age (mid-twenties at the start of the study) and general appearance gave us compatibility with most of the people we were observing.

We thus watched Dave and continued to develop our friendship with him. We also watched his friends and got to know a few of his more regular visitors. We continued to build friendly relations by doing, quite naturally, what Becker (1963), Polsky (1969), and Douglas (1972) had advocated for the early stages of field research: we gave them a chance to know us and form judgments about our trustworthiness by jointly pursuing those interests and activities which we had in common.

Then one day something happened which forced a breakthrough in the research. Dave had two guys visiting him from out of town and, after snorting quite a bit of cocaine, they turned their conversation to a trip they had just made from Mexico, where they piloted a load of marijuana back across the border in a small plane. Dave made a few efforts to shift the conversation to another subject, telling them to "button their lips," but they apparently thought that he was joking. They thought that anybody as close to Dave as we seemed to be undoubtedly knew the nature of his business. They made further allusions to his involvement in the operation and discussed the outcome of the sale. We could feel the wave of tension and awkwardness from Dave when this conversation began, as he looked toward us to see if we understood the implications of what was being said, but then he just shrugged it off as done. Later, after the two guys left, he discussed with us what happened. He admitted to us that he was a member of a smuggling crew and a major marijuana dealer on the side. He said that he knew he could trust us, but that it was his practice to say as little as possible to outsiders about his activities. This inadvertent slip, and Dave's subsequent opening up, were highly significant in forging our entry into Southwest County's drug world. From then on he was open in discussing the nature of his dealing and smuggling activities with us.

He was, it turned out, a member of a smuggling crew that was importing a ton of marijuana weekly and 40 kilos of cocaine every few months. During that first winter and spring, we observed Dave at work and also got to know the other members of his crew, including Ben, the smuggler himself. Ben was also very tall and broad shouldered, but his long black hair, now flecked with gray, bespoke his earlier membership in the hippie subculture. A large physical stature, we observed, was common to most of the male participants involved in this drug community. The women also had a unifying physical trait: they were extremely attractive and stylishly dressed. This included Dave's ex-wife, Jean, with whom he reconciled during the spring. We therefore became friendly with Jean and through her met a number of women ("dope chicks") who hung around the dealers and smugglers. As we continued to gain the friendship of Dave and Jean's associates we were progressively admitted into their inner circle and apprised of each person's dealing or smuggling role.

Once we realized the scope of Ben's and his associates' activities, we saw the enormous research potential in studying them. This scene was different from any analysis of drug trafficking that we had read in the sociological liter-ature because of the amounts they were dealing and the fact that they were importing it themselves. We decided that, if it was at all possible, we would capitalize on this situation, to "opportunistically" (Riemer 1977) take advan-tage of our prior expertise and of the knowledge, entrée, and rapport we had already developed with several key people in this setting. We therefore dis-cussed the idea of doing a study of the general subculture with Dave and sev-eral of his closest friends (now becoming our friends). We assured them of the anonymity, confidentiality, and innocuousness of our work. They were happy to reciprocate our friendship by being of help to our professional careers. In fact, they basked in the subsequent attention we gave their lives.

We began by turning first Dave, then others, into key informants and col-lecting their life histories in detail. We conducted a series of taped, depth in-terviews with an unstructured, open-ended format. We questioned them about such topics as their backgrounds, their recruitment into the occupation, the stages of their dealing careers, their relations with others, their motiva-tions, their lifestyle, and their general impressions about the community as a whole.

We continued to do taped interviews with key informants for the next six years until 1980, when we moved away from the area. After that, we occa-sionally did follow-up interviews when we returned for vacation visits. These later interviews focused on recording the continuing unfolding of events and included detailed probing into specific conceptual areas, such as dealing net-works, types of dealers, secrecy, trust, paranoia, reputation, the law, occupa-tional mobility, and occupational stratification. The number of taped interviews we did with each key informant varied, ranging between 10 and 30 hours of discussion.

Our relationship with Dave and the others thus took on an added dimen-sion—the research relationship. As Douglas (1976), Henslin (1972), and Wax (1952) have noted, research relationships involve some form of mutual ex-change. In our case, we offered everything that friendship could entail. We did routine favors for them in the course of our everyday lives, offered them insights and advice about their lives from the perspective of our more re-spectable position, wrote letters on their behalf to the authorities when they got in trouble, testified as character witnesses at their non-drug-related trials, and loaned them money when they were down and out. When Dave was arrested and brought to trial for check-kiting, we helped Jean organize his defense and raise the money to pay his fines. We spelled her in taking care of the children so that she could work on his behalf. When he was eventually sent to the state prison we maintained close ties with her and discussed our mutual efforts to buoy Dave up and secure his release. We also visited him in jail. During Dave's incarceration, however, Jean was courted by an old boyfriend and gave up her reconciliation with Dave. This proved to be another significant turning point in our research because, desperate for money, Jean looked up Dave's old dealing connections and went into the business herself. She did not stay with these marijuana dealers and smugglers for long, but soon moved into the cocaine business. Over the next several years her experiences in the world of cocaine dealing brought us into contact with a different group of people. While these people knew Dave and his associates (this was very common in the Southwest County dealing and smuggling community), they did not deal with them directly. We were thus able to gain access to a much wider and more diverse range of subjects than we would have had she not branched out on her own.

Dave's eventual release from prison three months later brought our involvement in the research to an even deeper level. He was broke and had nowhere to go. When he showed up on our doorstep, we took him in. We offered to let him stay with us until he was back on his feet again and could afford a place of his own. He lived with us for seven months, intimately sharing his daily experiences with us. During this time we witnessed, firsthand, his transformation from a scared ex-con who would never break the law again to a hard-working legitimate employee who only dealt to get money for his children's Christmas presents, to a full-time dealer with no pretensions at legitimate work. Both his process of changing attitudes and the community's gradual reacceptance of him proved very revealing.

We socialized with Dave, Jean, and other members of Southwest County's dealing and smuggling community on a near-daily basis, especially during the first four years of the research (before we had a child). We worked in their legitimate businesses, vacationed together, attended their weddings, and cared for their children. Throughout their relationship with us, several participants became co-opted to the researcher's perspective[1] and actively sought out instances of behavior which filled holes in the conceptualizations we were developing. Dave, for one, became so intrigued by our conceptual dilemmas that he undertook a "natural experiment" entirely on his own, offering an unlimited supply of drugs to a lower-level dealer to see if he could work up to higher levels of dealing, and what factors would enhance or impinge upon his upward mobility.

In addition to helping us directly through their own experiences, our key informants aided us in widening our circle of contacts. For instance, they let us know when someone in whom we might be interested was planning on dropping by, vouching for our trustworthiness and reliability as friends who could be included in business conversations. Several times we were even awakened in the night by phone calls informing us that someone had dropped by for a visit, should we want to "casually" drop over too. We rubbed the sleep from our eyes, dressed, and walked or drove over, feeling like sleuths out of a television series. We thus were able to snowball, through the active efforts of our key informants,[2] into an expanded study population. This was supplemented by our own efforts to cast a research net and befriend other dealers, moving from contact to contact slowly and carefully through the domino effect.

THE COVERT ROLE

The highly illegal nature of dealing in illicit drugs and dealers' and smugglers' general level of suspicion made the adoption of an overt research role highly sensitive and problematic. In discussing this issue with our key informants, they all agreed that we should be extremely discreet (for both our sakes and theirs). We carefully approached new individuals before we admitted that we were studying them. With many of these people, then, we took a covert posture in the research setting. As nonparticipants in the business activities which bound members together into the group, it was difficult to become fully accepted as peers. We therefore tried to establish some sort of peripheral, social membership in the general crowd, where we could be accepted as "wise" (Goffman 1963) individuals and granted a courtesy membership. This seemed an attainable goal, since we had begun our involvement by forming such relationships with our key informants. By being introduced to others in this wise rather than overt role, we were able to interact with people who would otherwise have shied away from us. Adopting a courtesy membership caused us to bear a courtesy stigma,[3] however, and we suffered since we, at times, had to disguise the nature of our research from both lay outsiders and academicians.

In our overt posture we showed interest in dealers' and smugglers' activities, encouraged them to talk about themselves (within limits, so as to avoid acting like narcs), and ran home to write field notes. This role offered us the advantage of gaining access to unapproachable people while avoiding researcher effects, but it prevented us from asking some necessary, probing questions and from tape recording conversations.[4] We therefore sought, at all times, to build toward a conversion to the overt role. We did this by working to develop their trust.

DEVELOPING TRUST

Like achieving entrée, the process of developing trust with members of unorganized deviant groups can be slow and difficult. In the absence of a formal structure separating members from outsiders, each individual must form his or her own judgment about whether new persons can be admitted to their confidence. No gatekeeper existed to smooth our path to being trusted, although our key informants acted in this role whenever they could by providing introductions and references. In addition, the unorganized nature of this group meant that we met people at different times and were constantly at different levels in our developing relationships with them. We were thus trusted more by some people than by others, in part because of their greater familiarity with us. But as Douglas (1976) has noted, just because someone knew us or even liked us did not automatically guarantee that they would trust us.

We actively tried to cultivate the trust of our respondents by tying them to us with favors. Small things, like offering the use of our phone, were followed with bigger favors, like offering the use of our car, and finally really meaningful favors, like offering the use of our home. Here we often trod a thin line, trying to ensure our personal safety while putting ourselves in enough of a risk position, along with our research subjects, so that they would trust us. While we were able to build a "web of trust" (Douglas 1976) with some members, we found that trust, in large part, was not a simple status to attain in the drug world. Johnson (1975) has pointed out that trust is not a one-time phenomenon, but an ongoing developmental process. From my experiences in this research I would add that it cannot be simply assumed to be a one-way process either, for it can be diminished, withdrawn, reinstated to varying degrees, and re-questioned at any point. Carey (1972) and Douglas (1972) have remarked on this waxing and waning process, but it was especially pronounced for us because our subjects used large amounts of cocaine over an extended period of time. This tended to make them alternately warm and cold to us. We thus lived through a series of ups and downs with the people we were trying to cultivate as research informants.

THE OVERT ROLE

After this initial covert phase, we began to feel that some new people trusted us. We tried to intuitively feel when the time was right to approach them and go overt. We used two means of approaching people to inform them that we were involved in a study of dealing and smuggling: direct and indirect. In some cases our key informants approached their friends or connections and, after vouching for our absolute trustworthiness, convinced these associates to talk to us. In other instances, we approached people directly, asking for their help with our project. We worked our way through a progression with these secondary contacts, first discussing the dealing scene overtly and later moving to taped life history interviews. Some people reacted well to us, but others responded skittishly, making appointments to do taped interviews only to break them as the day drew near, and going through fluctuating stages of being honest with us or putting up fronts about their dealing activities. This varied, for some, with their degree of active involvement in the business. During the times when they had quit dealing, they would tell us about their present and past activities, but when they became actively involved again, they would hide it from us.

This progression of covert to overt roles generated a number of tactical difficulties. The first was the problem of *coming on too fast* and blowing it. Early in
the research we had a dealer's old lady (we thought) all set up for the direct approach. We knew many dealers in common and had discussed many things tan-
gential to dealing with her without actually mentioning the subject. When we
asked her to do a taped interview of her bohemian lifestyle, she agreed without
hesitation. When the interview began, though, and she found out why we
were interested in her, she balked, gave us a lot of incoherent jumble, and ended
the session as quickly as possible. Even though she lived only three houses away
we never saw her again. We tried to move more slowly after that.

A second problem involved simultaneously *juggling our overt and covert roles*
with different people. This created the danger of getting our cover blown with
people who did not know about our research (Henslin 1972). It was very confusing to separate the people who knew about our study from those who did
not, especially in the minds of our informants. They would make occasional
veiled references in front of people, especially when loosened by intoxicants,
that made us extremely uncomfortable. We also frequently worried that our
snooping would someday be mistaken for police tactics. Fortunately, this never
happened.

CROSS-CHECKING

The hidden and conflictual nature of the drug dealing world made me feel the
need for extreme certainty about the reliability of my data. I therefore based
all my conclusions on independent sources and accounts that we carefully verified. First, we tested information against our own common sense and general
knowledge of the scene. We adopted a hard-nosed attitude of suspicion, assuming people were up to more than they would originally admit. We kept
our attention especially riveted on "reformed" dealers and smugglers who
were living better than they could outwardly afford, and were thereby able to
penetrate their public fronts.

Second, we checked out information against a variety of reliable sources.
Our own observations of the scene formed a primary reliable source, since
we were involved with many of the principals on a daily basis and knew exactly what they were doing. Having Dave live with us was a particular advantage because we could contrast his statements to us with what we could
clearly see was happening. Even after he moved out, we knew him so well
that we could generally tell when he was lying to us or, more commonly,
fooling himself with optimistic dreams. We also observed other dealers' and
smugglers' evasions and misperceptions about themselves and their activities.
These usually occurred when they broke their own rules by selling to people
they did not know, or when they commingled other people's money with
their own. We also cross-checked our data against independent, alternative
accounts. We were lucky, for this purpose, that Jean got reinvolved in the

drug world. By interviewing her, we gained additional insight into Dave's past, his early dealing and smuggling activities, and his ongoing involvement from another person's perspective. Jean (and her connections) also talked to us about Dave's associates, thereby helping us to validate or disprove their statements. We even used this pincer effect to verify information about people we had never directly interviewed. This occurred, for instance, with the tapes that Jack Douglas gave us from his earlier study. After doing our first round of taped interviews with Dave, we discovered that he knew the dealers Jack had interviewed. We were excited by the prospect of finding out what had happened to these people and if their earlier stories checked out. We therefore sent Dave to do some investigative work. Through some mutual friends he got back in touch with them and found out what they had been doing for the past several years.

Finally, wherever possible, we checked out accounts against hard facts: newspaper and magazine reports; arrest records; material possessions; and visible evidence. Throughout the research, we used all these cross-checking measures to evaluate the veracity of new information and to prod our respondents to be more accurate (by abandoning both their lies and their self-deceptions).[5]

After about four years of near-daily participant observation, we began to diminish our involvement in the research. This occurred gradually, as first pregnancy and then a child hindered our ability to follow the scene as intensely and spontaneously as we had before. In addition, after having a child, we were less willing to incur as many risks as we had before; we no longer felt free to make decisions based solely on our own welfare. We thus pulled back from what many have referred to as the "difficult hours and dangerous situations" inevitably present in field research on deviants (see Becker 1963; Carey 1972; Douglas 1972). We did, however, actively maintain close ties with research informants (those with whom we had gone overt), seeing them regularly and periodically doing follow-up interviews.

PROBLEMS AND ISSUES

Reflecting on the research process, I have isolated a number of issues which I believe merit additional discussion. These are rooted in experiences which have the potential for greater generic applicability.

The first is the *effect of drugs on the data-gathering process.* Carey (1972) has elaborated on some of the problems he encountered when trying to interview respondents who used amphetamines, while Wax (1952, 1957) has mentioned the difficulty of trying to record field notes while drinking sake. I found that marijuana and cocaine had nearly opposite effects from each other. The latter helped the interview process, while the former hindered it. Our attempts to interview respondents who were stoned on marijuana were unproductive for a number of reasons. The primary obstacle was the effects of the drug. Often, people became confused, sleepy, or involved in eating to varying degrees. This distracted them from our purpose. At times, people even simulated overreactions to marijuana to hide behind the drug's supposed disorienting influence and thereby avoid divulging information. Cocaine, in contrast, proved to be a research aid. The drug's warming and sociable influence opened people up, diminished their inhibitions, and generally increased their enthusiasm for both the interview experience and us.

A second problem I encountered involved *assuming risks while doing research*. As I noted earlier, dangerous situations are often generic to research on deviant behavior. We were most afraid of the people we studied. As Carey (1972), Henslin (1972), and Whyte (1955) have stated, members of deviant groups can become hostile toward a researcher if they think that they are being treated wrongfully. This could have happened at any time from a simple occurrence, such as a misunderstanding, or from something more serious, such as our covert posture being exposed. Because of the inordinate amount of drugs they consumed, drug dealers and smugglers were particularly volatile, capable of becoming malicious toward each other or us with little warning. They were also likely to behave erratically owing to the great risks they faced from the police and other dealers. These factors made them moody, and they vacillated between trusting us and being suspicious of us.

At various times we also had to protect our research tapes. We encountered several threats to our collection of taped interviews from people who had granted us these interviews. This made us anxious, since we had taken great pains to acquire these tapes and felt strongly about maintaining confidences entrusted to us by our informants. When threatened, we became extremely frightened and shifted the tapes between different hiding places. We even ventured forth one rainy night with our tapes packed in a suitcase to meet a person who was uninvolved in the research at a secret rendezvous so that he could guard the tapes for us.

We were fearful, lastly, of the police. We often worried about local police or drug agents discovering the nature of our study and confiscating or subpoenaing our tapes and field notes. Sociologists have no privileged relationship with their subjects that would enable us legally to withhold evidence from the authorities should they subpoena it.[6] For this reason we studiously avoided any publicity about the research, even holding back on publishing articles in scholarly journals until we were nearly ready to move out of the setting. The closest we came to being publicly exposed as drug researchers came when a former sociology graduate student (turned dealer, we had heard from inside sources) was arrested at the scene of a cocaine deal. His lawyer wanted us to testify about the dangers of doing drug-related research, since he was using his research status as his defense. Fortunately, the crisis was averted when his lawyer succeeded in suppressing evidence and had the case dismissed before the trial was to have begun. Had we been exposed, however, our respondents would have acquired guilt by association through their friendship with us.

Our fear of the police went beyond our concern for protecting our research subjects, however. We risked the danger of arrest ourselves through our own violations of the law. Many sociologists (Becker 1963; Carey 1972; Polsky 1969; Whyte 1955) have remarked that field researchers studying deviance must inevitably break the law in order to acquire valid participant observation data. This occurs in its most innocuous form from having "guilty knowledge": information about crimes that are committed. Being aware of major dealing and smuggling operations made us an accessory to their commission, since we failed to notify the police. We broke the law, secondly, through our "guilty observations," by being present at the scene of a crime and witnessing its occurrence (see also Carey 1972). We knew it was possible to get caught in a bust involving others, yet buying and selling was so pervasive that to leave every time it occurred would have been unnatural and highly suspicious. Sometimes drug transactions even occurred in our home, especially when Dave was living there, but we finally had to put a stop to that because we could not handle the anxiety. Lastly, we broke the law through our "guilty actions," by taking part in illegal behavior ourselves. Although we never dealt drugs (we were too scared to be seriously tempted), we consumed drugs and possessed them in small quantities. Quite frankly, it would have been impossible for a nonuser to have gained access to this group to gather the data presented here. This was the minimum involvement necessary to obtain even the courtesy membership we achieved. Some kind of illegal action was also found to be a necessary or helpful component of the research by Becker (1973), Carey (1972), Johnson (1975), Polsky (1969), and Whyte (1955).

Another methodological issue arose from the *cultural clash between our research subjects and ourselves*. While other sociologists have alluded to these kinds of differences (Humphreys 1970, Whyte 1955), few have discussed how the research relationships affected them. Relationships with research subjects are unique because they involve a bond of intimacy between persons who might not ordinarily associate together, or who might otherwise be no more than casual friends. When fieldworkers undertake a major project, they commit themselves to maintaining a long-term relationship with the people they study. However, as researchers try to get depth involvement, they are apt to come across fundamental differences in character, values, and attitudes between their subjects and themselves. In our case, we were most strongly confronted by differences in present versus future orientations, a desire for risk versus security, and feelings of spontaneity versus self-discipline. These differences often caused us great frustration. We repeatedly saw dealers act irrationally, setting themselves up for failure. We wrestled with our desire to point out their patterns of foolhardy behavior and offer advice, feeling competing pulls between our detached, observer role which advised us not to influence the natural setting, and our involved, participant role which called for us to offer friendly help whenever possible.[7]

Each time these differences struck us anew, we gained deeper insights into our core, existential selves. We suspended our own taken-for-granted feelings and were able to reflect on our culturally formed attitudes, character, and life choices from the perspective of the other. When comparing how we might act in situations faced by our respondents, we realized where our deepest priorities lay. These revelations had the effect of changing our self-conceptions:

whereas we, at one time, had thought of ourselves as what Rosenbaum (1981) has called "the hippest of non–addicts" (in this case nondealers), we were suddenly faced with being the straightest members of the crowd. Not only did we not deal, but we had a stable, long-lasting marriage and family life, and needed the security of a reliable monthly paycheck. Self-insights thus emerged as one of the unexpected outcomes of field research with members of a different cultural group.

The final issue I will discuss involved the various *ethical problems* which arose during this research. Many fieldworkers have encountered ethical dilemmas or pangs of guilt during the course of their research experiences (Carey 1972; Douglas 1976; Humphreys 1970; Johnson 1975; Klockars 1977, 1979; Rochford 1985). The researchers' role in the field makes this necessary because they can never fully align themselves with their subjects while maintaining their identity and personal commitment to the scientific community. Ethical dilemmas, then, are directly related to the amount of deception researchers use in gathering the data, and the degree to which they have accepted such acts as necessary and therefore neutralized them.

Throughout the research, we suffered from the burden of intimacies and confidences. Guarding secrets which had been told to us during taped interviews was not always easy or pleasant. Dealers occasionally revealed things about themselves or others that we had to pretend not to know when interacting with their close associates. This sometimes meant that we had to lie or build elaborate stories to cover for some people. Their fronts therefore became our fronts, and we had to weave our own web of deception to guard their performances. This became especially disturbing during the writing of the research report, as I was torn by conflicts between using details to enrich the data and glossing over description to guard confidences.[8]

Using the covert research role generated feelings of guilt, despite the fact that our key informants deemed it necessary, and thereby condoned it. Their own covert experiences were far more deeply entrenched than ours, being a part of their daily existence with non–drug world members. Despite the universal presence of covert behavior throughout the setting, we still felt a sense of betrayal every time we ran home to write research notes on observations we had made under the guise of innocent participants.

We also felt guilty about our efforts to manipulate people. While these were neither massive nor grave manipulations, they involved courting people to procure information about them. Our aggressively friendly postures were based on hidden ulterior motives: we did favors for people with the clear expectation that they could only pay us back with research assistance. Manipulation bothered us in two ways: immediately after it was done, and over the long run. At first, we felt awkward, phony, almost ashamed of ourselves, although we believed our rationalization that the end justified the means. Over the long run, though, our feelings were different. When friendship became intermingled with research goals, we feared that people would later look back on our actions and feel we were exploiting their friendship merely for the sake of our research project.

The last problem we encountered involved our feelings of whoring for data. At times, we felt that we were being exploited by others, that we were putting more into the relationship than they, that they were taking us for granted or using us. We felt that some people used a double standard in their relationship with us: they were allowed to lie to us, borrow money and not repay it, and take advantage of us, but we were at all times expected to behave honorably. This was undoubtedly an outgrowth of our initial research strategy where we did favors for people and expected little in return. But at times this led to our feeling bad. It made us feel like we were selling ourselves, our sincerity, and usually our true friendship, and not getting treated right in return.

CONCLUSIONS

The aggressive research strategy I employed was vital to this study. I could not just walk up to strangers and start hanging out with them as Liebow (1967) did, or be sponsored to a member of this group by a social service or reform organization as Whyte (1955) was, and expect to be accepted, let alone welcomed. Perhaps such a strategy might have worked with a group that had nothing to hide, but I doubt it. Our modern, pluralistic society is so filled with diverse subcultures whose interests compete or conflict with each other that each subculture has a set of knowledge which is reserved exclusively for insiders. In order to serve and prosper, they do not ordinarily show this side to just anyone. To obtain the kind of depth insight and information I needed, I had to become like the members in certain ways. They dealt only with people they knew and trusted, so I had to become known and trusted before I could reveal my true self and my research interests. Confronted with secrecy, danger, hidden alliances, misrepresentations, and unpredictable changes of intent, I had to use a delicate combination of overt and covert roles. Throughout, my deliberate cultivation of the norm of reciprocal exchange enabled me to trade my friendship for their knowledge, rather than waiting for the highly unlikely event that information would be delivered into my lap. I thus actively built a web of research contacts, used them to obtain highly sensitive data, and carefully checked them out to ensure validity.

Throughout this endeavor I profited greatly from the efforts of my husband, Peter, who served as an equal partner in this team field research project. It would have been impossible for me to examine this social world as an unattached female and not fall prey to sex role stereotyping which excluded women from business dealings. As a couple, our different genders allowed us to relate in different ways to both men and women (see Warren and Rasmussen 1977). We also protected each other when we entered the homes of dangerous characters, buoyed each other's initiative and courage, and kept the conversation going when one of us faltered. Conceptually, we helped each other keep a detached and analytical eye on the setting, provided multiperspectival insights, and corroborated, clarified, or (most revealingly) contradicted each other's observations and conclusions.

Finally, I feel strongly that to ensure accuracy, research on deviant groups must be conducted in the settings where it naturally occurs. As Polsky (1969: 115–16) has forcefully asserted:

> This means—there is no getting away from it—the study of career criminals *au natural,* in the field, the study of such criminals as they normally go about their work and play, the study of "uncaught" criminals and the study of others who in the past have been caught but are not caught at the time you study them. . . . Obviously we can no longer afford the convenient fiction that in studying criminals in their natural habitat, we would discover nothing really important that could not be discovered from criminals behind bars.

By studying criminals in their natural habitat I was able to see them in the full variability and complexity of their surrounding subculture, rather than within the artificial environment of a prison. I was thus able to learn about otherwise inaccessible dimensions of their lives, observing and analyzing firsthand the nature of their social organization, social stratification, lifestyle, and motivation.

NOTES

1. Gold (1958) discouraged this methodological strategy, cautioning against overly close friendship or intimacy with informants, lest they lose their ability to act as informants by becoming too much observers. Whyte (1955), in contrast, recommended the use of informants as research aides, not for helping in conceptualizing the data but for their assistance in locating data which supports, contradicts, or fills in the researcher's analysis of the setting.

2. See also Biernacki and Waldorf 1981; Douglas 1976; Henslin 1972; Hoffman 1980; McCall 1980; and West 1980 for discussions of "snowballing" through key informants.

3. See Kirby and Corzine 1981; Birenbaum 1970; and Henslin 1972 for more detailed discussion of the nature, problems, and strategies for dealing with courtesy stigmas.

4. We never considered secret tapings because, aside from the ethical problems involved, it always struck us as too dangerous.

5. See Douglas (1976) for a more detailed account of these procedures.

6. A recent court decision, where a federal judge ruled that a sociologist did not have to turn over his field notes to a grand jury investigating a suspicious fire at a restaurant where he worked, indicates that this situation may be changing (Fried 1984).

7. See Henslin 1972 and Douglas 1972, 1976 for further discussions of this dilemma and various solutions to it.

8. In some cases I resolved this by altering my descriptions of people and their actions as well as their names so that other members of the dealing and smuggling community would not recognize them. In doing this, however, I had to keep a primary concern for maintaining the sociological integrity of my data so that the generic conclusions I drew from them would be accurate. In places, then, where my attempts to conceal people's identities from people who know them have been inadequate, I hope that I caused them no embarrassment. See also Polsky 1969; Rainwater and Pittman 1967; and Humphreys 1970 for discussions of this problem.

REFERENCES

Adler, Patricia A., and Peter Adler. 1987. *Membership Roles in Field Research.* Beverly Hills, CA: Sage.

Becker, Howard. 1963. *Outsiders.* New York: Free Press.

Biernacki, Patrick, and Dan Waldorf. 1981. "Snowball sampling." *Sociological Methods and Research* 10: 141–63.

Birenbaum, Arnold. 1970. "On managing a courtesy stigma." *Journal of Health and Social Behavior* 11: 196–206.

Blumer, Herbert. 1969. *Symbolic Interactionism.* Englewood Cliffs, NJ: Prentice-Hall.

Carey, James T. 1972. "Problems of access and risk in observing drug scenes." In Jack D. Douglas, ed., *Research on Deviance,* pp. 71–92. New York: Random House.

Cummins, Marvin, et al. 1972. *Report of the Student Task Force on Heroin Use in Metropolitan Saint Louis.* Saint Louis: Washington University Social Science Institute.

Douglas, Jack D. 1972. "Observing deviance." In Jack D. Douglas, ed., *Research on Deviance,* pp. 3–34. New York: Random House.

———. 1976. *Investigative Social Research.* Beverly Hills, CA: Sage.

Fried, Joseph P. 1984. "Judge protects waiter's notes on fire inquiry." *New York Times,* April 8: 47.

Goffman, Erving. 1963. *Stigma.* Englewood Cliffs, NJ: Prentice-Hall.

Gold, Raymond. 1958. "Roles in sociological field observations." *Social Forces* 36: 217–23.

Henslin, James M. 1972. "Studying deviance in four settings: research experiences with cabbies, suicides, drug users and abortionees." In Jack D. Douglas, ed., *Research on Deviance,* pp. 35–70. New York: Random House.

Hoffman, Joan E. 1980. "Problems of access in the study of social elites and boards of directors." In William B. Shaffir, Robert A. Stebbins, and Allan Turowetz, eds., *Fieldwork Experience,* pp. 45–56. New York: St. Martin's.

Humphreys, Laud. 1970. *Tearoom Trade.* Chicago: Aldine.

Johnson, John M. 1975. *Doing Field Research.* New York: Free Press.

Kirby, Richard, and Jay Corzine. 1981. "The contagion of stigma." *Qualitative Sociology* 4: 3–20.

Klockars, Carl B. 1977. "Field ethics for the life history." In Robert Weppner, ed., *Street Ethnography,* pp. 201–26. Beverly Hills, CA: Sage.

———. 1979. "Dirty hands and deviant subjects." In Carl B. Klockars and Finnbarr W. O'Connor, eds., *Deviance and Decency,* pp. 261–82. Beverly Hills, CA: Sage.

Liebow, Elliott. 1967. *Tally's Corner.* Boston: Little, Brown.

McCall, Michal. 1980. "Who and where are the artists?" In William B. Shaffir, Robert A. Stebbins, and Allan Turowetz, eds., *Fieldwork Experience,* pp. 145–58. New York: St. Martin's.

Polsky, Ned. 1969. *Hustlers, Beats, and Others.* New York: Doubleday.

Rainwater, Lee R., and David J. Pittman. 1967. "Ethical problems in studying a politically sensitive and deviant community." *Social Problems* 14: 357–66.

Riemer, Jeffrey W. 1977. "Varieties of opportunistic research." *Urban Life* 5: 467–77.

Rochford, E. Burke, Jr. 1985. *Hare Krishna in America.* New Brunswick, NJ: Rutgers University Press.

Rosenbaum, Marsha. 1981. *Women on Heroin.* New Brunswick, NJ: Rutgers University Press.

Warren, Carol A. B., and Paul K. Rasmussen. 1977. "Sex and gender in field research." *Urban Life* 6: 349–69.

Wax, Rosalie. 1952. "Reciprocity as a field technique." *Human Organization* 11: 34–37.

———. 1957. "Twelve years later: An analysis of a field experience." *American Journal of Sociology* 63: 133–42.

West, W. Gordon. 1980. "Access to adolescent deviants and deviance." In William B. Shaffir, Robert A. Stebbins, and Allan Turowetz, eds., *Fieldwork Experience,* pp. 31–44. New York: St. Martin's.

Whyte, William F. 1955. *Street Corner Society*. Chicago: University of Chicago Press.

PART III

Constructing Deviance

As we noted in the General Introduction, the social constructionist perspective suggests that deviance should be regarded as lodged in a process of definition, rather than in some objective feature of an object, person, or act. This perspective therefore guides us to look at the process by which a society constructs definitions of deviance and applies them to specific groups of people associated with these objects or acts.

MORAL ENTREPRENEURS

The process of constructing and applying definitions of deviance can be understood as a **moral enterprise.** That is, it involves the constructions of moral meanings and the association of them with specific acts or conditions. The way people make deviance is similar to the way they manufacture anything else, but because deviance is an abstract concept rather than a tangible product, this process involves individuals drawing on the power and resources of organizations, institutions, agencies, symbols, ideas, communication, and audiences. Becker (1963) has suggested that we call the people involved in these activities **moral entrepreneurs.** The deviance-making enterprise has two facets: rule creating (without which there would be no deviant behavior) and rule enforcing (applying these rules to specific groups of people). We thus have two kinds of moral entrepreneurs: rule creators and rule enforcers.

Rule creating can be done by individuals acting either alone or in groups. Prominent individuals who have been influential in campaigning for definitions of deviance include Nancy Reagan and her "Just Say No" anti-drug campaign, former Surgeon General C. Everett Koop and his campaign to marginalize cigarette smoking, and Newt Gingrich and his campaign against welfare dependency. More commonly, however, individuals band together to use their collective energy and resources to change social definitions and create norms and rules. These groups of moral entrepreneurs represent interest groups that can be galvanized and activated into **pressure groups.** Rule creators ensure that our society is supplied with a constant stock of deviance and deviants by defining the behavior of others as immoral. They do this because they perceive threats in and feel fearful, distrustful, and suspicious of the behavior of these others. In so doing, they seek to transform private troubles into public issues and their private morality into the normative order.

Moral entrepreneurs manufacture public morality through a multi-stage process. Their first goal is to generate broad *awareness* of a problem; they must create a sense that certain conditions are problematic and pose a present or future potential danger to society. Because no rules exist to deal with the threatening condition, they construct the impression that these are necessary. In so doing, they draw on the testimonials of various "experts" in the field, such as scholars, doctors, eyewitnesses, ex-participants, and others with specific knowledge of the situation. These testimonials are disseminated to society via the media as "facts." Second, rule creators must bring about a **moral conversion,** convincing others of their views. With the problem outlined, they have to convert neutral parties and previous opponents into supporting partisans. Their successful conversion of others further legitimates their own beliefs. To effect a moral conversion, rule creators gain support from the endorsements of opinion leaders. These spokespersons need not have expert knowledge on any particular subject; they merely need to be liked and respected. Moral entrepreneurial groups thus often turn to athletes, actors, religious leaders, and media personalities for endorsements.

Once the public viewpoint has been swayed and a majority (or a vocal and powerful enough minority) of people have adopted a social definition, the issue may remain at the level of a norm or become elevated to the status of law through a legislative effort. In some cases both situations occur. For example, while the anti-smoking campaign has been successful in banning cigarette use in airplanes and various public places, it most effectively relies on normative informal sanctions.

Once norms or rules have been enacted, rule enforcers ensure that they are applied. In our society, this process often tends to be selective. Various individuals or groups have greater or lesser power to resist the enforcement of rules against them due to their socioeconomic, racial, religious, gender, political, or other status. Whole battles may begin anew over groups' strength to resist the enforcement of norms and laws, with this arena becoming once again a moral entrepreneurial combat zone.

The classic statement on moral entrepreneurial activities is included as the first reading in this section. Originally published in 1963 as part of Howard Becker's landmark book, *Outsiders,* this article outlines how moral entrepreneurs act as crusaders to transform certain behaviors into deviant acts. Becker reminds us that moral entrepreneurs believe that they are acting righteously to protect and preserve the moral code of society. Moral crusaders are typically people who have power over others and use this power to influence what decisions are made. Becker also looks at groups of people who have the power to enforce rules. These people, not necessarily members of elite groups, are charged with the job of reinforcing society's moral code. Thus, police, judges, psychiatrists, social workers, and teachers may be in the position to enforce their rules over others. Deviance, if it is conceptualized in this way, is a product of the domination of some groups (usually powerful ones) over others.

Following Becker's statement, we provide three examples of moral entrepreneurial crusades to illustrate this process. In Gerald Markle and Ronald Troyer's article, "Smoke Gets in Your Eyes: Cigarette Smoking as Deviant Behavior," the authors focus on efforts to marginalize this legitimate activity. Although their research took place during the 1970s, the issue has once again become salient due to recent moral crusades aimed at cigarette smoking, especially in public space. Medical and ethical arguments have been raised, during the era Markle and Troyer discuss and again currently, to sway public opinion and demonize public consumption. Markle and Troyer's work clearly identifies the opposing groups of moral entrepreneurs working to advance competing sides of the issue. A spate of claims has been put forward, pitting anti-smoking groups which have argued that secondhand smoke is toxic and that smokers should not be allowed to inflict their pollution onto others against opponents who have argued that the government should not legislate their morality. These issues illustrate the concern that frequently arises in deviance, when society must balance the right for individual freedoms (the desire to smoke) against the need for the common good (public health). Markle and Troyer's work can be set against the moral entrepreneurial groups operating today, with a coalition composed of the American Medical Association, The Public Health Service, The National Clearinghouse for Smoking and Health, and citizen groups such as GASP (Group Against Smoking Pollution) waging a moral war against the Tobacco Institute, the restaurant and tavern industry, and the U.S. Department of Agriculture, all three of which have vested interests in fighting a ban on smoking. This reading also shows that social power must be joined by effective argumentation and claims making to achieve a moral entrepreneurial victory.

Craig Reinarman then tackles moral attitudes toward illicit drugs in "The Social Construction of Drug Scares." He briefly offers a salient history of drug scares, the major players engineering them, and the social contexts that enhanced their growth. He then outlines seven factors common to drug scares that propel them toward realization. These enable him to dissect the essential processes at work in the rule creation and enforcement phases of drug scares, despite the contradictory cultural values of temperance and hedonistic consumptionism. From this article, we can see how drugs have been scapegoated to account for a wide array of social problems and to keep some groups down by defining their actions as deviant. It is clear that despite our society's views on the negative features associated with all illicit drugs, our moral entrepreneurial and enforcement efforts have been concentrated more stringently against the drugs used by members of the powerless lower class and minority racial groups.

Finally, Betsy Lucal's "The Problem with 'Battered Husbands'" raises the issue of a relatively new phenomenon: men who are beaten by their wives. The last several decades have witnessed a virtual explosion in concern about domestic violence. Domestic violence is currently seen as one of the major social problems in society, one that has been "defined up," as Krauthammer told us earlier, from previous definitions as a private concern. Lucal compares the highly successful campaign to define wife beating as deviant with the unsuccessful campaign to establish such a consciousness for battered husbands. She shows how the same act, spousal abuse, attained a prominent societal definition as deviant when committed against one spouse but not against the other. Lucal traces this difference to the activities of the moral entrepreneurs promoting these campaigns and to the rhetoric, media attention, social movement influences, and disparate gender images of men and women in society that affect social responsiveness to these arguments. Thus, unlike the case of battered women that had the women's movement fueling it, battered men have no such crusade. The media have not embraced this issue, and due to stereotypical gender issues in society, men are not seen as likely victims of domestic violence. For these reasons, the establishment of the construction of a deviant activity has been prevented.

DIFFERENTIAL SOCIAL POWER

Specific behavioral acts are not the only things that can be constructed as deviant; this definition can also be applied to a social status or lifestyle. When entire groups of people become relegated to a deviant status through their social condition (especially if it is ascribed through birth rather than voluntarily achieved), we see the force of inequality and differential social power in operation. This central tenet of labeling theory represents the fusion of the interactionist perspective with **conflict theory.** Based on the general writings of Karl Marx, conflict theorists believe that those who control the resources in society (politics, social status, wealth, religious beliefs, mobilization of the masses) have the ability to dominate, both materially and ideologically, the masses. Conflict theorists agree with interactionists that there is nothing inherently deviant in behavior but that social definitions, enacted on the basis of power relations, provide such definitions. A relatively small group of people thus has the ability to transform its definitions of reality into the norms that informally influence social behavior and the laws that formally dictate official behavior. Therefore, certain kinds of laws and enforcement are a product of political action by moral entrepreneurial interest groups that are connected to society's power base. Since Marx believed that society pits a dominant group against less forceful ones, deviance is created when the mighty use norms and rules to subjugate the weak. Conflict theory and labeling theory both advocate that deviance is constructed by groups who want to ensure that their strength and position are maintained. A simple way to do this is to define others' behavior as deviant. Thus, the relative deviance of conditions such as minority ethnic or racial status, feminine gender, lower social class, and homosexual orientation (as some of the following readings show), if taken in this light, can be seen to reflect the application of differential social power in our society. Individuals in these groups may find themselves discriminated against or blocked from the mainstream of society by virtue of this basic feature of their existence, unrelated to any particular situation or act. This application of the deviant label emphatically illustrates the role of power in the deviance-defining enterprise, for those positioned closer to the center of society, holding the greater social, economic, political, and moral resources, can turn the force of the deviant stigma onto others less fortunately placed. In so doing, they use the definition of deviance to reinforce their own favored position. This politicization of deviance and the power associated with its use serve to remind us that deviance is not a category inhabited only by those on the marginal outskirts of society: the exotics, erotics, and neurotics. Instead, any group can be pushed into this category by the exercise of another group's greater power.

A classic statement of conflict theory is presented in Richard Quinney's "The Social Reality of Crime." In a succinct look at the theory's major tenets, Quinney summarizes six basic premises. He describes how a dominant class creates legal definitions of human conduct. Crime exists because some behaviors conflict with the interests of the dominant class. Through class struggle and class conflict, crime is constructed, formulated, and applied so that less powerful groups are sub-dued and more powerful groups are strengthened.

William Chambliss's description of "The Saints and the Roughnecks" shows us how social class can operate to define some groups as deviant and others not. Although the former group actually engaged in more delinquent acts than the lat-ter, the Saints were perceived as "good boys" who were merely engaging in typi-cal adolescent hijinks. Because of their higher class background, their behavior was defined as socially normative, enabling the police, parents, teachers, and other parents to look the other way. On the other hand, the Roughnecks, who came from the "wrong side of the tracks," were perceived to be troublemakers, rabble-rousers, and delinquents. We see conflict and labeling theories in effect here since social class is the determinant of society's reactions. Behavior by teenagers from upstanding, middle-class families is tolerated, whereas similar behavior engaged in by lower-class youth is reinforced as deviant. Once again, labels are applied based on the basis of status, not on patterns of behavior.

In "The Police and the Black Male," Elijah Anderson gives us a glimpse into the perspective of the black man in the ghetto as he is handled by police: the agents of social control. For young black men living in America's inner cities, life may be made even more difficult by police who assume that if there is trouble in the neighborhood, it must be caused by these youngsters. Black men, walking alone at night or cruising a neighborhood in a car, may be stopped, harassed, ques-tioned, or beaten, even if they have done nothing wrong. Merely due to their as-cribed status as black, the activities of black men are scrutinized in different ways than are others in society. Drawing on conflict theory, we see how the deviant status is used by social control agents to subordinate less powerful groups.

Finally, Donald Sabo's "Homophobia in Sport" investigates one of the funda-mental bastions of American male culture, sport, and shows how this basis for male bonding and the development of masculine identity is founded not only on homophobia, the revulsion of homosexuality, but also on misogyny, the treat-ment of women as base sex objects and the disrespect for anything feminine. We see the process of social learning whereby the disrespect and denigration of ho-mosexuality are ingrained in young athletic participants, thereby becoming firmly established in traditional male culture. Almost nothing else can be more wound-ing to a young man than to call him a "faggot," a "sissy," or a "wimp." It is not unusual for coaches of athletic teams to shout that "you're playing like a bunch of girls." This reiterates society's negative label toward gays, giving them an inferior status. Heterosexual privilege is thus promoted at the expense of mutual consider-ation and respect. According to the conflict/labeling perspective, homosexuality is considered deviant only because the more powerful, traditional male system, with its roots in sport, has deemed it such.

7

Moral Entrepreneurs

HOWARD S. BECKER

RULE CREATORS

The prototype of the rule creator, but not the only variety as we shall see, is the crusading reformer. He is interested in the content of rules. The existing rules do not satisfy him because there is some evil which profoundly disturbs him. He feels that nothing can be right in the world until rules are made to correct it. He operates with an absolute ethic; what he sees is truly and totally evil with no qualification. Any means is justified to do away with it. The crusader is fervent and righteous, often self-righteous.

It is appropriate to think of reformers as crusaders because they typically believe that their mission is a holy one. The prohibitionist serves as an excellent example, as does the person who wants to suppress vice and sexual delinquency or the person who wants to do away with gambling.

These examples suggest that the moral crusader is a meddling busybody, interested in forcing his own morals on others. But this is a one-sided view. Many moral crusades have strong humanitarian overtones. The crusader is not only interested in seeing to it that other people do what he thinks [is] right. He believes that if they do what is right it will be good for them. Or he may feel that his reform will prevent certain kinds of exploitation of one person by another. Prohibitionists felt that they were not simply forcing their morals on others, but attempting to provide the conditions for a better way of life for people prevented by drink from realizing a truly good life. Abolitionists were not simply trying to prevent slave owners from doing the wrong thing; they were trying to help slaves achieve a better life. Because of the importance of the humanitarian motive, moral crusaders (despite their relatively single-minded devotion to their particular cause) often lend their support to other humanitarian crusades. Joseph Gusfield has pointed out that:

> The American temperance movement during the 19th century was a part of a general effort toward the improvement of the worth of the human being through improved morality as well as economic conditions. The mixture of the religious, the equalitarian, and the humanitarian was an outstanding facet of the moral reformism of many movements. Temperance

supporters formed a large segment of movements such as sabbatarianism, abolition, woman's rights, agrarianism, and humanitarian attempts to improve the lot of the poor. . . .

In its auxiliary interests the WCTU revealed a great concern for the improvement of the welfare of the lower classes. It was active in campaigns to secure penal reform, to shorten working hours and raise wages for workers, and to abolish child labor and in a number of other humanitarian and equalitarian activities. In the 1880s the WCTU worked to bring about legislation for the protection of working girls against the exploitation by men.[1]

As Gusfield says,[2] "Moral reformism of this type suggests the approach of a dominant class toward those less favorably situated in the economic and social structure." Moral crusaders typically want to help those beneath them to achieve a better status. That those beneath them do not always like the means proposed for their salvation is another matter. But this fact—that moral crusaders are typically dominated by those in the upper levels of the social structure—means that they add to the power they derive from the legitimacy of their moral position, the power they derive from their superior position in society.

Naturally, many moral crusades draw support from people whose motives are less pure than those of the crusader. Thus, some industrialists supported Prohibition because they felt it would provide them with a more manageable labor force.[3] Similarly, it is sometimes rumored that Nevada gambling interests support the opposition to attempts to legalize gambling in California because it would cut so heavily into their business, which depends in substantial measure on the population of Southern California.[4]

The moral crusader, however, is more concerned with ends than with means. When it comes to drawing up specific rules (typically in the form of legislation to be proposed to a state legislature or the Federal Congress), he frequently relies on the advice of experts. Lawyers, expert in the drawing of acceptable legislation, often play this role. Government bureaus in whose jurisdiction the problem falls may also have the necessary expertise, as did the Federal Bureau of Narcotics in the case of the marihuana problem.

As psychiatric ideology, however, becomes increasingly acceptable, a new expert has appeared—the psychiatrist. Sutherland, in his discussion of the natural history of sexual psychopath laws, pointed to the psychiatrist's influence.[5] He suggests the following as the conditions under which the sexual psychopath law, which provides that a person "who is diagnosed as a sexual psychopath may be confined for an indefinite period in a state hospital for the insane,"[6] will be passed.

First, these laws are customarily enacted after a state of fear has been aroused in a community by a few serious sex crimes committed in quick succession. This is illustrated in Indiana, where a law was passed following three or four sexual attacks in Indianapolis, with murder in two. Heads of families bought guns and watch dogs, and the supply of locks and chains in the hardware stores of the city was completely exhausted. . . .

A second element in the process of developing sexual psychopath laws is the agitated activity of the community in connection with the fear. The attention of the community is focused on sex crimes, and people in the most varied situations envisage dangers and see the need of and possibility for their control. . . .

The third phase in the development of those sexual psychopath laws has been the appointment of a committee. The committee gathers the many conflicting recommendations of persons and groups of persons, attempts to determine "facts," studies procedures in other states, and makes recommendations, which generally include bills for the legislature. Although the general fear usually subsides within a few days, a committee has the formal duty of following through until positive action is taken. Terror which does not result in a committee is much less likely to result in a law.[7]

In the case of sexual psychopath laws, there usually is no government agency charged with dealing in a specialized way with sexual deviations. Therefore, when the need for expert advice in drawing up legislation arises, people frequently turn to the professional group most closely associated with such problems:

> In some states, at the committee stage of the development of a sexual psychopath law, psychiatrists have played an important part. The psychiatrists, more than any others, have been the interest group in back of the laws. A committee of psychiatrists and neurologists in Chicago wrote the bill which became the sexual psychopath law of Illinois; the bill was sponsored by the Chicago Bar Association and by the state's attorney of Cook County and was enacted with little opposition in the next session of the State Legislature. In Minnesota all the members of the governor's committee except one were psychiatrists. In Wisconsin the Milwaukee Neuropsychiatric Society shared in pressing the Milwaukee Crime Commission for the enactment of a law. In Indiana the attorney-general's committee received from the American Psychiatric Association copies of all the sexual psychopath laws which had been enacted in other states.[8]

The influence of psychiatrists in other realms of the criminal law has increased in recent years.

In any case, what is important about this example is not that psychiatrists are becoming increasingly influential, but that the moral crusader, at some point in the development of his crusade, often requires the services of a professional who can draw up the appropriate rules in an appropriate form. The crusader himself is often not concerned with such details. Enough for him that the main point has been won; he leaves its implementation to others.

By leaving the drafting of the specific rule in the hands of others, the crusader opens the door for many unforeseen influences. For those who draft legislation for crusaders have their own interests, which may affect the legislation they prepare. It is likely that the sexual psychopath laws drawn by psychiatrists contain many features never intended by the citizens who spearheaded the drives to "do something about sex crimes," features which do however reflect the professional interests of organized psychiatry.

RULE ENFORCERS

The most obvious consequence of a successful crusade is the creation of a new set of rules. With the creation of a new set of rules we often find that a new set of enforcement agencies and officials is established. Sometimes, of course, existing agencies take over the administration of the new rule, but more frequently a new set of rule enforcers is created. The passage of the Harrison Act presaged the creation of the Federal Narcotics Bureau, just as the passage of the Eighteenth Amendment led to the creation of police agencies charged with enforcing the Prohibition Laws.

With the establishment of organizations of rule enforcers, the crusade becomes institutionalized. What started out as a drive to convince the world of the moral necessity of a new rule finally becomes an organization devoted to the enforcement of the rule. Just as radical political movements turn into organized political parties and lusty evangelical sects become staid religious denominations, the final outcome of the moral crusade is a police force. To understand, therefore, how the rules creating a new class of outsiders are applied to particular people we must understand the motives and interests of police, the rule enforcers.

Although some policemen undoubtedly have a kind of crusading interest in stamping out evil, it is probably much more typical for the policeman to have a certain detached and objective view of his job. He is not so much concerned with the content of any particular rule as he is with the fact that it is his job to enforce the rule. When the rules are changed, he punishes what was once acceptable behavior just as he ceases to punish behavior that has been made legitimate by a change in the rules. The enforcer, then, may not be interested in the content of the rule as such, but only in the fact that the existence of the rule provides him with a job, a profession, and a *raison d'être*.

Since the enforcement of certain rules provides justification for his way of life, the enforcer has two interests which condition his enforcement activity: first, he must justify the existence of his position and, second, he must win the respect of those he deals with.

These interests are not peculiar to rule enforcers. Members of all occupations feel the need to justify their work and win the respect of others. Musicians would like to do this but have difficulty finding ways of successfully impressing their worth on customers. Janitors fail to win their tenants' respect, but develop an ideology which stresses the quasi-professional responsibility they have to keep confidential the intimate knowledge of tenants they acquire in the course of their work.[9] Physicians, lawyers, and other professionals, more successful in winning the respect of clients, develop elaborate mechanisms for maintaining a properly respectful relationship.

In justifying the existence of his position, the rule enforcer faces a double problem. On the one hand, he must demonstrate to others that the problem still exists: the rules he is supposed to enforce have some point, because infractions occur. On the other hand, he must show that his attempts at enforcement are effective and worthwhile, that the evil he is supposed to deal with is in fact being dealt with adequately. Therefore, enforcement organizations, particularly when they are seeking funds, typically oscillate between two kinds of claims. First, they say that by reason of their efforts the problem they deal with is approaching solution. But, in the same breath, they say the problem is perhaps worse than ever (though through no fault of their own) and requires renewed and increased effort to keep it under control. Enforcement officials can be more vehement than anyone else in their insistence that the problem they are supposed to deal with is still with us, in fact is more with us than ever before. In making these claims, enforcement officials provide good reason for continuing the existence of the position they occupy.

We may also note that enforcement officials and agencies are inclined to take a pessimistic view of human nature. If they do not actually believe in original sin, they at least like to dwell on the difficulties in getting people to abide by the rules, on the characteristics of human nature that lead people toward evil. They are skeptical of attempts to reform rule-breakers.

The skeptical and pessimistic outlook of the rule enforcer, of course, is reinforced by his daily experience. He sees, as he goes about his work, the evidence that the problem is still with us. He sees the people who continually repeat offenses, thus definitely branding themselves in his eyes as outsiders. Yet it is not too great a stretch of the imagination to suppose that one of the underlying reasons for the enforcer's pessimism about human nature and the possibilities of reform is the fact that if human nature were perfectible and people could be permanently reformed, his job would come to an end.

In the same way, a rule enforcer is likely to believe that it is necessary for the people he deals with to respect him. If they do not, it will be very difficult to do his job; his feeling of security in his work will be lost. Therefore, a good deal of enforcement activity is devoted not to the actual enforcement of rules, but to coercing respect from the people the enforcer deals with. This means that one may be labeled as deviant not because he has actually broken a rule, but because he has shown disrespect to the enforcer of the rule.

Westley's study of policemen in a small industrial city furnishes a good example of this phenomenon. In his interview, he asked policemen, "When do you think a policeman is justified in roughing a man up?" He found that "at least 37% of the men believed that it was legitimate to use violence to coerce respect."[10] He gives some illuminating quotations from his interviews:

> Well, there are cases. For example, when you stop a fellow for a routine questioning, say a wise guy, and he starts talking back to you and telling you you are no good and that sort of thing. You know you can take a man in on a disorderly conduct charge, but you can practically never make it stick. So what you do in a case like that is to egg the guy on until

he makes a remark where you can justifiably slap him and, then, if he fights back, you can call it resisting arrest.

Well, a prisoner deserves to be hit when he goes to the point where he tries to put you below him.

You've gotta get rough when a man's language becomes very bad, when he is trying to make a fool of you in front of everybody else. I think most policemen try to treat people in a nice way, but usually you have to talk pretty rough. That's the only way to set a man down, to make him show a little respect.[11]

What Westley describes is the use of an illegal means of coercing respect from others. Clearly, when a rule enforcer has the option of enforcing a rule or not, the difference in what he does may be caused by the attitude of the offender toward him. If the offender is properly respectful, the enforcer may smooth the situation over. If the offender is disrespectful, then sanctions may be visited on him. Westley has shown that this differential tends to operate in the case of traffic offenses, where the policeman's discretion is perhaps at a maximum.[12] But it probably operates in other areas as well.

Ordinarily, the rule enforcer has a great deal of discretion in many areas, if only because his resources are not sufficient to cope with the volume of rule-breaking he is supposed to deal with. This means that he cannot tackle everything at once and to this extent must temporize with evil. He cannot do the whole job and knows it. He takes his time, on the assumption that the problems he deals with will be around for a long while. He establishes priorities, dealing with things in their turn, handling the most pressing problems immediately and leaving others for later. His attitude toward his work, in short, is professional. He lacks the naive moral fervor characteristic of the rule creator.

If the enforcer is not going to tackle every case he knows of at once, he must have a basis for deciding when to enforce the rule, which persons committing which acts to label as deviant. One criterion for selecting people is the "fix." Some people have sufficient political influence or know-how to be able to ward off attempts at enforcement, if not at the time of apprehension then at a later stage in the process. Very often, this function is professionalized; someone performs the job on a full-time basis, available to anyone who wants to hire him. A professional thief described fixers this way:

There is in every large city a regular fixer for professional thieves. He has no agents and does not solicit and seldom takes any case except that of a professional thief, just as they seldom go to anyone except him. This centralized and monopolistic system of fixing for professional thieves is found in practically all of the large cities and many of the small ones.[13]

Since it is mainly professional thieves who know about the fixer and his operations, the consequence of this criterion for selecting people to apply the rules to is that amateurs tend to be caught, convicted, and labeled deviant much more frequently than professionals. As the professional thief notes:

You can tell by the way the case is handled in court when the fix is in. When the copper is not very certain he has the right man, or the testimony of the copper and the complainant does not agree, or the prosecutor goes easy on the defendant, or the judge is arrogant in his decisions, you can always be sure that someone has got the work in. This does not happen in many cases of theft, for there is one case of a professional to twenty-five or thirty amateurs who know nothing about the fix. These amateurs get the hard end of the deal every time. The coppers bawl out about the thieves, no one holds up his testimony, the judge delivers an oration, and all of them get credit for stopping a crime wave. When the professional hears the case immediately preceding his own, he will think, "He should have got ninety years. It's the damn amateurs who cause all the heat in the stores." Or else he thinks, "Isn't it a damn shame for that copper to send that kid away for a pair of hose, and in a few minutes he will agree to a small fine for me for stealing a fur coat?" But if the coppers did not send the amateurs away to strengthen their records of convictions, they could not sandwich in the professionals whom they turn loose.[14]

Enforcers of rules, since they have no stake in the content of particular rules themselves, often develop their own private evaluation of the importance of various kinds of rules and infractions of them. This set of priorities may differ considerably from those held by the general public. For instance, drug users typically believe (and a few policemen have personally confirmed it to me) that police do not consider the use of marihuana to be as important a problem or as dangerous a practice as the use of opiate drugs. Police base this conclusion on the fact that, in their experience, opiate users commit other crimes (such as theft or prostitution) in order to get drugs, while marihuana users do not.

Enforcers then, responding to the pressures of their own work situation, enforce rules and create outsiders in a selective way. Whether a person who commits a deviant act is in fact labeled a deviant depends on many things extraneous to his actual behavior: whether the enforcement official feels that at this time he must make some show of doing his job in order to justify his position, whether the misbehaver shows proper deference to the enforcer, whether the "fix" has been put in, and where the kind of act he has committed stands on the enforcer's list of priorities.

The professional enforcer's lack of fervor and routine approach to dealing with evil may get him into trouble with the rule creator. The rule creator, as we have said, is concerned with the content of the rules that interest him. He sees them as the means by which evil can be stamped out. He does not understand the enforcer's long-range approach to the same problems and cannot see why all the evil that is apparent cannot be stamped out at once.

When the person interested in the content of a rule realizes or has called to his attention the fact that enforcers are dealing selectively with the evil that concerns him, his righteous wrath may be aroused. The professional is denounced for viewing the evil too lightly, for failing to do his duty. The moral entrepreneur, at whose insistence the rule was made, arises again to say that the outcome of the last crusade has not been satisfactory or that the gains once made have been whittled away and lost.

NOTES

1. Joseph R. Gusfield, "Social Structure and Moral Reform: A Study of the Woman's Christian Temperance Union," *American Journal of Sociology*, LXI (November, 1955), 223.

2. *Ibid.*

3. See Raymond G. McCarthy, editor, *Drinking and Intoxication* (New Haven and New York: Yale Center of Alcohol Studies and The Free Press of Glencoe, 1959), pp. 395–396.

4. This is suggested in Oscar Lewis, *Sagebrush Casinos: The Story of Legal Gambling in Nevada* (New York: Doubleday and Co., 1953), pp. 223–234.

5. Edwin H. Sutherland, "The Diffusion of Sexual Psychopath Laws," *American Journal of Sociology*, LVI (September, 1950), 142–148.

6. *Ibid.*, p. 142.

7. *Ibid.*, pp. 143–145.

8. *Ibid.*, pp. 145–146.

9. See Ray Gold, "Janitors Versus Tenants: A Status-Income Dilemma," *American Journal of Sociology*, LVII (March, 1952), 486–493.

10. William A. Westley, "Violence and the Police," *American Journal of Sociology*, LIX (July, 1953), 39.

11. *Ibid.*

12. See William A. Westley, "The Police: A Sociological Study of Law, Custom, and Morality" (unpublished Ph.D. dissertation, University of Chicago, Department of Sociology, 1951).

13. Edwin H. Sutherland (editor), *The Professional Thief* (Chicago: University of Chicago Press, 1937), pp. 87–88.

14. *Ibid.*, pp. 91–92.

8

Smoke Gets in Your Eyes

Cigarette Smoking as Deviant Behavior

GERALD E. MARKLE AND RONALD J. TROYER

For the past fifteen years most sociological studies of deviance have focused on individuals and their subcultures and neglected the process of the social definition and redefinition of deviant categories:[1] how, under what conditions, and by whom are deviant categories created, maintained or abolished? Our investigation of cigarette smoking will address these questions.

Our methodology will follow Bloor's (1976) "strong program" in the sociology of knowledge. It will be causal, focusing on the forces which bring about collective social definitions. It will be impartial with respect to truth or falsity, rationality or irrationality, success or failure. And our analysis will be symmetrical: the same types of cause should explain true or false beliefs. Too often, according to Bloor, scholars seek to explain the causes of error or deviation while assuming that logic, rationality and truth are their explanation.

Gusfield's (1963) work on the American temperance movement, particularly its creation of deviant and criminal labels, is our exemplar. His analysis is causal, focusing on status politics and the symbolic nature of the crusade. This focus remains unchanged, though the initial winner of the conflict became the eventual loser. Moreover, the validity of his argument is independent of truth claims about alcohol; whether the drug is actually an aphrodisiac, a depressant or a tool of the devil is largely irrelevant. Finally, his investigation is symmetrical because it considers the motives and actions of both sides, and indeed the interaction of dominant and deviant groups, necessary to understand the conflict.

From Gusfield's work, we identify two dynamic processes by which deviant categories may be created or redefined. In the assimilative model, the violator admits deviance, and thus the reformer views that person with pity and sympathy:

> While legislation may be sought . . . the major activities are efforts to persuade the sufferer to remake his habits and customs. The orientation of the movement . . . is toward the welfare of the potential abstainer by his conversion to the habits of abstinence (1963: 69).

In the coercive model, on the other hand, the reformer's exhortations "fall on deaf and angry ears. The expected homage is not paid; the act of deference is absent" (1963: 69). The violator must be approached and engaged in political and legal battle as an enemy. As its name implies, the temperance movement began in the assimilative mode, but understanding and education failed to produce penitence. It was coercive measures in the form of power politics that eventually led to prohibition.

Cigarette smoking, as a social phenomenon, provides an opportunity to examine the creation of deviant categories. Like alcohol and other drugs which are common to everyday life, cigarette smoking has a long history of changing normative definitions. In an earlier article (Nuehring and Markle, 1974), we demonstrated that cigarette smoking was reemerging as a deviant behavior, and argued that this reemergence was a status battle between non-medical (primarily political and economic) interests. Furthermore, recent and significant changes in smoking attitudes and behaviors seemed to reflect those social conflicts. We characterized the anticigarette rules and legislation of the late 1960s and early 1970s as assimilative, but speculated that those forces might soon turn to coercive tactics. In this paper we continue that analysis and show that the antismoking movement has kept growing and now has coercive, as well as assimilative, tactics. Looking at both sides of the controversy, we examine the role of vested interests in this process. In showing the conflict between pro- and antismoking forces we also show how the image of the smoker has become more deviant in the mid- and late-1970s.

PROSMOKING FORCES

There are enormous vested interests in maintaining cigarette smoking as an approved behavior. In this section we consider the involvement, particularly that relating to finances, of three groups: the tobacco industry, advertising and government. In some cases (e.g., tobacco manufacturers) we show specific political linkages between vested interest and behavior; for others (e.g., the federal government) we show contrary and ambiguous linkages; most often, however, we merely correlate behavioral outcomes with financial interests.

Tobacco farming plays an important role in the American economy. In 1977 there were some 400,000 farms in the United States which harvested 2.3 billion dollars worth of tobacco (U.S. Department of Agriculture, 1978a: 10). Tobacco is currently the fifth largest crop for the entire United States, representing 2.5 percent of the total cash receipts from all farm commodities; moreover tobacco may be the best farm investment of all crops, averaging $2,375 per acre.[2] In three states (North Carolina, Kentucky and South Carolina) tobacco is a major agricultural product (USDA, 1978a: 17; Tobacco Institute, 1978).

Tobacco manufacturing interests are much more concentrated than farming interests. In 1977 R. J. Reynolds alone accounted for one-third of all sales in the domestic cigarette market. Its 1977 sales and earnings reached "record levels": sales were $6.36 billion, up 10.6 percent from 1976; net earnings were $423 million, up 19.9 percent from 1976. Return on the average common stockholder's equity was 18.8 percent. Interestingly, only 44 percent of all net sales came from domestic tobacco: international sales of tobacco added another 20 percent while transportation, energy, foods and beverages (e.g., Hawaiian Punch, ChunKing food) and aluminum products made up the balance (R. J. Reynolds, 1978). Other major tobacco manufacturers have also diversified their operations. For example, the Miller Brewing Company is owned by Phillip Morris and accounts for 26 percent of the operating revenues of the parent company (Phillip Morris, 1978).

In order to promote and defend these interests, the eleven companies which manufacture cigarettes have created the Tobacco Institute, located in Washington, D.C. The Institute seeks to "promote a better public understanding of the tobacco industry and its place in the nation's economy" (Encyclopedia of Associations, 1978). Much of the activity of the Institute has been in the political arena, channeling tens of thousands of dollars into the election campaigns of national leaders. A former congressman from North Carolina heads the Institute and employs three lobbyists with extensive political experience. Senator Kennedy is quoted as claiming that: "hour for hour and dollar for dollar, they're probably the most effective lobby on Capitol Hill" (Jensen, 1978).

Tobacco companies and their executives have also acted on their own to further the interests of cigarettes. The National Information Center on Political Finance places direct and indirect contributions by executives of the six largest tobacco companies to 1972 campaigns at $278,000. And in May 1976, three directors at R. J. Reynolds resigned when it was disclosed the company had illegally funneled $65,000 to $90,000 in corporate funds to domestic political campaigns (Jensen, 1976: 1).

The role of the federal government in the tobacco controversy is ambiguous and contrary. Certainly it appears to have a vested interest in promoting tobacco sales, since it collected $2.3 billion in cigarette excise taxes in 1977.[3] Moreover, the U.S. leads the world in tobacco exports: in 1977 the net balance of tobacco trade contributed $1.33 billion to the U.S. balance of payments (Tobacco Institute, 1978).

Within the Federal government, congressional representatives of major tobacco states and the U.S. Department of Agriculture have influenced tobacco production. Since 1933, the USDA has controlled supplies, supported prices, subsidized exports, and provided marketing assistance to tobacco farmers. The importance and impact of this program may be assessed by its size: as of March 1, 1978, 621 million pounds of flue-cured tobacco, valued at $629 million, were held under guaranteed government loan (U.S. Department of Agriculture, 1978b). Since tobacco allotments are tied to specific farms, they increase land values: in 1967 the program added as much as $6,000 to an acre of burley tobacco (Shuffett and Hoskins, 1969) while by 1975 an allotment was probably worth $10,000 per acre (Mann, 1975).

The financial impact of the cigarette controversy has filtered through the American economy. Aside from the direct combatants, the sector with the greatest vested interest in the dispute is the advertising industry and media. After cigarette advertising was banned from the broadcast media in 1971, cigarette sales increased for three successive years. This irony has an explanation: with the ban on cigarette commercials, powerful antismoking commercials—with their enormous public impact—were sharply curtailed. To fill this void, advertisers switched to the print media; by 1971 manufacturers had tripled their expenditures for newspaper, magazine and outdoor advertising (Morris, 1972: 43), and in 1975 R. J. Reynolds spent $31.7 million to advertise its top-selling Winston cigarettes (Advertising Age, 1976). The total expenditures are vast: in 1977 the tobacco companies spent an estimated $400 million on advertising (Jensen, 1978).

The strength of the prosmoking forces was vividly demonstrated during the battle over Proposition 5 in California. One month before the election a poll showed Californians favoring the measure by a 58–38 margin. But by election eve the situation had changed dramatically. As Walter Cronkite (1978) reported:

> Californians are used to expensive campaigns for their votes on ballot initiatives. But they've never been wooed so lavishly to vote against anything as they are on Proposition 5, the measure regulating their indoor smoking. . . . The opponents have blitzed the airwaves with a nearly $5 million campaign, more than the combined expenditures of all the candidates for governor. Most of the money is coming from four giants of the tobacco industry. Two of them, Philip Morris and R. J. Reynolds, have spent well over a million dollars apiece. . . . The campaign has turned around Proposition 5 from an almost sure winner in August to a 16 point underdog in one poll this week. And it's become an issue in public debate.

With tobacco interests eventually spending $5.6 million compared to an outlay of $578,000 by their opponents (Rood, 1978), Proposition 5 was handily defeated.

ANTISMOKING FORCES

There is no strong well-financed national antismoking group but there are many interest groups with diverse motives who are opposed to cigarette smoking. In this section we consider the involvement of the federal government, voluntary action citizen groups, and private entrepreneurs in the antismoking movement.

While some departments of the federal government have promoted tobacco interests, others have been at the forefront of the antismoking campaign. The Public Health Service, for example, has been actively involved in anticigarette activities for a number of years. One of its offices, The National

Clearinghouse for Smoking and Health, publishes and distributes materials urging people to quit smoking. Its efforts have been somewhat limited in recent years because of budget reductions and relegation to virtual obscurity. From its inception in 1966 to 1972, the budget of the Clearinghouse was $2.5 million per year. In 1973 the allocation dropped to $900,000 and in 1974 it was moved out of Washington to the Center for Disease Control in Atlanta with no budget of its own. However, HEW Secretary Califano has reinvigorated the office. He changed its name to the Office for Smoking and Health, moved it back to Washington, and requested $6 million for its 1978 budget.

Federal regulatory agencies such as the Civil Aeronautics Board, the Federal Trade Commission and the Interstate Commerce Commission have all issued rulings favorable to the antismoking position. Not only have these agencies been willing to act, but the rulings have become increasingly coercive, such as the increasingly stringent restrictions on smoking on interstate transportation. The reasons for this regulatory militancy are not clear. However we previously (Nuehring and Markle, 1974) speculated that internal organizational needs for role definition and power influenced agency behavior.

An additional reason for the antismoking actions by the regulatory agencies appears to be the pressure applied by antismoking citizen groups. ASH (Action on Smoking and Health), for example, has filed numerous petitions with agencies seeking anticigarette decisions. Headed by John Banzhaf, the man who persuaded the FCC to issue its "equal time" ruling on anticigarette commercials in 1967, ASH serves as the legal arm of the antismoking movement. Among its activities in 1977, for example, ASH filed two petitions with the CAB, one each with the FAA and the FDA, convinced the Food and Drug Administration that a special warning about smoking be included in birth control pill packages, and assisted in a law suit filed by a nonsmoker (ASH, 1978).

Other citizen groups exerting pressure on government bodies include GASP (Group Against Smokers Pollution), and NIC (National Interagency Council on Smoking). GASP, which has members in every state, promotes the rights of nonsmokers and seeks antismoking laws and regulations (GASP, n.d.). GASP chapters, for example, have been involved in promoting the California antismoking referendum and the New Jersey attempt to ban smoking in public places. The NIC has concentrated more on education against smoking but does favor prohibition of cigarette advertising (Encyclopedia of Associations, 1978: 801).

Health organizations such as the American Cancer Society, the American Heart Association and the American Lung Association initially endorsed assimilative reform of the smoker. Recently, however, they have approved more coercive tactics. Since 1974, the American Cancer Society, for example, has called for elimination of all cigarette advertising (Brody, 1974), an end to the tobacco subsidy program (New York Times, 1977), state and local restriction of smoking in public places, and strict enforcement of the ban on cigarette sales to minors (Brody, 1978). And all of those health groups appear to have rejuvenated their antismoking campaign by again making antismoking spots available to the mass media.[4]

Other vested interests benefiting from and sometimes encouraging the antismoking movement include profit-making enterprises. A check of the *1977–78 Books in Print,* for example, reveals 17 monographs with stop or quit smoking in the title. Insurance companies have found that they can profit as well as attract new customers by offering discounts as high as 20 and 25 percent to nonsmokers (Cole, 1973). Organized crime has benefited from the antismoking movement support of "sin taxes." The vast tax differential, for example 26¢ a pack in New York to 2¢ a pack in North Carolina, has made cigarette smuggling a lucrative business estimated to net $800 million annually (*Time,* 1977).

Several commercial enterprises, including Smoke No More, The Schick Treatment, Damon Hypothesis and Fresh Start, are in the quit smoking business (Condon, 1976). The two largest organizations are Smokewatchers, with franchises in 17 states, and SmokEnders, which claims to have graduated 100,000 people from its course. Founded in 1969 SmokEnders claims that smokers can "unlearn their habit . . . comfortably, intelligently, pleasantly, even joyfully" (SmokEnders, 1977). In fact an independent retrospective study claims that SmokEnders has a 70 percent success rate (Condon, 1976). It is difficult to evaluate SmokEnders methods: everyone in the program signs a statement that he or she will refrain from talking about the course and will not teach SmokEnders' course without the company's permission. What is known is that these methods are profitable: in 1976 the company grossed $5 million, up from $4.1 million in 1975 (Condon, 1976).

EMERGING CONFRONTATIONS

As antismoking forces shift from an assimilative attack on the product to a coercive attack on the smoker, there have been recent attempts to label the smoker as a psychological misfit. As a psychiatrist in a letter to the *New York Times* has proposed:

> What we need is a national campaign that results in stigmatization rather than glorification of the smoker. This, in my opinion, would be the most effective way of reducing the number of smokers and confining their smoking to the privacy of their homes (Gardner, 1978).

Behavioral scientists have contributed to this stigmatization, perhaps inadvertently, with studies exploring the personality configuration of the smoker. Perhaps most influential has been Eysenck's (1965) contention that smokers are more extroverted and neurotic than nonsmokers. Rae (1975) and Cherry and Kierman (1976) claimed support for this theory in their studies of British college students and adults. In the United States, Smith (1970) reported smokers scoring higher on extroversion.

More negative evaluative characteristics have been attributed to smokers by some authors. Studies of both adults and college students indicated that smokers scored lower on agreeableness and strength of character, and higher on antisocial tendencies, impulsiveness, crudeness and orality; in addition, they were more externally oriented and happy-go-lucky (Smith, 1969, 1970). Another study of college students reported that smokers demonstrated personality traits of defiance, impulsivity and danger seeking; showed more oral preoccupations and stress; and had perceptions of having experienced minimal warmth, protection and affection while growing up (Jacobs and Spilkin, 1971). Reynolds and Nichols (1976) claimed that the smokers, among the 885 students they examined, were less well-adjusted and more likely to engage in antisocial activities. Finally, studies in both the United States (Borland and Rudolph, 1975) and Canada (Hanley and Robinson, 1976) purport that smokers are academic underachievers.

Some authors have directly questioned the mental health of smokers (Smith, 1970; Srole and Fischer, 1973); Fisher (1976), for example, claimed that female smokers are preoccupied with power because of penis envy. Dr. Jerome H. Jaffe, the top White House drug abuse official in the Nixon Administration, suggested that people who smoke more than a pack of cigarettes a day be described as suffering from a "compulsive smoking disorder" (Brody, 1975: 38). And a new term—*compulsive smoking syndrome*—has been proposed as a disorder to be listed in the *Diagnostic and Statistical Manual* of the American Psychiatric Association.

These attempts to stigmatize the smoker have not gone unchallenged. For years the industry has portrayed the smoker as a desirable person (e.g., the "Marlboro Man"). And recent and heavy advertising in soft pornographic magazines[5] is clearly an attempt to pair smoking with erogeneous behavior. Nationally syndicated columnist William Safire (1974) has also protested that government attempts to lower the social status of smokers is an abuse of power. Other citizens have similarly complained:

> I am particularly sick of being issued the lousiest seat on a plane (although I have paid the same) because I might want a smoke during flight; of getting "kitchen" tables in the non-smoking sections of restaurants (although my bill will be just as high) (Murphy, 1978).

As attempts to restrict and label the smoker continue, and as nonsmokers become more militant, interpersonal relations between users and abstainers may become strained. In fact, one study found that nonsmokers rated persons more negatively when they smoked and that smokers greatly underestimated the extent to which their smoking was considered discourteous (Bleda and Sandman, 1977). If this study of Army and Air Force personnel is indicative of the general population, further confrontations and conflicts will follow.[6] As the Secretary of Health, Education and Welfare has noted:

> . . . the etiquette of smoking has changed, slowly but perceptibly. Once the smoker asked, "Would you like a cigarette?" Today the question is, "Do you mind if I smoke?" And more and more nonsmokers are finding the courage to answer with a polite but emphatic, "Yes, I do mind" (Califano, 1978).

CONCLUSION

Our first paper concluded with the statement: "Activities of the past few years and the next few years may show that, once again, the sale and possession of cigarettes is emerging as a deviant behavior" (Nuehring and Markle, 1974: 525). In fact in the mid- and late-1970s the antismoking movement has become more active, more politically effective and more coercive. Cigarette smokers, who in increasing numbers see themselves as deviant actors, have been labeled as drug addicts and neurotics as well as air polluters and fire hazards. Smokers who do not accept these labels face increasing confrontations with militant nonsmokers. These battles may be interpersonal or highly organized political campaigns such as the 1978 vote in California—The Clean Indoor Air Act—which would have banned smoking in nearly all public facilities and private businesses.

Most commentators interpret the smoking controversy from a medical framework. Scientists have demonstrated that smoking causes cancer and a host of other diseases. Therefore abstention is seen as a rational response to a health danger, while smoking is labeled as an irrational behavior. We do not deny that the health issue has played a role in the controversy. However, we have shown that since the early 1900s, and particularly since the 1950s, people have been informed about the health risks of smoking. To think of smoking behavior as irrational leads to a scientific dead end. Our task is to understand the phenomenon, not to place it beyond the scope of systematic investigation.

We might attribute the smoking controversy, and its resultant coercive regulations and laws, to the defensive behavior of the nonsmoker. Convinced that smoking is a health hazard for all people, abstainers are simply asserting their right to health and long life. This explanation is fine as far as it goes. But it does not account for the recent escalation of the smoking controversy. We maintain that individual opposition to smoking is only effective, or even allowable, within a collective context. Only be examining the institutional and organizational forces involved in the dispute can the dynamics of the controversy be appreciated.

Thus we maintain that sociopolitical rather than health factors have been decisive in shaping the smoking controversy. Increasingly smokers and nonsmokers may be seen as members of two status groups in conflict. At stake is the collective conferral of legitimacy and consequent prestige. If the norms of the nonsmoker are officially endorsed, the smoker clearly will drop in status ranking.

Enormous vested interests are attempting to define and redefine the social status of the smoker. For years cigarette advertising has glorified the smoker. Recent increases in smoking among teenagers, particularly women, demonstrate the vigor of this campaign. On the other hand the antismoking movement has inspired a multitude of new laws and regulations which restrict the smoker's habit. Because strictures are rarely enforced their actual impact on public smoking is difficult to assess. However the important effect of these measures is that they constantly stigmatize the smoker.

Our investigation of cigarette smoking is a case study of how deviant categories are created. But our findings can be linked to those on a variety of other social problems. For example, the antismoking movement has much in common with the temperance movement. Both issues involve human health, but both social movements have strong political and symbolic components. Both were initially characterized by assimilative tactics which later developed into coercive tactics. The battles over smoking are not as identified with clear regional, ethnic and religious boundaries as were those over drinking. However the link between vested interests and public opinion is easier to document in the smoking controversy than it was for temperance.

Smoking, like many other emerging social problems, is characterized by a conflict of vested interests over the collective definition of individual behavior. We have shown how each side of the conflict has tried, and continues to try, to get its definition as the accepted public definition. Whether cigarette smoking again becomes more desirable or more deviant is contingent on the outcome of the current status battles between pro and antismoking interests.

NOTES

1. Recent exceptions are studies on deviant drinking (Levine, 1978; Schneider, 1978), hyperactivity (Conrad, 1975), child abuse (Pfohl, 1977), and rape (Rose, 1977). See Spector and Kitsuse (1977) for a theoretical rationale of this type of study.

2. This compares very favorably to such crops as corn and wheat which yield an average of less than $400 per acre. Despite these data, tobacco farmers are not a wealthy group. Accoring to Mann (1975: 89): "Comparing commercial tobacco farms with other types of commercial farms, we find that tobacco farms are smaller in size (128 acres vs. 530) and have a lower value of land and buildings ($36,000 vs. $104,000), and lower value of farm products sold ($10,470 vs. $25,680)." In fact, some writers have argued that most tobacco farmers could not survive without the high return from this crop and that the allotment program should be considered as "an important social welfare program in the traditionally poor south" (Bradford and Infanger, 1978).

3. Other governmental units also benefit from the smoking habit. In 1977, state and local governments collected approximately $4 billion through taxes on tobacco products (U.S. Department of Agriculture, 1978a: 26).

4. The antismoking campaigns waged by these health groups show some ambivalence. Epstein (1978), for example, points out that the American Cancer Society has endorsed several restrictive antismoking measures but never launched the well-organized lobbying effort needed to insure their enactment and implementation.

5. In 1976, cigarette advertising accounted for an astounding 38 percent of all advertising revenue for *Oui* and 21 percent for *Playboy* (*ASH Newsletter,* 1977).

6. When, on a Western Airlines flight, a man complained about cigarette smoke coming from the smoking section, six smokers, including two in the nonsmoking section, lit up and began to blow smoke at him (Salpukas, 1975). Nonsmokers are showing similar militance. Several incidents on commuter trains are reported including a nonsmoker breathing garlic on other commuters who ignored the no smoking sign, and that several nonsmokers have equipped themselves with battery operated fans to blow smoke back in smokers' faces (1975).

REFERENCES

Advertising Age. 1976. "Costs of cigarette advertising: 1975–1968." *Advertising Age* 47 (November 22): 36, 38.

ASH (Action on Smoking and Health). 1975. "Smoking and health: History of the battle, 1964–1975." Washington, DC: ASH.

————. 1978. "History of the war against smoking: 1964–1978." Washington, DC: ASH.

ASH Newsletter. 1977. "News you should know." *ASH* 7 (January–February): 6.

————. 1978. "Senate hearings on health bill and teenage smoking." *ASH* 8 (May–June): 3.

Bleda, Paul R., and Paul H. Sandman. 1977. "In smokes way: Socioemotional reactions to another's smoking." *Journal of Applied Psychology* 62 (4): 452–458.

Bloor, David. 1976. *Knowledge and Social Imagery.* London: Routledge and Kegan Paul.

Borland, Barry L., and Joseph P. Rudolph. 1975. "Relative effects of low socio-economic status, parental smoking and poor scholastic perfomance on smoking among high school students." *Social Science and Medicine* 9 (1): 27–30.

Bradford, Garnett L., and Craig L. Infanger. 1978. "The tobacco industry and health: Understanding a contradiction." *Intellect* 106 (February): 318–321.

Brody, Jane E. 1974. "Cancer society steps up drive on growing smoking problems." *New York Times,* February 8: 32.

————. "Heavy smoking called disorder." *New York Times,* June 5: 38.

————. "Massive drive urged to combat smoking." *New York Times,* February 1: A-10.

Califano, Joseph A., Jr. 1978. Address before the National Interagency Council on Smoking and Health, Shoreham Hotel, Washington, D.C., January 11.

Cherry, Nicola, and Kathy Kierman. 1976. "Personality scores and smoking behaviour." *British Journal of Social Preventive Medicine* 30 (2): 123–31.

Cole, Robert J. 1973. "Insurance discounts for nonsmokers spread from life to care policies." *New York Times,* November 1: 63.

Condon, James C. 1976. "Tobacco is a dirty word—they loved it." *New York Times,* December 5, Section III: 5.

Congressional Quarterly Weekly Report. 1978. "Text of President Carter's January 12, News Conference." 36 (January 21): 138–141. Washington, D.C.: Congressional Quarterly, Inc.

Conrad, Peter. 1975. "The discovery of hyperkinesis: Notes on the medicalization of deviant behavior." *Social Problems* 23 (October): 12–21.

Cronkite, Walter. 1978. *CBS-TV Evening News,* November 2.

Dullea, Georgia. 1975. "And as the gaspers of the world unite, nonsmokers become fashionably cool." *New York Times,* November 5: 51.

Encyclopedia of Associations. 1978. *Encyclopedia of Associations,* 12th edition, Volume 11. Detroit: Gale Research Company.

Eysenck, Hans Jurgen. 1965. *Smoking, Health, and Personality.* Basic Books: New York.

Epstein, Samuel. 1978. *The Politics of Cancer.* San Francisco: Sierra Club Books.

Fisher, Jerid M. 1976. "Sex differences in smoking dynamics." *Journal of Health and Social Behavior* 17 (2): 145–150.

Gardner, Richard A. 1977. "Letter to the editor." *New York Times,* July 30: 18.

GASP (Group Against Smokers' Pollution). n.d. "Does tobacco smoke turn you off?" GASP: College Park, Maryland.

Glasnick, Jack. 1974. "Letter to the editor." *New York Times,* April 6: 30.

Gusfield, Joseph R. 1963. *Symbolic Crusade: Status Politics and the American*

Temperance Movement. Urbana, IL: University of Illinois Press.

Hanley, J. A., and J. C. Robinson. 1976. "Cigarette smoking and the young: A national survey." *Canadian Medical Association Journal* 114 (March 20): 511–517.

Jacobs, Martin A., and Aron Z. Spilkin. 1971. "Personality patterns associated with heavy cigarette smoking in male college students." *Journal of Consulting and Clinical Psychology* 37 (3): 428–432.

Jensen, Michael C. 1976. "3 directors out at R. J. Reynolds." *New York Times,* May 29: 1.

———. 1978. "Tobacco: A potent lobby." *New York Times,* February 19, Section III: 1.

Lear, Martha Weinman. 1974. "All the warnings, gone up in smoke." *New York Times,* March 10: 18–19, 86, 91.

Levine, Harry Gene. 1978. "The discovery of addiction: Changing conceptions of habitual drunkenness in America." *Journal of Studies on Alcohol* 39 (January): 143–174.

Mann, Charles Kellogg. 1975. *Tobacco: The Ants and the Elephants.* Salt Lake City: Olympus Publishing Company.

Morris, John D. 1972. "Cigarette ads up in publications." *New York Times,* January 19: 43.

Murphy, Phyllis. 1978. "Letters to the editor." *Christian Science Monitor,* February 10: 19.

New York Times. 1977. "Cancer group opens drive to end tobacco subsidy." January 15: A-8.

Newsweek. 1978. "The new issues." *Newsweek,* October 2: 56–61.

Nuehring, Elaine, and Gerald E. Markle. 1974. "Nicotine and norms: The re-emergence of a deviant behavior." *Social Problems* 21 (April): 513–526.

Pfohl, Stephen J. 1977. "The discovery of child abuse." *Social Problems* 24 (February): 310–323.

Phillip Morris Incorporated. 1978. *Annual Report—1977.* New York: Phillip Morris Incorporated.

Rae, Gordon. 1975. "Extraversion, neuroticism and cigarette smoking." *British Journal of Social and Clinical Psychology* 14 (November): 429–430.

Reynolds, Carl, and Robert Nichols. 1976. "Personality and behavioral correlates of cigarette smoking: One-year follow-up." *Psychological Reports* 38 (1): 251–258.

R. J. Reynolds Industries, Inc. 1978. *1977 Annual Report.* Winston-Salem, North Carolina: R. J. Reynolds Industries.

Rood, W. B. 1978. "Tobacco firms pour more money into proposition 5 fight." *Los Angeles Times,* November 5: Part I, 3.

Rose, Vicki McNickle. 1977. "The rise of the rape problem." Pages 167–195 in Armand L. Mauss and Julie Camile Wolfe (eds.), *This Land of Promises.* Philadelphia: J.B. Lippincott.

Safire, William. 1974. "On puffery." *New York Times,* May 16: 41.

Schneider, Joseph W. 1978. "Deviant drinking as disease: Alcoholism as a social accomplishment." *Social Problems* 25 (4): 361–372.

Shuffett, D. M., and J. Hoskins. 1969. "Capitalization of burley tobacco allotment rights." *American Journal of Agricultural Ecomonics* 51 (May): 471–474.

Simon, William E., and Louis H. Primavera. 1976. "The personality of the cigarette smoker: Some empirical data." *The International Journal of the Addictions* 11 (1): 81–94.

Smith, Gene M. 1969. "Relations between personality and smoking behavior in preadult subjects." *Journal of Consulting and Clinical Psychology* 33: 710–715.

———. "Personality and smoking: A review of the empirical literature." In William A. Hunt (ed.), *Learning Mechanisms in Smoking.* Chicago: Aldine.

SmokEnders. 1977. "The easy way." Phillipsburg, New Jersey: Smok-Enders, Inc.

Spector, Malcolm, and John I. Kitsuse. 1977. *Constructing Social Problems.*

Menlo Park, CA: Cummings Publishing Company.

Srole, Leo, and Anita Kassen Fischer. 1973. "The social epidemiology of smoking behavior 1953 and 1970: The Midtown Manhattan study." *Social Science and Medicine* 7 (May): 341–358.

Time. 1977. "The Mafia: Big, bad, and booming." *Time,* May 16: 32–42.

Tobacco Institute. 1978. "Tobacco industry profile." Tobacco Institute: Washington, D. C.

U.S. Department of Agriculture. 1969. *Annual Report on Tobacco Statistics, 1968.* Statistical Bulletin No. 450, Agricultural Marketing Service. Washington, DC: Government Printing Office.

———. 1971. *Annual Report on Tobacco Statistics, 1970.* Statistical Bulletin No. 454, Agricultural Marketing Service. Washington, DC: Government Printing Office.

———. 1978a. *Annual Report on Tobacco Statistics, 1977.* Statistical Bulletin No. 605, Agricultural Marketing Service. Washington, DC: Government Printing Office.

———. 1978b. "ASCS commodity fact sheet: Flue cured tobacco." Agricultural Stabilization and Conservation Service. Washington, DC: Government Printing Office.

U.S. Department of Health, Education and Welfare. 1973. *Adult Use of Tobacco—1970.* Washington, DC: Government Printing Office.

———. 1976a. "Adult use of Tobacco—1975: Summary." Washington, DC.

———. 1976b. "Adult use of Tobacco—1975: Tables." Washington, DC.

———. 1976c. "Adult use of Tobacco—1975: Methodology." Washington, DC.

Warner, Kenneth E. 1978. "Possible increases in the underreporting of cigarette consumption." *Journal of the American Statistical Association* 73 (June): 314–318.

9

The Social Construction
of Drug Scares

CRAIG REINARMAN

D rug "wars," anti-drug crusades, and other periods of marked public concern about drugs are never merely reactions to the various troubles people can have with drugs. These drug scares are recurring cultural and political phenomena *in their own right* and must, therefore, be understood sociologically on their own terms. It is important to understand why people ingest drugs and why some of them develop problems that have something to do with having ingested them. But the premise of this chapter is that it is equally important to understand patterns of acute societal concern about drug use and drug problems. This seems especially so for U.S. society, which has had *recurring* anti-drug crusades and a *history* of repressive anti-drug laws.

Many well-intentioned drug policy reform efforts in the U.S. have come face to face with staid and stubborn sentiments against consciousness–altering substances. The repeated failures of such reform efforts cannot be explained solely in terms of ill-informed or manipulative leaders. Something deeper is involved, something woven into the very fabric of American culture, something which explains why claims that some drug is the cause of much of what is wrong with the world are *believed* so often by so many. The origins and nature of the *appeal* of anti-drug claims must be confronted if we are ever to understand how "drug problems" are constructed in the U.S. such that more enlightened and effective drug policies have been so difficult to achieve.

In this chapter I take a step in this direction. First, I summarize briefly some of the major periods of anti-drug sentiment in the U.S. Second, I draw from them the basic ingredients of which drug scares and drug laws are made. Third, I offer a beginning interpretation of these scares and laws based on those broad features of American culture that make *self-control* continuously problematic.

DRUG SCARES AND DRUG LAWS

What I have called drug scares (Reinarman and Levine, 1989a) have been a recurring feature of U.S. society for 200 years. They are relatively autonomous

Reprinted by permission of Craig Reinarman.

from whatever drug-related problems exist or are said to exist.[1] I call them "scares" because, like Red Scares, they are a form of moral panic ideologically constructed so as to construe one or another chemical bogeyman, à la "communists," as the core cause of a wide array of pre-existing public problems.

The first and most significant drug scare was over drink. Temperance movement leaders constructed this scare beginning in the late 18th and early 19th century. It reached its formal end with the passage of Prohibition in 1919.[2] As Gusfield showed in his classic book *Symbolic Crusade* (1963), there was far more to the battle against booze than long-standing drinking problems. Temperance crusaders tended to be native born, middle-class, non-urban Protestants who felt threatened by the working-class, Catholic immigrants who were filling up America's cities during industrialization.[3] The latter were what Gusfield termed "unrepentant deviants" in that they continued their long-standing drinking practices despite middle-class W.A.S.P. norms against them. The battle over booze was the terrain on which was fought a cornucopia of cultural conflicts, particularly over whose morality would be the dominant morality in America.

In the course of this century-long struggle, the often wild claims of Temperance leaders appealed to millions of middle-class people seeking explanations for the pressing social and economic problems of industrializing America. Many corporate supporters of Prohibition threw their financial and ideological weight behind the Anti-Saloon League and other Temperance and Prohibitionist groups because they felt that traditional working-class drinking practices interfered with the new rhythms of the factory, and thus with productivity and profits (Rumbarger, 1989). To the Temperance crusaders' fear of the bar room as a breeding ground of all sorts of tragic immorality, Prohibitionists added the idea of the saloon as an alien, subversive place where unionists organized and where leftists and anarchists found recruits (Levine, 1984).

This convergence of claims and interests rendered alcohol a scapegoat for most of the nation's poverty, crime, moral degeneracy, "broken" families, illegitimacy, unemployment, and personal and business failure—problems whose sources lay in broader economic and political forces. This scare climaxed in the first two decades of this century, a tumultuous period rife with class, racial, cultural, and political conflict brought on by the wrenching changes of industrialization, immigration, and urbanization (Levine, 1984; Levine and Reinarman, 1991).

America's first real drug law was San Francisco's anti-opium den ordinance of 1875. The context of the campaign for this law shared many features with the context of the Temperance movement. Opiates had long been widely and legally available without a prescription in hundreds of medicines (Brecher, 1972; Musto, 1973; Courtwright, 1982; cf. Baumohl, 1992), so neither opiate use nor addiction was really the issue. This campaign focused almost exclusively on what was called the "Mongolian vice" of opium *smoking* by Chinese immigrants (and white "fellow travelers") in dens (Baumohl, 1992). Chinese immigrants came to California as "coolie" labor to build the railroad and dig the gold mines. A small minority of them brought along the practice of smoking opium—a practice originally brought to China by British and American traders in the 19th century. When the railroad was completed and the gold dried up, a decade-long depression ensued. In a tight labor market, Chinese immigrants were a target. The white Workingman's Party fomented racial ha- *completed* tred of the low-wage "coolies" with whom they now had to compete for *w/ Chinese* work. The first law against opium smoking was only one of many laws enacted to harass and control Chinese workers (Morgan, 1978).

By calling attention to this broader political-economic context I do not wish to slight the specifics of the local political-economic context. In addition to the Workingman's Party, downtown businessmen formed merchant associations and urban families formed improvement associations, both of which fought for more than two decades to reduce the impact of San Francisco's vice districts on the order and health of the central business district and on family neighborhoods (Baumohl, 1992).

In this sense, the anti-opium den ordinance was not the clear and direct result of a sudden drug scare alone. The law was passed against a specific form of drug use engaged in by a disreputable group that had come to be seen as threatening in lean economic times. But it passed easily because this new threat was understood against the broader historical backdrop of long-standing local concerns about various vices as threats to public health, public morals, and public order. Moreover, the focus of attention were dens where it was suspected that whites came into intimate contact with "filthy, idolatrous" Chinese (see Baumohl, 1992). Some local law enforcement leaders, for example, complained that Chinese men were using this vice to seduce white women into sexual slavery (Morgan, 1978). Whatever the hazards of opium smoking, its initial criminalization in San Francisco had to do with both a general context of recession, class conflict, and racism, and with specific local interests in the control of vice and the prevention of miscegenation.

A nationwide scare focusing on opiates and cocaine began in the early *opiates* 20th century. These drugs had been widely used for years, but were first crim- *cocaine +* inalized when the addict population began to shift from predominantly white, middle-class, middle-aged women to young, working-class males, African-Americans in particular. This scare led to the Harrison Narcotics Act of 1914, *44's* the first federal anti-drug law (see Duster, 1970).

Many different moral entrepreneurs guided its passage over a six-year campaign: State Department diplomats seeking a drug treaty as a means of expanding trade with China, trade which they felt was crucial for pulling the economy out of recession; the medical and pharmaceutical professions whose interests were threatened by self-medication with unregulated proprietary tonics, many of which contained cocaine or opiates; reformers seeking to control what they saw as the deviance of immigrants and Southern Blacks who were migrating off the farms; and a pliant press which routinely linked drug use with prostitutes, criminals, transient workers (e.g., the Wobblies), and African-Americans (Musto, 1973). In order to gain the support of Southern Congressmen for a new federal law that might infringe on "states' rights," State Department officials and other crusaders repeatedly spread unsubstantiated suspicions, repeated in the press, that, e.g., cocaine induced African-American men to rape white women (Musto, 1973: 6–10, 67). In short, there was more to this drug scare, too, than mere drug problems.

In the Great Depression, Harry Anslinger of the Federal Narcotics Bureau pushed Congress for a federal law against marijuana. He claimed it was a "killer weed" and he spread stories to the press suggesting that it induced violence—especially among Mexican-Americans. Although there was no evidence that marijuana was widely used, much less that it had any untoward effects, his crusade resulted in its criminalization in 1937—and not incidentally a turnaround in his Bureau's fiscal fortunes (Dickson, 1968). In this case, a new drug law was put in place by a militant moral-bureaucratic entrepreneur who played on racial fears and manipulated a press willing to repeat even his most absurd claims in a context of class conflict during the Depression (Becker, 1963). While there was not a marked scare at the time, Anslinger's claims were never contested in Congress because they played upon racial fears and widely held Victorian values against taking drugs solely for pleasure.

In the drug scare of the 1960s, political and moral leaders somehow reconceptualized this same "killer weed" as the "drop out drug" that was leading America's youth to rebellion and ruin (Himmelstein, 1983). Bio-medical scientists also published uncontrolled, retrospective studies of very small numbers of cases suggesting that, in addition to poisoning the minds and morals of youth, LSD produced broken chromosomes and thus genetic damage (Cohen et al., 1967). These studies were soon shown to be seriously misleading if not meaningless (Tjio et al., 1969), but not before the press, politicians, the medical profession, and the National Institute of Mental Health used them to promote a scare (Weil, 1972: 44–46).

I suggest that the reason even supposedly hard-headed scientists were drawn into such propaganda was that dominant groups felt the country was at war—and not merely with Vietnam. In this scare, there was not so much a "dangerous class" or threatening racial group as multi-faceted political and cultural conflict, particularly between generations, which gave rise to the perception that middle-class youth who rejected conventional values were a dangerous threat.[4] This scare resulted in the Comprehensive Drug Abuse Control Act of 1970, which criminalized more forms of drug use and subjected users to harsher penalties.

Most recently we have seen the crack scare, which began in earnest *not* when the prevalence of cocaine use quadrupled in the late 1970s, nor even when thousands of users began to smoke it in the more potent and dangerous form of freebase. Indeed, when this scare was launched, crack was unknown outside of a few neighborhoods in a handful of major cities (Reinarman and Levine, 1989a) and the prevalence of illicit drug use had been dropping for several years (National Institute on Drug Use, 1990). Rather, this most recent scare began in 1986 when freebase cocaine was renamed crack (or "rock") and sold in precooked, inexpensive units on ghetto streetcorners (Reinarman and Levine, 1989b). Once politicians and the media linked this new form of cocaine use to the inner-city, minority poor, a new drug scare was underway and the solution became more prison cells rather than more treatment slots.

The same sorts of wild claims and Draconian policy proposals of Temperance and Prohibition leaders resurfaced in the crack scare. Politicians have so outdone each other in getting "tough on drugs" that each year since crack came on the scene in 1986 they have passed more repressive laws providing billions more for law enforcement, longer sentences, and more drug offenses punishable by death. One result is that the U.S. now has more people in prison than any industrialized nation in the world—about half of them for drug offenses, the majority of whom are racial minorities.

In each of these periods more repressive drug laws were passed on the grounds that they would reduce drug use and drug problems. I have found no evidence that any scare actually accomplished those ends, but they did greatly expand the quantity and quality of social control, particularly over subordinate groups perceived as dangerous or threatening. Reading across these historical episodes one can abstract a recipe for drug scares and repressive drug laws that contains the following *seven ingredients:*

1. A Kernel of Truth Humans have ingested fermented beverages at least since human civilization moved from hunting and gathering to primitive agriculture thousands of years ago (Levine, forthcoming). The pharmacopoeia has expanded exponentially since then. So, in virtually all cultures and historical epochs, there has been sufficient ingestion of consciousness-altering chemicals to provide some basis for some people to claim that it is a problem.

2. Media Magnification In each of the episodes I have summarized and many others, the mass media has engaged in what I call the *routinization of caricature*—rhetorically recrafting worst cases into typical cases and the episodic into the epidemic. The media dramatize drug problems, as they do other problems, in the course of their routine news-generating and sales-promoting procedures (see Brecher, 1972: 321–34; Reinarman and Duskin, 1992; and Molotch and Lester, 1974).

3. Politico-Moral Entrepreneurs I have added the prefix "politico" to Becker's (1963) seminal concept of moral entrepreneur in order to emphasize the fact that the most prominent and powerful moral entrepreneurs in drug scares are often political elites. Otherwise, I employ the term just as he intended: to denote the *enterprise,* the work, of those who create (or enforce) a rule against what they see as a social evil.[5]

In the history of drug problems in the U.S., these entrepreneurs call attention to drug using behavior and define it as a threat about which "something must be done." They also serve as the media's primary source of sound bites on the dangers of this or that drug. In all the scares I have noted, these entrepreneurs had interests of their own (often financial) which had little to do with drugs. Political elites typically find drugs a functional demon in that (like "outside agitators") drugs allow them to deflect attention from other, more systemic sources of public problems for which they would otherwise have to take some responsibility. Unlike almost every other political issue, however, to be "tough on drugs" in American political culture allows a leader to take a firm stand without risking votes or campaign contributions.

4. Professional Interest Groups In each drug scare and during the passage of each drug law, various professional interests contended over what Gusfield (1981: 10–15) calls the "ownership" of drug problems—"the ability to create and influence the public definition of a problem" (1981: 10), and thus to define what should be done about it. These groups have included industrialists, churches, the American Medical Association, the American Pharmaceutical Association, various law enforcement agencies, scientists, and most recently the treatment industry and groups of those former addicts converted to disease ideology.[6] These groups claim for themselves, by virtue of their specialized forms of knowledge, the legitimacy and authority to name what is wrong and to prescribe the solution, usually garnering resources as a result.

5. Historical Context of Conflict This trinity of the media, moral entrepreneurs, and professional interests typically interact in such a way as to inflate the extant "kernel of truth" about drug use. But this interaction does not by itself give rise to drug scares or drug laws without underlying conflicts which make drugs into functional villains. Although Temperance crusaders persuaded millions to pledge abstinence, they campaigned for years without achieving alcohol control laws. However, in the tumultuous period leading up to Prohibition, there were revolutions in Russia and Mexico, World War I, massive immigration and impoverishment, and socialist, anarchist, and labor movements, to say nothing of increases in routine problems such as crime. I submit that all this conflict made for a level of cultural anxiety that provided fertile ideological soil for Prohibition. In each of the other scares, similar conflicts—economic, political, cultural, class, racial, or a combination—provided a context in which claims makers could viably construe certain classes of drug users as a threat.

6. Linking a Form of Drug Use to a "Dangerous Class" Drug scares are never about drugs per se, because drugs are inanimate objects without social consequence until they are ingested by humans. Rather, drug scares are about the use of a drug by particular groups of people who are, typically, *already* perceived by powerful groups as some kind of threat (see Duster, 1970; Himmelstein, 1978). It was not so much alcohol problems *per se* that most animated the drive for Prohibition but the behavior and morality of what dominant groups saw as the "dangerous class" of urban, immigrant, Catholic, working-class drinkers (Gusfield, 1963; Rumbarger, 1989). It was *Chinese* opium smoking dens, not the more widespread use of other opiates, that prompted California's first drug law in the 1870s. It was only when smokable cocaine found its way to the African-American and Latino underclass that it made headlines and prompted calls for a drug war. In each case, politico-moral entrepreneurs were able to construct a "drug problem" by linking a substance to a group of users perceived by the powerful as disreputable, dangerous, or otherwise threatening.

7. Scapegoating a Drug for a Wide Array of Public Problems The final ingredient is scapegoating, i.e., blaming a drug or its alleged effects on a group of its users for a variety of preexisting social ills that are typically only indirectly associated with it. Scapegoating may be the most crucial element because it gives great explanatory power and thus broader resonance to claims about the horrors of drugs (particularly in the conflictual historical contexts in which drug scares tend to occur).

Scapegoating was abundant in each of the cases noted previously. To listen to Temperance crusaders, for example, one might have believed that without alcohol use, America would be a land of infinite economic progress with no poverty, crime, mental illness, or even sex outside marriage. To listen to leaders of organized medicine and the government in the 1960s, one might have surmised that without marijuana and LSD there would have been neither conflict between youth and their parents nor opposition to the Vietnam War. And to believe politicians and the media in the past 6 years is to believe that without the scourge of crack the inner cities and the so-called underclass would, if not disappear, at least be far less scarred by poverty, violence, and crime. There is no historical evidence supporting any of this.

In short, drugs are richly functional scapegoats. They provide elites with fig leaves to place over unsightly social ills that are endemic to the social system over which they preside. And they provide the public with a restricted aperture of attribution in which only a chemical bogeyman or the lone deviants who ingest it are seen as the cause of a cornucopia of complex problems.

TOWARD A CULTURALLY-SPECIFIC
THEORY OF DRUG SCARES

Various forms of drug use have been and are widespread in almost all societies comparable to ours. A few of them have experienced limited drug scares, usually around alcohol decades ago. However, drug scares have been *far* less common in other societies, and never as virulent as they have been in the U.S. (Brecher, 1972; Levine, 1992; MacAndrew and Edgerton, 1969). There has never been a time or place in human history without drunkenness, for example, but in *most* times and places drunkenness has not been nearly as problematic as it has been in the U.S. since the late 18th century (Levine, forthcoming). Moreover, in comparable industrial democracies, drug laws are generally less repressive. Why then do claims about the horrors of this or that consciousness-altering chemical have such unusual power in American culture?

Drug scares and other periods of acute public concern about drug use are not just discrete, unrelated episodes. There is a historical pattern in the U.S.

that cannot be understood in terms of the moral values and perceptions of individual anti-drug crusaders alone. I have suggested that these crusaders have
benefitted in various ways from their crusades. For example, making claims
about how a drug is damaging society can help elites increase the social control of groups perceived as threatening (Duster, 1970), establish one class's
moral code as dominant (Gusfield, 1963), bolster a bureaucracy's sagging fiscal
fortunes (Dickson, 1968), or mobilize voter support (Reinarman and Levine,
1989a, b). However, the recurring character of pharmaco-phobia in U.S. history suggests that there is something about our *culture* which makes citizens
more vulnerable to anti-drug crusaders' attempts to demonize drugs. Thus, an
answer to the question of America's unusual vulnerability to drug scares must
address why the scapegoating of consciousness-altering substances regularly
resonates with or appeals to substantial portions of the population.

There are three basic parts to my answer. The first is that claims about the
evils of drugs are especially viable in American culture in part because they
provide a welcome *vocabulary of attribution* (cf. Mills, 1940). Armed with
"DRUGS" as a generic scapegoat, citizens gain the cognitive satisfaction of
having a folk devil on which to blame a range of bizarre behaviors or other
conditions they find troubling but difficult to explain in other terms. This
much may be true of a number of other societies, but I hypothesize that this is
particularly so in the U.S. because in our political culture individualistic explanations for problems are so much more common than social explanations.

Second, claims about the evils of drugs provide an especially serviceable
vocabulary of attribution in the U.S. in part because our society developed
from a *temperance culture* (Levine, 1992). American society was forged in the
fires of ascetic Protestantism and industrial capitalism, both of which demand
self-control. U.S. society has long been characterized as the land of the individual "self-made man." In such a land, self-control has had extraordinary importance. For the middle-class Protestants who settled, defined, and still dominate
the U.S., self-control was both central to religious world views and a charac-
terological necessity for economic survival and success in the capitalist market
(Weber, 1930 [1985]). With Levine (1992), I hypothesize that in a culture in
which self-control is inordinately important, drug-induced altered states of
consciousness are especially likely to be experienced as "loss of control," and
thus to be inordinately feared.[7]

Drunkenness and other forms of drug use have, of course, been present
everywhere in the industrialized world. But temperance cultures tend to arise
only when industrial capitalism unfolds upon a cultural terrain deeply imbued
with the Protestant ethic.[8] This means that only the U.S., England, Canada,
and parts of Scandanavia have Temperance cultures, the U.S. being the most
extreme case.

It may be objected that the influence of such a Temperance culture was
strongest in the 19th and early 20th century and that its grip on the American
zeitgeist has been loosened by the forces of modernity and now, many say, postmodernity. The third part of my answer, however, is that on the foundation of

a Temperance culture, advanced capitalism has built a *postmodern, mass consumption culture* that exacerbates the problem of self-control in new ways.

Early in the 20th century, Henry Ford pioneered the idea that by raising wages he could simultaneously quell worker protests and increase market demand for mass-produced goods. This mass consumption strategy became central to modern American society and one of the reasons for our economic success (Marcuse, 1964; Aronowitz, 1973; Ewen, 1976; Bell, 1978). Our economy is now so fundamentally predicated upon mass consumption that theorists as diverse as Daniel Bell and Herbert Marcuse have observed that we live in a mass consumption culture. Bell (1978), for example, notes that while the Protestant work ethic and deferred gratification may still hold sway in the workplace, Madison Avenue, the media, and malls have inculcated a new indulgence ethic in the leisure sphere in which pleasure-seeking and immediate gratification reign.

Thus, our economy and society have come to depend upon the constant cultivation of new "needs," the production of new desires. Not only the hardware of social life such as food, clothing, and shelter but also the software of the self—excitement, entertainment, even eroticism—have become mass consumption commodities. This means that our society offers an increasing number of incentives for indulgence—more ways to lose self-control—and a decreasing number of countervailing reasons for retaining it.

In short, drug scares continue to occur in American society in part because people must constantly manage the contradiction between a Temperance culture that insists on self-control and a mass consumption culture which renders self-control continuously problematic. In addition to helping explain the recurrence of drug scares, I think this contradiction helps account for why in the last dozen years millions of Americans have joined 12-Step groups, more than 100 of which have nothing whatsoever to do with ingesting a drug (Reinarman, forthcoming). "Addiction," or the generalized loss of self-control, has become the meta-metaphor for a staggering array of human troubles. And, of course, we also seem to have a staggering array of politicians and other moral entrepreneurs who take advantage of such cultural contradictions to blame new chemical bogeymen for our society's ills.

NOTES

1. In this regard, for example, Robin Room wisely observes "that we are living at a historic moment when the rate of (alcohol) dependence as a cognitive and existential experience is rising, although the rate of alcohol consumption and of heavy drinking is falling." He draws from this a more general hypothesis about "long waves" of drinking and societal reactions to them: "[I]n periods of increased questioning of drinking and heavy drinking, the trends in the two forms of dependence, psychological and physical, will tend to run in opposite directions. Conversely, in periods of a "wettening" of sentiments, with the curve of alcohol consumption beginning to rise, we may expect the rate of physical dependence . . . to rise while the rate of dependence as a cognitive experience falls" (1991: 154).

2.   I say "formal end" because Temperance ideology is not merely alive and well in the War on Drugs but is being applied to all manner of human troubles in the burgeoning 12-Step Movement (Reinarman, forthcoming).

3.   From Jim Baumohl I have learned that while the Temperance movement attracted most of its supporters from these groups, it also found supporters among many others (e.g., labor, the Irish, Catholics, former drunkards, women), each of which had its own reading of and folded its own agenda into the movement.

4.   This historical sketch of drug scares is obviously not exhaustive. Readers interested in other scares should see, e.g., Brecher's encyclopedic work *Licit and Illicit Drugs* (1972), especially the chapter on glue sniffing, which illustrates how the media actually created a new drug problem by writing hysterical stories about it. There was also a PCP scare in the 1970s in which law enforcement officials claimed that the growing use of this horse tranquilizer was a severe threat because it made users so violent and gave them such super-human strength that stun guns were necessary. This, too, turned out to be unfounded and the "angel dust" scare was short-lived (see Feldman et al., 1979). The best analysis of how new drugs themselves can lead to panic reactions among users is Becker (1967).

5.   Becker wisely warns against the "one-sided view" that sees such crusaders as merely imposing their morality on others. Moral entrepreneurs, he notes, do operate "with an absolute ethic," are "fervent and righteous," and will use "any means" necessary to "do away with" what they see as "totally evil." However, they also "typically believe that their mission is a holy one," that if people do what they want it "will be good for them." Thus, as in the case of abolitionists, the crusades of moral entrepreneurs often "have strong humanitarian overtones" (1963: 147–8). This is no less true for those whose moral enterprise promotes drug scares. My analysis, however, concerns the character and consequences of their efforts, not their motives.

6.   As Gusfield notes, such ownership sometimes shifts over time, e.g., with alcohol problems, from religion to criminal law to medical science. With other drug problems, the shift in ownership has been away from medical science toward criminal law. The most insightful treatment of the medicalization of alcohol/drug problems is Peele (1989).

7.   See Baumohl's (1990) important and erudite analysis of how the human will was valorized in the therapeutic temperance thought of 19th-century inebriate homes.

8.   The third central feature of Temperance cultures identified by Levine (1992), which I will not dwell on, is predominance of spirits drinking, i.e., more concentrated alcohol than wine or beer and thus greater likelihood of drunkenness.

REFERENCES

Aronowitz, Stanley. 1973. *False Promises: The Shaping of American Working Class Consciousness.* New York: McGraw-Hill.

Baumohl, Jim. 1990. "Inebriate Institutions in North America, 1840–1920." *British Journal of Addiction* 85: 1187–1204.

Baumohl, Jim. 1992. "The 'Dope Fiend's Paradise' Revisited: Notes from Research in Progress on Drug Law Enforcement in San Francisco, 1875–1915." *Drinking and Drug Practices Surveyor* 24: 3–12.

Becker, Howard S. 1963. *Outsiders: Studies in the Sociology of Deviance.* Glencoe, IL: Free Press.

———. 1967. "History, Culture, and Subjective Experience: An Exploration of the Social Bases of Drug-Induced Experiences." *Journal of Health and Social Behavior* 8: 162–176.

Bell, Daniel. 1978. *The Cultural Contradictions of Capitalism*. New York: Basic Books.

Brecher, Edward M. 1972. *Licit and Illicit Drugs*. Boston: Little Brown.

Cohen, M. M., K. Hirshorn, and W. A. Frosch. 1967. "In Vivo and in Vitro Chromosomal Damage Induced by LSD-25." *New England Journal of Medicine* 227: 1043.

Courtwright, David. 1982. *Dark Paradise: Opiate Addiction in America Before 1940*. Cambridge, MA: Harvard University Press.

Dickson, Donald. 1968. "Bureaucracy and Morality." *Social Problems*. 16: 143–156.

Duster, Troy. 1970. *The Legislation of Morality: Law, Drugs, and Moral Judgement*. New York: Free Press.

Ewen, Stuart. 1976. *Captains of Consciousness: Advertising and the Social Roots of Consumer Culture*. New York: McGraw-Hill.

Feldman, Harvey W., Michael H. Agar, and George M. Beschner. 1979. *Angel Dust*. Lexington, MA: Lexington Books.

Gusfield, Joseph R. 1963. *Symbolic Crusade: Status Politics and the American Temperance Movement*. Urbana: University of Illinois Press.

———. 1981. *The Culture of Public Problems: Drinking-Driving and the Symbolic Order*. Chicago: University of Chicago Press.

Himmelstein, Jerome. 1978. "Drug Politics Theory." *Journal of Drug Issues*. 8.

———. 1983. *The Strange Career of Marihuana*. Westport, CT: Greenwood Press.

Levine, Harry Gene. 1984. "The Alcohol Problem in America: From Temperance to Alcoholism." *British Journal of Addiction*. 84: 109–119.

———. 1992. "Temperance Cultures: Concern About Alcohol Problems in Nordic and English-Speaking Cultures." In G. Edwards et al., Eds., *The Nature of Alcohol and Drug Related Problems*. New York: Oxford University Press.

———. forthcoming. *Drunkenness and Civilization*. New York: Basic Books.

Levine, Harry Gene, and Craig Reinarman. 1991. "From Prohibition to Regulation: Lessons from Alcohol Policy for Drug Policy." *Milbank Quarterly*. 69: 461–494.

MacAndrew, Craig, and Robert Edgerton. 1969. *Drunken Comportment*. Chicago: Aldine.

Marcuse, Herbert. 1964. *One-Dimensional Man: Studies in the Ideology of Advanced Industrial Society*. Boston: Beacon Press.

Mills, C. Wright. 1940. "Situated Actions and Vocabularies of Motive." *American Sociological Review* 5: 904–913.

Molotch, Harvey, and Marilyn Lester. 1974. "News as Purposive Behavior: On the Strategic Uses of Routine Events, Accidents, and Scandals." *American Sociological Review* 39: 101–112.

Morgan, Patricia. 1978. "The Legislation of Drug Law: Economic Crisis and Social Control." *Journal of Drug Issues* 8: 53–62.

Musto, David. 1973. *The American Disease: Origins of Narcotic Control*. New Haven, CT: Yale University Press.

National Institute on Drug Abuse. 1990. *National Household Survey on Drug Abuse: Main Findings 1990*. Washington, DC: U.S. Department of Health and Human Services.

Peele, Stanton. 1989. *The Diseasing of America: Addiction Treatment Out of Control*. Lexington, MA: Lexington Books.

Reinarman, Craig. forthcoming. "The 12-Step Movement and Advanced Capitalist Culture: Notes on the Politics of Self-Control in Postmodernity." In B. Epstein, R. Flacks, and M. Darnovsky, Eds., *Contemporary Social Movements and Cultural Politics*. New York: Oxford University Press.

Reinarman, Craig, and Ceres Duskin. 1992. "Dominant Ideology and Drugs in the Media." *International Journal on Drug Policy* 3: 6–15.

Reinarman, Craig, and Harry Gene Levine. 1989a. "Crack in Context: Politics and Media in the Making of a Drug Scare." *Contemporary Drug Problems.* 16: 535–577.

———. 1989b. "The Crack Attack: Politics and Media in America's Latest Drug Scare." In Joel Best, Ed., *Images of Issues: Typifying Contemporary Social Problems,* pp. 115–137. New York: Aldine de Gruyter.

Room, Robin G. W. 1991. "Cultural Changes in Drinking and Trends in Alcohol Problems Indicators: Recent U.S. Experience." In Walter B. Clark and Michael E. Hilton, Eds., *Alcohol in America: Drinking Practices and Problems,* pp. 149–162. Albany: State University of New York Press.

Rumbarger, John J. 1989. *Profits, Power, and Prohibition: Alcohol Reform and the Industrializing of America, 1800–1930.* Albany: State University of New York Press.

Tijo, J. H., W. N. Pahnke, and A. A. Kurland. 1969. "LSD and Chromosomes: A Controlled Experiment." *Journal of the American Medical Association.* 210: 849.

Weber, Max. 1985 (1930). *The Protestant Ethic and the Spirit of Capitalism.* London: Unwin.

Weil, Andrew. 1972. *The Natural Mind.* Boston: Houghton Mifflin.

10

The Problem with
"Battered Husbands"

BETSY LUCAL

N o topic in the sociological study of family violence has been more controversial in recent years than that of husband battering. Since the publication of Steinmetz's article, "The Battered Husband Syndrome," in
1977, disagreement and polemic have reigned. Any attempt to discuss the issues surrounding the idea of battered husbands enters the researcher into an
emotionally charged and hotly contested debate. According to Straus, this debate has indeed been heated. He reports that his public presentations on the
issue have been obstructed by "booing, shouting and picketing" and that
Steinmetz has been subjected to a "letter-writing campaign opposing her promotion," "phone calls threatening her and her family, and a bomb threat at a
conference where she spoke" (Straus 1992, pp. 225–226).

It has been a classic debate, filled with claims and counterclaims. Controversy has surrounded the issue of the respective rates of wife and husband battering, of what those rates truly mean, and of what society's response should be.

Much of the debate has centered around the question of whether there really are very many battered husbands out there. The question of whether battered husbands are or are not a social problem worthy of support has revolved
around this issue of rates. On one side, there are researchers who cite numerous studies they believe show that, as Straus (1993, p. 67) puts it, "physical assaults by wives" are "a major social problem." Straus, in fact, claims that more
than 30 studies show a "rate of assault by women on male partners that is about
the same as the rate of assault of men on female partners" (1993, p. 70; also
see Straus 1992). On the other side are those researchers who, questioning the
validity of those findings, argue that the abuse of husbands is not a major problem (e.g., Dobash and Dobash 1981; Berk, Berk, Loseke, and Rauma 1983;
Pagelow 1984, 1985; Saunders 1986, 1988a; Kurz 1989, 1993; Dobash, Dobash,
Wilson, and Daly 1992). Interestingly, Gelles (1994), who has authored studies that others (such as Straus) have interpreted as confirming that there are as
many battered husbands as battered wives, recently published an editorial in

which he estimates that there are about 100,000 battered men in the United States each year, compared to 2–4 million battered women (also see Gelles 1979, in which he questions the battered husband phenomenon).

From a deviant behavior perspective, at issue are the process of "deviance-making," the events that lead to a phenomenon being identified as deviant; and reactions to deviance, the form and types of responses to attempts at deviance-making (Terry and Steffensmeier 1988). In this article, I compare the successful construction of the deviant label "battered wives" to the failed construction of the label "battered husbands" by describing the process involved in the construction of those issues as social problems. As I show here, perhaps the most important reason that "battered husbands" have not been labeled successfully is our traditional gender images of women and men. It is virtually impossible to construct a social problem when, as in the case of battered husbands, it is difficult to apply the deviant label.

Since the claims being made about battered husbands are, at least implicitly, based on comparisons to battered wives, it is reasonable to compare the claims and counterclaims made about these two issues to attempt to account for the success of one and the failure of the other as a social problem.[1] In this article, I trace the development of the battered husbands issue and compare it to the battered wives issue in an attempt to understand why it is that battered husbands have not become a social problem.

THEORETICAL FRAMEWORK

Spector and Kitsuse (1977) provided the first explication of the constructionist perspective on social problems (though their argument is basically similar to Blumer's 1971 framework which analyzed social problems as collective behavior). From this perspective, social problems are claims-making activities rather than conditions. The main analytical task is to account for the development, character, and continuation of claims-making and responding endeavors. Whether the condition exists "objectively" is of no concern here because "social problems are what people think they are" (Spector and Kitsuse 1977, p. 73). It is the definitional process that needs to be analyzed, with the focus being on the *viability* rather than the validity of claims and on the responses to those claims (Schneider 1984).

The social constructionist framework provides a basis for assessing claims about battered husbands and for providing evidence as to why they have not become a social problem. Using this perspective also presents a challenge. The simplest way to dismiss battered husbands as a potential social problem is to demonstrate, or at least to argue, that there are not (m)any of them out there. From a strict social constructionist perspective (Best 1989), however, such numbers do not matter to the construction of a social problem. That is, the actual existence of wives and husbands who are battered by their spouses is, in large part, irrelevant. Because it has often been acknowledged that it is quite difficult to collect accurate and complete data about the incidence of family violence, starting from a position that makes no judgment about the validity of claims (i.e., one that puts aside the controversy over the relative numbers of battered wives and husbands) is reasonable. Therefore, while the incidence debate is clearly crucial to the issue of husband battering, it is not the focus of this article. Instead, I assess the claims and counterclaims that have been made in the attempts to establish "battered wives" and "battered husbands" as social problems. I compare the components of the successful construction of battered wives as a social problem to the ineffective attempts to do the same for battered husbands.

While countless researchers have studied successful social problem construction, very little research has been done on failed attempts to create social problems (but see Ball and Lilly 1984; Weitzer 1991). This paper compares a successful and an unsuccessful problem in the same substantive area by analyzing the factors that led to their different outcomes. There are three main factors that contributed to these outcomes: social movements, the type and extent of academic and mass media attention to the issues, and gender images of women and men. The first two factors are related to the process of deviance-making, while the third is also tied to the process of applying a deviant label.

METHODS

In attempting to account for the failure of battered husbands as a social problem, I compared the claims-making process and other developments concerning this issue with those that successfully constructed battered wives as a social problem. This application of the case-oriented comparative method (Ragin 1987; Neuman 1994) provides a means for answering such questions as: What are the different combinations of conditions that are associated with these two processes and outcomes? How do these different conditions fit together to produce differing outcomes? The goal of the analysis is to identify the differences that are responsible for contradictory outcomes. Here, that involves identifying the conditions that led to battered wives becoming a social problem and battered husbands, thus far, failing to do so. After identifying the broad set of factors associated with the success of the movement to make battered wives a social problem and fitting them into the "ideal" process of social problem construction, it is possible to see how those same factors have related to the failure of battered husbands as a social problem.

In addition to the justifications cited above, I chose to compare battered wives and battered husbands for two reasons. Because they occur in similar types of relationships (marital or quasi-marital) and among people with similar statuses (spouses or peer intimates), a comparison of responses to claims about these two types of abuse provides a way to look at how other factors led to differing outcomes. That is, given these two similarities between the issues, what led to one becoming a social problem and the other failing?

The analysis considers the development over time of the issues of wife and husband battering. The 1974 publication of Erin Pizzey's *Scream Quietly or the Neighbours Will Hear* was taken as the turning point between traditional attention to the issue of wives who were being beaten by their husbands and the beginning of claims about a new problem called "battered wives." Academic literature appearing after this year is the major basis of the analysis. For battered husbands, the analysis begins with Steinmetz's 1977 article, "The Battered Husband Syndrome," which was published in *Victimology.* These dates also provide anchoring points for the analysis of mass media attention to the two issues.

Research and writing addressing the issue of battered wives has proliferated since it was first raised. By using this literature to identify the factors that resulted in successful claims–making on behalf of abused wives, it is possible to distinguish those factors that have been absent, or otherwise unsuccessful, in the process of attempting to make abused husbands, about whom much less has been written, a social problem. Because this is a strict constructionist analysis, my interest is in factors that are related to the viability of claims about the two groups. The questions are: What are the factors that helped battered wives succeed as a social problem? How did these factors work against claims about battered husbands?

My sample is neither random nor systematic. I simply attempted to include as many sources as possible, in most cases, finding them through snowball sampling (Neuman 1994). It was far easier to analyze all of the available work on battered husbands than to claim to have done so with battered wives. Because of the sheer number of books and articles written on the latter, it would have been impossible to include them all. However, I believe I have included all the major works on the subject. Because so many materials I used supported the same conclusions (i.e., I was not finding any new information), I believe my analysis is adequate and appropriate.

The major part of this analysis is based on thematic content analysis (Krippendorff 1980) of secondary data in the form of existing literature on battered husbands and battered wives. Using Best (1989), Spector and Kitsuse (1977), Tierney (1982), and Studer (1984) as a basis for deciding what to look for in articles and books about wife and husband battering, I collected data on the two issues. Claims, their organization, and responses to them were analyzed in terms of the themes, concepts, and processes suggested in the literature on the social construction of social problems. The analysis consisted of collecting information for the existing literature that related to the process of attempting to construct these two issues as social problems.

ANALYSIS

The battered wives issue has been a successful competitor in the social problems marketplace. The success of this issue as a social problem can be attributed to the combination of a variety of factors: (a) organizational factors, as exemplified by the feminist movement in general and the battered women's movement in particular; (b) the proliferation of social science research and literature, as well as continued popular media attention; and (c) a stereotypic image of women that leads to their identification as "appropriate" (Dobash and Dobash 1977, p. 426) and/or acceptable victims. These same factors have worked *against* the construction of battered husbands as a social problem. There has been no social movement, no organized response of any kind, on behalf of battered husbands. Social science responses, in large part, have been contradictory; sustained mass media attention has been lacking. Finally, if gender images make the identification and definition of battered wives easier, they make similar perceptions of battered husbands all the more difficult.

The Social Movements

Much has been written about the success of the feminist movement and the battered women's movement in constructing abused wives as a social problem. Together the two movements have been crucial to the success of this issue.

The feminist movement provided the context for the discovery of wife battering by declaring that what happened to women and men in the privacy of their homes was political, by identifying violence against women as a widespread problem in contemporary society, and by questioning the legitimacy of husbands' violence toward their wives (Schechter 1982; Greenblat 1985). The definition of violence between women and men as the result of social relationships of power and domination provided for the redefinition of marital violence as a social problem (Murray 1988). This process of redefinition was not limited to the United States; it has also been documented in Canada (Walker 1990) and in Great Britain (Dobash and Dobash 1987, 1992).

According to Tierney (1982), the success of the battered women's movement in the United States is attributable to three factors related to the organization of the movement itself. The preexisting organizational base for the movement (the feminist movement and its allies), the movement's flexibility (both national and local organizations with compatibly different foci) and the existence of incentives for professionals to support the movement (also see Johnson 1981; Loseke and Cahill 1984; Studer 1984) all contributed to the success of the movement for battered women.

Another factor that probably contributed to the success of the movement was the involvement of battered wives. Especially early on in the movement, women who received assistance from shelters later joined their staffs (Schechter 1982; Rodriguez 1988). While this feminist type of shelter was not the only, or even the most prevalent (Johnson 1981; Ferraro and Johnson 1985), kind in existence, the involvement of formerly battered wives in activism enhanced the visibility and credibility of the cause.

The lack of a social movement, while not dooming a potential social problem to failure, does make construction more difficult: There is no organized effort to promote the issue as a problem. Because there was no men's movement when the claims about battered husbands were first made, there was no existing organizational basis of support for the issue. There were no co-optable social networks for advocates of battered husbands to turn to for resources. There were no incentives for other professionals to get involved in the issue: Large numbers of battered husbands did not come forward seeking services. The lack of visibility of battered husbands also meant that they could not promote their own cause as members of a movement. When viewed in light of the importance of social movement involvement in the construction of battered wives as a social problem, this lack of movement support can be seen to have played a significant role in the failure of battered husbands as a social problem.

How did the new men's movement of the 1980s and 1990s affect this issue? This movement has provided neither a voice nor resources for advocates of battered husbands. Its attention has focused elsewhere, on everything from child custody and the constraints of the traditional male role (Doyle 1989; Renzetti and Curran 1993) to a "romantic assertion of primitive masculinity in all its innocent strength and virtue" (Adler, Springen, Glick, and Gordon 1991, p. 49). It has not focused on male victims of spouse abuse. Additionally, those men involved in domestic violence activism have focused their energies on supporting the battered women's movement and on working with men who batter (Doyle 1989; Kimmel 1989).

One additional social movement factor is also a contributor to the lack of viability of claims about battered husbands. Because of fears about losing funding for shelters and of taking attention away from their clients, activists from the battered women's movement have been in the forefront of denials of the existence of battered husbands (see, e.g., Fields and Kirchner 1978; Dobash and Dobash 1992). They contend that since wife battering is a severe and widespread problem, it deserves attention. Battered husbands, on the other hand, given their dubious status, should not be allowed to take attention and funding away from battered wives.

All of these social movement factors worked to influence the process of deviance-making and social problem construction. In the case of battered wives, the result was the successful construction and application of a deviant label. For battered husbands, the result was an aborted attempt at construction. As I show in the next section, academic and mass media attention to the issues also contributed to their different outcomes.

Social Science and Mass Media Attention

The increase in social science research and literature since the 1970s has also supported the subsistence of battered wives as a social problem (Studer 1984).[2] Harry (1987) argues that this social science literature has both accompanied and buttressed the rise of battered women as a social problem. Kurz (1989) also points to the role of researchers who provided statistical evidence of the extent of the abuse in making wife battering a social problem. (This points to the importance of considering the role of incidence in the construction of a problem.) A survey of professional attention to the issue showed that approximately 380 articles on battered wives were indexed in *Sociological Abstracts* between 1974 and the first quarter of 1994.[3]

The extent of attention given to battered husbands by social scientists has been significantly less and, at least as importantly, of a different type. A review of the index of sociology journal articles produced just three articles on husband battering. (See Note 2.) Additionally, an examination of the sociological literature on the subject showed that the majority of attention to Steinmetz's claims has been negative (see, e.g., Pleck, Pleck, Grossman, and Bart 1977; Fields and Kirchner 1978; Dobash and Dobash 1981; Berk et al. 1983; Pagelow 1984, 1985; Saunders 1986, 1988a, 1988b; Kurz 1989, 1993; Bograd 1990; Brush 1990; Dobash, Dobash, Daly and Wilson 1992).

There are, though, notable exceptions to this trend. Straus, in particular, has supported and put forth claims about the existence of battered husbands (Gelles and Straus 1988; Stets and Straus 1990; Straus 1993). It is, however, important to note that Straus's most recent forays into this debate do not use the politically explosive "battered husbands" term: He refers instead to "physical assaults by wives" (Straus 1993, p. 67). This choice of words is also evidence of the difficulty that exists in attempting to apply the "battered husbands" label. Others have also supported the acceptance of battered husbands as a social problem (Landley and Levy 1977; McNeely and Robinson-Simpson 1987; Flynn 1990; McNeely and Mann 1990). The main theme of claims about battered husbands is that the problem is equal in incidence to that of battered wives.

The result of the profound disagreement about the existence of battered husbands (which stems largely from theoretical and methodological differences between the two camps) has been adamant support for the issue among a small group of researchers and fierce denial of its existence among a somewhat larger group. For example, their theoretical differences are such that they are arguing past one another. The belief that there are a significant number of abused husbands originates from the theoretical position that we live in a violent society and that, as such, wives can be expected to be as violent as husbands (e.g., Straus, Gelles, and Steinmetz 1980; Straus and Gelles 1986; McNeely and Robinson-Simpson 1987; McNeely and Mann 1990; Straus 1992). The argument that there are not many battered husbands grows out of the feminist contention that we live in a male-dominated society that is at least partly founded on and perpetuated by men's violence against women (e.g., Dobash and Dobash 1981, 1992; Kurz 1989, 1993; Bograd 1990). In other words, one side insists that there is a group of people to whom the label "battered husbands" can be applied, while the other argues just as vehemently that no such group can exist. However, while this issue has been hotly debated among researchers, it has rarely made it out of academic circles.

Once again, while professional/scientific attention and responses to claims may not be necessary to the construction of a social problem, they either enhance or diminish its credibility. Though there has by no means been a uniform response to wife battering, the responses *have* been sustained and persistent. Attention to battered husbands, on the other hand, has been sporadic and often negative. The small number of articles on battered husbands suggests that they did not catch on with professionals the way battered wives did, making their chances of being promoted as a social problem slim.

Like professional attention, popular media attention to battered wives has been helpful in establishing and maintaining it as a social problem. Tierney (1982) suggests that violence against wives was a good subject for the media because it was a "new" problem for the public; it was controversial, mixing violence and social relevance; and it provided a focal point for the discussion of issues such as feminism, inequality, and family life in the United States (also see Studer 1984; Schneider 1985). An examination of coverage of battered wives in popular magazines (using the *Reader's Guide to Periodical Literature*) shows that coverage has waxed and waned since 1974. The number of articles appearing in a year has ranged from 1 in 1975/76 and 1981/82 to 21 in 1991. Between 1974 and 1992, 151 articles appeared.[4]

Battered husbands have also received attention from the mass media. As Pagelow points out, "The very idea of husband battering seemed to titillate the collective imagination of the mass media" (1984, p. 268). After an initial flurry of attention to the issue, however, it virtually disappeared from the mass media. Between 1977 and 1992, *Reader's Guide* lists only three magazine articles on husband battering.[4,5]

Blumer (1971) argued that social science researchers should take the lead in defining social problems from the public, which seems to have worked out in the case of battered wives. The issue was raised at the grass-roots level and then taken up by researchers (many of whom were also activists). Neither academics nor the public (in the form of the popular press attention), however, have taken up the cause of battered husbands in a way that would lead it to take on the status of a social problem. The popular press rarely makes claims about battered husbands; academics are busy disagreeing among themselves about their very existence. This lack of media attention to battered husbands has hindered the attempt to construct and apply a deviant label.

Professional and mass media attention to the issue of battered wives has been instrumental in its creation and continuation as an identified social problem. Along with social movement/organizational factors, this attention has been crucial to the construction of the social problem called "battered wives." The *lack* of these two factors has been of considerable importance to the failure of battered husbands. Both of these factors can be related to gender images. For battered wives to become a social problem, wife beating had to come to be seen as something that was problematic, and it had to be labeled as deviance. Stereotypical gender images of women and men were essential to this process. Conversely, these same gender images have worked against the definition of battered husbands as a social problem.

Gender Images

If women are seen as passive, dependent, and weak, then it is easy to accept their identification as potential victims of strong, assertive men but difficult to identify them as batterers. If men are accepted as being aggressive and/or

sometimes out of control, or as likely to use violence to assert their dominance and to get what they want, then wives are going to be beaten; it is difficult to imagine husbands being beaten. To reverse the traditional scenarios
contradicts gender norms. In this way, stereotypical images of women and
men influence what kind of claims can be made about them and what kind of
labels can be applied to them. It is here that attempts at deviance-making for
battered husbands are most compromised.

As Straus points out, "To be violent is not unmasculine. But to be physically violent *is* unfeminine according to contemporary American standards"
(1977, p. 448). Violence by men is supported, in many contexts, by societal
norms. There is a long tradition of husbands being permitted to beat their
wives to make them obey—to keep them in line (see, e.g., Davidson 1977).
There is no similar tradition of wives beating their husbands (Steinmetz 1977).[6]

To recognize battered husbands as a significant social problem requires at
least implicit acknowledgement that men can be victims of acts perpetrated by
women. While it may be easy for us to see men as potential victims of violent
acts committed by other men, it is difficult to imagine them being victimized
by women. "Victim" connotes a weak, passive person—an image antithetical
to masculinity. We, as a society, are not inclined to see men that way. Similarly,
in this context, men likely would be reticent about naming themselves as victims of women's violence. This results in difficulties in attempting to apply a
label of "battered husbands" to men—whether or not they have been abused
by their wives. Reluctance to "believe in" battered husbands then becomes a
matter of subscribing to traditional images of women and men. It is also a
matter of questioning the prevalence of female aggression and assertion of
dominance (through violence) in our society (Dobash et al. 1992), which may
be based not on stereotypes but on everyday experiences and/or the influence
of feminism.

More specifically, dependency has been a key notion in identifying victims
of family violence. Children, the elderly, and women fit more easily into this
category than adult men do. Our image of what it means to be a man mili-
tates against the inclusion of an image of husbands in positions of dependency
(physical, economic, etc.) that lead to victimization by their wives. The power
of this image of dependency and the difficulty of fitting men into it is evidenced by efforts of husband battering supporters to depict them as dependent (O'Toole and Webster 1988). For example, "A wife need not be an
Amazon to abuse her husband. Sometimes a woman is physically stronger than
her husband because the man is sick, handicapped, or much older than his
wife" (Langley and Levy 1977, p. 190). Since the dependent husband is an
exceptional image, however, these efforts to establish husband battering as a
social problem, and to apply a deviant label, have failed.

The fact that, because of gender images, men do not make good victims
of domestic violence has impeded the construction of a battered husbands
problem by making it difficult to attach such a label to men. Combined with
the other two factors, this element has led to the failure of battered husbands
to be considered a social problem. Particularly when viewed in the context of

the successful construction of battered wives, the operation of these factors appears to have compromised the viability of battered husbands as a social problem.

DISCUSSION AND CONCLUSIONS

In this article, I have argued that if we set aside the debate about possible quantitative differences between battered wives and battered husbands, the issue of abused husbands still faces obstacles to becoming a social problem. That is, even if Straus and other writers' assessment of the data is correct, battered husbands are not yet a social problem. Because no social movement has organized around the issue and because there has been little social science or mass media attention to it (relative to battered wives), husband battering has remained marginal to public and professional consciousness. Gender images of women and men, however, have been the most detrimental to the construction of battered husbands as a social problem. Because men do not make good victims, it has been difficult to present them as likely to be abused (especially compared to women, the elderly, and children). That is, it has been difficult to construct and apply a label of "battered husbands" because of social movement, mass media attention, and gender images factors.

Given these obstacles to the construction of battered husbands as a social problem, it is possible that even a resolution of the debate about numbers will not lead to agreement about the existence of the problem. In the end, it may be beliefs that husband and wife battering are qualitatively different—as suggested by my argument about gender images—that will determine the fate of the issue. That is, it may be the difficulty of making the label "battered husbands" stick that makes the most difference in constructing battered husbands as a social problem. For example, a recent study of college students found that they tended to take wife battering more seriously, and to see it as more violent, than husband battering (Harris and Cook 1994). Additionally, issues of differences in seriousness (Dobash and Dobash 1981; Fields and Kirchner 1978; Straus 1980, 1992; Straus et al. 1980; Berk et al. 1983; McLeod 1984; Pagelow 1984; Steinmetz 1987; Steinmetz and Lucca 1988), or whether women are more likely to use violence in self-defense (Straus 1980, 1993; Straus et al. 1980; Dobash and Dobash 1981; Walker and Brown 1985; Saunders 1988a, 1988b; Stets and Straus 1990), and of the failure of researchers supporting the battered husbands issue to account for the context of the violence (Gelles 1979; Saunders 1986; Kurz 1989, 1993; Dobash et al. 1992) also remain to be resolved.

As sociological attention to family violence increases, the lack of attention to a battered husbands problem becomes all the more obvious. Although numerous books and articles have been written about wife battering, child abuse and neglect, and elder abuse and neglect, hardly any work has been done on the husband battering issue. No one has studied a sample of battered husbands.

Most of the energy addressing the issue has been devoted to debating whether battered husbands exist at all. Because both sides are arguing past each other, given their different theoretical orientations, it is unlikely that the debate will be resolved any time soon. For now, because of the issues raised in this article, husband battering will retain its marginal status as a social issue.

NOTES

1.   For detailed information about the history of the battered wives issue see Schechter (1982), Tierney (1982) and Studer (1984). Pagelow (1984) and Lucal (1991) provide a history of the battered husbands issue.

2.   Not everyone believes that research attention has helped the battered women's movement. (See, e.g., Adams, Jackson, and Lauby 1988.)

3.   For battered wives, I searched "sociofile," the CD-ROM version of *Sociological Abstracts,* using the following key words: wife battering, battered wives, battered women, and abused wives. For battered husbands, I searched on: husband battering, battered husbands, battered men, and abused husbands.

4.   I searched on the following key words for battered wives: abused women and wife beating or abuse; for battered husbands: abused men or husband beating or abuse.

5.   Although only magazines were studied for this research, it is interesting to note the wide discrepancy between the portrayal of battered husbands and battered wives in television and movies. Battered wives appear fairly frequently on the small screen, either in movies (e.g., "The Burning Bed") or on talk shows. Feature films about battered wives also exist (e.g., "Sleeping with the Enemy"). Battered husbands have also appeared but much less frequently: on the small screen, most recently on "Donahue" in 1991, on "Sally Jessy Raphael" in 1993 and in a made-for-TV movie, "Men Don't Tell," that same year. They have not been the subject of feature films. Additionally, in June 1994, in the midst of his highly publicized encounter with police over charges that he murdered his former wife, O. J. Simpson reportedly claimed that he sometimes felt like a battered husband or boyfriend. A former football player, however, does not make a good victim; nothing has come of his claim. Instead, attention refocused on wife battering.

6.   Straus (1992, 1993) maintains that just the opposite is true. He reports that "There seems to be an implicit cultural norm permitting or encouraging minor assaults by wives in certain circumstances" (1993, p. 78). His argument is that physical assaults by wives are a major social problem because they lead to wife battering.

REFERENCES

1988. "Family Violence Research: Aid or Obstacle to the Battered Women's Movement?" *Response to the Victimization of Women and Children* 11: 14–16.

Adler, J., K. Springen, D. Glick, and J. Gordon. 1991. "Drums, Sweat and Tears." *Newsweek* (24 June): 46–51.

Ball, R. A., and J. R. Lilly. 1984. "When Is a 'Problem' Not a Problem?: Deflection Activities in a Clandestine Motel." In *Studies in the Sociology of Social Problems,* edited by J. W. Schneider and J. I. Kitsuse (pp. 14–39). Norwood, NJ: Ablex.

"The Battered Husbands." 1978. *Time.* Vol. 111, no. 12 (March 20).

Berk, R. A., S. F. Berk, D. R. Loseke, and D. Rauma. 1983. "Mutual Combat and Other Family Violence Myths." In *The Dark Side of Families,* edited by D. Finkelhor, R. J. Gelles, G. T. Hotaling, and M. A. Straus (pp. 197–212). Beverly Hills: Sage.

Best, J. (ed.) 1989. *Images of Issues.* New York: Aldine de Gruyter.

Blumer, H. 1971. "Social Problems as Collective Behavior." *Social Problems* 18: 298–306.

Bograd, M. 1990. "Why We Need Gender to Understand Human Violence." *Journal of Interpersonal Violence* 5: 132–35.

Brush, L. D. 1990. "Violent Acts and Injurious Outcomes in Married Couples: Methodological Issues in the National Survey of Families and Households." *Gender and Society* 4: 56–67.

Davidson, T. 1977. "Wifebeating: A Recurring Phenomenon Throughout History." In *Battered Women,* edited by M. Roy (pp. 2–23). New York: Van Nostrand Rinehold.

Dobash, R. E., and R. P. Dobash. 1977. "Wives: The 'Appropriate' Victims of Marital Violence." *Victimology* 2: 426–42.

———. 1981. "Social Science and Social Action: The Case of Wife Beating." *Journal of Family Issues* 2: 439–70.

———. 1987. "The Responses of the British and American Women's Movements to Violence Against Women." In *Women, Violence and Social Control,* edited by J. Hanmer and M. Maynard (pp. 169–179). Atlantic Highlands: Humanities Press International.

———. 1992. *Women, Violence and Social Change.* London: Routledge.

Dobash, R. P., R. E. Dobash, M. Wilson, and M. Daly. 1992. "The Myth of Sexual Symmetry in Marital Violence." *Social Problems* 39: 71–91.

———. 1993. "Marital Violence Is Not Symmetrical: A Response to Campbell." *SSSP Newsletter* 24 (Fall): 26–30.

Doyle, J. A. 1989. *The Male Experience* (Second edition). Dubuque: Brown.

Ferraro, K. J., and J. M. Johnson. 1985. "The New Underground Railroad." *Studies in Symbolic Interaction* 6: 377–86.

Fields, M. D., and R. M. Kirchner. 1978. "Battered Women Are Still in Need: A Reply to Steinmetz." *Victimology* 3: 216–26.

Flynn, C. P. 1990. "Relationship Violence by Women: Issues and Implications." *Family Relations* 39: 194–8.

Gelles, R. J. 1979. "The Myth of the Battered Husband and New Facts about Family Violence." *Ms.* 8 (October): 65–6, 71–2.

———. 1994. "Battered Men: Few in Number, Very Much Alone." *The Plain Dealer* (March 6, 1994): 7C.

Gelles, R. J., and M. A. Straus. 1988. *Intimate Violence.* New York: Simon and Schuster.

Greenblatt, C. S. 1985. "'Don't Hit Your Wife . . . Unless . . .': Preliminary Findings on Normative Support for the Use of Physical Force by Husbands." *Victimology* 10: 221–41.

Harris, R. J., and C. A. Cook. 1994. "Attributions About Spouse Abuse: It Matters Who the Batterers and Victims Are." *Sex Roles* 30: 553–65.

Hatty, S. 1987. "Woman Battering as a Social Problem: The Denial of Injury." *Australian and New Zealand Journal of Sociology* 23: 36–46.

Johnson, J. M. 1981. "Program Enterprise and Official Cooptation in the Battered Women's Shelter Movement." *American Behavioral Scientist* 24: 827–42.

Kimmel, M. S. 1989. "From Pedestals to Partners: Men's Responses to Feminism." In *Women: A Feminist Perspective* (Fourth edition), edited by J. Freeman (pp. 581–94). Mountain View: Mayfield.

Krippendorff, K. 1980. *Content Analysis: An Introduction to Its Methodology.* Beverly Hills: Sage.

Kurz, D. 1989. "Social Science Perspectives on Wife Abuse: Current Dilemmas and Future Directions." *Gender and Society* 3: 489–505.

———. 1993. "Physical Assaults by Husbands: A Major Social Problem." In *Current Controversies on Family Violence,* edited by R. J. Gelles and D. R. Loseke (pp. 88–103). Newbury Park: Sage.

Langley, R., and R. C. Levy. 1977. *Wife Beating: The Silent Crisis.* New York: Dutton.

Loseke, D. R., and S. E. Cahill. 1984. "The Social Construction of Deviance: Experts on Battered Women." *Social Problems* 31: 296–310.

Lucal, E. M. 1991. *Of Social Movements and Gender Images: Why Battered Husbands Aren't a Social Problem and Battered Wives Are.* M.A. Thesis. Department of Sociology, Kent State University, Kent, OH.

McLeod, M. 1984. "Women Against Men: An Examination of Domestic Violence Based on an Analysis of Official Data and National Victimization Data." *Justice Quarterly* 1: 171–93.

McNeely, R. L., and C. R. Mann. 1990. "Domestic Violence Is a Human Issue." *Journal of Interpersonal Violence* 5: 129–32.

McNeely, R. L., and G. Robinson-Simpson. 1987. "The Truth About Domestic Violence: A Falsely Framed Issue." *Social Work* 32: 485–90.

Murray, S. B. 1988. "The Unhappy Marriage of Theory and Practice: An Analysis of a Battered Women's Shelter." *NWSA Journal* 1: 75–92.

Neuman, W. L. 1994. *Social Research Methods* (Second edition). Boston: Allyn and Bacon.

O'Toole, R., and S. Webster. 1988. "Differentiation of Family Mistreatment: Similarities and Differences by Status of the Victim." *Deviant Behavior* 9: 347–68.

Pagelow, M. D. 1984. *Woman Battering.* Beverly Hills: Sage.

———. 1985. "The 'Battered Husband Syndrome': Social Problem or Much Ado About Little?" In *Marital Violence,* edited by N. Johnson (pp. 172–95). London: Routledge and Kegan Paul.

Pizzey, E. 1974. *Scream Quietly or the Neighbours Will Hear.* Short Hills, England: Ridley Enslow.

Pleck, E., J. H. Pleck, M. Grossman, and P. Bart. 1977. "The Battered Data Syndrome: A Reply to Steinmetz' Article." *Victimology* 2: 680–83.

Ragin, C. C. 1987. *The Comparative Method.* Berkeley: University of California.

Renzetti, C. M., and D. J. Curran. 1992. *Women, Men and Society* (Second edition). Boston: Allyn and Bacon.

Rodriguez, N. M. 1988. "Transcending Bureaucracy: Feminist Politics at a Shelter for Battered Women." *Gender and Society* 2: 214–27.

Saunders, D. G. 1986. "When Battered Women Use Violence: Husband-Abuse or Self-Defense?" *Victims and Violence* 1: 47–60.

———. 1988a. "Wife Abuse, Husband Abuse, or Mutual Combat?: A Feminist Perspective on the Empirical Findings." In *Feminist Perspectives on Wife Abuse,* edited by K. Yllo and M. Bograd (pp. 90–113). Beverly Hills: Sage.

———. 1988b. "Other 'Truths' About Domestic Violence: A Reply to McNeely and Robinson-Simpson." *Social Work* 33: 179–83.

Schechter, S. 1982. *Women and Male Violence: The Visions and Struggles of the Battered Women's Movement*. Boston: South End.

Schneider, J. W. 1984. "Introduction." In *Studies in the Sociology of Social Problems,* edited by J. W. Schneider and J. I. Kitsuse (pp. vii–xx). Norwood, NJ: Ablex.

———. 1985. "Social Problems Theory: The Constructionist View." *Annual Review of Sociology* 11: 209–29.

Spector, M., and J. I. Kitsuse. 1977. *Constructing Social Problems*. Menlo Park: Cummings.

Steinmetz, S. 1977. "The Battered Husband Syndrome." *Victimology* 2: 499–509.

———. 1987. "Family Violence: Past, Present and Future." In *Handbook of Marriage and the Family,* edited by M. B. Sussman and S. K. Steinmetz (pp. 725–65). New York: Plenum.

Steinmetz, S. K., and J. S. Lucca. 1988. "Husband Battering." In *Handbook of Family Violence,* edited by V. B. van Hasselt, R. L. Morrison, A. S. Bellack, and M. Hersen (pp. 233–46). New York: Plenum.

Stets, J. E., and M. A. Straus. 1990. "Gender Differences in Reporting Marital Violence and Psychological Consequences." In *Physical Violence in American Families,* edited by M. A. Straus and R. J. Gelles (pp. 151–65). New Brunswick: Transaction.

Straus, M. A. 1977. "Wife Beating: How Common and Why?" *Victimology* 2: 443–58.

———. 1980. "Victims and Aggressors in Marital Violence." *American Behavioral Scientist* 23: 681–704.

———. 1992. "Sociological Research and Social Policy: The Case of Family Violence." *Sociological Forum* 7: 211–37.

———. 1993. "Physical Assaults by Wives: A Major Social Problem." In *Current Controversies on Family Violence,* edited by R. J. Gelles and D. R. Loseke (pp. 67–87). Newbury Park: Sage.

Straus, M. A., and R. J. Gelles. 1986. "Societal Change and Change in Family Violence from 1975 to 1985 as Revealed by Two National Surveys." *Journal of Marriage and the Family* 48: 465–79.

Straus, M. A., Gelles, R. J., and S. K. Steinmetz. 1980. *Behind Closed Doors.* New York: Anchor.

Studer, M. 1984. "Wife Beating as a Social Problem: The Process of Definition." *International Journal of Women's Studies* 7: 412–22.

Terry, R. M., and D. J. Steffensmeier. 1988. "Conceptual and Theoretical Issues in the Study of Deviance." *Deviant Behavior* 9: 55–76.

Tierney, K. J. 1982. "The Battered Woman Movement and the Creation of the Wife Beating Problem." *Social Problems* 29: 207–20.

Walker, G. A. 1990. *Family Violence and the Women's Movement.* Toronto: University of Toronto Press.

Walker, L. E. A., and A. Brown. 1985. "Gender and Victimization by Intimates." *Journal of Personality* 53: 179–95.

Weitzer, R. 1991. "Prostitutes' Rights in the United States: The Failure of a Movement." *Sociological Quarterly* 32: 23–41.

11

The Social Reality of Crime

RICHARD QUINNEY

A theory that helps us begin to examine the legal order critically is the one I call the *social reality of crime*. Applying this theory, we think of crime as it is affected by the dynamics that mold the society's social, economic, and political structure. First, we recognize how criminal law fits into capitalist society. The legal order gives reality to the crime problem in the United States. Everything that makes up crime's social reality, including the application of criminal law, the behavior patterns of those who are defined as criminal, and the construction of an ideology of crime, is related to the established legal order. The social reality of crime is constructed on conflict in our society.

The theory of the social reality of crime is formulated as follows.

I. The Official Definition of Crime: Crime as a legal definition of human conduct is created by agents of the dominant class in a politically organized society.

The essential starting point is a definition of crime that itself is based on the legal definition. Crime, as *officially* determined, is a *definition* of behavior that is conferred on some people by those in power. Agents of the law (such as legislators, police, prosecutors, and judges) are responsible for formulating and administering criminal law. Upon *formulation* and *application* of these definitions of crime, persons, and behaviors become criminal.

Crime, according to this first proposition, is not inherent in behavior, but is a judgment made by some about the actions and characteristics of others. This proposition allows us to focus on the formulation and administration of the criminal law as it applies to the behaviors that become defined as criminal. Crime is seen as a result of the class-dynamic process that culminate in defining persons and behaviors as criminal. It follows, then, that the greater the number of definitions of crime that are formulated and applied, the greater the amount of crime.

II. Formulating Definitions of Crime: Definitions of crime are composed of behaviors that conflict with the interests of the dominant class.

Definitions of crime are formulated according to the interests of those who have the power to translate their interests into public policy. Those definitions

From Richard Quinney, *Criminology* (Boston: Little, Brown, 1975), pp. 37–41. Reprinted by permission of the author.

are ultimately incorporated into the criminal law. Furthermore, definitions of crime in a society change as the interests of the dominant class change. In other words, those who are able to have their interests represented in public policy regulate the formulation of definitions of crime.

The powerful interests are reflected not only in the definitions of crime and the kinds of penal sanctions attached to them, but also in the *legal policies* on handling those defined as criminals. Procedural rules are created for enforcing and administering the criminal law. Policies are also established on programs for treating and punishing the criminally defined and programs for controlling and preventing crime. From the initial definitions of crime to the subsequent procedures, correctional and penal programs, and policies for controlling and preventing crime, those who have the power regulate the behavior of those without power.

III. Applying Definitions of Crime: Definitions of crime are applied by the class that has the power to shape the enforcement and administration of criminal law.

The dominant interests intervene in all the stages at which definitions of crime are created. Because class interests cannot be effectively protected merely by formulating criminal law, the law must be enforced and administered. The interests of the powerful, therefore, also operate where the definitions of crime reach the *application* stage. As Vold has argued, crime is "political behavior and the criminal becomes in fact a member of a 'minority group' without sufficient public support to dominate the control of the police power of the state." Those whose interests conflict with the ones represented in the law must either change their behavior or possibly find it defined as criminal.

The probability that definitions of crime will be applied varies according to how much the behaviors of the powerless conflict with the interests of those in power. Law enforcement efforts and judicial activity are likely to increase when the interests of the dominant class are threatened. Fluctuations and variations in applying definitions of crime reflect shifts in class relations.

Obviously, the criminal law is not applied directly by those in power; its enforcement and administration are delegated to authorized *legal agents*. Because the groups responsible for creating the definitions of crime are physically separated from the groups that have the authority to enforce and administer law, local conditions determine how the definitions will be applied. In particular, communities vary in their expectations of law enforcement and the administration of justice. The application of definitions is also influenced by the visibility of offenses in a community and by the public's norms about reporting possible violations. And especially important in enforcing and administering the criminal law are the legal agents' occupational organization and ideology.

The probability that these definitions will be applied depends on the actions of the legal agents who have the authority to enforce and administer the law. A definition of crime is applied depending on their evaluation. Turk has argued that during "criminalization," a criminal label may be affixed to people because of real or fancied attributes: "Indeed, a person is evaluated, either favorably or unfavorably, not because he *does* something, or even because he *is* something, but because others react to their perceptions of him as offensive or inoffensive." Evaluation by the definers is affected by the way in which the suspect handles the situation, but ultimately the legal agents' evaluations and subsequent decisions are the crucial factors in determining the criminality of human acts. As legal agents evaluate more behaviors and persons as worthy of being defined as crimes, the probability that definitions of crime will be applied grows.

IV. How Behavior Patterns Develop in Relation to Definitions of Crime: Behavior patterns are structured in relation to definitions of crime, and within this context people engage in actions that have relative probabilities of being defined as criminal.

Although behavior varies, all behaviors are similar in that they represent patterns within society. All persons—whether they create definitions of crime or are the objects of these definitions—act in reference to *normative systems* learned in relative social and cultural settings. Because it is not the quality of the behavior but the action taken against the behavior that gives it the character of criminality, that which is defined as criminal is relative to the behavior patterns of the class that formulates and applies definitions. Consequently, people whose behavior patterns are not represented when the definitions of crime are formulated and applied are more likely to act in ways that will be defined as criminal than those who formulate and apply the definitions.

Once behavior patterns become established with some regularity within the segments of society, individuals have a framework for creating *personal action patterns.* These continually develop for each person as he moves from one experience to another. Specific action patterns give behavior an individual substance in relation to the definitions of crime.

People construct their own patterns of action in participating with others. It follows, then, that the probability that persons will develop action patterns with a high potential for being defined as criminal depends on (1) structured opportunities, (2) learning experiences, (3) interpersonal associations and identifications, and (4) self-conceptions. Throughout the experiences, each person creates a conception of self as a human social being. Thus prepared, he behaves according to the anticipated consequences of his actions.

In the experiences shared by the definers of crime and the criminally defined, personal-action patterns develop among the latter because they are so defined. After they have had continued experience in being defined as criminal, they learn to manipulate the application of criminal definitions.

Furthermore, those who have been defined as criminal begin to conceive of themselves as criminal. As they adjust to the definitions imposed on them, they learn to play the criminal role. As a result of others' reactions, therefore, people may develop personal-action patterns that increase the likelihood of their being defined as criminal in the future. That is, increased experience with definitions of crime increases the probability of their developing actions that may be subsequently defined as criminal.

Thus, both the definers of crime and the criminally defined are involved in reciprocal action patterns. The personal-action patterns of both the definers and the defined are shaped by their common, continued, and related experiences. The fate of each is bound to that of the other.

V. Constructing an Ideology of Crime: An ideology of crime is constructed and diffused by the dominant class to secure its hegemony.

This ideology is created in the kinds of ideas people are exposed to, the manner in which they select information to fit the world they are shaping, and their way of interpreting this information. People behave in reference to the *social meanings* they attach to their experiences.

Among the conceptions that develop in a society are those relating to what people regard as crime. The concept of crime must of course be accompanied by ideas about the nature of crime. Images develop about the relevance of crime, the offender's characteristics, the appropriate reaction to crime, and the relation of crime to the social order. These conceptions are constructed by communication, and, in fact, an ideology of crime depends on the portrayal of crime in all personal and mass communication. This ideology is thus diffused throughout the society.

One of the most concrete ways by which an ideology of crime is formed and transmitted is the official investigation of crime. The President's Commission on Law Enforcement and Administration of Justice is the best contemporary example of the state's role in shaping an ideology of crime. Not only are we as citizens more aware of crime today because of the President's Commission, but official policy on crime has been established in a crime bill, the Omnibus Crime Control and Safe Streets Act of 1968. The crime bill, itself a reaction to the growing fears of class conflict in American society, creates an image of a severe crime problem and, in so doing, threatens to negate some of our basic constitutional guarantees in the name of controlling crime.

Consequently, the conceptions that are most critical in actually formulating and applying the definitions of crime are those held by the dominant class. These conceptions are certain to be incorporated into the social reality of crime. The more the government acts in reference to crime, the more probable it is that definitions of crime will be created and that behavior patterns will develop in opposition to those definitions. The formulation of definitions of crime, their application, and the development of behavior patterns in relation to the definitions, are thus joined in full circle by the construction of an ideological hegemony toward crime.

VI. Constructing the Social Reality of Crime: The social reality of crime is constructed by the formulation and application of definitions of crime, the development of behavior patterns in relation to these definitions, and the construction of an ideology of crime.

The first five propositions are collected here into a final composition proposition. The theory of the social reality of crime, accordingly, postulates creating a series of phenomena that increase the probability of crime. The result, holistically, is the social reality of crime.

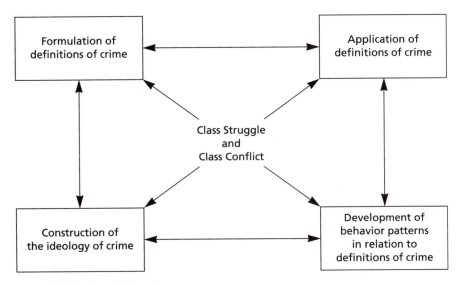

The Social Reality of Crime

Because the first proposition of the theory is a definition and the sixth is a composite, the body of the theory consists of the four middle propositions. These form a model of crime's social reality. The model, as diagrammed, relates the proposition units into a theoretical system (see figure above). Each unit is related to the others. The theory is thus a system of interacting developmental propositions. The phenomena denoted in the propositions and their relationships culminate in what is regarded as the amount and character of crime at any time—that is, in the social reality of crime.

The theory of the social reality of crime as I have formulated it is inspired by a change that is occurring in our view of the world. This change, pervading all levels of society, pertains to the world that we all construct and from which, at the same time, we pretend to separate ourselves in our human experiences. For the study of crime, a revision in thought has directed attention to the criminal process: All relevant phenomena contribute to creating definitions of crime, development of behaviors by those involved in criminal-defining situations, and constructing an ideology of crime. The result is the social reality of crime that is constantly being constructed in society.

12

The Saints and the Roughnecks

WILLIAM J. CHAMBLISS

Eight promising young men—children of good, stable, white, upper-middle-class families, active in school affairs, good pre-college students—were some of the most delinquent boys at Hanibal High School. While community residents and parents knew that these boys occasionally sowed a few wild oats, they were totally unaware that sowing wild oats completely occupied the daily routine of these young men. The Saints were constantly occupied with truancy, drinking, wild driving, petty theft, and vandalism. Yet not one was officially arrested for any misdeed during the two years I observed them.

This record was particularly surprising in light of my observations during the same two years of another gang of Hanibal High School students, six lower-class white boys known as the Roughnecks. The Roughnecks were constantly in trouble with police and community even though their rate of delinquency was about equal with that of the Saints. What was the cause of this disparity? the result? The following consideration of the activities, social class, and community perceptions of both gangs may provide some answers.

THE SAINTS FROM MONDAY TO FRIDAY

The Saints' principal daily concern was with getting out of school as early as possible. The boys managed to get out of school with minimum danger that they would be accused of playing hookey through an elaborate procedure for obtaining "legitimate" release from class. The most common procedure was for one boy to obtain the release of another by fabricating a meeting of some committee, program, or recognized club. Charles might raise his hand in his 9:00 chemistry class and asked to be excused—a euphemism for going to the bathroom. Charles would go to Ed's math class and inform the teacher that Ed was needed for a 9:30 rehearsal of the drama club play. The math teacher would recognize Ed and Charles as "good students" involved in numerous school activities and would permit Ed to leave at 9:30. Charles would return to his class, and Ed would go to Tom's English class to obtain his release. Tom would engineer Charles's escape. The strategy would continue until as many

From "The Saints and the Roughnecks," William J. Chambliss, *Society*, V. 11, No. 1, pp. 24–31.

of the Saints as possible were freed. After a stealthy trip to the car (which had been parked in a strategic spot), the boys were off for a day of fun.

Over the two years I observed the Saints, this pattern was repeated nearly every day. There were variations on the theme, but in one form or another, the boys used this procedure for getting out of class and then off the school grounds. Rarely did all eight of the Saints manage to leave school at the same time. The average number avoiding school on the days I observed them was five.

Having escaped from the concrete corridors the boys usually went either to a pool hall on the other (lower-class) side of town or to a cafe in the suburbs. Both places were out of the way of people the boys were likely to know (family or school officials), and both provided a source of entertainment. The pool hall entertainment was the generally rough atmosphere, the occasional hustler, the sometimes drunk proprietor, and, of course, the game of pool. The cafe's entertainment was provided by the owner. The boys would "accidentally" knock a glass on the floor or spill cola on the counter—not all the time, but enough to be sporting. They would also bend spoons, put salt in sugar bowls, and generally tease whoever was working in the cafe. The owner had opened the cafe recently and was dependent on the boys' business which was, in fact, substantial since between the horsing around and the teasing they bought food and drinks.

THE SAINTS ON WEEKENDS

On weekends the automobile was even more critical than during the week, for on weekends the Saints went to Big Town—a large city with a population of over a million 25 miles from Hanibal. Every Friday and Saturday night most of the Saints would meet between 8:00 and 8:30 and would go into Big Town. Big Town activities included drinking heavily in taverns or nightclubs, driving drunkenly through the streets, and committing acts of vandalism and playing pranks.

By midnight on Fridays and Saturdays the Saints were usually thoroughly high, and one or two of them were often so drunk they had to be carried to the cars. Then the boys drove around town, calling obscenities to women and girls; occasionally trying (unsuccessfully so far as I could tell) to pick girls up; and driving recklessly through red lights and at high speeds with their lights out. Occasionally they played "chicken." One boy would climb out the back window of the car and across the roof to the driver's side of the car while the car was moving at high speed (between 40 and 50 miles an hour); then the driver would move over and the boy who had just crawled across the car roof would take the driver's seat.

Searching for "fair game" for a prank was the boys' principal activity after they left the tavern. The boys would drive alongside a foot patrolman and ask directions to some street. If the policeman leaned on the car in the course of answering the question, the driver would speed away, causing him to lose his balance. The Saints were careful to play this prank only in an area where they were not going to spend much time and where they could quickly disappear around a corner to avoid having their license plate number taken.

Construction sites and road repair areas were the special province of the Saints' mischief. A soon-to-be-repaired hole in the road inevitably invited the Saints to remove lanterns and wooden barricades and put them in the car, leaving the hole unprotected. The boys would find a safe vantage point and wait for an unsuspecting motorist to drive into the hole. Often, though not always, the boys would go up to the motorist and commiserate with him about the dreadful way the city protected its citizenry.

Leaving the scene of the open hole and the motorist, the boys would then go searching for an appropriate place to erect the stolen barricade. An "appropriate place" was often a spot on a highway near a curve in the road where the barricade would not be seen by an oncoming motorist. The boys would wait to watch an unsuspecting motorist attempt to stop and (usually) crash into the wooden barricade. With saintly bearing the boys might offer help and understanding.

A stolen lantern might well find its way onto the back of a police car or hang from a street lamp. Once a lantern served as a prop for a reenactment of the "midnight ride of Paul Revere" until the "play," which was taking place at 2:00 A.M. in the center of a main street of Big Town, was interrupted by a police car several blocks away. The boys ran, leaving the lanterns on the street, and managed to avoid being apprehended.

Abandoned houses, especially if they were located in out-of-the-way places, were fair game for destruction and spontaneous vandalism. The boys would break windows, remove furniture to the yard and tear it apart, urinate on the walls, and scrawl obscenities inside.

Through all the pranks, drinking, and reckless driving the boys managed miraculously to avoid being stopped by police. Only twice in two years was I aware that they had been stopped by a Big City policeman. Once was for speeding (which they did every time they drove whether they were drunk or sober), and the driver managed to convince the policeman that it was simply an error. The second time they were stopped they had just left a nightclub and were walking through an alley. Aaron stopped to urinate and the boys began making obscene remarks. A foot patrolman came into the alley, lectured the boys, and sent them home. Before the boys got to the car one began talking in a loud voice again. The policeman, who had followed them down the alley, arrested this boy for disturbing the peace and took him to the police station where the other Saints gathered. After paying a $5.00 fine, and with the assurance that there would be no permanent record of the arrest, the boy was released.

The boys had a spirit of frivolity and fun about their escapades. They did not view what they were engaged in as "delinquency," though it surely was by any reasonable definition of that word. They simply viewed themselves as having a little fun and who, they would ask, was really hurt by it? The answer had to be no one, although this fact remains one of the most difficult things to explain about the gang's behavior. Unlikely though it seems, in two years of drinking, driving, carousing, and vandalism no one was seriously injured as a result of the Saints' activities.

THE SAINTS IN SCHOOL

The Saints were highly successful in school. The average grade for the group was "B," with two of the boys having close to a straight "A" average. Almost all of the boys were popular and many of them held offices in the school. One of the boys was vice-president of the student body one year. Six of the boys played on athletic teams.

At the end of their senior year, the student body selected ten seniors for special recognition as the "school wheels"; four of the ten were Saints. Teachers and school officials saw no problem with any of these boys and anticipated that they would all "make something of themselves."

How the boys managed to maintain this impression is surprising in view of their actual behavior while in school. Their technique for covering truancy was so successful that teachers did not even realize that the boys were absent from school much of the time. Occasionally, of course, the system would backfire and then the boy was on his own. A boy who was caught would be most contrite, would plead guilty and ask for mercy. He inevitably got the mercy he sought.

Cheating on examinations was rampant, even to the point of orally communicating answers to exams as well as looking at one another's papers. Since none of the group studied, and since they were primarily dependent on one another for help, it is surprising that grades were so high. Teachers contributed to the deception in their admitted inclination to give these boys (and presumably others like them) the benefit of the doubt. When asked how the boys did in school, and when pressed on specific examinations, teachers might admit that they were disappointed in John's performance, but would quickly add that they "knew that he was capable of doing better," so John was given a higher grade than he had actually earned. How often this happened is impossible to know. During the time that I observed the group, I never saw any of the boys take homework home. Teachers may have been "understanding" very regularly.

One exception to the gang's generally good performance was Jerry, who had a "C" average in his junior year, experienced disaster the next year, and failed to graduate. Jerry had always been a little more nonchalant than the others about the liberties he took in school. Rather than wait for someone to come get him from class, he would offer his own excuse and leave. Although he probably did not miss any more classes than most of the others in the group, he did not take the requisite pains to cover his absences. Jerry was the only Saint whom I ever heard talk back to a teacher. Although teachers often called him a "cut up" or a "smart kid," they never referred to him as a troublemaker or as a kid headed for trouble. It seems likely, then, that Jerry's failure his senior year and his mediocre performance his junior year were consequences of his not playing the game the proper way (possibly because he was disturbed by his parents' divorce). His teachers regarded him as "immature" and not quite ready to get out of high school.

THE POLICE AND THE SAINTS

The local police saw the Saints as good boys who were among the leaders of the youth in the community. Rarely, the boys might be stopped in town for speeding or for running a stop sign. When this happened the boys were always polite, contrite, and pled for mercy. As in school, they received the mercy they asked for. None ever received a ticket or was taken into the precinct by the local police.

The situation in Big City, where the boys engaged in most of their delinquency, was only slightly different. The police there did not know the boys at all, although occasionally the boys were stopped by a patrolman. Once they were caught taking a lantern from a construction site. Another time they were stopped for running a stop sign, and on several occasions they were stopped for speeding. Their behavior was as before: contrite, polite, and penitent. The urban police, like the local police, accepted their demeanor as sincere. More important, the urban police were convinced that these were good boys just out for a lark.

THE ROUGHNECKS

Hanibal townspeople never perceived the Saints' high level of delinquency. The Saints were good boys who just went in for an occasional prank. After all, they were well dressed, well mannered, and had nice cars. The Roughnecks were a different story. Although the two gangs of boys were the same age, and both groups engaged in an equal amount of wild-oat sowing, everyone agreed that the not-so-well-dressed, not-so-well-mannered, not-so-rich boys were heading for trouble. Townspeople would say, "You can see the gang members at the drugstore, night after night, leaning against the storefront (sometimes drunk) or slouching around inside buying cokes, reading magazines, and probably stealing old Mr. Wall blind. When they are outside and girls walk by, even respectable girls, these boys make suggestive remarks. Sometimes their remarks are downright lewd."

From the community's viewpoint, the real indication that these kids were in for trouble was that they were constantly involved with the police. Some of them had been picked up for stealing, mostly small stuff, of course, "but still it's stealing small stuff that leads to big time crimes." "Too bad," people said. "Too bad that these boys couldn't behave like the other kids in town: stay out of trouble, be polite to adults, and look to their future."

The community's impression of the degree to which this group of six boys (ranging in age from 16 to 19) engaged in delinquency was somewhat distorted. In some ways the gang was more delinquent than the community thought; in other ways they were less.

The fighting activities of the group were fairly readily and accurately perceived by almost everyone. At least once a month, the boys would get into some sort of fight, although most fights were scraps between members of the group or involved only one member of the group and some peripheral hanger-on. Only three times in the period of observation did the group fight together: once against a gang from across town, once against two blacks, and once against a group of boys from another school. For the first two fights the group went out "looking for trouble"—and they found it both times. The third fight followed a football game and began spontaneously with an argument on the football field between one of the Roughnecks and a member of the opposition's football team.

Jack had a particular propensity for fighting and was involved in most of the brawls. He was a prime mover of the escalation of arguments into fights.

More serious than fighting, had the community been aware of it, was theft. Although almost everyone was aware that the boys occasionally stole things, they did not realize the extent of the activity. Petty stealing was a frequent event for the Roughnecks. Sometimes they stole as a group and coordinated their efforts; other times they stole in pairs. Rarely did they steal alone.

The thefts ranged from very small things like paperback books, comics, and ballpoint pens to expensive items like watches. The nature of the thefts varied from time to time. The gang would go through a period of systematically shoplifting items from automobiles or school lockers. Types of thievery varied with the whim of the gang. Some forms of thievery were more profitable than others, but all thefts were for profit, not just thrills.

Roughnecks siphoned gasoline from cars as often as they had access to an automobile, which was not very often. Unlike the Saints, who owned their own cars, the Roughnecks would have to borrow their parents' cars, an event which occurred only eight or nine times a year. The boys claimed to have stolen cars for joy rides from time to time.

Ron committed the most serious of the group's offenses. With an unidentified associate the boy attempted to burglarize a gasoline station. Although this station had been robbed twice previously in the same month, Ron denied any involvement in either of the other thefts. When Ron and his accomplice approached the station, the owner was hiding in the bushes beside the station. He fired both barrels of a double-barreled shotgun at the boys. Ron was severely injured; the other boy ran away and was never caught. Though he remained in critical condition for several months, Ron finally recovered and served six months of the following year in reform school. Upon release from reform school, Ron was put back a grade in school, and began running around with a different gang of boys. The Roughnecks considered the new gang less delinquent than themselves, and during the following year Ron had no more trouble with the police.

The Roughnecks, then, engaged mainly in three types of delinquency: theft, drinking, and fighting. Although community members perceived that this gang of kids was delinquent, they mistakenly believed that their illegal activities were primarily drinking, fighting, and being a nuisance to passersby. Drinking was limited among the gang members, although it did occur, and theft was much more prevalent than anyone realized.

Drinking would doubtless have been more prevalent had the boys had ready access to liquor. Since they rarely had automobiles at their disposal, they could not travel very far, and the bars in town would not serve them. Most of the boys had little money, and this, too, inhibited their purchase of alcohol. Their major source of liquor was a local drunk who would buy them a fifth if they would give him enough extra to buy himself a pint of whiskey or a bottle of wine.

The community's perception of drinking as prevalent stemmed from the fact that it was the most obvious delinquency the boys engaged in. When one of the boys had been drinking, even a casual observer seeing him on the corner would suspect that he was high.

There was a high level of mutual distrust and dislike between the Roughnecks and the police. The boys felt very strongly that the police were unfair and corrupt. Some evidence existed that the boys were correct in their perception.

The main source of the boys' dislike for the police undoubtedly stemmed from the fact that the police would sporadically harass the group. From the standpoint of the boys, these acts of occasional enforcement of the law were whimsical and uncalled for. It made no sense to them, for example, that the police would come to the corner occasionally and threaten them with arrest for loitering when the night before the boys had been out siphoning gasoline from cars and the police had been nowhere in sight. To the boys, the police were stupid on the one hand, for not being where they should have been and catching the boys in a serious offense, and unfair on the other hand, for trumping up "loitering" charges against them.

From the viewpoint of the police, the situation was quite different. They knew, with all the confidence necessary to be a policeman, that these boys were engaged in criminal activities. They knew this partly from occasionally catching them, mostly from circumstantial evidence ("the boys were around when those tires were slashed"), and partly because the police shared the view of the community in general that this was a bad bunch of boys. The best the police could hope to do was to be sensitive to the fact that these boys were engaged in illegal acts and arrest them whenever there was some evidence that they had been involved. Whether or not the boys had in fact committed a particular act in a particular way was not especially important. The police had a broader view: their job was to stamp out these kids' crimes; the tactics were not as important as the end result.

Over the period that the group was under observation, each member was arrested at least once. Several of the boys were arrested a number of times and spent at least one night in jail. While most were never taken to court, two of the boys were sentenced to six months' incarceration in boys' schools.

THE ROUGHNECKS IN SCHOOL

The Roughnecks' behavior in school was not particularly disruptive. During school hours they did not all hang around together, but tended instead to spend most of their time with one or two other members of the gang who were their special buddies. Although every member of the gang attempted to avoid school as much as possible, they were not particularly successful and most of them attended school with surprising regularity. They considered school a burden—something to be gotten through with a minimum of conflict. If they were "bugged" by a particular teacher, it could lead to trouble. One of the boys, Al, once threatened to beat up a teacher and, according to the other boys, the teacher hid under a desk to escape him.

Teachers saw the boys the way the general community did, as heading for trouble, as being uninterested in making something of themselves. Some were also seen as being incapable of meeting the academic standards of the school. Most of the teachers expressed concern for this group of boys and were willing to pass them despite poor performance, in the belief that failing them would only aggravate the problem.

The group of boys had a grade point average just slightly above "C." No one in the group failed a grade, and no one had better than a "C" average. They were very consistent in their achievement or, at least, the teachers were consistent in their perception of the boys' achievement.

Two of the boys were good football players. Herb was acknowledged to be the best player in the school and Jack was almost as good. Both boys were criticized for their failure to abide by training rules, for refusing to come to practice as often as they should, and for not playing their best during practice. What they lacked in sportsmanship they made up for in skill, apparently, and played every game no matter how poorly they had performed in practice or how many practice sessions they had missed.

TWO QUESTIONS

Why did the community, the school, and the police react to the Saints as though they were good, upstanding, nondelinquent youths with bright futures but to the Roughnecks as though they were tough, young criminals who were headed for trouble? Why did the Roughnecks and the Saints in fact have quite different careers after high school—careers which, by and large, lived up to the expectations of the community?

The most obvious explanation for the differences in the community's and law enforcement agencies' reactions to the two gangs is that one group of boys was "more delinquent" than the other. Which group *was* more delinquent? The answer to this question will determine in part how we explain the differential responses to these groups by the members of the community and, particularly, by law enforcement and school officials.

In sheer number of illegal acts, the Saints were the more delinquent. They were truant from school for at least part of the day almost every day of the week. In addition, their drinking and vandalism occurred with surprising regularity. The Roughnecks, in contrast, engaged sporadically in delinquent episodes. While these episodes were frequent, they certainly did not occur on a daily or even a weekly basis.

The difference in frequency of offenses was probably caused by the Roughnecks' inability to obtain liquor and to manipulate legitimate excuses from school. Since the Roughnecks had less money that the Saints, and teachers carefully supervised their school activities, the Roughnecks' hearts may have been as black as the Saints', but their misdeeds were not nearly as frequent.

There are really no clear-cut criteria by which to measure qualitative differences in antisocial behavior. The most important dimension of the difference is generally referred to as the "seriousness" of the offenses.

If seriousness encompasses the relative economic costs of delinquent acts, then some assessment can be made. The Roughnecks probably stole an average of about $5.00 worth of goods a week. Some weeks the figure was considerably higher, but these times must be balanced against long periods when almost nothing was stolen.

The Saints were more continuously engaged in delinquency but their acts were not for the most part costly to property. Only their vandalism and occasional theft of gasoline would so qualify. Perhaps once or twice a month they would siphon a tankful of gas. The other costly items were street signs, construction lanterns, and the like. All of these acts combined probably did not quite average $5.00 a week, partly because much of the stolen equipment was abandoned and presumably could be recovered. The difference in cost of stolen property between the two groups was trivial, but the Roughnecks probably had a slightly more expensive set of activities than did the Saints.

Another meaning of seriousness is the potential threat of physical harm to members of the community and to the boys themselves. The Roughnecks were more prone to physical violence; they not only welcomed an opportunity to fight; they went seeking it. In addition, they fought among themselves frequently. Although the fighting never included deadly weapons, it was still a menace, however minor, to the physical safety of those involved.

The Saints never fought. They avoided physical conflict both inside and outside the group. At the same time, though, the Saints frequently endangered their own and other people's lives. They did so almost every time they drove a car, especially if they had been drinking. Sober, their driving was risky; under the influence of alcohol it was horrendous. In addition, the Saints endangered the lives of others with their pranks. Street excavations left unmarked were a very serious hazard.

Evaluating the relative seriousness of the two gangs' activities is difficult. The community reacted as though the behavior of the Roughnecks was a problem, and they reacted as though the behavior of the Saints was not. But the members of the community were ignorant of the array of delinquent acts that characterized the Saints' behavior. Although concerned citizens were unaware of much of the Roughnecks' behavior as well, they were much better informed about the Roughnecks' involvement in delinquency than they were about the Saints'.

Visibility

Differential treatment of the two gangs resulted in part because one gang was infinitely more visible than the other. This differential visibility was a direct function of the economic standing of the families. The Saints had access to automobiles and were able to remove themselves from the sight of the community. In as routine a decision as to where to go to have a milkshake after school, the Saints stayed away from the mainstream of community life. Lacking transportation, the Roughnecks could not make it to the edge of town. The center of town was the only practical place for them to meet since their homes were scattered throughout the town and any noncentral meeting place put an undue hardship on some members. Through necessity the Roughnecks congregated in a crowded area where everyone in the community passed frequently, including teachers and law enforcement officers. They could easily see the Roughnecks hanging around the drugstore.

The Roughnecks, of course, made themselves even more visible by making remarks to passersby and by occasionally getting into fights on the corner. Meanwhile, just as regularly, the Saints were either at the cafe on one edge of town or in the pool hall at the other edge of town. Without any particular realization that they were making themselves inconspicuous, the Saints were able to hide their time-wasting. Not only were they removed from the mainstream of traffic, but they were almost always inside a building.

On their escapades the Saints were also relatively invisible, since they left Hanibal and travelled to Big City. Here, too, they were mobile, roaming the city, rarely going to the same area twice.

Demeanor

To the notion of visibility must be added the difference in the responses of group members to outside intervention with their activities. If one of the Saints was confronted with an accusing policeman, even if he felt he was truly innocent of a wrongdoing, his demeanor was apologetic and penitent. A Roughneck's attitude was almost the polar opposite. When confronted with a threatening adult authority, even one who tried to be pleasant, the Roughneck's hostility and disdain were clearly observable. Sometimes he might attempt to put up a veneer of respect, but it was thin and was not accepted as sincere by the authority.

School was no different from the community at large. The Saints could manipulate the system by feigning compliance with the school norms. The availability of cars at school meant that once free from the immediate sight of the teacher, the boys could disappear rapidly. And this escape was well enough planned that no administrator or teacher was nearby when the boys left. A Roughneck who wished to escape for a few hours was in a bind. If it were possible to get free from class, downtown was still a mile away, and even if he arrived there, he was still very visible. Truancy for the Roughnecks meant almost certain detection, while the Saints enjoyed almost complete immunity from sanctions.

Bias

Community members were not aware of the transgressions of the Saints. Even if the Saints had been less discreet, their favorite delinquencies would have been perceived as less serious than those of the Roughnecks.

In the eyes of the police and school officials, a boy who drinks in an alley and stands intoxicated on the street corner is committing a more serious offense than is a boy who drinks to inebriation in a nightclub or a tavern and drives around afterwards in a car. Similarly, a boy who steals a wallet from a store will be viewed as having committed a more serious offense than a boy who steals a lantern from a construction site.

Perceptual bias also operates with respect to the demeanor of the boys in the two groups when they are confronted by adults. It is not simply that adults dislike the posture affected by boys of the Roughneck ilk; more important is the conviction that the posture adopted by the Roughnecks is an indication of their devotion and commitment to deviance as a way of life. The posture becomes a cue, just as the type of the offense is a cue, to the degree to which the known transgressions are indicators of the youths' potential for other problems.

Visibility, demeanor, and bias are surface variables which explain the day-to-day operations of the police. Why do these surface variables operate as they do? Why did the police choose to disregard the Saints' delinquencies while breathing down the backs of the Roughnecks?

The answer lies in the class structure of American society and the control of legal institutions by those at the top of the class structure. Obviously, no representative of the upper class drew up the operational chart for the police which led them to look in the ghettoes and on streetcorners—which led them to see the demeanor of lower-class youth as troublesome and that of upper middle-class youth as tolerable. Rather, the procedure simply developed from experience—experience with irate and influential upper-middle-class parents insisting that their son's vandalism was simply a prank and his drunkenness only a momentary "sowing of wild oats"—experience with cooperative or indifferent, powerless, lower-class parents who acquiesced to the law's definition of their son's behavior.

ADULT CAREERS OF THE
SAINTS AND THE ROUGHNECKS

The community's confidence in the potential of the Saints and the Roughnecks apparently was justified. If anything the community members underestimated the degree to which these youngsters would turn out "good" or "bad."

Seven of the eight members of the Saints went on to college immediately after high school. Five of the boys graduated from college in four years. The sixth one finished college after two years in the army, and the seventh spent four years in the air force before returning to college and receiving a B.A. degree. Of these seven college graduates, three went on for advanced degrees. One finished law school and is now active in state politics, one finished medical school and is practicing near Hanibal, and one boy is now working for a Ph.D. The other four college graduates entered submanagerial, managerial, or executive training positions with larger firms.

The only Saint who did not complete college was Jerry. Jerry had failed to graduate from high school with the other Saints. During his second senior year, after the other Saints had gone on to college, Jerry began to hang around with what several teachers described as a "rough crowd"—the gang that was heir apparent to the Roughnecks. At the end of his second senior year, when he did graduate from high school, Jerry took a job as a used car salesman, got married, and quickly had a child. Although he made several abortive attempts to go to college by attending night school, when I last saw him (ten years after high school) Jerry was unemployed and had been living on unemployment for almost a year. His wife worked as a waitress.

Some of the Roughnecks have lived up to community expectations. A number of them were headed for trouble. A few were not.

Jack and Herb were the athletes among the Roughnecks and their athletic prowess paid off handsomely. Both boys received unsolicited athletic scholarships to college. After Herb received his scholarship (near the end of his senior year), he apparently did an about-face. His demeanor became very similar to that of the Saints. Although he remained a member in good standing of the Roughnecks, he stopped participating in most activities and did not hang on the corner as often.

Jack did not change. If anything, he became more prone to fighting. He even made excuses for accepting the scholarship. He told the other gang members that the school had guaranteed him a "C" average if he would come to play football—an idea that seems far-fetched, even in this day of highly competitive recruiting.

During the summer after graduation from high school, Jack attempted suicide by jumping from a tall building. The jump would certainly have killed most people trying it, but Jack survived. He entered college in the fall and played four years of football. He and Herb graduated in four years, and both are teaching and coaching in high schools. They are married and have stable families. If anything, Jack appears to have a more prestigious position in the community than does Herb, though both are well respected and secure in their positions.

Two of the boys never finished high school. Tommy left at the end of his junior year and went to another state. That summer he was arrested and placed on probation on a manslaughter charge. Three years later he was arrested for murder; he pleaded guilty to second degree murder and is serving a 30-year sentence in the state penitentiary.

Al, the other boy who did not finish high school, also left the state in his senior year. He is serving a life sentence in a state penitentiary for first degree murder.

Wes is a small-time gambler. He finished high school and "bummed around." After several years he made contact with a bookmaker who employed him as a runner. Later he acquired his own area and has been working it ever since. His position among the bookmakers is almost identical to the position he had in the gang; he is always around but no one is really aware of him. He makes no trouble and he does not get into any. Steady, reliable, capable of keeping his mouth closed, he plays the game by the rules, even though the game is an illegal one.

That leaves only Ron. Some of his former friends reported that they had heard he was "driving a truck up north," but no one could provide any concrete information.

REINFORCEMENT

The community responded to the Roughnecks as boys in trouble, and the boys agreed with that perception. Their pattern of deviancy was reinforced, and breaking away from it became increasingly unlikely. Once the boys acquired an image of themselves as deviants, they selected new friends who affirmed that self-image. As that self-conception became more firmly entrenched, they also became willing to try new and more extreme deviances. With their growing alienation came freer expression of disrespect and hostility for representatives of the legitimate society. This disrespect increased the community's negativism, perpetuating the entire process of commitment to deviance. Lack of a commitment to deviance works the same way. In either case, the process will perpetuate itself unless some event (like a scholarship to college or a sudden failure) external to the established relationship intervenes. For two of the Roughnecks (Herb and Jack), receiving college athletic scholarships created new relations and culminated in a break with the established pattern of deviance. In the case of one of the Saints (Jerry), his parents' divorce and his failing to graduate from high school changed some of his other relations. Being held back in school for a year and losing his place among the Saints had sufficient impact on Jerry to alter his self-image and virtually to assure that he would not go on to college as his peers did. Although the experiments of life can rarely be reversed, it seems likely in view of the behavior of the other boys who did not enjoy this special treatment by the school that Jerry, too, would have "become something" had he graduated as anticipated. For Herb and Jack outside intervention worked to their advantage; for Jerry it was his undoing.

Selective perception and labeling—finding, processing, and punishing some kinds of criminality and not others—means that visible, poor, nonmobile, outspoken, undiplomatic "tough" kids will be noticed, whether their actions are seriously delinquent or not. Other kids, who have established a reputation for being bright (even though underachieving), disciplined and in-volved in respectable activities, who are mobile and monied, will be invisible when they deviate from sanctioned activities. They'll sow their wild oats—perhaps even wider and thicker than their lower-class cohorts—but they won't be noticed. When it's time to leave adolescence most will follow the expected path, settling into the ways of the middle class, remembering fondly the delin-quent but unnoticed fling of their youth. The Roughnecks and others like them may turn around, too. It is more likely that their noticeable deviance will have been so reinforced by police and community that their lives will be effectively channeled into careers consistent with their adolescent background.

13

The Police and the Black Male

ELIJAH ANDERSON

The police, in the Village-Northton [neighborhood] as elsewhere, represent society's formal, legitimate means of social control.[1] Their role includes protecting law-abiding citizens from those who are not law-abiding, by preventing crime and by apprehending likely criminals. Precisely how the police fulfill the public's expectations is strongly related to how they view the neighborhood and the people who live there. On the streets, color-coding often works to confuse race, age, class, gender, incivility, and criminality, and it expresses itself most concretely in the person of the anonymous black male. In doing their job, the police often become willing parties to this general color-coding of the public environment, and related distinctions, particularly those of skin color and gender, come to convey definite meanings. Although such coding may make the work of the police more manageable, it may also fit well with their own presuppositions regarding race and class relations, thus shaping officers' perceptions of crime "in the city." Moreover, the anonymous black male is usually an ambiguous figure who arouses the utmost caution and is generally considered dangerous until he proves he is not. . . .

There are some who charge—. . . perhaps with good reason—that the police are primarily agents of the middle class who are working to make the area more hospitable to middle-class people at the expense of the lower classes. It is obvious that the police assume whites in the community are at least middle class and are trustworthy on the streets. Hence the police may be seen primarily as protecting "law-abiding" middle-class whites against anonymous "criminal" black males.

To be white is to be seen by the police—at least superficially—as an ally, eligible for consideration and for much more deferential treatment than that accorded blacks in general. This attitude may be grounded in the backgrounds of the police themselves.[2] Many have grown up in Eastern City's "ethnic" neighborhoods. They may serve what they perceive as their own class and neighborhood interests, which often translates as keeping blacks "in their place"—away from neighborhoods that are socially defined as "white." In trying to do their job, the police appear to engage in an informal policy of monitoring young black men as a means of controlling crime, and often they seem to go beyond the bounds of duty. The following field note shows what pressures and racism young black men in the Village may endure at the hands of the police:

Elijah Anderson, *Streetwise* (Chicago: The University of Chicago Press, 1990). Reprinted by permission of the publisher and the author.

At 8:30 on a Thursday evening in June I saw a police car stopped on a side street near the Village. Beside the car stood a policeman with a young black man. I pulled up behind the police car and waited to see what would happen. When the policeman released the young man, I got out of my car and asked the youth for an interview.

"So what did he say to you when they stopped you? What was the problem?" I asked. "I was just coming around the corner, and he stopped me, asked me what was my name, and all that. And what I had in my bag. And where I was coming from. Where I lived, you know, all the basic stuff, I guess. Then he searched me down and, you know, asked me who were the supposedly tough guys around here? That's about it. I couldn't tell him who they are. How do I know? Other gang members could, but I'm not from a gang, you know. But he tried to put me in a gang bag, though." "How old are you?" I asked. "I'm seventeen, I'll be eighteen next month." "Did he give any reason for stopping you?' "No, he didn't. He just wanted my address, where I lived, where I was coming from, that kind of thing. I don't have no police record or nothin'. I guess he stopped me on principle, 'cause I'm black." "How does that make you feel?" I asked. "Well, it doesn't bother me too much, you know, as long as I know that I hadn't done nothin', but I guess it just happens around here. They just stop young black guys and ask 'em questions, you know. What can you do?"

On the streets late at night, the average young black man is suspicious of others he encounters, and he is particularly wary of the police. If he is dressed in the uniform of the "gangster," such as a black leather jacket, sneakers, and a "gangster cap," if he is carrying a radio or a suspicious bag (which may be confiscated), or if he is moving too fast or too slow, the police may stop him. As part of the routine, they search him and make him sit in the police car while they run a check to see whether there is a "detainer" on him. If there is nothing, he is allowed to go on his way. After this ordeal the youth is often left afraid, sometimes shaking, and uncertain about the area he had previously taken for granted. He is upset in part because he is painfully aware of how close he has come to being in "big trouble." He knows of other youths who have gotten into a "world of trouble" simply by being on the streets at the wrong time or when the police were pursuing a criminal. In these circumstances, particularly at night, it is relatively easy for one black man to be mistaken for another. Over the years, while walking through the neighborhood I have on occasion been stopped and questioned by police chasing a mugger, but after explaining myself I was released.

Many youths, however, have reason to fear such mistaken identity or harassment, since they might be jailed, if only for a short time, and would have to post bail money and pay legal fees to extricate themselves from the mess (Anderson 1986). When law-abiding blacks are ensnared by the criminal justice system, the scenario may proceed as follows. A young man is arbitrarily stopped by the police and questioned. If he cannot effectively negotiate with the officer(s), he may be accused of a crime and arrested. To resolve this situation he needs financial resources, which for him are in short supply. If he does not have money for an attorney, which often happens, he is left to a public defender who may be more interested in going along with the court system than in fighting for a poor black person. Without legal support, he may well wind up "doing time" even if he is innocent of the charges brought against him. The next time he is stopped for questioning he will have a record, which will make detention all the more likely.

Because the young black man is aware of many cases when an "innocent" black person was wrongly accused and detained, he develops an "attitude" toward the police. The street word for police is "the man," signifying a certain machismo, power, and authority. He becomes concerned when he notices "the man" in the community or when the police focus on him because he is outside his own neighborhood. The youth knows, or soon finds out, that he exists in a legally precarious state. Hence he is motivated to avoid the police, and his public life becomes severely circumscribed.

To obtain fair treatment when confronted by the police, the young man may wage a campaign for social regard so intense that at times it borders on obsequiousness. As one streetwise black youth said: "If you show a cop that you nice and not a smartass, they be nice to you. They talk to you like the man you are. You gonna get ignorant like a little kid, they gonna get ignorant with you." Young black males often are particularly deferential toward the police even when they are completely within their rights and have done nothing wrong. Most often this is not out of blind acceptance or respect for the "law," but because they know the police can cause them hardship. When confronted or arrested, they adopt a particular style of behavior to get on the policeman's good side. Some simply "go limp" or politely ask, "What seems to be the trouble, officer?" This pose requires a deference that is in sharp contrast with the youth's more usual image, but many seem to take it in stride or not even to realize it. Because they are concerned primarily with staying out of trouble, and because they perceive the police as arbitrary in their use of power, many defer in an equally arbitrary way. Because of these pressures, however, black youths tend to be especially mindful of the police and, when they are around, to watch their own behavior in public. Many have come to expect harassment and are inured to it; they simply tolerate it as part of living in the Village-Northton.

After a certain age, say twenty-four, a black man may no longer be stopped so often, but he continues to be the object of policy scrutiny. As one twenty-seven-year-old black college graduate speculated:

> I think they see me with my little bag with papers in it. They see me with penny loafers on. I have a tie on, some days. They don't stop me so much now. See, it depends on the circumstances. If something goes down, and they hear that the guy had on a big black coat, I may be the one. But when I was younger, they could just stop me, carte blanche, any old time. Name taken, searched, and this went on endlessly. From the time I was about twelve until I was sixteen or seventeen, endlessly, endlessly. And I come from a lower-middle-class black neighborhood, OK, that borders a white neighborhood. One neighborhood is all black, and one is all

white. OK, just because we were so close to that neighborhood, we were stopped endlessly. And it happened even more when we went up into a suburban community. When we would ride up and out to the suburbs, we were stopped every time we did it.

If it happened today, now that I'm older, I would really be upset. In the old days when I was younger, I didn't know any better. You just expected it, you knew it was gonna happen. Cops would come up, "What you doing, where you coming from?" Say things to you. They might even call you nigger.

Such scrutiny and harassment by local police makes black youths see them as a problem to get beyond, to deal with, and their attempts affect their overall behavior. To avoid encounters with "the man," some streetwise young men camouflage themselves, giving up the urban uniform and emblems that identify them as "legitimate" objects of police attention. They may adopt a more conventional presentation of self, wearing chinos, sweat suits, and generally more conservative dress. Some youths have been known to "ditch" a favorite jacket if they see others wearing one like it, because wearing it increases their chances of being mistaken for someone else who may have committed a crime.

But such strategies do not always work over the long run and must be constantly modified. For instance, because so many young ghetto blacks have begun to wear Fila and Adidas sweat suits as status symbols, such dress has become incorporated into the public image generally associated with young black males. These athletic suits, particularly the more expensive and colorful ones, along with high-priced sneakers, have become the leisure dress of successful drug dealers, and other youths will often mimic their wardrobe to "go for bad" in the quest for local esteem. Hence what was once a "square" mark of distinction approximating the conventions of the wider culture has been adopted by a neighborhood group devalued by that same culture. As we saw earlier, the young black male enjoys a certain power over fashion: whatever the collective peer group embraces can become "hip" in a manner the wider society may not desire (see Goffman 1963). These same styles then attract the attention of the agents of social control.

THE IDENTIFICATION CARD

Law-abiding black people, particularly those of the middle class, set out to approximate middle-class whites in styles of self-presentation in public, including dress and bearing. Such middle-class emblems, often viewed as "square," are not usually embraced by young working-class blacks. Instead, their connections with and claims on the institutions of the wider society seem to be symbolized by the identification card. The common identification card associates its holder with a firm, a corporation, a school, a union, or some other institution of substance and influence. Such a card, particularly from a prominent establishment, puts the police and others on notice that the youth is "somebody," thus creating an important distinction between a black man who can claim a connection with the wider society and one who is summarily judged as "deviant." Although blacks who are established in the middle class might take such cards for granted, many lower-class blacks, who continue to find it necessary to campaign for civil rights denied them because of skin color, believe that carrying an identification card brings them better treatment than is meted out to their less fortunate brothers and sisters. For them this link to the wider society, though often tenuous, is psychically and socially important. The young college graduate continues:

> I know [how] I used to feel when I was enrolled in college last year, when I had an ID card. I used to hear stories about the blacks getting stopped over by the dental school, people having trouble sometimes. I would see that all the time. Young black male being stopped by the police. Young black male in handcuffs. But I knew that because I had that ID card that I would not be mistaken for just somebody snatching a pocketbook, or just somebody being where maybe I wasn't expected to be. See, even though I was intimidated by the campus police—I mean, the first time I walked into the security office to get my ID they all gave me the double-take to see if I was somebody they were looking for. See, after I got the card, I was like, well, they can think that now, but I have this [ID card]. Like, see, late at night when I be walking around, and the cops be checking me out, giving me the looks, you know. I mean, I know guys, students, who were getting stopped all the time, sometimes by the same officer, even though they had the ID. And even they would say, "Hey, I got the ID, so why was I stopped?"

The cardholder may believe he can no longer be treated summarily by the police, that he is no longer likely to be taken as a "no count," to be prejudicially confused with that class of blacks "who are always causing trouble on the trolley." Furthermore, there is a firm belief that if the police stop a person who has a card, they cannot "do away with him without somebody coming to his defense." This concern should not be underestimated. Young black men trade stories about mistreatment at the hands of the police; a common one involves policemen who transport youths into rival gang territories and release them, telling them to get home the best way they can. From the youth's perspective, the card signifies a certain status in circumstances where little recognition was formerly available.

"DOWNTOWN" POLICE AND LOCAL POLICE

In attempting to manage the police—and by implication to manage themselves—some black youths have developed a working connection of the police in certain public areas of the Village-Northton. Those who spend a good amount of their time on these corners, and thus observing the police, have come to distinguish between the "downtown" police and the "regular" local police.

The local police are the ones who spend time in the area; normally they drive around in patrol cars, often one officer to a car. These officers usually make a kind of working peace with the young men on the streets; for example, they know the names of some of them and may even befriend a young boy. Thus they offer an image of the police department different from that displayed by the "downtown" police. The downtown police are distant, impersonal, and often actively looking for "trouble." They are known to swoop down arbitrarily on gatherings of black youths standing on a street corner; they might punch them around, call them names, and administer other kinds of abuse, apparently for sport. A young Northton man gave the following narrative about his experiences with the police.

> And I happen to live in a violent part. There's a real difference between the violence level in the Village and the violence level in Northton. In the nighttime it's more dangerous over there.
>
> It's so bad now, they got downtown cops over there now. They doin' a good job bringin' the highway patrol over there. Regular cops don't like that. You can tell that. They even try to emphasize to us the certain category. Highway patrol come up, he leave, they say somethin' about it. "We can do our job over here." We call [downtown police] Nazis. They about six feet eight, seven feet. We walkin', they jump out. "You run, and we'll blow your nigger brains out." I hate bein' called a nigger. I want to say somethin' but get myself in trouble.
>
> When a cop do somethin', nothing happen to 'em. They come from downtown. From what I heard some of 'em don't even wear their real badge numbers. So you have to put up with that. Just keep your mouth shut when they stop you, that's all. Forget about questions, get against the wall, just obey 'em. "Put all that out right there"—might get rough with you now. They snatch you by the shirt, throw you against the wall, pat you hard, and grab you by the arms, and say, "Get outta here." They call you nigger this and little black this, and things like that. I take that. Some of the fellas get mad. It's a whole different world.
>
> Yeah, they lookin' for trouble. They gotta look for trouble when you got five, eight police cars together and they laughin' and talkin', start teasin' people. One night we were at a bar, we read in the paper that the downtown cops comin' to straighten things out. Same night, three police cars, downtown cops with their boots on, they pull the sticks out, beatin' around the corner, chase into bars. My friend Todd, one of 'em grabbed him and knocked the shit out of him. He punched 'im, a little short white guy. They start a riot. Cops started that shit. Everybody start seein' how wrong the cops was—they start throwin' bricks and bottles, cussin' 'em out. They lock my boy up; they had to let him go. He was just standin' on the corner, they snatch him like that.
>
> One time one of 'em took a gun and began hittin' people. My boy had a little hickie from that. He didn't know who the cop was, because there was no such thing as a badge number. They have phony badge numbers. You can tell they're tougher, the way they dress, plus they're

bigger. They have boots, trooper pants, blond hair, blue eyes, even black [eyes]. And they seven feet tall, and six foot six inches and six foot eight inches. Big! They are the rough cops. You don't get smart with them or they beat the shit out of you *in front of everybody,* they don't care.

We call 'em Nazis. Even the blacks among them. They ride along with 'em. They stand there and watch a white cop beat your brains out. What takes me out is the next day you don't see 'em. Never see 'em again, go down there, come back, and they ride right back downtown, come back, do their little dirty work, go back downtown, and put their real badges on. You see 'em with a forty-five or fifty-five number: "Ain't no such number here, I'm sorry, son." Plus, they got unmarked cars. No sense takin' 'em to court. But when that happened at that bar, another black cop from the sixteenth [local] district, ridin' a real car, came back and said, "Why don't y'all go on over to the sixteenth district and file a complaint? Them musclin' cops was wrong. Beatin' people." So about ten people went over there; sixteenth district knew nothin' about it. They come in unmarked cars, they must have been downtown cops. Some of 'em do it. Some of 'em are off duty, on their way home. District commander told us they do that. They have a patrol over there, but them cops from down-town have control of them cops. Have bigger ranks and bigger guns. They carry .357s and regular cops carry little .38s. Downtown cops are all around. They carry magnums.

Two cars the other night. We sittin' on the steps playing cards. Some-body called the cops. We turn around and see four regular police cars and two highway police cars. We drinkin' beer and playin' cards. Police get out and say you're gamblin'. We say we got nothin' but cards here, we got no money. They said all right, got back in their cars, and drove away. Downtown cops dressed up like troopers. That's intimidation. Damn!

You call a cop, they don't come. My boy got shot, we had to take him to the hospital ourselves. A cop said, "You know who did it?" We said no. He said, "Well, I hope he dies if y'all don't say nothin'." What he say that for? My boy said, "I hope your mother die," he told the cop right to his face. And I was grabbin' another cop, and he made a complaint about that. There were a lot of witnesses. Even the nurse behind the counter said the cop had no business saying nothin' like that. He said it loud, "I hope he dies." Nothin' like that should be comin' from a cop.

Such behavior by formal agents of social control may reduce the crime rate, but it raises questions about social justice and civil rights. Many of the old-time liberal white residents of the Village view the police with some am-bivalence. They want their streets and homes defended, but many are con-vinced that the police manhandle "kids" and mete out an arbitrary form of "justice." These feelings make many of them reluctant to call the police when they are needed, and they may even be less than completely cooperative after a crime has been committed. They know that far too often the police simply "go out and pick up some poor black kid." Yet they do cooperate, if ambiva-lently, with these agents of social control.

In an effort to gain some balance in the emerging picture of the police in the Village-Northton, I interviewed local officers. The following edited conversation with Officer George Dickens (white) helps place in context the fears and concerns of local residents, including black males:

I'm sympathetic with the people who live in this neighborhood [the Village-Northton], who I feel are victims of drugs. There are a tremendous number of decent, hardworking people who are just trying to live their life in peace and quiet, not cause any problems for their neighbors, not cause any problems for themselves. They just go about their own business and don't bother anyone. The drug situation as it exists in Northton today causes them untold problems. And some of the young kids are involved in one way or another with this drug culture. As a result, they're gonna come into conflict even with the police they respect and have some rapport with.

We just went out last week on Thursday and locked up ten young men on Cherry Street, because over a period of about a week, we had undercover police officers making drug buys from those young men. This was very well documented and detailed. They were videotaped selling the drugs. And as a result, right now, if you walk down Cherry Street, it's pretty much a ghost town; there's nobody out. [Before, Cherry Street was notorious for drug traffic.] Not only were people buying drugs there, but it was a very active street. There's been some shock value as a result of all those arrests at one time.

Now, there's two reactions to that. The [television] reporters went out and interviewed some people who said, "Aw, the police overreacted, they locked up innocent people. It was terrible, it was harassment." One of the neighbors from Cherry Street called me on Thursday, and she was outraged. Because she said, "Officer, it's not fair. We've been working with the district for well over a year trying to solve some of the problems on Cherry Street." But most of the neighbors were thrilled that the police came and locked all those kids up. So you're getting two conflicting reactions here. One from the people that live there that just wanta be left alone, alright? Who are really being harassed by the drug trade and everything that's involved in it. And then you have a reaction from the people that are in one way or another either indirectly connected or directly connected, where they say, "You know, if a young man is selling drugs, to him that's a job." And if he gets arrested, he's out of a job. The family's lost their income. So they're not gonna pretty much want anybody to come in there to make arrests. So you've got contradicting elements of the community there. My philosophy is that we're going to try to make Northton livable. If that means we have to arrest some of the residents of Northton, that's what we have to do.

You talk to Tyrone Pitts, you know the group that they formed was formed because of a reaction to complaints against one of the officers of how the teenagers were being harassed. And it turned out that basically what he [the officer] was doing was harassing drug dealers. When North-ton against Drugs actually formed and seemed to jell, they developed a close working relationship with the police here. For that reason, they felt the officer was doing his job.

I've been here eighteen months. I've seen this neighborhood go from . . . let me say, this is the only place I've ever worked where I've seen a rapport between the police department and the general community like the one we have right now. I've never seen it any place else before coming here. And I'm not gonna claim credit because this happened while I hap-pened to be here. I think a lot of different factors were involved. I think the community was ready to work with the police because of the terrible situation in reference to crack. My favorite expression when talking about crack is "crack changed everything." Crack changed the rules of how the police and the community have to interact with each other. Crack changed the rules about how the criminal justice system is gonna work, whether it works well or poorly. Crack is causing the prisons to be over-crowded. Crack is gonna cause the people that do drug rehabilitation to be overworked. It's gonna cause a wide variety of things. And I think the reason the rapport between the police and the community in Northton developed at the time it did is very simply that drugs to a certain extent made many areas in this city unlivable.

In effect the officer is saying that the residents, regardless of former atti-tudes, are now inclined to be more sympathetic with the police and to work with them. And at the same time, the police are more inclined to work with the residents. Thus, not only are the police and the black residents of North-ton working together, but different groups in the Village and Northton are working with each other against drugs. In effect, law-abiding citizens are com-ing together, regardless of race, ethnicity, and class. He continues:

Both of us [police and the community] are willing to say, "Look, let's try to help each other." The nice thing about what was started here is that it's spreading to the rest of the city. If we don't work together, this prob-lem is gonna devour us. It's gonna eat us alive. It's a state of emergency, more or less.

In the past there was significant negative feeling among young black men about the "downtown" cops coming into the community and harassing them. In large part these feelings continue to run strong, though many young men appear to "know the score" and to be resigned to their situation, accommo-dating and attempting to live with it. But as the general community feels under attack, some residents are willing to forgo certain legal and civil rights and un-dergo personal inconvenience in hopes of obtaining a sense of law and order. The officer continues:

Today we don't have too many complaints about police harassment in the community. Historically there were these complaints, and in almost any minority neighborhood in Eastern City where I ever worked there was more or less a feeling of that [harassment]. It wasn't just Northton; it was a feeling that the police were the enemy. I can honestly say that for the first time in my career I don't feel that people look at me like I'm the enemy. And it feels nice; it feels real good not to be the enemy, ha-ha. I think we [the police] realize that a lot of problems here [in the Village-Northton] are related to drugs. I think the neighborhood realizes that too. And it's a matter of "Who are we gonna be angry with? Are we gonna be angry with the police because we feel like they're this army of occupation, or are we gonna argue with these people who are selling drugs to our kids and shooting up our neighborhoods and generally causing havoc in the area? Who deserves the anger more?" And I think, to a large extent, people of the Village-Northton decided it was the drug dealers and not the police.

I would say there are probably isolated incidents where the police would stop a male in an area where there is a lot of drugs, and this guy may be perfectly innocent, not guilty of doing anything at all. And yet he's stopped by the police because he's specifically in that area, on that street corner where we know drugs are going hog wild. So there may be isolated incidents of that. At the same time, I'd say I know for a fact that our complaints against police in this division, the whole division, were down about 45 percent. If there are complaints, if there are instances of abuse by the police, I would expect that our complaints would be going up. But they're not; they're dropping.

Such is the dilemma many Villagers face when they must report a crime or deal in some direct way with the police. Stories about police prejudice against blacks are often traded at Village get-togethers. Cynicism about the effectiveness of the police mixed with community suspicion of their behavior toward blacks keeps middle-class Villagers from embracing the notion that they must rely heavily on the formal means of social control to maintain even the minimum freedom of movement they enjoy on the streets.

Many residents of the Village, especially those who see themselves as the "old guard" or "old-timers," who were around during the good old days when antiwar and antiracist protest was a major concern, sigh and turn their heads when they see the criminal justice system operating in the ways described here. They express hope that "things will work out," that tensions will ease, that crime will decrease and police behavior will improve. Yet as incivility and crime become increasing problems in the neighborhood, whites become less tolerant of anonymous blacks and more inclined to embrace the police as their heroes.

Such criminal and social justice issues, crystallized on the streets, strain relations between the newcomers and many of the old guard, but in the present context of drug-related crime and violence in the Village-Northton, many of the old-timers are adopting a "law and order" approach to crime and public safety, laying blame more directly on those they see as responsible for such crimes, though they retain some ambivalence. Newcomers can share such feelings with an increasing number of old-time "liberal" residents. As one middle-aged white woman who has lived in the Village for fifteen years said:

When I call the police, they respond. I've got no complaints. They are fine for me. I know they sometimes mistreat black males. But let's face it, most of the crime is committed by them, and so they can simply tolerate more scrutiny. But that's them.

Gentrifiers and the local old-timers who join them, and some traditional residents continue to fear, care more for their own safety and well-being than for the rights of young blacks accused of wrong-doing. Yet reliance on the police, even by an increasing number of former liberals, may be traced to a general feeling of oppression at the hands of street criminals, whom many believe are most often black. As these feelings intensify and as more yuppies and students inhabit the area and press the local government for services, especially police protection, the police may be required to "ride herd" more stringently on the youthful black population. Thus young black males are often singled out as the "bad" element in an otherwise healthy diversity, and the tensions between the lower-class black ghetto and the middle- and upper-class white community increase rather than diminish.

NOTES

1.   See Rubinstein (1973); Wilson (1978); Fogelson (1977); Reiss (1971); Bittner (1967); Banton (1964).

2.   For an illuminating typology of police work that draws a distinction between "fraternal" and "professional" codes of behavior, see Wilson (1968).

REFERENCES

Anderson, Elijah. 1986. "Of old heads and young boys: Notes on the urban black experience." Unpublished paper commissioned by the National Research Council, Committee on the Status of Black Americans.

Banton, Michael. 1964. *The policeman and the community*. New York: Basic Books.

Bittner, Egon. 1967. The police on Skid Row. *American Sociological Review* 32 (October): 699–715.

Fogelson, Robert. 1977. *Big city police*. Cambridge: Harvard University Press.

Goffman, Erving. 1963. *Behavior in public places*. New York: Free Press.

Reiss, Albert J. 1971. *The police and the public*. New Haven: Yale University Press.

Rubinstein, Jonathan. 1973. *City police*. New York: Farrar, Straus and Giroux.

Wilson, James Q. 1968. "The police and the delinquent in two cities." In *Controlling delinquents*, ed. Stanton Wheeler. New York: John Wiley.

14

Homophobia in Sport

DONALD F. SABO

Feelings about gay issues in sports run so deep because homophobia—irrational fear or hatred of gay men, lesbians, and bisexuals—is so prevalent. You know you're homophobic if you get anxious and afraid when you think you may be perceived as gay or lesbian by others. Becoming anxious or repulsed when you find yourself attracted to a person of your own sex, or being afraid that you have homo- or bisexual tendencies, are also signs of homophobia. Homophobic thoughts and feelings are a pervasive part of our culture and gender politics.

Like other forms of prejudice, homophobia feeds on stereotypes and promotes unfair discrimination. Key to the controversy surrounding gays and lesbians in sports is the potential of homosexuality to undermine institutionalized male dominance and sexist beliefs. Ironically, many so-called gay issues in sports are more fundamentally gender issues and sex equality issues.

LEARNING HOMOPHOBIC MASCULINITY

I was recently the guest speaker at an athletic banquet at a parochial middle school. Parents, grandparents, coaches, school administrators, and young athletes had gathered together in the church basement cafeteria to celebrate their athletic dreams and achievements. The kids, who ranged from the fourth through the eighth grades, were absolutely beautiful, and the room seemed to almost tilt back and forth from the force of their collective energy. The room hushed, however, as they filed up one by one to receive accolades and laminated plastic trophies that were, in their sparkling eyes, as good as gold.

After the awards ceremony, a father of a fourth-grade boy told me that his son had been called a lesbian by a friend. His son had replied indignantly, "I can't be a lesbian, stupid. You have to be a girl to be a lesbian." The anecdote illustrates two lessons about how homophobia filters into a boy's developing identity. First, males are not born homophobes, they gradually learn homophobic sentiments and ideas as they grow up. . . . Second, the flowering of homophobic beliefs and emotions in young males is intricately tied to their development of gender identity. While growing up, boys internalize various

From Michael A. Messner and Donald F. Sabo, *Sex, Violence and Power in Sports,* Crossing Press, 1994. Reprinted by permission of Donald F. Sabo and the publisher.

cultural messages about masculinity. They learn to behave "like men," which means not to behave like women. They are told—don't be a sissy or a wimp. Keep a stiff upper lip. Big boys don't cry. Take it like a man. Be independent; try no to depend on others. Be tough and aggressive, and keep your feelings to yourself. Homophobia is yet another message sent to boys across the American cultural air waves. This message says that, like oil and water, homosexuality and masculinity do not mix. As boys begin to view homosexuality as a "negation of masculinity," as Bob Connell (1992, 736) puts it, they come to equate homophobia with masculinity. Gregory Herek (1986, 563) explains that "to be 'a man' in contemporary American society is to be homophobic— that is, to be hostile toward homosexual persons in general and gay men in particular."

In athletics, the lessons of homophobia are learned and acted out in various ways, of which teasing and ridiculing are probably the most common. The razzing and joking that go on in most locker rooms communicate that homosexuals are inferior, silly, sick, or disgusting. The terms fag and faggot may be used to mock, insult, or aggravate a teammate. Lack of toughness, open displays of sympathy, or other behavior that is considered feminine might also provoke ridicule.

Some coaches use homophobia as a motivational device. Playing on the gender and sexual insecurities of adolescent athletes, coaches use the threat of homosexual stigmatization to muster allegiance to themselves or esprit d'corps among the ranks. Such ploys work because they reflect and feed the anti-gay sentiment that already exists in the locker-room subculture. An incident I remember from a high-school football practice illustrates this dynamic. A sophomore named Brian, a big lug but rather flabby, lacked the physical strength and the "killer instinct" that we were taught to believe was necessary to be a good player. One hot August afternoon, Coach "Sleepy Joe" Shumock decided to teach poor Brian "how to block, once and for all." He lined up the entire defensive team and made Brian block each one of us, one after the other. All the while, Coach taunted him:

> How many sisters you got at home, Brian? Is it six or seven? How long did it take your mother to find out you were a boy, Brian? When did you stop wearing dresses like your sisters, Brian? Maybe Brian would like to bake cookies for us tomorrow, boys. You're soft, Brian, maybe too soft for this team. What do you think, boys, is Brian too soft for the team?

The ordeal went on for a least ten minutes, until Brian collapsed, exhausted and in tears. Coach had won. I felt sorry for Brian; he may not have been "an animal" or a good player, but he was out there sweating and beating himself up with the rest of us. I realized, though, that being "soft" was to be avoided at all costs. I ultimately sided with the coach and the "team." I identified "up" the male hierarchy in solidarity with the team rather than "down" with Brian's vulnerability and suffering. I now understand that the coach's onslaught of homophobic messages stayed with me long after Brian's tears had dried in the August heat.

The "homophobe persona" is another vehicle through which homophobia gets translated into social action. Some young men act out the role of a gay-hater as a way of constructing their manly identity and building a reputation among their male peers. They usually do this by publicly ridiculing gay men, but they may also physically confront gays. Gay-bashing, the physical beating, even killing, of gays, is an extreme form of this behavior. Gay-haters' conscious or unconscious strategy is to become the "big man on campus" by being the big homophobe on campus. Teammates or fraternity brothers in effect become the audience to which the homophobe plays out the theme "I am a real man because I hate gays" or "I'll prove I'm heterosexual by hassling gays and lesbians." The belief that some sports are inherently more masculine than others also promotes homophobia among athletes. Jim Estep, a college professor and a former captain of the U.S. Naval Academy fencing team, recalled that whereas "football, basketball, and lacrosse were considered macho sports, fencing was looked down upon as being effeminate. There was the added inference that, if fencing was effeminate, then the participants in it might by gay."

This belief was communicated among students, Jim knew, even though nobody ever actually accused him or his teammates of being a homosexual. Once, when it was announced at the evening meal that Jim had earned second place in an eastern United States intercollegiate fencing tournament, some of the men at his table hooted and waved limp wrists in his face. And at the regular meetings of team captains from various sports, those from the more prestigious "macho" sports ignored him.

> My voice wasn't heard. My sport was viewed as less important and less macho than theirs. In their minds, football and basketball were about masculine traits like size, strength, and force. Fencing, in contrast, was more about strategy and subtlety, and these traits were considered feminine and inferior.

In summary, homophobia is not inborn but learned. As a boy grows up, homophobic sentiments and attitudes become grafted to his developing personality and gender identity. A boy's inner sense of manliness develops in juxtaposition to his sense of what it means to be gay or feminine. And yet, there is more to homophobia than sexual preference or conformity to gender stereotypes. Like other forms of prejudice, homophobia gets acted out and perpetuated in our relationships with other people. Homophobia derives from and reinforces a wide range of power relations among individuals and groups. These relations, in turn, are tied to systems of inequality.

Organizational/
Institutional Deviance

W e have just looked at how some categories of people and behavior become defined as deviant. Yet interactionist theory, as we noted in the General Introduction, suggests that a deviant label floating around abstractly in society is not meaningful unless it gets attached to people. Groups in society not only work to create definitions of deviance, but also create situations in which deviance occurs and is labeled. In this section we will examine the role of organizations and institutions in creating a context in which deviance is evoked and shaped.

CREATING DEVIANCE

Society is not made up solely of individuals taking care of their own business and needs. It is filled with many organizations and institutions, such as hospitals, businesses, police forces, courts, prisons, social welfare agencies, corporations, and even the government. People who work within these organizations are guided, constrained, or pressured to act in ways that help the organization accomplish its goals. Organizational goals may range from performing complex operations to simply staying financially or politically afloat. In our intricate society, organizations sometimes have difficulty achieving these mandates through legitimate means. Whether out of convenience or through necessity, organizations often create a climate in which deviance becomes accepted as the best or only way to accomplish the tasks they deem desirable or vital. In these contexts, people's deviant behavior is not committed to further their own ends, but those of the organization's. Schrager and Short (1978: 411–12) have offered the following definition of organizational crime:

> Organizational crimes are illegal acts of omission or commission of an individual or a group of individuals in a legitimate formal organization in accordance with the operative goals of the organization, which have a serious physical or economic impact on employees, customers, or the general public.

Organizational crimes directed against employees may consist of unsafe working conditions, against consumers they may take the form of hazardous products, and against the general public they may entail environmental pollution, dangerous public construction, and risks to public safety. These result from conditions in which the individuals who work for organizations perceive that they will benefit the organization more by allowing dangerous laxity than by enforcing stringent safety standards, and in which the regulatory enforcement is careless or permissive.

Tom Barker and David Carter illustrate the organizational constraints fostering deviance among law enforcement officers in "'Fluffing Up the Evidence and Covering Your Ass': On Police Lying." They outline a taxonomy of police lying, ranging from organizationally beneficial lying that is accepted and even taught in the academy, to lying that is tolerated as an organizationally necessary evil although considered morally wrong, to deviant lies that violate policies and procedures for legitimate or illegitimate purposes. Barker and Carter carefully document the police norms that excuse and justify lying so that this act is redefined as the accomplishment of institutional goals rather than irresponsible and illegal individual manipulation.

William Chambliss then shows the pervasiveness of the institutional mandate for deviance at even greater heights in examining the activities of the most elite group in society: the federal government. Although government agents and agencies have the power to make and enforce laws, Chambliss shows in "State-Organized Crime" that they do not always seek to further their goals by legitimate means. Partly due to their power and position, they hold themselves separate from ordinary citizens and above the law. As their position frames their interests, they justify the necessity of violating the norms and laws that they expect others to follow. This includes the smuggling of drugs (Vietnam) and arms (Contragate); murder, assassinations, and coups d'etat (both domestic and international); and other activities such as spying on citizens, including opposing political parties (Watergate). Therefore, we see that deviance is found not only on the margins of society but also within its mainstream and among its elites.

LABELING DEVIANCE

When not pursuing their own deviant activities, organization members often apply the deviant label to others. Many societal institutions, especially those in the medical and criminal justice arenas, are charged with the mandate of processing potential deviants. In so doing, they establish specific rules and procedures for handling people and assigning them to various categories of deviance. As in many bureaucracies, these rules and procedures tend to become organized, stabilized, and systematized, often functioning informally as organizational ends in themselves. This practice aids organizations in processing large numbers of people in a smooth and efficient manner, thus establishing and maintaining order while protecting themselves from outside criticism.

Organizations at work want to handle tasks in routine ways with a minimum of variation. People delegated to intake processing will assess individuals and assign them to bureaucratic categories based on **typifications** they hold about models of deviants and deviant behavior that are rooted in the numerous cases they encounter. Differences among individuals must then be minimized or neutralized to facilitate moving people through institutional mechanisms. Therefore, most institutions incorporate some way of "stripping down" individuals' self-conceptions and replacing them with ones useful to the organization. These procedures range in severity and intensity according to the bureaucratic organization involved, culminating in what Erving Goffman (1961) has called **total institutions.** These organizations, such as prisons, military boot camps, and monasteries, demand the total submission of clients, isolate clients from outside populations, and operate for the benefit of a small, elite group rather than for the good of the large group of clients. They typically deindividualize people through a haircut, clothing change, seizure of possessions, and loss of rights, subjecting them to individual and collective humiliations and degradations. Although all these experiences foster the smooth functioning of deviance-processing organizations, the result is a deep rooting of the deviant label.

Erving Goffman's "The Moral Career of the Mental Patient" offers us some insight into a classic total institution: a state mental hospital. Goffman considers the needs of this organization, how they result in its handling of people, and the consequences for individuals so treated. He examines the effects of people's betrayals by others as they are involuntarily committed and the mortifying **degradation ceremonies** they undergo once admitted. Restricted in movement and freedom and punished for breaking rules designed to ease the accomplishment of staff goals, clients are told that these treatments are for their "needs." These experiences frame and shape people's changing conceptions of self.

In "Bad Blood: The Moral Stigmatization of Paid Plasma Donors," Martin Kretzmann outlines the way that commercial blood plasma centers label the clients who sell them plasma as deviant. Although donating blood and other services may be considered altruistic, selling it renders the client discreditable, and this definition is forced onto donors by the policies and practices of the center's administrative and medical staff. Kretzmann compares the treatment of paid and unpaid donors, the assumptions implicitly made about their status and the worth of their blood, and the implications this has for the identity labels that clients assume.

The judicial labeling process is the focus of Lisa Frohmann's "Discrediting Victims' Allegations of Sexual Assault: Prosecutorial Accounts of Case Rejections." Frohmann discusses the organizational dynamics of the prosecutor's office, focusing on the ways in which rape cases and rape victims are "typified," with those displaying discrepant (that is, individualized) features discredited. Complainants are made to feel deviant and devalued if their cases do not mold to the norm, and their cases are dropped from further legal pursuit. This case is particularly interesting because it is the rape victims, not the offenders, who are made to feel disvalued by institutional processing assumptions and procedures.

Therefore, individuals operating within institutional frameworks are likely to be affected by the organizational dynamics shaping these frameworks. They are either spurred to deviance or labeled as deviant by their bureaucratic mechanisms.

15

"Fluffing Up the Evidence and Covering Your Ass"

On Police Lying

TOM BARKER AND DAVID CARTER

Lying and other deceptive practices are an integral part of the police officer's working environment. At first blush, one's reaction to this statement might be rather forthright. Police officers should not lie. If you can't trust your local police who can you trust. However, as with most issues the matter is not that simple.

We are all aware that police officers create false identities for undercover operations. We know that they make false promises to hostage takers and kidnappers. We also know officers will strain the truth in order to spare the feelings of a crime victim and his/her loved ones. Police officers are trained to lie and be deceptive in these law enforcement practices. They are also trained to use techniques of interrogation which require deception and even outright lying. Police officers learn much of this in the police academy. Where they are also warned about the impropriety of perjury and the need to record all incidents fully and accurately in all official reports. The recruit learns that all rules and regulations must be obeyed. He/she learns of the danger of lying to internal affairs or a supervisor. The recruit is told to be truthful in his dealings with the noncriminal element of the public in that mutual trust is an important element in police community relations. Once the recruit leaves the academy—and in some departments where officers work in the field before attending rookie school—the officer soon learns from his/her peers that police lying is the norm under certain circumstances.

Our purpose is to discuss the patterns of lying which might occur in a police organization, the circumstances under which they occur, and the possible consequences of police lying.

From "'Fluffing Up the Evidence and Covering Your Ass': Some Conceptual Notes on Police Lying," Tom Barker and David Carter, *Deviant Behavior*, V. 11, No. 1, Taylor & Francis, Inc., Washington, D.C., pp. 61–73, 1990. Reprinted with permission.

TAXONOMY OF POLICE LIES

Accepted Lying

Certain forms of police lying and/or deception are an accepted part of police officers' working environment. The lies told in this category are accepted by the police organization because they fulfill a defined police purpose. Administrators and individual police officers believe that certain lies are necessary to control crime and "arrest the guilty." In these instances, the organization will freely admit the intent to lie and define the acts as legitimate policing strategies. On face value, most would agree with the police that lies in this category are acceptable and necessary. However, a troubling and difficult question is "to what extent, if at all is it *proper* for law enforcement officials to employ trickery and deceit as part of their law enforcement practices" (Skolnick, 1982, italics added)? As we shall see, the answer to this question is not so easy. Acceptable lies may be very functional for the police but are they always proper, moral, ethical, and legal?

The most readily apparent patterns of "accepted" police lying are the deceptive practices that law enforcement officers believe are necessary to perform undercover operations or detect other forms of secret and consensual crimes. Police officers engaged in these activities must not only conceal their true identity but they must talk, act, and dress out of character, fabricating all kinds of stories in order to perform these duties. One could hardly imagine that FBI Special Agent Joseph Pistone could have operated for six years in the Mafia without the substantial number of lies he had to tell (Pistone, 1987). However, the overwhelming majority of undercover operations are not as glamorous nor as dangerous as working six years with the Mafia or other organized crime groups. The most common police undercover operations occur in routine vice operations dealing with prostitution, bootlegging, gambling, narcotics, bribery of public officials (e.g. ABSCAM, MILAB, BRILAB), and sting operations.

These deceptive practices in undercover operations are not only acceptable to the law enforcement community but considered necessary for undercover operations to be effective. Nevertheless, such activities are not without problems. The "Dirty Harry" problem in police work raises the question as to what extent morally good police practices warrant or justify ethically, politically, or legally suspect means to achieve law enforcement objectives (Klockars, 1980). Marx also raises the issue that many of the tactics used by law enforcement officers in such recent undercover operations as ABSCAM, MILAB, and BRILAB, police run fencing or sting operations and anti-crime decoy squads may have lost sight of "the profound difference between carrying out an investigation to determine if a suspect is, in fact, breaking the law, and carrying it out to determine if an individual can be induced to break the law" (Marx, 1985: 106). One Congressman involved in the ABSCAM case refused the first offer of a cash bribe to only later accept the money after federal agents, concluding that he was an alcoholic, gave him liquor (Marx, 1985: 104).

Encouraging the commission of a crime may be a legally accepted police practice when the officer acts as a willing victim or his/her actions facilitate the commission of a crime which was going to be committed in the first place. However, it is possible for "encouragement" to lead the suspect to raise the defense of entrapment. According to *Black's Law Dictionary* entrapment is "the act of officers or agents of the government in inducing a person to commit a crime not contemplated by him, for the purpose of instituting a criminal prosecution against him" (277). For the defense of entrapment to prevail the defendant must show that the officer or his/her agent has gone beyond providing the encouragement and opportunity for the commission of a crime and through trickery, fraud, or other deception has induced the suspect to commit a crime. This defense is raised far more times than it is successful because the current legal criterion to determine entrapment is what is known as the subjective test.

In the subjective test the predisposition of the offender rather than the objective methods of the police is the key factor in determining entrapment (Skolnick, 1982; Marx, 1985; Stitt and James, 1985). This makes it extremely difficult for a defendant with a criminal record to claim that he/she would not have committed the crime except for the actions of the officer. The "objective test" of entrapment raised by a minority of the Supreme Court has focused on the nature of the police conduct rather than the predisposition of the offender (Stitt and James, 1985). For example, the objective test probably would examine whether the production of crack by a police organization to use in undercover drug arrests is proper and legal. . . .

Police administrators are well aware of the possibility that the entire organization may be labeled deviant because of the deviant acts of its members. The "rotten apple theory" of police corruption has often been used as an impression management technique by police administrators who are aware of this possibility (Barker, 1977). It is easier to explain police deviance as a result of individual aberrations than admit the possibility of systemic problems and invite public scrutiny. However, candor and public scrutiny may be the best way to insure that corruption and other forms of police deviance do not occur or continue in an organization (see Cooper and Belair, 1978).

Thus, accepted lies are those which the organization views as having a viable role in police operations. The criteria for the lie to be accepted is:

It must be in furtherance of a legitimate organizational purpose.

There must be a clear relationship between the need to deceive and the accomplishment of an organizational purpose.

The nature of the deception must be one wherein officers and the management structure acknowledge that deception will better serve the public interest than the truth.

The ethical standing of the deception and the issue of law appear to be collateral concerns.

Tolerated Lying

A second category of police lies are those which are recognized as "lies" by the police organization but are tolerated as "necessary evils." Police administrators will admit to deception or "not exactly telling the whole truth" when confronted with the facts. These types of situational or "white" lies are truly in the gray area of propriety and the police can provide logical rationales for their use. When viewed from an ethical standpoint they may be "wrong" but from the police perspective they are necessary (i.e., tolerated) to achieve organizational objectives or deal with what Goldstein has termed the basic problems of police work (Goldstein, 1977: 9).

The basic problems of police work arise from the mythology surrounding police work; e.g., statutes usually require and the public expects the police to enforce all the laws all the time, the public holds the police responsible for preventing crime and apprehending all criminals, the public views the police as being capable of handling all emergencies, etc. (Goldstein, 1977). Most police administrators will not publicly admit that they do not have the resources, the training, or the authority to do some of the duties that the pubic expects. In fact, many police administrators and police officers lacking the education and insight into police work would be hard pressed to explain police work, particularly discretionary decision making to outside groups. They therefore resort to lies and deception to support police practices.

Police administrators often deny that their departments practice anything less than full enforcement of all laws rather than attempt to explain the basis for police discretionary decisions and selective enforcement. We continually attempt to deal with social problems through the use of criminal sanctions and law enforcement personnel. Mandatory sentencing for all offenders committing certain felony and misdemeanor offenses is often seen as a panacea for these offenses. For example, in recent years many politically active groups such as Mothers Against Drunk Drivers (M.A.D.D.) have pressured legislators for stronger laws with mandatory enforcement in drunk driving cases. However, their sentiment in cases not involving accidents may not be shared by the general public (Formby and Smykla, 1984) or the police. One can only speculate as to the number of discretionary decisions still being made by police officers in DUI offenses in departments where full enforcement is the official policy. One of the authors learned of an individual who had two DUI offenses reduced and asked a police supervisor about it.

> *Barker:* The chief said that all DUI suspects are charged and those over the legal blood alcohol level never have the charge reduced. In fact, he said this at a M.A.D.D. meeting. Yet, I heard that (——) had two DUI offenses reduced.
>
> *Supervisor:* That is true, Tom. However, (——) is helping us with some drug cases. M.A.D.D. may not understand but they do not have to make drug busts.

The point of note is that the police, in response to political pressure, make a policy on DUI cases and vow that the policy will be followed. However, in this case, that vow was not true. The police made a discretionary judgement that the assistance of the DUI offender in drug investigations was of greater importance than a DUI prosecution. Thus, this policy deviation was tantamount to a lie to the M.A.D.D. membership—a lie tolerated by the police department.

The public also expects the police to handle any disorderly or emergency situations. The American public believes that one of the methods for handling any problem is "calling the cops" (Bittner, 1972). However, in many of these order maintenance situations the police do not have the authority, resources, or training to deal with the problem. They often face a situation "where something must be done now" yet an arrest is not legally possible or would be more disruptive. The officer is forced to reach into a bag of tricks for a method of dealing with the crisis. Lying to the suspects or the complainants is often that method. For example, . . . in domestic disturbances police officers face volatile situations where the necessary conditions for an arrest often are not present. Frequently there is a misdemeanor where the officer does not have a warrant, an offense has not been committed in his/her presence, and the incident occurred in a private residence. However, the officer may feel that something must be done. Consequently, the officer may lie and threaten to arrest one or both combatants, or take one of the parties out to the street or the patrol car to discuss the incident and arrest them for disorderly conduct or public intoxication when they reach public property. Another option is to make an arrest and lie about the circumstances in order to make the arrest appear legal. Obviously, the latter strategy will not be a tolerated pattern of lying. It would fall into the pattern of deviant lying to be discussed later.

Officers soon learn that the interrogation stage of an arrest is an area where certain lies are tolerated and even taught to police officers. The now famous Miranda v. Arizona case decided by the U.S. Supreme Court in 1966 quoted excerpts from Inbau and Reid's *Criminal Interrogation and Confession* text to show that the police used deception and psychologically coercive methods in their interrogation of suspects (George, 1966: 155–266). The latest edition of this same text gives examples of deceptive and lying practices for . . . skilled interrogators to engage in (Inbau, Reid, Buckley, 1986).

As an illustration of these techniques . . . the book notes that an effective means to interrogate multiple suspects of a crime is "playing one offender against the other." In this regard it is suggested that the "interrogator may merely intimate to one offender that the other has confessed, or else the interrogator may actually *tell* the offender so" (emphasis added, p. 132).

This attitude, which is borne in the frustrations of many officers, sets a dangerous precedent for attitudes related to civil liberties. When law enforcement officers begin to tolerate lies because it serves their ends—regardless of the constitutional and ethical implications of those lies—then fundamental elements of civil rights are threatened.

Deviant Police Lies

The last example raises the possibility of the third category of police lying—deviant lies. After all "he (the suspect) can lie through his teeth. Why not us?" Deviant police lies are those which violate substantive or procedural law and/or police department rules and regulations. The deviant lies which violate substantive or procedural law are improper and should not be permitted. However, organization members (including supervisors), and other actors in the criminal justice system are often aware of their occurrence. Noted defense attorney Alan Dershowitz states that police lying is well known by actors in the criminal justice system. He clearly illustrates these as the "Rules of the Justice Game." In part, the rules include:

Rule IV: Almost all police lie about whether they violated the Constitution in order to convict guilty defendants.

Rule V: All prosecutors, judges, and defense attorneys are aware of Rule IV.

Rule VI: Many prosecutors implicitly encourage police to lie about whether they violated the Constitution in order to convict guilty defendants.

Rule VII: All judges are aware of Rule VI.

Rule VIII: Most trial judges pretend to believe police officers who they know are lying.

Rule IX: All appellate judges are aware of Rule VIII, yet many pretend to believe the trial judges who pretend to believe the lying police officers. (Dershowitz, 11983: xxi–xxii).

This may be an extreme position. However, other criminal defense attorneys believe that the police will lie in court. In fact, one study concluded that "the possibility of police perjury is a part of the working reality of criminal defense attorneys" (Kittel, 1986: 20). Fifty-seven percent of 277 attorneys surveyed in this study believed that police perjury takes place very often or often (Kittel, 1986: 16). Police officers themselves have reported that they believe their fellow officers will lie in court (Barker, 1978). An English barrister believes that police officers have perjured themselves on an average of three out of ten trials (Wolchover, 1986).

As part of the research for this chapter, one of the authors asked an Internal Affairs (IA) investigator of a major U.S. police department about officer lying:

Carter: During the course of IA investigations, do you detect officers lying to you?

IA Investigator: Yes, all the time. They'll lie about anything, everything.

Carter: Why is that?

IA Investigator: To tell me what I want to hear. To help get them out of trouble. To make themselves feel better—rationalizing I guess. They're so used to lying on the job, I guess it becomes second nature.

An analysis of deviant lies reveals that the intent of the officer in telling deviant lies may be either in support of perceived legitimate goals, or illegitimate goals.

Deviant Lies in Support of Perceived Legitimate Goals The deviant lies told by the officer to achieve perceived legitimate goals usually occur to "put criminals in jail," prevent crime, and perform various other policing responsibilities. The police officer believes that because of his/her unique experiences in dealing with criminals and the public he/she knows the guilt or innocence of those they arrest (Manning, 1978). Frequently, officers feel this way independently of any legal standards. However, the final determination of guilt or innocence is in the Judicial System. The officer(s), convinced that the suspect is factually guilty of the offense, may believe that necessary elements of legal guilt are lacking, e.g., no probable cause for a "stop," no Miranda warning, not enough narcotics for a felony offense, etc. Therefore, the officer feels that he/she must supply the missing elements. One police officer told one of the authors that it is often necessary to "fluff up the evidence" to get a search warrant or insure conviction. The officer will attest to facts, statements, or evidence which never occurred or occurred in a different fashion. Obviously when he/she does this under oath, perjury has then been committed. Once a matter of record, the perjury must continue for the officer to avoid facing disciplinary action and even criminal prosecution.

These lies are rationalized by the officer because they are necessary to ensure that criminals do not get off on "technicalities." A central reason for these deviant lies is officer frustration. There is frustration with the criminal justice system because of the inability of courts and corrections to handle large caseloads. Frustration with routinized practices of plea negotiations and intricate criminal procedures which the officer may not fully understand. The officer sees the victims of crimes and has difficulty in reconciling the harm done to them with the wide array of due process protections afforded to defendants. Nevertheless, the officer has fallen into "the avenging angel syndrome" where the end justifies the means. The officer can easily rationalize lying and perjury to accomplish what is perceived to be the "right thing." The officer's views are short-sighted and provincial. There is no recognition that such behavior is a threat to civil liberties and that perjury is as fundamentally improper as the criminal behavior of the accused.

Deviant Lies in Support of Illegitimate Goals Lies in this category are told to effect an act of corruption or to protect the officer from organizational discipline or civil and/or criminal liability. Deviant lies may be manifest in police perjury as the officer misrepresents material elements of an arrest or search in order to "fix" a criminal prosecution for a monetary reward. Lying and/or perjury in court is an absolute necessity in departments where corrupt acts occur on a regular basis. Sooner or later every police officer who engages in corrupt acts or observes corrupt acts on the part of other officers will face the possibility of having to lie under oath to protect him/herself or fellow officers.

Skolnick has suggested that perjury and corruption are both systematic forms of police deviance which occur for the same sort of reason: "police know that other police are on the take and police know that other police are perjuring themselves" (Skolnick, 1982: 42).

It is also possible that other forms of police deviance will lead to deviant lying. For example, the officer who commits an act of police brutality may have to lie on the report to his/her supervisor and during testimony to avoid the possibility of criminal sanction, a civil lawsuit, or department charges. The officer who has sex on duty, or sleeps or drinks on duty may have to lie to a supervisor or internal affairs to avoid department discipline. The officer who causes an injury or death to a suspect which is not strictly according to law or police policy may have to lie to protect himself or his fellow officers from criminal and/or civil liability.

Other implications . . . were clear: officers were willing to lie during the sworn deposition to protect themselves and others. They would swear to the truth of facts which were plainly manufactured for their protection. More-over, their remorse was not that they lied, but that they got caught in misconduct. Similarly, a police chief in West Virginia recently told a federal judge he lied to investigators in order to cover up for four officers accused of beating handcuffed prisoners (*Law Enforcement News,* March 15, 1989, p. 2). Again, the illegitimate goal of "protection" surfaces as a motive for lying.

The typical police bureaucracy is a complex organization with a myriad of rules and regulations. The informal organization, including many supervisors, overlooks these rules until someone decides to "nail someone." Given the plethora of rules and regulations in most large urban police departments it is virtually impossible to work a shift without violating one. It may be common practice to eat a free meal, leave one's beat for personal reasons, not wear one's hat when out of the car, to live outside the city limits, etc. All of these acts may be forbidden by a policy, rule, or regulation. When a supervisor decides to discipline an officer for violating one of these acts, the officer, and often fellow officers, may resort to lies to protect themselves and each other. After all, such "minor" lies are inherent in the "Blue Code." Manning observes that rule enforcement by police supervisors represents a mock bureaucracy where ritualistic and punitive enforcement is applied after the fact (Manning, 1978). The consequences of these seemingly understandable lies can be disastrous when discovered. The officer(s) may be suspended, reduced in rank, or dismissed. The same organization where members routinely engage in acceptable, tolerated, and deviant lying practices can take on a very moralistic attitude when it discovers that one of its own has told a lie to avoid internal discipline. Nevertheless, the lies told in these examples are told in support of the illegitimate goal of avoiding departmental discipline.

CONCLUSION

The effects of lying, even those which are acceptable or tolerated, are multi-fold. Lies can and do create distrust within the organization. When the public learns members of the police lie or engage in deceptive practices this can undermine citizen confidence in the police. As we have seen some police lies violate citizens' civil rights and others are told to cover up civil rights violations. Police lying contributes to police misconduct and corruption and undermines the organization's discipline system. Furthermore, deviant police lies undermine the effectiveness of the criminal justice system.

REFERENCES

Barker, T. 1977. "Peer Group Support for Police Organizational Deviance." *Criminology* 15(3): 353–366.

———. 1978. "An Empirical Study of Police Deviance Other Than Corruption." *Journal of Police Science and Administration.* 6(3): 264–272.

Bittner, E. 1972. *The Functions of the Police in Modern Society.* Washington, DC: U.S. Government Printing Office.

Birmingham Post-Herald. 1989. "Sheriff's Chemist Makes Crack." April 19: B2.

Black, H. C. 1983. *Black's Law Dictionary,* Abridged Fifth Edition. St. Paul, MN: West Publishing Co.

Cooper, G. R., and Robert R. Belair. 1978. *Privacy and Security of Criminal History Information: Privacy and the Media.* U.S. Department of Justice. Washington, DC: U.S. Government Printing Office.

Dershowitz, A. M. 1983. *The Best Defense.* New York: Vintage Books.

Formby, W. A., and John O. Smykla. 1984. "Attitudes and Perception Toward Drinking and Driving: A Simulation of Citizen Awareness." *Journal of Police Science and Administration.* 12(4): 379–384.

George, J. B. 1966. *Constitutional Limitations on Evidence in Criminal Cases.* Ann Arbor, MI: Institute of Continuing Legal Education.

Goldstein, H. 1977. *Policing a Free Society.* Cambridge, MA: Ballenger Publishing Company.

Inbau, F. E., J. E. Reid, and Joseph P. Buckley. 1986. *Criminal Interrogation and Confessions,* 3rd ed. Baltimore: Williams and Wilkins.

Kittel, N. G. 1986. "Police Perjury: Criminal Defense Attorneys' Perspective." *American Journal of Criminal Justice* XI(1): 11–22.

Klockars, C. B. 1980. "The Dirty Harry Problem." *The Annals,* 452: 33–47 (November).

Law Enforcement News. 1989. March 15: 2

Manning, P. K. 1978. "Lying, Secrecy and Social Control." In Peter K. Manning and John Van Maanen (Eds.), *Policing: A View from the Street.* Santa Monica, CA: Goodyear Publishing Co., 238–255.

Marx, G. T. 1985. "Who Really Gets Stung? Some Issues Raised by the New Police Undercover Work." In F. A. Elliston and Michael Feldberg (Eds.), *Moral Issues in Police Work.* Totowa, NJ: Rowman and Allanheld.

New York Times. 1989. "Dead Officer, Dropped Charges: A Scandal in Boston." March 20: K9.

Pistone, J. D. 1987. *Donnie Brasco: My Undercover Life in the Mafia.* New York: Nail Books.

Skolnick, J. 1982. "Deception by Police." *Criminal Justice Ethics.* 1(2): 40–54.

Stitt, B. G., and Gene G. James. 1985. "Entrapment: An Ethical Analysis." In F. A. Elliston and Michael Feldberg (Eds.), *Moral Issues in Police Work.* Totowa, NJ: Rowman and Allanheld.

Wolchover, D. 1986. "Police Perjury in London." *New Law Journal:* 180–184 (February).

16

State-Organized Crime

WILLIAM J. CHAMBLISS

STATE-ORGANIZED CRIME DEFINED

The most important type of criminality organized by the state consists of acts defined by law as criminal and committed by state officials in the pursuit of their job as representatives of the state. Examples include a state's complicity in piracy, smuggling, assassinations, criminal conspiracies, acting as an accessory before or after the fact, and violating laws that limit their activities. In the latter category would be included the use of illegal methods of spying on citizens, diverting funds in ways prohibited by law (e.g., illegal campaign contributions, selling arms to countries prohibited by law, and supporting terrorist activities).

State-organized crime does not include criminal acts that benefit only individual officeholders, such as the acceptance of bribes or the illegal use of violence by the police against individuals, unless such acts violate existing criminal law and are official policy. For example, the current policies of torture and random violence by the police in South Africa are incorporated under the category of state-organized crime because, apparently, those practices are both state policy and in violation of existing South African law. On the other hand, the excessive use of violence by the police in urban ghettoes is not state-organized crime for it lacks the necessary institutionalized policy of the state. . . .

SMUGGLING

Smuggling occurs when a government has successfully cornered the market on some commodity or when it seeks to keep a commodity of another nation from crossing its borders. In the annals of crime, everything from sheep to people, wool to wine, gold to drugs, and even ideas, has been prohibited for either export or import. Paradoxically, whatever is prohibited, it is at the expense of one group of people for the benefit of another. Thus, the laws that prohibit the import or export of a commodity inevitably face a built-in resistance. Some part of the population will always want to either possess or to dis-

From "State-Organized Crime," William J. Chambliss, *Criminology* 27(2), 1989.
Reprinted by permission of the American Society of Criminology and the author.

tribute the prohibited goods. At times, the state finds itself in the position of having its own interests served by violating precisely the same laws passed to prohibit the export or import of the goods it has defined as illegal.

Narcotics and the Vietnam War

Sometime around the eighth century, Turkish traders discovered a market for opium in Southeast Asia (Chambliss, 1977; McCoy, 1973). Portuguese traders several centuries later found a thriving business in opium trafficking conducted by small ships sailing between trading ports in the area. One of the prizes of Portuguese piracy was the opium that was taken from local traders and exchanged for tea, spices, and pottery. Several centuries later, when the French colonized Indochina, the traffic in opium was a thriving business. The French joined the drug traffickers and licensed opium dens throughout Indochina. With the profits from those licenses, the French supported 50% of the cost of their colonial government (McCoy, 1973: 27).

When the Communists began threatening French rule in Indochina, the French government used the opium profits to finance the war. It also used cooperation with the hill tribes who controlled opium production as a means of ensuring the allegiance of the hill tribes in the war against the Communists (McCoy, 1973).

The French were defeated in Vietnam and withdrew, only to be replaced by the United States. The United States inherited the dependence on opium profits and the cooperation of the hill tribes, who in turn depended on being allowed to continue growing and shipping opium. The CIA went a step further than the French and provided the opium-growing feudal lords in the mountains of Vietnam, Laos, Cambodia, and Thailand with transportation for their opium via Air America, the CIA airline in Vietnam.

Air America regularly transported bundles of opium from airstrips in Laos, Cambodia, and Burma to Saigon and Hong Kong (Chambliss, 1977: 56). An American stationed at Long Cheng, the secret CIA military base in northern Laos during the war, observed:

> . . . so long as the Meo leadership could keep their wards in the boondocks fighting and dying in the name of, for these unfortunates anyway, some nebulous cause . . . the Meo leadership [was paid off] in the form of a carte-blanche to exploit U.S.-supplied airplanes and communication gear to the end of greatly streamlining the opium operations. . . . (Chambliss, 1977: 56)

This report was confirmed by Laotian Army General Ouane Rattikone, who told me in an interview in 1974 that he was the principal overseer of the shipment of opium out of the Golden Triangle via Air America. U.S. law did not permit the CIA or any of its agents to engage in the smuggling of opium.

After France withdrew from Vietnam and left the protection of democracy to the United States, the French intelligence service that preceded the

CIA in managing the opium smuggling in Asia continued to support part of its clandestine operations through drug trafficking (Kruger, 1980). Although those operations are shrouded in secrecy, the evidence is very strong that the French intelligence agencies helped to organize the movement of opium through the Middle East (especially Morocco) after their revenue from opium from Southeast Asia was cut off.

In 1969 Michael Hand, a former Green Beret and one of the CIA agents stationed at Long Cheng when Air America was shipping opium, moved to Australia, ostensibly as a private citizen. On arriving in Australia, Hand entered into a business partnership with an Australian national, Frank Nugan. In 1976 they established the Nugan Hand Bank in Sydney (Commonwealth of New South Wales, 1982a, 1982b). The Nugan Hand Bank began as a storefront operation with minimal capital investment, but almost immediately it boasted deposits of over $25 million. The rapid growth of the bank resulted from large deposits of secret funds made by narcotics and arms smugglers and large deposits from the CIA (Nihill, 1982).

In addition to the records from the bank that suggest the CIA was using the bank as a conduit for its funds, the bank's connection to the CIA and other U.S. intelligence agencies is evidenced by the people who formed the directors and principal officers of the bank, including the following:

- Admiral Earl F. Yates, president of the Nugan Hand Bank was, during the Vietnam War, chief of staff for strategic planning of U.S. forces in Asia and the Pacific.

- General Edwin F. Black, president of Nugan Hand's Hawaii branch, was commander of U.S. troops in Thailand during the Vietnam War and, after the war, assistant army chief of staff for the Pacific.

- General Erle Cocke, Jr., head of the Nugan Hand Washington, D.C., office.

- George Farris, worked in the Nugan Hand Hong Kong and Washington, D.C. offices. Farris was a military intelligence specialist who worked in a special forces training base in the Pacific.

- Bernie Houghton, Nugan Hand's representative in Saudi Arabia. Houghton was also a U.S. naval intelligence undercover agent.

- Thomas Clines, director of training in the CIA's clandestine service, was a London operative for Nugan Hand who helped in the takeover of a London-based bank and was stationed at Long Cheng with Michael Hand and Theodore S. Shackley during the Vietnam War.

- Dale Holmgreen, former flight service manager in Vietnam for Civil Air Transport, which became Air America. He was on the board of directors of Nugan Hand and ran the bank's Taiwan office.

- Walter McDonald, an economist and former deputy director of CIA for economic research, was a specialist in petroleum. He became a consultant to Nugan Hand and served as head of its Annapolis, Maryland, branch.

- General Roy Manor, who ran the Nugan Hand Philippine office, was a Vietnam veteran who helped coordinate the aborted attempt to rescue the Iranian hostages, chief of staff for the U.S. Pacific command, and the U.S. government's liaison officer to Philippine President Ferdinand Marcos.

On the board of directors of the parent company formed by Michael Hand that preceded the Nugan Hand Bank were Grant Walters, Robert Peterson, David M. Houton, and Spencer Smith, all of whom listed their address as c/o Air America, Army Post Office, San Francisco, California.

Also working through the Nugan Hand Bank was Edwin F. Wilson, a CIA agent involved in smuggling arms to the Middle East and later sentenced to prison by a U.S. court for smuggling illegal arms to Libya. Edwin Wilson's associate in Mideast arms shipments was Theodore Shackley, head of the Miami, Florida, CIA station.* In 1973, when William Colby was made director of Central Intelligence, Shackley replaced him as head of covert operations for the Far East; on his retirement from the CIA William Colby became Nugan Hand's lawyer.

In the late 1970s the bank experienced financial difficulties, which led to the death of Frank Nugan. He was found dead of a shotgun blast in his Mercedes Benz on a remote road outside Sydney. The official explanation was suicide, but some investigators speculated that he might have been murdered. In any event, Nugan's death created a major banking scandal and culminated in a government investigation. The investigation revealed that millions of dollars were unaccounted for in the bank's records and that the bank was serving as a money-laundering operation for narcotics smugglers and as a conduit through which the CIA was financing gun smuggling and other illegal operations throughout the world. These operations included illegally smuggling arms to South Africa and the Middle East. There was also evidence that the CIA used the Nugan Hand Bank to pay for political campaigns that slandered politicians, including Australia's Prime Minister Witham (Kwitny, 1987).

Michael Hand tried desperately to cover up the operations of the bank. Hundreds of documents were destroyed before investigators could get into the bank. Despite Hand's efforts, the scandal mushroomed and eventually Hand was forced to flee Australia. He managed this, while under indictment for a rash of felonies, with the aid of a CIA official who flew to Australia with a false passport and accompanied him out of the country. Hand's father, who lives in New York, denies knowing anything about his son's whereabouts.

Thus, the evidence uncovered by the government investigation in Australia linked high-level CIA officials to a bank in Sydney that was responsible for financing and laundering money for a significant part of the narcotics trafficking originating in Southeast Asia (Commonwealth of New South Wales,

---

*It was Shackley who, along with Rafael "Chi Chi" Quintero, a Cuban-American, forged the plot to assassinate Fidel Castro by using organized-crime figures Santo Trafficatnte, Jr., and Sam Giancana.

1982b; Owen, 1983). It also linked the CIA to arms smuggling and illegal involvement in the democratic processes of a friendly nation. Other investigations reveal that the events in Australia were but part of a worldwide involvement in narcotics and arms smuggling by the CIA and French intelligence (Hougan, 1978; Kruger, 1980; Owen, 1983).

Arms Smuggling

One of the most important forms of state-organized crime today is arms smuggling. To a significant extent, U.S. involvement in narcotics smuggling after the Vietnam War can be understood as a means of funding the purchase of military weapons for nations and insurgent groups that could not be funded legally through congressional allocations or for which U.S. law prohibited support (NARMIC, 1984).

In violation of U.S. law, members of the National Security Council (NSC), the Department of Defense, and the CIA carried out a plan to sell millions of dollars worth of arms to Iran and use profits from those sales to support the Contras in Nicaragua (Senate Hearings, 1986). The Boland amendment, effective in 1985, prohibited any U.S. official from directly or indirectly assisting the Contras. To circumvent the law, a group of intelligence and military officials established a "secret team" of U.S. operatives, including Lt. Colonel Oliver North, Theodore Shackley, Thomas Clines, and Maj. General Richard Secord, among others (testimony before U.S. Senate, 1986). Shackley and Clines, as noted, were CIA agents in Long Cheng; along with Michael Hand they ran the secret war in Laos, which was financed in part from profits from opium smuggling. Shackley and Clines had also been involved in the 1961 invasion of Cuba and were instrumental in hiring organized-crime figures in an attempt to assassinate Fidel Castro.

Senator Daniel Inouye of Hawaii claims that this "secret government within our government" waging war in Third World countries was part of the Reagan doctrine (the *Guardian,* July 29, 1987). Whether President Reagan or then Vice President Bush was aware of the operations is yet to be established. What cannot be doubted in the face of overwhelming evidence in testimony before the Senate and from court documents is that this group of officials of the state oversaw and coordinated the distribution and sale of weapons to Iran and to the Contras in Nicaragua. These acts were in direct violation of the Illegal Arms Export Control Act, which made the sale of arms to Iran unlawful, and the Boland amendment, which made it a criminal act to supply the Contras with arms or funds.

The weapons that were sold to Iran were obtained by the CIA through the Pentagon. Secretary of Defense Caspar Weinberger ordered the transfer of weapons from Army stocks to the CIA without the knowledge of Congress four times in 1986. The arms were then transferred to middlemen, such as Iranian arms dealer Yaacov Nimrodi, exiled Iranian arms dealer Manucher Ghorbanifar, and Saudi Arabian businessman Adnan Khashoggi. Weapons were also flown directly to the Contras, and funds from the sale of weapons were diverted to support Contra warfare. There is also considerable evidence that this "secret team," along with other military and CIA officials, cooperated with narcotics smuggling in Latin America in order to fund the Contras in Nicaragua.

In 1986, the Reagan administration admitted that Adolfo Chamorro's Contra group, which was supported by the CIA, was helping a Colombian drug trafficker transport drugs into the United States. Chamorro was arrested in April 1986 for his involvement (Potter and Bullington, 1987: 54). Testimony in several trials of major drug traffickers in the past 5 years has revealed innumerable instances in which drugs were flown from Central America into the United States with the cooperation of military and CIA personnel. These reports have also been confirmed by military personnel and private citizens who testified that they saw drugs being loaded on planes in Central America and unloaded at military bases in the United States. Pilots who flew planes with arms to the Contras report returning with planes carrying drugs.

At the same time that the United States was illegally supplying the Nicaraguan Contras with arms purchased, at least in part, with profits from the sale of illegal drugs, the administration launched a campaign against the Sandanistas for their alleged involvement in drug trafficking. Twice during his weekly radio shows in 1986, President Reagan accused the Sandanistas of smuggling drugs. Barry Seal, an informant and pilot for the Drug Enforcement Administration (DEA), was ordered by members of the CIA and DEA to photograph the Sandanistas loading a plane. During a televised speech in March 1986, Reagan showed the picture that Seal took and said that it showed Sandanista officials loading a plane with drugs for shipment to the United States. After the photo was displayed, Congress appropriated $100 million in aid for the Contras. Seal later admitted to reporters that the photograph he took was a plane being loaded with crates that did not contain drugs. He also told reporters that he was aware of the drug smuggling activities of the Contra network and a Colombian cocaine syndicate. For his candor, Seal was murdered in February 1987. Shortly after his murder, the DEA issued a "low key clarification" regarding the validity of the photograph, admitting that there was no evidence that the plane was being loaded with drugs.

Other testimony linking the CIA and U.S. military officials to complicity in drug trafficking includes the testimony of John Stockwell, a former high-ranking CIA official, who claims that drug smuggling and the CIA were essential components in the private campaign for the Contras. Corroboration for these assertions comes also from George Morales, one of the largest drug traffickers in South America, who testified that he was approached by the CIA in 1984 to fly weapons into Nicaragua. Morales claims that the CIA opened up an airstrip in Costa Rica and gave the pilots information on how to avoid radar traps. According to Morales, he flew 20 shipments of weapons into Costa Rica in 1984 and 1985. In return, the CIA helped him to smuggle thousands of kilos of cocaine into the United States. Morales alone channeled $250,000 quarterly to Contra leader Adolfo Chamorro from his trafficking activity. A pilot for Morales, Gary Betzner, substantiated Morales's claims and admitted flying 4,000 pounds of arms into Costa Rica and 500 kilos of cocaine to Lakeland, Florida, on his return trips. From 1985 to 1987, the CIA arranged 50 to 100 flights using U.S. airports that did not undergo inspection.

The destination of the flights by Morales and Betzner was a hidden airstrip on the ranch of John Hull. Hull, an admitted CIA agent, was a primary player in Oliver North's plan to aid the Contras. Hull's activities were closely monitored by Robert Owen, a key player in the Contra supply network. Owen established the Institute for Democracy, Education, and Assistance, which raised money to buy arms for the Contras and which, in October 1985, was asked by Congress to distribute $50,000 in "humanitarian aid" to the Contras. Owen worked for Oliver North in coordinating illegal aid to the Contras and setting up the airstrip on the ranch of John Hull.

According to an article in the *Nation,* Oliver North's network of operatives and mercenaries had been linked to the largest drug cartel in South America since 1983. The DEA estimates that Colombian Jorge Ochoa Vasquez, the "kingpin" of the Medellin drug empire, is responsible for supplying 70% to 80% of the cocaine that enters the United States every year. Ochoa was taken into custody by Spanish police in October 1984 when a verbal order was sent by the U.S. Embassy in Madrid for his arrest. The embassy specified that Officer Cos-Gayon, who had undergone training with the DEA, should make the arrest. Other members of the Madrid Judicial Police were connected to the DEA and North's arms smuggling network. Ochoa's lawyers informed him that the United States would alter his extradition if he agreed to implicate the Sandanista government in drug trafficking. Ochoa refused and spent 20 months in jail before returning to Colombia. The Spanish courts ruled that the United States was trying to use Ochoa to discredit Nicaragua and released him (the *Nation,* September 5, 1987).

There are other links between the U.S. government and the Medellin cartel. Jose Blandon, General Noriega's former chief advisor, claims that DEA operations have protected the drug empire in the past and that the DEA paid Noriega $4.7 million for his silence. Blandon also testified in Senate committee hearings that Panama's bases were used as training camps for the Contras in exchange for "economic" support from the United States. Finally, Blandon contends that the CIA gave Panamanian leaders intelligence documents about U.S. senators and aides; the CIA denies these charges (the *Christian Science Monitor,* February 11, 1988: 3).

Other evidence of the interrelationship among drug trafficking, the CIA, the NSC, and aid to the Contras includes the following:

- In January 1983, two Contra leaders in Costa Rica persuaded the Justice Department to return over $36,000 in drug profits to drug dealers Julio Zavala and Carlos Cabezas for aid to the Contras (Potter and Bullington, 1987: 22).

- Michael Palmer, a drug dealer in Miami, testified that the State Department's Nicaraguan humanitarian assistance office contracted with his company, Vortex Sales and Leasing, to take humanitarian aid to the Contras. Palmer claims that he smuggled $40 million in marijuana to the United States between 1977 and 1985 (the *Guardian*, March 20, 1988: 3).

- During House and Senate hearings in 1986, it was revealed that a major DEA investigation of the Medellin drug cartel of Colombia, which was expected to culminate in the arrest of several leaders of the cartel, was compromised when someone in the White House leaked the story of the investigation to the *Washington Times* (a conservative newspaper in Washington, D.C.), which published the story on July 17, 1984. According to DEA Administrator John Lawn, the leak destroyed what was "probably one of the most significant operations in DEA history" (Sharkey, 1988: 24).

- When Honduran General Jose Buseo, who was described by the Justice Department as an "international terrorist," was indicted for conspiring to murder the president of Honduras in a plot financed by profits from cocaine smuggling, Oliver North and officials from the Department of Defense and the CIA pressured the Justice Department to be lenient with General Buseo. In a memo disclosed by the Iran-Contra committee, North stated that if Buseo was not protected "he will break his long-standing silence about the Nic[araguan] resistance and other sensitive operations" (Sharkey, 1988: 27).

On first blush, it seems odd that government agencies and officials would engage in such wholesale disregard of the law. As a first step in building an explanation for these and other forms of state-organized crime, let us try to understand why officials of the CIA, the NSC, and the Department of Defense would be willing to commit criminal acts in pursuit of other goals.

WHY?

Why would government officials from the NSC, the Defense Department, the State Department, and the CIA become involved in smuggling arms and narcotics, money laundering, assassinations, and other criminal activities? The answer lies in the structural contradictions that inhere in nation-states (Chambliss, 1980).

As Weber, Marx, and Gramsci pointed out, no state can survive without establishing legitimacy. The law is a fundamental cornerstone in creating legitimacy and an illusion (at least) of social order. It claims universal principles that demand some behaviors and prohibit others. The protection of property and personal security are obligations assumed by states everywhere both as a means of legitimizing the state's franchise on violence and as a means of protecting commercial interests (Chambliss and Seidman, 1982).

The threat posed by smuggling to both personal security and property interests makes laws prohibiting smuggling essential. Under some circumstances, however, such laws contradict other interests of the state. This contradiction prepares the ground for state-organized crime as a solution to the conflicts and dilemmas posed by the simultaneous existence of contradictory "legitimate" goals.

The military-intelligence establishment in the United States is resolutely committed to fighting the spread of "communism" throughout the world. This mission is not new but has prevailed since the 1800s. Congress and the presidency are not consistent in their support for the money and policies thought by the front-line warriors to be necessary to accomplish their lofty goals. As a result, programs under way are sometimes undermined by a lack of funding and even by laws that prohibit their continuation (such as the passage of laws prohibiting support for the Contras). Officials of government agencies adversely affected by political changes are thus placed squarely in a dilemma: If they comply with the legal limitations on their activities they sacrifice their mission. The dilemma is heightened by the fact that they can anticipate future policy changes that will reinstate their resources and their freedom. When that time comes, however, programs adversely affected will be difficult if not impossible to re-create.

A number of events that occurred between 1960 and 1980 left the military and the CIA with badly tarnished images. Those events and political changes underscored their vulnerability. The CIA lost considerable political clout with elected officials when its planned invasion of Cuba (the infamous Bay of Pigs invasion) was a complete disaster. Perhaps as never before in its history, the United States showed itself vulnerable to the resistance of a small nation. The CIA was blamed for this fiasco even though it was President Kennedy's decision to go ahead with the plans that he inherited from the previous administration. To add to the agency's problems, the complicity between it and ITT to invade Chile and overthrow the Allende government was yet another scar (see below), as was the involvement of the CIA in narcotics smuggling in Vietnam.

These and other political realities led to a serious breach between Presidents Kennedy, Johnson, Nixon, and Carter and the CIA. During President Nixon's tenure in the White House, one of the CIA's top men, James Angleton, referred to Nixon's national security advisor, Henry Kissinger (who became secretary of state), as "objectively, a Soviet Agent" (Hougan, 1984: 75). Another top agent of the CIA, James McCord (later implicated in the Watergate burglary), wrote a secret letter to his superior, General Paul Gaynor, in January 1973 in which he said:

> When the hundreds of dedicated fine men and women of the CIA no longer write intelligence summaries and reports with integrity, without fear of political recrimination—when their fine Director [Richard Helms] is being summarily discharged in order to make way for a politician who will write or rewrite intelligence the way the politicians want them written, instead of the way truth and best judgment dictates, our nation is in the deepest of trouble and freedom itself was never so imperiled. Nazi Germany rose and fell under exactly the same philosophy of governmental operation. (Hougan, 1984: 26–27)

McCord (1974: 60) spoke for many of the top military and intelligence officers in the United States when he wrote in his autobiography: "I believed that the whole future of the nation was at stake." These views show the depth of feeling toward the dangers of political "interference" with what is generally accepted in the military-intelligence establishment as their mission (Goulden, 1984).

When Jimmy Carter was elected president, he appointed Admiral Stansfield Turner as director of Central Intelligence. At the outset, Turner made it clear that he and the president did not share the agency's view that they were conducting their mission properly (Goulden, 1984; Turner, 1985). Turner insisted on centralizing power in the director's office and on overseeing clandestine and covert operations. He met with a great deal of resistance. Against considerable opposition from within the agency, he reduced the size of the covert operation section from 1,200 to 400 agents. Agency people still refer to this as the "Halloween massacre."

Old hands at the CIA do not think their work is dispensable. They believe zealously, protectively, and one is tempted to say, with religious fervor, that the work they are doing is essential for the salvation of humankind. With threats from both Republican and Democratic administrations, the agency sought alternative sources of revenue to carry out its mission. The alternative was already in place with the connections to the international narcotics traffic, arms smuggling, the existence of secret corporations incorporated in foreign countries (such as Panama), and the established links to banks for the laundering of money for covert operations.

STATE-ORGANIZED
ASSASSINATIONS AND MURDER

Assassination plots and political murders are usually associated in people's minds with military dictatorships and European monarchies. The practice of assassination, however, is not limited to unique historical events but has become a tool of international politics that involves modern nation-states of many different types.

In the 1960s a French intelligence agency hired Christian David to assassinate the Moroccan leader Ben Barka (Hougan, 1978: 204–207). Christian David was one of those international "spooks" with connections to the DEA, the CIA, and international arms smugglers, such as Robert Vesco.

In 1953, the CIA organized and supervised a coup d'etat in Iran that overthrew the democratically elected government of Mohammed Mossadegh, who had become unpopular with the United States when he nationalized foreign-owned oil companies. The CIA's coup replaced Mossadegh with Reza Shah Pahlevi, who denationalized the oil companies and with CIA guidance established one of the most vicious secret intelligence organizations in the world: SAVAK. In the years to follow, the shah and CIA-trained agents of SAVAK murdered thousands of Iranian citizens. They arrested almost 1,500 people monthly, most of whom were subjected to inhuman torture and punishments without trial. Not only were SAVAK agents trained by the CIA, but there is evidence that they were instructed in techniques of torture (Hersh, 1979: 13).

In 1970 the CIA repeated the practice of overthrowing democratically elected governments that were not completely favorable to U.S. investments. When Salvador Allende was elected president of Chile, the CIA organized a coup that overthrew Allende, during which he was murdered, along with the head of the military, General Rene Schneider. Following Allende's overthrow, the CIA trained agents for the Chilean secret service (DINA). DINA set up a team of assassins who could "travel anywhere in the world . . . to carry out sanctions including assassinations" (Dinges and Landau, 1980: 239). One of the assassinations carried out by DINA was the murder of Orlando Letellier, Allende's ambassador to the United States and his former minister of defense. Letellier was killed when a car bomb blew up his car on Embassy Row in Washington, D.C. (Dinges and Landau, 1982).

Other bloody coups known to have been planned, organized, and executed by U.S. agents include coups in Guatemala, Nicaragua, the Dominican Republic, and Vietnam. American involvement in those coups was never legally authorized. The murders, assassinations, and terrorist acts that accompany coups are criminal acts by law, both in the United States and in the country in which they take place.

More recent examples of murder and assassination for which government officials are responsible include the death of 80 people in Beirut, Lebanon, when a car bomb exploded on May 8, 1985. The bomb was set by a Lebanese counterterrorist unit working with the CIA. Senator Daniel Moynihan has said that when he was vice president of the Senate Intelligence Committee, President Reagan ordered the CIA to form a small antiterrorist effort in the Mideast. Two sources said that the CIA was working with the group that planted the bomb to kill the Shiite leader Hussein Fadallah (the *New York Times,* May 13, 1985).

A host of terrorist plans and activities connected with the attempt to overthrow the Nicaraguan government, including several murders and assassinations, were exposed in an affidavit filed by free-lance reporters Tony Avirgan and Martha Honey. They began investigating Contra activities after Avirgan was injured in an attempt on the life of Contra leader Eden Pastora. In 1986, Honey and Avirgan filed a complaint with the U.S. District Court in Miami charging John Hull, Robert Owen, Theodore Shackley, Thomas Clines, Chi Chi Quintero, Maj. General Richard Secord, and others working for the CIA in Central America with criminal conspiracy and the smuggling of cocaine to aid the Nicaraguan rebels.

A criminal conspiracy in which the CIA admits participating is the publication of a manual, *Psychological Operation in Guerilla Warfare,* which was distributed to the people in Nicaragua. The manual describes how the people should proceed to commit murder, sabotage, vandalism, and violent acts in order to undermine the government. Encouraging or instigating such crimes is not only a violation of U.S. law, it was also prohibited by Reagan's executive order of 1981, which forbade any U.S. participation in foreign assassinations.

The CIA is not alone in hatching criminal conspiracies. The DEA organized a "Special Operations Group," which was responsible for working out plans to assassinate political and business leaders in foreign countries who were involved in drug trafficking. The head of this group was a former CIA agent, Lou Conein (also known as "Black Luigi"). George Crile wrote in the *Washington Post* (June 13, 1976):

> When you get down to it, Conein was organizing an assassination program. He was frustrated by the big-time operators who were just too insulated to get to. . . . Meetings were held to decide whom to target and what method of assassination to employ.

Crile's findings were also supported by the investigative journalist Jim Hougan (1978: 132).

It is a crime to conspire to commit murder. The official record, including testimony by participants in three conspiracies before the U.S. Congress and in court, make it abundantly clear that the crime of conspiring to commit murder is not infrequent in the intelligence agencies of the United States and other countries.

It is also a crime to cover up criminal acts, but there are innumerable examples of instances in which the CIA and the FBI conspired to interfere with the criminal prosecution of drug dealers, murderers, and assassins. In the death of Letellier, mentioned earlier, the FBI and the CIA refused to cooperate with the prosecution of the DINA agents who murdered Letellier (Dinges and Landau, 1980: 208–209). Those agencies were also involved in the cover-up of the criminal activities of a Cuban exile, Ricardo (Monkey) Morales. While an employee of the FBI and the CIA, Morales planted a bomb on an Air Cubana flight from Venezuela, which killed 73 people. The Miami police confirmed Morales's claim that he was acting under orders from the CIA (Lernoux, 1984: 188). In fact, Morales, who was arrested for overseeing the shipment of 10 tons of marijuana, admitted to being a CIA contract agent who conducted bombings, murders, and assassinations. He was himself killed in a bar after he made public his work with the CIA and the FBI.

Colonel Muammar Qaddafi, like Fidel Castro, has been the target of a number of assassination attempts and conspiracies by the U.S. government. One plot, the *Washington Post* reported, included an effort to "lure [Qaddafi] into some foreign adventure or terrorist exploit that would give a growing number of Qaddafi opponents in the Libyan military a chance to seize power, or such a foreign adventure might give one of Qaddafi's neighbors, such as Algeria or Egypt, a justification for responding to Qaddafi militarily" (the *Washington Post*, April 14, 1986). The CIA recommended "stimulating" Qaddafi's fall "by encouraging disaffected elements in the Libyan army who could be spurred to assassination attempts" (the *Guardian*, November 20, 1985: 6).

Opposition to government policies can be a very risky business, as the ecology group Greenpeace discovered when it opposed French nuclear testing in the Pacific. In the fall of 1985 the French government planned a series of atomic tests in the South Pacific. Greenpeace sent its flagship to New Zealand with instructions to sail into the area where the atomic testing was scheduled to occur. Before the ship could arrive at the scene, however, the French secret service located the ship in the harbor and blew it up. The blast from the bomb killed one of the crew.

OTHER STATE-ORGANIZED CRIMES

Every agency of government is restricted by law in certain fundamental ways. Yet structural pressures exist that can push agencies to go beyond their legal limits. The CIA, for example, is not permitted to engage in domestic intelligence. Despite this, the CIA has opened and photographed the mail of over 1 million private citizens (Rockefeller Report, 1975: 101–115), illegally entered people's homes, and conducted domestic surveillance through electronic devices (Parenti, 1983: 170–171).

Agencies of the government also cannot legally conduct experiments on human subjects that violate civil rights or endanger the lives of the subjects. But the CIA conducted experiments on unknowing subjects by hiring prostitutes to administer drugs to their clients. CIA-trained medical doctors and psychologists observed the effects of the drugs through a two-way mirror in expensive apartments furnished to the prostitutes by the CIA. At least one of the victims of these experiments died and others suffered considerable trauma (Anderson and Whitten, 1976; Crewdson and Thomas, 1977; Jacobs, 1977a, 1977b).

The most flagrant violation of civil rights by federal agencies is the FBI's counterintelligence program, known as COINTELPRO. This program was designed to disrupt, harass, and discredit groups that the FBI decided were in some way "un-American." Such groups included the American Civil Liberties Union, antiwar movements, civil rights organizations, and a host of other legally constituted political groups whose views opposed some of the policies of the United States (Church Committee, 1976). With the exposure of COINTELPRO, the group was disbanded. There is evidence, however, that the illegal surveillance of U.S. citizens did not stop with the abolition of COINTELPRO but continues today (Klein, 1988). . . .

CONCLUSION

. . . We need to explore different political, economic, and social systems in varying historical periods to discover why some forms of social organization are more likely to create state-organized crimes than others. We need to explore the possibility that some types of state agencies are more prone to engaging in criminality than others. It seems likely, for example, that state agencies whose activities can be hidden from scrutiny are more likely to engage in criminal acts than those whose record is public. This principle may also apply to whole nation-states: the more open the society, the less likely it is that state-organized crime will become institutionalized.

There are also important parallels between state-organized criminality and the criminality of police and law enforcement agencies generally. Local police departments that find it more useful to cooperate with criminal syndicates than to combat them are responding to their own particular contradictions, conflicts, and dilemmas (Chambliss, 1988). An exploration of the theoretical implications of these similarities could yield some important findings.

The issue of state-organized crime raises again the question of how crime should be defined to be scientifically useful. For the purposes of this analysis, I have accepted the conventional criminological definition of crime as acts that are in violation of the criminal law. This definition has obvious limitations (see Schwendinger and Schwendinger, 1975), and the study of state-organized crime may facilitate the development of a more useful definition by underlying the interrelationship between crime and the legal process. At the very least, the study of state-organized crime serves as a reminder that crime is a political phenomenon and must be analyzed accordingly.

REFERENCES

Anderson, Jack, and Lee Whitten. 1976. The CIA's "sex squad." *Washington Post,* June 22: B13.

Chambliss, William J. 1977. Markets, profits, labor and smack. *Contemporary Crises* 1: 53–57.

———. 1980. On lawmaking. *British Journal of Law and Society* 6: 149–172.

———. 1988. *On the Take: From Petty Crooks to Presidents.* Revised ed. Bloomington: Indiana University Press.

Chambliss, William J., and Robert B. Seidman. 1982. *Law, Order and Power.* Rev. ed. Reading, MA: Addison-Wesley.

Church Committee. 1976. *Intelligence Activities and the Rights of Americans.* Washington, DC: Government Printing Office.

Commonwealth of New South Wales. 1982a. New South Wales Joint Task Force on Drug Trafficking. Federal Parliament Report. Sydney: Government of New South Wales.

———. 1982b. Preliminary Report of the Royal Commission to Investigate the Nugan Hand Bank Failure. Federal Parliament Report. Sydney: Government of New South Wales.

Crewdson, John M., and Jo Thomas. 1977. Abuses in testing of drugs by CIA to be panel focus. *New York Times,* September 20.

Dinges, John, and Saul Landau. 1980. *Assassination on Embassy Row.* New York: McGraw-Hill.

———. 1982. The CIA's link to Chile's plot. *Nation,* June 12: 712–713.

Goulden, Joseph C. 1984. *Death Merchant: The Brutal True Story of Edwin P. Wilson.* New York: Simon and Schuster.

Hersh, Seymour. 1979. Ex-analyst says CIA rejected warning on Shah. *New York Times,* January 7: A10. Cited in Piers Beirne and James Messerschmidt, *Criminology.* New York: Harcourt Brace Jovanovich, forthcoming.

Hougan, Jim. 1978. *Spooks: The Haunting of America—The Private Use of Secret Agents.* New York: William Morrow.

———. 1984. *Secret Agenda: Watergate, Deep Throat, and the CIA.* New York: Random House.

Jacobs, John. 1977a. The diaries of a CIA operative. *Washington Post,* September 5: 1.

————. 1977b. Turner cites 149 drug-test projects. *Washington Post,* August 4: 1.

Klein, Lloyd. 1988. *Big Brother Is Still Watching You.* Paper presented at the annual meetings of the American Society of Criminology, Chicago, November 12.

Kruger, Henrik. 1980. *The Great Heroin Coup.* Boston: South End Press.

Kwitny, Jonathan. 1987. *The Crimes of Patriots.* New York: W. W. Norton.

Lernoux, Penny. 1984. The Miami connection. *Nation,* February 18: 186–198.

McCord, James W., Jr. 1974. *A Piece of Tape.* Rockville, MD: Washington Media Services.

McCoy, Alfred W. 1973. *The Politics of Heroin in Southeast Asia.* New York: Harper & Row.

NARMIC. 1984. Military Exports to South Africa: A Research Report on the Arms Embargo. Philadelphia: American Friends Service Committee.

Nihill, Grant. 1982. Bank links to spies, drugs. *Advertiser,* November 10: 1.

Owen, John. 1983. *Sleight of Hand: The $25 Million Nugan Hand Bank Scandal.* Sydney: Calporteur Press.

Parenti, Michael. 1983. *Democracy for the Few.* New York: St. Martin's.

Potter, Gary W., and Bruce Bullington. 1987. *Drug Trafficking and the Contras: A Case Study of State-Organized Crime.* Paper presented at annual meeting of the American Society of Criminology, Montreal.

Rockefeller Report. 1975. Report to the President by the Commission on CIA Activities Within the United States. Washington, DC: Government Printing Office.

Schwendinger, Herman, and Julia Schwendinger. 1975. Defenders of order or guardians of human rights. Issue in *Criminology* 7: 72–81.

Senate Hearings. 1986. Senate Select Committee on Assassination, Alleged Assassination Plots Involving Foreign Leaders. Interim Report of the Senate Select Committee to Study Governmental Operations with Respect to Intelligence Activities. 94th Cong., 1st sess., November 20. Washington, DC: Government Printing Office.

Sharkey, Jacqueline. 1988. The Contra-drug trade-off. *Common Cause,* September–October: 23–33.

Turner, Stansfield. 1985. *Secrecy and Democracy: The CIA in Transition.* New York: Houghton Mifflin.

U.S. Department of State. 1985. Revolution Beyond Our Border: Information on Central America. State Department Report N 132. Washington, DC: U.S. Department of State.

17

The Moral Career
of the Mental Patient

ERVING GOFFMAN

Traditionally the term *career* has been reserved for those who expect to enjoy the rises laid out within a respectable profession. The term is coming to be used, however, in a broadened sense to refer to any social strand of any person's course through life. The perspective of natural history is taken: unique outcomes are neglected in favor of such changes over time as are basic and common to the members of a social category, although occurring independently to each of them. Such a career is not a thing that can be brilliant or disappointing; it can no more be a success than a failure. In this light, I want to consider the mental patient, drawing mainly upon data collected during a year's participant observation of patient social life in a public mental hospital,[1] wherein an attempt was made to take the patient's point of view.

One value of the concept of career is its two-sidedness. One side is linked to internal matters held dearly and closely, such as image of self and felt identity; the other side concerns official position, jural relations, and style of life, and is part of a publicly accessible institutional complex. The concept of career, then, allows one to move back and forth between the personal and the public, between the self and its significant society, without having overly to rely for data upon what the person says he thinks he imagines himself to be.

This paper, then, is an exercise in the institutional approach to the study of self. The main concern will be with the *moral* aspects of career—that is, the regular sequence of changes that career entails in the person's self and in his framework of imagery for judging himself and others.[2]

The category "mental patient" itself will be understood in one strictly sociological sense. In this perspective, the psychiatric view of a person becomes significant only in so far as this view itself alters his social fate—an alteration which seems to become fundamental in our society when, and only when, the person is put through the process of hospitalization.[3] I therefore exclude certain neighboring categories: the undiscovered candidates who would be judged "sick" by psychiatric standards but who never come to be viewed as such by themselves or others, although they may cause everyone a great deal of trouble;[4] the office patient whom a psychiatrist feels he can handle with

From "The Moral Career of the Mental Patient," Erving Goffman, *Psychiatry* 22, pp. 123–142, 1959. Reprinted by permission of the Guilford Press.

drugs or shock on the outside; the mental client who engages in psychothera-
peutic relationships. And I include anyone, however robust in temperament,
who somehow gets caught up in the heavy machinery of mental hospital ser-
vicing. In this way the effects of being treated as a mental patient can be kept
quite distinct from the effects upon a person's life of traits a clinician would
view as psychopathological.[5] Persons who become mental hospital patients
vary widely in the kind and degree of illness that a psychiatrist would impute
to them, and in the attributes by which laymen would describe them. But
once started on the way, they are confronted by some importantly similar circumstances and respond to these in some importantly similar ways. Since these
similarities do not come from mental illness, they would seem to occur in
spite of it. It is thus a tribute to the power of social forces that the uniform
status of mental patient cannot only assure an aggregate of persons a common
fate and eventually, because of this, a common character, but that this social
reworking can be done upon what is perhaps the most obstinate diversity of
human materials that can be brought together by society. Here there lacks
only the frequent forming of a protective group-life by ex-patients to illustrate
in full the classic cycle of response by which deviant subgroupings are psycho-
dynamically formed in society.

This general sociological perspective is heavily reinforced by one key finding of sociologically oriented students in mental hospital research. As has been
repeatedly shown in the study of nonliterate societies, the awesomeness, dis-
tastefulness, and barbarity of a foreign culture can decrease in the degree that
the student becomes familiar with the point of view to life that is taken by his
subjects. Similarly, the student of mental hospitals can discover that the crazi-
ness or "sick behavior" claimed for the mental patient is by and large a product
of the claimant's social distance from the situation that the patient is in, and is
not primarily a product of mental illness. Whatever the refinements of the various patients' psychiatric diagnoses, and whatever the special ways in which social life on the "inside" is unique, the researcher can find that he is participating
in a community not significantly different from any other he has studied.[6] Of
course, while restricting himself to the off-ward grounds community of paroled
patients, he may feel, as some patients do, that life in the locked wards is bizarre;
and while on a locked admissions or convalescent ward, he may feel that
chronic "back" wards are socially crazy places. But he need only move his
sphere of sympathetic participation to the "worst" ward in the hospital, and
this too can come into social focus as a place with a livable and continuously
meaningful social world. This in no way denies that he will find a minority in
any ward or patient group that continues to seem quite beyond the capacity to
follow rules of social organization, or that the orderly fulfillment of normative
expectations in patient society is partly made possible by strategic measures that
have somehow come to be institutionalized in mental hospitals.

The career of the mental patient falls popularly and naturalistically into
three main phases: the period prior to entering the hospital, which I shall call
the *prepatient phase;* the period in the hospital, the *inpatient phase;* the period

after discharge from the hospital, should this occur, namely, the *ex-patient phase*.[7] This paper will deal only with the first two phases.

THE PREPATIENT PHASE

A relatively small group of prepatients come into the mental hospital willingly, because of their own idea of what will be good for them, or because of whole-hearted agreement with the relevant members of their family. Presumably these recruits have found themselves acting in a way which is evidence to them that they are losing their minds or losing control of themselves. This view of oneself would seem to be one of the most pervasively threatening things that can happen to the self in our society, especially since that it is likely to occur at a time when the person is in any case sufficiently troubled to exhibit the kind of symptom which he himself can see. As Sullivan described it,

> What we discover in the self-system of a person undergoing schizophrenic changes or schizophrenic processes, is then, in its simplest form, an extremely fear-marked puzzlement, consisting of the use of rather generalized and anything but exquisitely refined referential processes in an attempt to cope with what is essentially a failure at being human—a failure at being anything that one could respect as worth being.[8]

Coupled with the person's disintegrative re-evaluation of himself will be the new, almost equally pervasive circumstance of attempting to conceal from others what he takes to be the new fundamental facts about himself, and attempting to discover whether others too have discovered them.[9] Here I want to stress that perception of losing one's mind is based on culturally derived and socially engrained stereotypes as to the significance of symptoms such as hearing voices, losing temporal and spatial orientation, and sensing that one is being followed, and that many of the most spectacular and convincing of these symptoms in some instances psychiatrically signify merely a temporary emotional upset in a stressful situation, however terrifying to the person at the time. Similarly, the anxiety consequent upon this perception of oneself, and the strategies devised to reduce this anxiety, are not a product of abnormal psychology, but would be exhibited by any person socialized into our culture who came to conceive of himself as someone losing his mind. Interestingly, subcultures in American society apparently differ in the amount of ready imagery and encouragement they supply for such self-views, leading to differential rates of *self*-referral; the capacity to take this disintegrative view of oneself without psychiatric prompting seems to be one of the questionable cultural privileges of the upper classes.[10]

For the person who has come to see himself—with whatever justification—as mentally unbalanced, entrance to the mental hospital can sometimes bring relief, perhaps in part because of the sudden transformation in the structure of his basic social situations; instead of being to himself a questionable person trying to maintain a role as a full one, he can become an officially questioned person known to himself to be not so questionable as that. In other cases, hospitalization can make matters worse for the willing patient, confirming by the objective situation what has theretofore been a matter of the private experience of self.

Once the willing prepatient enters the hospital, he may go through the same routine of experiences as do those who enter unwillingly. In any case, it is the latter that I mainly want to consider, since in America at present these are by far the more numerous kind.[11] Their approach to the institution takes one of three classic forms: they come because they have been implored by their family or threatened with the abrogation of family ties unless they go "willingly"; they come by force under police escort; they come under misapprehension purposely induced by others, this last restricted mainly to youthful prepatients.

The prepatient's career may be seen in terms of an extrusory model; he starts out with relationships and rights, and ends up, at the beginning of his hospital stay, with hardly any of either. The moral aspects of this career, then, typically begin with the experience of abandonment, disloyalty, and embitterment. This is the case even though to others it may be obvious that he was in need of treatment, and even though in the hospital he may soon come to agree.

The case histories of most mental patients document offense against some arrangement for face-to-face living—a domestic establishment, a work place, a semipublic organization such as a church or store, a public region such as a street or park. Often there is also a record of some *complainant,* some figure who takes that action against the offender which eventually leads to this hospitalization. This may not be the person who makes the first move, but it is the person who makes what turns out to be the first effective move. Here is the *social* beginning of the patient's career, regardless of where one might locate the psychological beginning of his mental illness.

The kinds of offenses which lead to hospitalization are felt to differ in nature from those which lead to other extrusory consequences—to imprisonment, divorce, loss of job, disownment, regional exile, noninstitutional psychiatric treatment, and so forth. But little seems known about these differentiating factors; and when one studies actual commitments, alternate outcomes frequently appear to have been possible. It seems true, moreover, that for every offense that leads to an effective complaint, there are many psychiatrically similar ones that never do. No action is taken; or action is taken which leads to other extrusory outcomes; or ineffective action is taken, leading to the mere pacifying or putting off of the person who complains. Thus, as Clausen and Yarrow have nicely shown, even offenders who are eventually hospitalized are likely to have had a long series of ineffective actions taken against them.[12]

Separating those offenses which could have been used as grounds for hospitalizing the offender from those that are so used, one finds a vast number of what students of occupation call career contingencies.[13] Some of these contingencies in the mental patient's career have been suggested, if not explored, such as socio-economic status, visibility of the offense, proximity to a mental hospital, amount of treatment facilities available, community regard for the type of treatment given in available hospitals, and so on.[14] For information about other contingencies one must rely on atrocity tales: a psychotic man is tolerated by his wife until she finds herself a boyfriend, or by his adult children until they move from a house to an apartment; an alcoholic is sent to a mental hospital because the jail is full, and a drug addict because he declines to avail himself of psychiatric treatment on the outside; a rebellious adolescent daughter can no longer be managed at home because she now threatens to have an open affair with an unsuitable companion; and so on. Correspondingly there is an equally important set of contingencies causing the person to bypass this fate. And should the person enter the hospital, still another set of contingencies will help determine when he is to obtain a discharge—such as the desire of his family for his return, the availability of a "manageable" job, and so on. The society's official view is that inmates of mental hospitals are there primarily because they are suffering from mental illness. However, in the degree that the "mentally ill" outside hospitals numerically approach or surpass those inside hospitals, one could say that mental patients *distinctively* suffer not from mental illness, but from contingencies.

Career contingencies occur in conjunction with a second feature of the prepatient's career—the *circuit of agents*—and agencies—that participate fatefully in his passage from civilian to patient status.[15] Here is an instance of that increasingly important class of social system whose elements are agents and agencies, which are brought into systemic connection through having to take up and send on the same persons. Some of these agent-roles will be cited now, with the understanding that in any concrete circuit a role may be filled more than once, and a single person may fill more than one of them.

First is the *next-of-relation*—the person whom the prepatient sees as the most available of those upon whom he should be able to most depend in times of trouble; in this instance the last to doubt his sanity and the first to have done everything to save him from the fate which, it transpires, he has been approaching. The patient's next-of-relation is usually his next of kin; the special term is introduced because he need not be. Second is the *complainant,* the person who retrospectively appears to have started the person on his way to the hospital. Third are the *mediators*—the sequence of agents and agencies to which the prepatient is referred and through which he is relayed and processed on his way to the hospital. Here are included police, clergy, general medical practitioners, office psychiatrists, personnel in public clinics, lawyers, social service workers, school teachers, and so on. One of these agents will have the legal mandate to sanction commitment and will exercise it, and so those agents who precede him in the process will be involved in something whose outcome is not yet settled. When the mediators retire from the scene, the prepatient has become an inpatient, and the significant agent has become the hospital administrator.

While the complainant usually takes action in a lay capacity as a citizen, an employer, a neighbor, or a kinsman, mediators tend to be specialists and differ from those they serve in significant ways. They have experience in handling trouble, and some professional distance from what they handle. Except in the case of policemen, and perhaps some clergy, they tend to be more psychiatrically oriented than the lay public, and will see the need for treatment at times when the public does not.[16]

An interesting feature of these roles is the functional effects of their interdigitation. For example, the feelings of the patient will be influenced by whether or not the person who fills the role of complainant also has the role of next-of-relation—an embarrassing combination more prevalent, apparently, in the higher classes than in the lower.[17] Some of these emergent effects will be considered now.[18]

In the prepatient's progress from home to the hospital he may participate as a third person in what he may come to experience as a kind of *alienative coalition*. His next-of-relation presses him into coming to "talk things over" with a medical practitioner, an office psychiatrist, or some other counselor. Disinclination on his part may be met by threatening him with desertion, disownment, or other legal action, or by stressing the joint and explorative nature of the interview. But typically the next-of-relation will have set the interview up, in the sense of selecting the professional, arranging for time, telling the professional something about the case, and so on. This move effectively tends to establish the next-of-relation as the responsible person to whom pertinent findings can be divulged, while effectively establishing the other as the patient. The prepatient often goes to the interview with the understanding that he is going as an equal of someone who is so bound together with him that a third person could not come between them in fundamental matters; this, after all, is one way in which close relationships are defined in our society. Upon arrival at the office the prepatient suddenly finds that he and his next-of-relation have not been accorded the same roles, and apparently that a prior understanding between the professional and the next-of-relation has been put in operation against him. In the extreme but common case the professional first sees the prepatient alone, in the role of examiner and diagnostician, and then sees the next-of-relation alone, in the role of advisor, while carefully avoiding talking things over seriously with them both together.[19] And even in those nonconsultative cases where public officials must forcibly extract a person from a family that wants to tolerate him, the next-of-relation is likely to be induced to "go along" with the official action, so that even here the prepatient may feel that an alienative coalition has been formed against him.

The moral experience of being third man in such a coalition is likely to embitter the prepatient, especially since his troubles have already probably led to some estrangement from his next-of-relation. After he enters the hospital, continued visits by his next-of-relation can give the patient the "insight" that his own best interests were being served. But the initial visits may temporarily strengthen his feeling of abandonment; he is likely to beg his visitor to get him out or at least to get him more privileges and to sympathize with the monstrousness of his plight—to which the visitor ordinarily can respond only by trying to maintain a hopeful note, by not "hearing" the requests, or by assuring the patient that the medical authorities know about these things and are doing what is medically best. The visitor then nonchalantly goes back into a world that the patient has learned is incredibly thick with freedom and privileges, causing the patient to feel that his next-of-relation is merely adding a pious gloss to a clear case of traitorous desertion.

The depth to which the patient may feel betrayed by his next-of-relation seems to be increased by the fact that another witnesses his betrayal—a factor which is apparently significant in many three-party situations. An offended person may well act forbearantly and accommodatively toward an offender when the two are alone, choosing peace ahead of justice. The presence of a witness, however, seems to add something to the implications of the offense. For then it is beyond the power of the offended and offender to forget about, erase, or suppress what has happened; the offense has become a public social fact.[20] When the witness is a mental health commission, as is sometimes the case, the witnessed betrayal can verge on a "degradation ceremony."[21] In such circumstances, the offended patient may feel that some kind of extensive reparative action is required before witnesses, if his honor and social weight are to be restored.

Two other aspects of sensed betrayal should be mentioned. First, those who suggest the possibility of another's entering a mental hospital are not likely to provide a realistic picture of how in fact it may strike him when he arrives. Often he is told that he will get required medical treatment and a rest, and may well be out in a few months or so. In some cases they may thus be concealing what they know, but I think, in general, they will be telling what they see as the truth. For here there is a quite relevant difference between patients and mediating professionals; mediators, more so than the public at large, may conceive of mental hospitals as short-term medical establishments where required rest and attention can be voluntarily obtained, and not as places of coerced exile. When the prepatient finally arrives he is likely to learn quite quickly, quite differently. He then finds that the information given him about life in the hospital has had the effect of his having put up less resistance to entering than he now sees he would have put up had he known the facts. Whatever the intentions of those who participated in his transition from person to patient, he may sense they have in effect "conned" him into his present predicament.

I am suggesting that the prepatient starts out with at least a portion of the rights, liberties, and satisfactions of the civilian and ends up on a psychiatric ward stripped of almost everything. The question here is *how* this stripping is managed. This is the second aspect of betrayal I want to consider.

As the prepatient may see it, the circuit of significant figures can function as a kind of *betrayal funnel*. Passage from person to patient may be effected through a series of linked stages, each managed by a different agent. While each stage tends to bring a sharp decrease in adult free status, each agent may try to maintain the fiction that no further decrease will occur. He may even manage to turn the prepatient over to the next agent while sustaining this note. Further, through words, cues, and gestures, the prepatient is implicitly asked by the current agent to join with him in sustaining a running line of polite small talk that tactfully avoids the administrative facts of the situation, becoming, with each stage, progressively more at odds with these facts. The spouse would rather not have to cry to get the prepatient to visit a psychiatrist; psychiatrists would rather not have a scene when the prepatient learns that he and his spouse are being seen separately and in different ways; the police infrequently bring a prepatient to the hospital in a strait jacket, finding it much easier all around to give him a cigarette, some kindly words, and freedom to relax in the back seat of the patrol car; and finally, the admitting psychiatrist finds he can do his work better in the relative quiet and luxury of the "admission suite" where, as an incidental consequence, the notion can survive that a mental hospital is indeed a comforting place. If the prepatient heeds all of these implied requests and is reasonably decent about the whole thing, he can travel the whole circuit from home to hospital without forcing anyone to look directly at what is happening or to deal with the raw emotion that his situation might well cause him to express. His showing consideration for those who are moving him toward the hospital allows them to show consideration for him, with the joint result that these interactions can be sustained with some of the protective harmony characteristic of ordinary face-to-face dealings. But should the new patient cast his mind back over the sequence of steps leading to hospitalization, he may feel that everyone's *current* comfort was being busily sustained while his long-range welfare was being undermined. This realization may constitute a moral experience that further separates him for the time from the people on the outside.[22]

I would now like to look at the circuit of career agents from the point of view of the agents themselves. Mediators in the person's transition from civil to patient status—as well as his keepers, once he is in the hospital—have an interest in establishing a responsible next-of-relation as the patient's deputy or *guardian;* should there be no obvious candidate for the role, someone may be sought out and pressed into it. Thus while a person is gradually being transformed into a patient, a next-of-relation is gradually being transformed into a guardian. With a guardian on the scene, the whole transition process can be kept tidy. He is likely to be familiar with the prepatient's civil involvements and business, and can tie up loose ends that might otherwise be left to entangle the hospital. Some of the prepatient's abrogated civil rights can be transferred to him, thus helping to sustain the legal fiction that while the prepatient does not actually have his rights he somehow actually has not lost them.

Inpatients commonly sense, at least for a time, that hospitalization is a massive unjust deprivation, and sometimes succeed in convincing a few persons on the outside that this is the case. It often turns out to be useful, then, for those identified with inflicting these deprivations, however justifiably, to be able to point to the cooperation and agreement of someone whose relationship to the patient places him above suspicion, firmly defining him as the person most likely to have the patient's personal interest at heart. If the guardian is satisfied with what is happening to the new inpatient, the world ought to be.[23]

Now it would seem that the greater the legitimate personal stake one party has in another, the better he can take the role of guardian to the other. But the structural arrangements in society which lead to the acknowledged merging of two persons' interests lead to additional consequences. For the person to whom the patient turns for help—for protection against such threats as involuntary commitment—is just the person to whom the mediators and hospital administrators logically turn for authorization. It is understandable, then, that some patients will come to sense, at least for a time, that the closeness of a relationship tells nothing of its trustworthiness.

There are still other functional effects emerging from this complement of roles. If and when the next-of-relation appeals to mediators for help in the trouble he is having with the prepatient, hospitalization may not, in fact, be in his mind. He may not even perceive the prepatient as mentally sick, or, if he does, he may not consistently hold to this view.[24] It is the circuit of mediators, with their greater psychiatric sophistication and their belief in the medical character of mental hospitals, that will often define the situation for the next-of-relation, assuring him that hospitalization is a possible solution and a good one, that it involves no betrayal, but is rather a medical action taken in the best interests of the prepatient. Here the next-of-relation may learn that doing his duty to the prepatient may cause the prepatient to distrust and even hate him for the time. But the fact that this course of action may have had to be pointed out and prescribed by professionals, and be defined by them as a moral duty, relieves the next-of-relation of some of the guilt he may feel.[25] It is a poignant fact that an adult son or daughter may be pressed into the role of mediator, so that the hostility that might otherwise be directed against the spouse is passed on to the child.[26]

Once the prepatient is in the hospital, the same guilt-carrying function may become a significant part of the staff's job in regard to the next-of-relation.[27] These reasons for feeling that he himself has not betrayed the patient, even though the patient may then think so, can later provide the next-of-relation with a defensible line to take when visiting the patient in the hospital and a basis for hoping that the relationship can be reestablished after its hospital moratorium. And of course this position, when sensed by the patient, can provide him with excuses for the next-of-relation, when and if he comes to look for them.[28]

Thus while the next-of-relation can perform important functions for the mediators and hospital administrators, they in turn can perform important functions for him. One finds, then, an emergent unintended exchange or reciprocation of functions, these functions themselves being often unintended.

The final point I want to consider about the prepatient's moral career is its peculiarly *retroactive* character. Until a person actually arrives at the hospital there usually seems no way of knowing for sure that he is destined to do so, given the determinative role of career contingencies. And until the point of hospitalization is reached, he or others may not conceive of him as a person who is becoming a mental patient. However, since he will be held against his will in the hospital, his next-of-relation and the hospital staff will be in great need of a rationale for the hardships they are sponsoring. The medical elements of the staff will also need evidence that they are still in the trade they were trained for. These problems are eased, no doubt unintentionally, by the case-history construction that is placed on the patient's past life, this having the effect of demonstrating that all along he had been becoming sick, that he finally became very sick, and that if he had not been hospitalized much worse things would have happened to him—all of which, of course, may be true. Incidentally, if the patient wants to make sense out of his stay in the hospital, and, as already suggested, keep alive the possibility of once again conceiving of his next-of-relation as a decent, well-meaning person, then he too will have reason to believe some of this psychiatric work-up of his past.

Here is a very ticklish point for the sociology of careers. An important aspect of every career is the view the person constructs when he looks backward over his progress; in a sense, however, the whole of the prepatient career derives from this reconstruction. The fact of having had a prepatient career, starting with an effective complaint, becomes an important part of the mental patient's orientation, but this part can begin to be played only after hospitalization proves that what he had been having, but no longer has, is a career as a prepatient.

THE INPATIENT PHASE

The last step in the prepatient's career can involve his realization—justified or not—that he has been deserted by society and turned out of relationships by those closest to him. Interestingly enough, the patient, especially a first admission, may manage to keep himself from coming to the end of this trail, even though in fact he is now in a locked mental hospital ward. On entering the hospital, he may very strongly feel the desire not to be known to anyone as a person who could possibly be reduced to these present circumstances, or as a person who conducted himself in the way he did prior to commitment. Consequently, he may avoid talking to anyone, may stay by himself when possible, and may even be "out of contact" or "manic" so as to avoid ratifying any interaction that presses a politely reciprocal role upon him and opens him up to what he has become in the eyes of others. When the next-of-relation makes an effort to visit, he may be rejected by mutism, or by the patient's refusal to enter the visiting room, these strategies sometimes suggesting that the patient still clings to a remnant of relatedness to those who made up his past, and is protecting this remnant from the final destructiveness of dealing with the new people that they have become.[29]

Usually the patient comes to give up this taxing effort at anonymity, at not-hereness, and begins to present himself for conventional social interaction to the hospital community. Thereafter he withdraws only in special ways—by always using his nickname, by signing his contribution to the patient weekly with his initial only, or by using the innocuous "cover" address tactfully provided by some hospitals; or he withdraws only at special times, when, say, a flock of nursing students makes a passing tour of the ward, or when, paroled to the hospital grounds, he suddenly sees he is about to cross the path of a civilian he happens to know from home. Sometimes this making of oneself available is called "settling down" by the attendants. It marks a new stand openly taken and supported by the patient, and resembles the "coming out" process that occurs in other groupings.[30]

Once the prepatient begins to settle down, the main outlines of his fate tend to follow those of a whole class of segregated establishments—jails, concentration camps, monasteries, work camps, and so on—in which the inmate spends the whole round of life on the grounds, and marches through his regimented day in the immediate company of a group of persons of his own institutional status.[31]

Like the neophyte in many of these "total institutions," the new inpatient finds himself cleanly stripped of many of his accustomed affirmations, satisfactions, and defenses, and is subjected to a rather full set of mortifying experiences: restriction of free movement; communal living; diffuse authority of a whole echelon of people; and so on. Here one begins to learn about the limited extent to which a conception of oneself can be sustained when the usual setting of supports for it are suddenly removed.

While undergoing these humbling moral experiences, the inpatient learns to orient himself in terms of the "ward system."[32] In public mental hospitals this usually consists of a series of graded living arrangements built around wards, administrative units called services, and parole statuses. The "worst" level involves often nothing but wooden benches to sit on, some quite indifferent food, and a small piece of room to sleep in. The "best" level may involve a room of one's own, ground and town privileges, contacts with staff that are relatively undamaging, and what is seen as good food and ample recreational facilities. For disobeying the pervasive house rules, the inmate will receive stringent punishments expressed in terms of loss of privileges; for obedience he will eventually be allowed to reacquire some of the minor satisfactions he took for granted on the outside.

The institutionalization of these radically different levels of living throws light on the implications for self of social settings. And this in turn affirms that the self arises not merely out of its possessor's interactions with significant others, but also out of the arrangements that are evolved in an organization for its members.

There are some settings which the person easily discounts as an expression or extension of him. When a tourist goes slumming, he may take pleasure in the situation not because it is a reflection of him but because it so assuredly is not. There are other settings, such as living rooms, which the person manages on his own and employs to influence in a favorable direction other persons' views of him. And there are still other settings, such as a work place, which express the employee's occupational status, but over which he has no final control, this being exerted, however tactfully, by his employer. Mental hospitals provide an extreme instance of this latter possibility. And this is due not merely to their uniquely degraded living levels, but also to the unique way in which significance for self is made explicit to the patient, piercingly, persistently, and thoroughly. Once lodged on a given ward, the patient is firmly instructed that the restrictions and deprivations he encounters are not due to such things as tradition or economy—and hence dissociable from self—but are intentional parts of his treatment, part of his need at the time, and therefore an expression of the state that his self has fallen to. Having every reason to initiate requests for better conditions, he is told that when the staff feels he is "able to manage" or will be "comfortable with" a higher ward level, then appropriate action will be taken. In short, assignment to a given ward is presented not as a reward or punishment, but as an expression of his general level of social functioning, his status as a person. Given the fact that the worst ward levels provide a round of life that inpatients with organic brain damage can easily manage, and that these quite limited human beings are present to prove it, one can appreciate some of the mirroring effects of the hospital.[33]

The ward system, then, is an extreme instance of how the physical facts of an establishment can be explicitly employed to frame the conception a person takes of himself. In addition, the official psychiatric mandate of mental hospitals gives rise to even more direct, even more blatant, attacks upon the inmate's view of himself. The more "medical" and the more progressive a mental hospital is—the more it attempts to be therapeutic and not merely custodial—the more he may be confronted by high-ranking staff arguing that his past has been a failure, that the cause of this has been within himself, that his attitude to life is wrong, and that if he wants to be a person he will have to change his way of dealing with people and his conceptions of himself. Often the moral value of these verbal assaults will be brought home to him by requiring him to practice taking this psychiatric view of himself in arranged confessional periods, whether in private sessions or group psychotherapy.

Now a general point may be made about the moral career of inpatients which has bearing on many moral careers. Given the stage that any person has reached in a career, one typically finds that he constructs an image of his life course—past, present, and future—which selects, abstracts, and distorts in such a way as to provide him with a view of himself that he can usefully expound in current situations. Quite generally, the person's line concerning self defensively brings him into appropriate alignment with the basic values of his society, and so may be called an *apologia*. If the person can manage to present a view of his current situation which shows the operation of favorable personal qualities in the past and a favorable destiny awaiting him, it may be called a *success story*. If the facts of a person's past and present are extremely dismal, then about the best he can do is to show that he is not responsible for what has become of him, and the term *sad tale* is appropriate. Interestingly enough, the more the person's past forces him out of apparent alignment with central moral values, the more often he seems compelled to tell his sad tale in any company in which he finds himself. Perhaps he partly responds to the need he feels in others of not having their sense of proper life courses affronted. In any case, it is among convicts, "wino's," and prostitutes that one seems to obtain sad tales the most readily.[34] It is the vicissitudes of the mental patient's sad tale that I want to consider now.

In the mental hospital, the setting and the house rules press home to the patient that he is, after all, a mental case who has suffered some kind of social collapse on the outside, having failed in some overall way, and that here he is of little social weight, being hardly capable of acting like a full-fledged person at all. These humiliations are likely to be most keenly felt by middle-class patients, since their previous condition of life little immunizes them against such affronts; but all patients feel some downgrading. Just as any normal member of his outside subculture would do, the patient often responds to this situation by attempting to assert a sad tale proving that he is not "sick," that the "little trouble" he did get into was really somebody else's fault, that his past life course had some honor and rectitude, and that the hospital is therefore unjust in forcing the status of mental patient upon him. This self-respecting tendency is heavily institutionalized within the patient society where opening social contacts typically involve the participants' volunteering information about their current ward location and length of stay so far, but not the reasons for their stay—such interaction being conducted in the manner of small talk on the outside.[35] With greater familiarity, each patient usually volunteers relatively acceptable reasons for his hospitalization, at the same time accepting without open immediate question the lines offered by other patients. Such stories as the following are given and overtly accepted.

> I was going to night school to get a M.A. degree, and holding down a job in addition, and the load got too much for me.

> The others here are sick mentally but I'm suffering from a bad nervous system and that is what is giving me these phobias.

> I got here by mistake because of a diabetes diagnosis, and I'll leave in a couple of days. [The patient had been in seven weeks.]

> I failed as a child, and later with my wife I reached out for dependency.

> My trouble is that I can't work. That's what I'm in for. I had two jobs with a good home and all the money I wanted.[36]

The patient sometimes reinforces these stories by an optimistic definition of his occupational status: A man who managed to obtain an audition as a radio announcer styles himself a radio announcer; another who worked for some months as a copy boy and was then given a job as a reporter on a large trade journal, but fired after three weeks, defines himself as a reporter.

A whole social role in the patient community may be constructed on the basis of these reciprocally sustained fictions. For these face-to-face niceties tend to be qualified by behind-the-back gossip that comes only a degree closer to the "objective" facts. Here, of course, one can see a classic social function of informal networks of equals: they serve as one another's audience for self-supporting tales—tales that are somewhat more solid than pure fantasy and somewhat thinner than the facts.

But the patient's *apologia* is called forth in a unique setting, for few settings could be so destructive of self-stories except, of course, those stories already constructed along psychiatric lines. And this destructiveness rests on more than the official sheet of paper which attests that the patient is of unsound mind, a danger to himself and others—an attestation, incidentally, which seems to cut deeply into the patient's pride, and into the possibility of his having any.

Certainly the degrading conditions of the hospital setting belie many of the self-stories that are presented by patients; and the very fact of being in the mental hospital is evidence against these tales. And of course, there is not always sufficient patient solidarity to prevent patient discrediting patient, just as there is not always a sufficient number of "professionalized" attendants to prevent attendant discrediting patient. As one patient informant repeatedly suggested to a fellow patient:

If you're so smart, how come you got your ass in here?

The mental hospital setting, however, is more treacherous still. Staff has much to gain through discreditings of the patient's story—whatever the felt reason for such discreditings. If the custodial faction in the hospital is to succeed in managing his daily round without complaint or trouble from him, then it will prove useful to be able to point out to him that the claims about himself upon which he rationalizes his demands are false, that he is not what he is claiming to be, and that in fact he is a failure as a person. If the psychiatric faction is to impress upon him its views about his personal make-up, then they must be able to show in detail how their version of his past and their version of his character hold up much better than his own.[37] If both the custodial and psychiatric factions are to get him to cooperate in the various psychiatric treatments, then it will prove useful to disabuse him of *his* view of their purposes, and cause him to appreciate that they know what they are doing, and are doing what is best for him. In brief, the difficulties caused by a patient are closely tied to his version of what has been happening to him, and if cooperation is to be secured, it helps if this version is discredited. The patient must "insightfully" come to take, or effect to take, the hospital's view of himself.

NOTES

1.   The study was conducted during 1955–56 under the auspices of the Laboratory of Social-environmental Studies of the National Institute of Mental Health. I am grateful to the Laboratory Chief, John A. Clausen, and to Dr. Winfred Overholser, Superintendent, and the late Dr. Jay Hoffman, then First Assistant Physician of Saint Elizabeth's Hospital, Washington, D.C., for the ideal cooperation they freely provided. A preliminary report is contained in Goffman, "Interpersonal Persuasion," pp. 117–193; in *Group Processes: Transactions of the Third Conference*, edited by Bertram Schaffner: New York, Josiah Macy, Jr. Foundation, 1957. A shorter version of this paper was presented at the Annual Meeting of the American Sociological Society, Washington, D.C., August 1957.

2.   Material on moral career can be found in early social anthropological work on ceremonies of status transition, and in classic social psychological descriptions of those spectacular changes in one's view of self that can accompany participation in social movements and sects. Recently new kinds of relevant data have been suggested by psychiatric interest in the problem of "identity" and sociological studies of work careers and "adult socialization."

3.   This point has recently been made by Elaine and John Cumming, *Closed Ranks;* Cambridge, Commonwealth Fund, Harvard Univ. Press, 1957; pp. 101–102. "Clinical experience supports the impression that many people define mental illness as 'That condition for which a person is treated in a mental hospital.' . . . Mental illness, it seems, is a condition which afflicts people who must go to a mental institution, but until they do almost anything they do is normal." Leila Deasy has pointed out to me the correspondence here with the situation in white collar crime. Of those who are detected in this activity, only the ones who do not manage to avoid going to prison find themselves accorded the social role of the criminal.

4.   Case records in mental hospitals are just now coming to be exploited to show the incredible amount of trouble a person may cause for himself and others before anyone begins to think about him psychiatrically, let alone take psychiatric action against him. See John A. Clausen and Marian Radke Yarrow, "Paths to the Mental Hospital," *J. Social Issues* (1955) 11: 25–32; August B. Hollingshead and Frederick C. Redlich, *Social Class and Mental Illness;* New York, Wiley, 1958: pp. 173–174.

5.   An illustration of how this perspective may be taken to all forms of deviancy may be found in Edwin Lemert, *Social Pathology;* New York, McGraw-Hill, 1951; see especially pp. 74–76. A specific application to mental defectives may be found in Stewart E. Perry, "Some Theoretic Problems of Mental Deficiency and Their Action Implications," *Psychiatry* (1954) 17: 45–73; see especially p. 68.

6.   Conscientious objectors who voluntarily went to jail sometimes arrived at the same conclusion regarding criminal inmates. See, for example, Alfred Hassler, *Diary of a Self-made Convict;* Chicago, Regnery, 1954; p. 74.

7.   This simple picture is complicated by the somewhat special experience of roughly a third of ex-patients—namely, readmission to the hospital, this being the recidivist or "repatient" phase.

8.   Harry Stack Sullivan, *Clinical Studies in Psychiatry;* edited by Helen Swick Perry, Mary Ladd Gawel, and Martha Gibbon; New York, Norton, 1956; pp. 184–185.

9.   This moral experience can be contrasted with that of a person learning to become a marihuana addict, whose discovery that he can be "high" and still "op" effectively without being detected apparently leads to a new level of use. See Howard S. Becker, "Marihuana Use and Social Control," *Social Problems* (1955) 3: 35–44; see especially pp. 40–41.

10. See note 4: Hollingshead and Redlich, p. 187, Table 6, where relative frequency is given of self-referral by social class grouping.

11. The distinction employed here between willing and unwilling patients cuts
across the legal one, of voluntary and
committed, since some persons who are
glad to come to the mental hospital may
be legally committed, and of those who
come only because of strong familial pressure, some may sign themselves in as voluntary patients.

12. Clausen and Yarrow; see note 4.

13. An explicit application of this notion
to the field of mental health may be found
in Edwin M. Lemert, "Legal Commitment and Social Control," *Sociology and
Social Research* (1946) 30: 370–378.

14. For example, Jerome K. Meyers and
Leslie Schaffer, "Social Stratification and
Psychiatric Practice: A Study of an Outpa-
tient Clinic," *Amer. Sociological Rev.* (1954)
19: 307–310, Lemert, see note 5; pp.
402–403. *Patients in Mental Institutions,*
1941; Washington, D.C., Department of
Commerce, Bureau of Census, 1941; p. 2.

15. For one circuit of agents and its bear-
ing on career contingencies, see Oswald
Hall, "The Stages of a Medical Career,"
Amer. J. Sociology (1948) 53: 227–336.

16. See Cumming, note 3; p. 92.

17. Hollingshead and Redlich, note 4;
p. 187.

18. For an analysis of some of these circuit
implications for the inpatient, see Leila C.
Deasy and Olive W. Quinn, "The Wife
of the Mental Patient and the Hospital
Psychiatrist," *J. Social Issues* (1955) 11:
49–60. An interesting illustration of this
kind of analysis may also be found in Alan
G. Gowman, "Blindness and the Role of
Companion," *Social Problems* (1956) 4:
68–75. A general statement may be found
in Robert Merton, "The Role Set: Problems in Sociological Theory," *British J.
Sociology* (1957) 8: 106–120.

19. I have one case record of a man who
claims he thought *he* was taking his wife
to see the psychiatrist, not realizing until
too late that his wife had made the
arrangements.

20. A paraphrase from Kurt Riezler, "The
Social Psychology of Shame," *Amer. J.
Sociology* (1943) 48: 458.

21. See Harold Garfinkel, "Conditions of
Successful Degradation Ceremonies,"
Amer. J. Sociology (1956) 61: 420–424.

22. Concentration camp practices provide
a good example of the function of the
betrayal funnel in inducing cooperation
and reducing struggle and fuss, although
here the mediators could not be said to be
acting in the best interests of the inmates.
Police picking up persons from their
homes would sometimes joke good-na-
turedly and to offer to wait while coffee
was being served. Gas chambers were fit-
ted out like delousing rooms, and victims
taking off their clothes were told to note
where they were leaving them. The sick,
aged, weak, or insane who were selected
for extermination were sometimes driven
away in Red Cross ambulances to camps
referred to by terms such as "observation
hospital." See David Boder, *I Did Not
Interview the Dead;* Urbana, Univ. of Illinois Press, 1949; p. 81; and Elie A.
Cohen, *Human Behavior in the Concentration Camp;* London, Cape, 1954; pp. 32,
37, 107.

23. Interviews collected by the Clausen
group at NIMH suggest that when a wife
comes to be a guardian, the responsibility
may disrupt previous distance from in-
laws, leading either to a new supportive
coalition with them or to a marked withdrawal from them.

24. For an analysis of these nonpsychiatric
kinds of perception, see Marian Radke
Yarrow, Charlotte Green Schwartz, Har-
riet S. Murphy, and Leila Calhoun Deasy,
"The Psychological Meaning of Mental
Illness in the Family," *J. Social Issues*
(1955) 11: 12–24; Charlotte Green
Schwartz, "Perspectives on Deviance:
Wives' Definitions of their Husbands'
Mental Illness," *Psychiatry* (1957) 20:
275–291.

25. This guilt-carrying function is found,
of course, in other role-complexes. Thus,
when a middle-class couple engages in the
process of legal separation or divorce, each
of their lawyers usually takes the position
that his job is to acquaint his client with all
of the potential claims and rights, pressing
his client into demanding these, in spite of
any nicety of feelings about the rights and

honorableness of the ex-partner. The client, in all good faith, can then say to self and to the ex-partner that the demands are being made only because the lawyer insists it is best to do so.

26. Recorded in the Clausen data.

27. This point is made by Cumming, see note 3; p. 129.

28. There is an interesting contrast here with the moral career of the tuberculosis patient. I am told by Julius Roth that tuberculosis patients are likely to come to the hospital willingly, agreeing with their next-of-relation about treatment. Later in their hospital career, when they learn how long they yet have to stay and how depriving and irrational some of the hospital rulings are, they may seek to leave, be advised against this by the staff and by relatives, and only then begin to feel betrayed.

29. The inmate's initial strategy of holding himself aloof from ratifying contact may partly account for the relative lack of group-information among inmates in public mental hospitals, a connection that has been suggested to me by William R. Smith. The desire to avoid personal bonds that would give license to the asking of biographical questions could also be a factor. In mental hospitals, of course, as in prisoner camps, the staff may consciously break up incipient group-formation in order to avoid collective rebellious action and other ward disturbances.

30. A comparable coming out occurs in the homosexual world, when a person finally comes frankly to present himself to a "gay" gathering not as a tourist but as someone who is "available." See Evelyn Hooker, "A Preliminary Examination of Group Behavior of Homosexuals," *J. Psychology* (1956) 42: 217–225; especially p. 221. A good fictionalized treatment may be found in James Baldwin's *Giovanni's Room;* New York, Dial, 1956; pp. 41–63. A familiar instance of the coming out process is no doubt to be found among prepubertal children at the moment one of these actors sidles *back* into a room that had been left in an angered huff and injured *amour-propre.* The phrase itself

presumably derives from a *rite-de-passage* ceremony once arranged by upper-class mothers for their daughters. Interestingly enough, in large mental hospitals the patient sometimes symbolizes a complete coming out by his first active participation in the hospital-wide patient dance.

31. See Goffman, "Characteristics of Total Institutions," pp. 43–84; in *Proceedings of the Symposium of Preventive and Social Psychiatry;* Washington, D.C., Walter Reed Army Institute of Research, 1958.

32. A good description of the ward system may be found in Ivan Belknap, *Human Problems of a State Mental Hospital;* New York, McGraw-Hill, 1956; see especially p. 164.

33. Here is one way in which mental hospitals can be worse than concentration camps and prisons as places in which to "do" time; in the latter, self-insulation from the symbolic implications of the settings may be easier. In fact, self-insulation from hospital settings may be so difficult that patients have to employ devices for this which staff interpret as psychotic symptoms.

34. In regard to convicts, see Anthony Heckstall-Smith, *Eighteen Months;* London, Wingate, 1954; pp. 52–53. For "winos's" see the discussion in Howard G. Bain, "A Sociological Analysis of the Chicago Skid-Row Lifeway"; unpublished M.A. thesis, Dept. of Sociology, pp. 141–146. Bain's neglected thesis is a useful source of material on moral careers.

Apparently one of the occupational hazards of prostitution is that clients and other professional contacts sometimes persist in expressing sympathy by asking for a defensible dramatic explanation for the fall from grace. In having to bother to have a sad tale ready, perhaps the prostitute is more to be pitied than damned. Good examples of prostitute sad tales may be found in Sir Henry Mayhew, "Those That Will Not Work," pp. 210–272; in his *London Labour and the London Poor,* Vol. 4; London, Griffin, Bohn, and Cox, 1862. For a contemporary source, see *Women of the Streets,* edited by C. H. Rolph; London, Zecker and Warburg, 1955; especially p. 6. "Almost always, however, after a few comments on the police, the girl would begin to explain how it was that she was in the life, usually in terms of self-justification." Lately, of course, the psychological expert has helped out the profession in the construction of wholly remarkable sad tales. See, for example, Harold Greenwald, *Call Girl;* New York, Ballantine, 1958.

35. A similar self-protecting rule has been observed in prisons. Thus, Hassler, see note 6, in describing a conversation with a fellow-prisoner: "He didn't say much about why he was sentenced, and I didn't ask him, that being the accepted behavior in prison" (p. 76). A novelistic version for the mental hospital may be found in J. Kerkhoff, *How Thin the Veil: A Newspaperman's Story of His Own Mental Crack-up and Recovery;* New York, Greenberg, 1952; p. 27.

36. From the writer's field notes of informal interaction with patients, transcribed as near verbatim as he was able.

37. The process of examining a person psychiatrically and then altering or reducing his status in consequence is known in hospital and prison parlance as *bugging,* the assumption being that once you come to the attention of the testers you either will automatically be labeled crazy or the process of testing itself will make you crazy. Thus psychiatric staff are sometimes seen not as *discovering* whether you are sick, but as *making* you sick; and "Don't bug me, man," can mean, "Don't pester me to the point where I'll get upset." Sheldon Messenger has suggested to me that this meaning of bugging is related to the other colloquial meaning, of wiring a room with a secret microphone to collect information usable for discrediting the speaker.

18

Bad Blood

The Moral Stigmatization of
Paid Plasma Donors

MARTIN J. KRETZMANN

Commercial Blood Plasma Centers are familiar establishments in the urban landscape, yet they have rarely been investigated. The present article reports on an ethnographic study on one such center in which I participated as a paid donor. Within these settings, donors undergo a plasmapheresis procedure which extracts the commercially valuable component, or plasma, from blood and returns the red cells to the donor. In exchange for each bag of plasma, donors receive a small cash payment. In a sense, this is an eminently economic-rational transaction and serves the vital purpose of maintaining a regular source of blood products for hospitals and the pharmaceutical industry. The social meaning attached to selling plasma, however, carries a persistent and nonrational cultural imagery of negative stereotypes and moral stigma.

From the perspective of conventional social sensibilities and morality, paid plasma donors are often typified as "dope addicts, liars, degenerates, unemployed derelicts . . . bums, the faceless, the undernourished and unwashed" (Titmuss 1971, 114–15). As Espeland (1984) and Espeland and Clemens (1989) pointed out, the social meaning of an act, such as sex, military service, or giving blood, may be dramatically transformed when economic interests are introduced. Selling plasma is stigmatized because it is based on economic self-interest. More specifically, this stigma is associated with the common notion that this is a source of "easy money" for people who are not motivated to make an honest living. As with the bum who asks for spare change, we wonder where the money will go.

The treatment of paid donors is in sharp contrast to the relationship of trust and deference accorded to unpaid blood donors who in a sense become "addicted to altruism" (Piliavin, Callero, and Evans 1982). Plasma donors are always suspect. The plasma center staff I observed automatically assumed that anyone needing to sell plasma was concealing information which would reveal

From "Bad Blood: The Moral Stigmatization of Paid Plasma Donors," Martin J. Kretzmann, *Journal of Contemporary Ethnography,* pp. 416–441, 1992. Reprinted by permission of Sage Publications, Inc.

that their plasma is dangerous or otherwise unsuitable for medical purposes. Unlike nonprofit blood banks, interactions between staff and paid donors were marked by an absence of trust and a social-moral devaluation of donors.

Similar to the orientation of emergency-ward staff in the urban hospital studied by Roth (1972), the plasma center staff felt that their clients "were not to be trusted and did not have to be treated politely" (p. 842). In short, plasma center staff imputed to donors the discredited identity of "bad blood." Paid donors were regarded by the staff primarily in terms of the doubtful quality of their plasma, that is, the likelihood that the donors would knowingly or un-knowingly transmit blood-related diseases to recipients. This suspicion was based on the moral judgment of their clientele as morally unworthy.

METHOD AND SETTING

My purpose in participating as a paid donor was to subjectively experience the setting of the urban plasma center from the point of view of a new donor. The decision to adopt a covert participant observer role followed naturally from this purpose, and it was my intention that I be handled by the staff no differently from the average donor. Lacking any previous insider knowledge of plasma centers, I entered the setting cold. Although the average new donor comes to a plasma center with no clear idea of what to expect, it is often not what they expect.

I found the covert role to be particularly appropriate in this situation be-cause the staff routinely assumed that donors were concealing information in order to pass the physical screening. I observed cases where this was true, and for some, passing is perhaps a normal part of being a paid donor. Over a pe-riod of 8 weeks, I spent between 5 and 8 hours per week in the setting, with normally two visits per week. I modeled my appearance and behavior after the other young male donors. My attire consisted of old jeans, work shirt, and leather jacket. I attempted to give an impression of being a bit "scruffy" and unshaven but still respectable. Like other new donors, I attempted to conform to the unfamiliar formal rules and informal norms in the plasma center. In most parts of the setting, this restricted my activity to passive observation be-cause, indeed, passivity and compliance were expected of donors at all times.

The plasma center I studied was located in a large city in the Rocky Moun-tain region. This center, like many, was a standardized plasmapheresis program operated by a national pharmaceutical firm. As is common for urban plasma centers, it was located in the poorer area of the city where the disadvantaged are concentrated. . . .

The interior of the center was divided into three major areas: a waiting area for donors in front furnished with rows of metal folding chairs, a larger rear section off to one side where the actual plasmapheresis took place, and between these areas an enclosed counter area restricted to staff. All dealings between staff and donors in the waiting area, including the routine physical screening, were performed across this counter. Access between each of these areas was through waist-high swinging doors.

An 8 × 12 ft. glass-enclosed smoking booth was situated in one corner of the waiting area. Due to the number of nonsmokers who frequented the booth, I soon recognized that this provided donors with a backstage region. Here, they could talk more freely and relax to some extent, although donors had to be alert to their names being called by the staff at any time. The majority of my interactions with other donors took place in the confines of this booth where for the price of a bummed cigarette, a conversation was easily initiated or, in many cases, simply overheard.

The large plasmapheresis area was equipped with special vinyl couches and an immaculately white plasmapheresis machine the size of a dishwasher beside each. This area was run by a separate female staff of phlebotomists with a male supervisor. The area was divided into a number of subsections by signs on poles that designated the "yellow," "blue," or "orange" and so on sections. After an initial routine screening by the staff in the waiting area, donors were ordered to report to one of these sections. On completing the hour-long plasmapheresis procedure, donors returned to the front counter with their "bag" of yellowish plasma. A staff member there would sign them out and pay them from a roll of currency kept in a lab coat pocket.

DONOR SCREENING AND THE
PLASMAPHERESIS PROCEDURE

Plasma centers are faced with two organizational priorities: (a) to produce a regular supply of plasma and (b) to take routine precautions against contamination of this supply, primarily from hepatitis and other blood-related diseases. To ensure a steady supply of donors, various economic incentives were used. To safeguard against contamination, physical screening procedures dictated by strict Food and Drug Administration (FDA) regulations were followed.

The safe assumption for the staff when processing donors was that he or she was concealing important medical information (see Titmuss 1971, 77, 151). Because the staff did not rely on the truthfulness of donors, they watched the donors carefully for nonverbal clues of drug use or other medical conditions that would disqualify them from donating. For example, one of the first questions directed to me as a potential new donor was "Do you shoot drugs?" The way in which the staff member looked into my eyes as she asked this blunt question told me that it would be my *reaction* to the question rather than the response which was of greater interest.

The routine screening procedure took place in two steps. After signing in on a clipboard at the front counter, there was usually a half hour delay or more until again being called up to the counter. The first step was recording the donor's weight, temperature, and hematocrit (red blood cell count). The weight scale was located on the donor side of the counter, requiring the staff member to lean across the counter to adjust the weights. Donors were obliged to remove jackets and coats and drop them to the floor while being weighed. In my first screening, I unthinkingly placed my jacket on the counter, and experienced momentary panic when I realized I had violated a basic rule of etiquette. The staff's counter was not free space.

After this initial step, an invisible dye was applied to one fingernail that marked the donor as "one of theirs." A staff member explained to me that this was the routine procedure among plasma centers in the city to prevent donors from going to another center. Some donors attempt to donate more frequently than the FDA regulations allow in order to earn more money. A quick check under ultraviolet light revealed who the donor belonged to; thus cheaters could easily be spotted.

Following a delay of up to an hour, the second screening step was to record blood pressure, pulse, temperature, and a check of the lymph glands and arms for signs of infection. If passed, a staff member then rapidly read off a long list of questions concerning medical symptoms and drug use from a standard form. These questions were required by the FDA and neither the staff nor the donors appeared to take them very seriously. The forms were quickly checked off by the staff and then initialed by the donor. The formal paperwork and rules were strictly adhered to, but as Espeland (1984) pointed out, rules and objective criteria were only followed "on paper" for the sake of appearance (pp. 142–43).

But this does not imply that the screening was ineffective. Instead, determining donor suitability was accomplished by an informal process of spotting those physical symptoms or nonverbal clues that might indicate drug abuse or illness. This ended the standard screening and marked the beginning of another long wait until being called up to donate. The plasmapheresis took an additional hour to complete.

Because I was a new donor, an additional screening step including urinalysis and an examination by a staff physician was required before the second step. Each new donor was instructed to fill a clear plastic cup in the restroom and return it to the front counter. Donors were then required to hold out their cups at the counter in full view of those in the waiting area while it was tested by the staff. Women were particularly embarrassed by this, as was evident in their pained and awkward expressions as they carried their cups. After this, we were told to take a seat until our names were called.

Periodically, a staff physician would come into the waiting area and call the name of a new donor for examination. These exams took place in a small back office. This turned out to be more of an interrogation than a physical exam. There were many questions about personal habits, social life, and drug and alcohol use as well as medical problems. After my response to each question, the physician, an older woman, made brief notes on her chart but never seemed satisfied with my answers to her questions as my fieldnotes indicate:

> Many questions about drug use, both direct and indirect, looking for any symptoms. Makes short notes on the chart. Looks in my eyes each time

and pauses before reacting to my answers. Often repeats "Are you sure? Never?" Never stops during the pauses, searches her notes, thinking. Goes back to things I said earlier, such as sleeping on two pillows. . . . Makes many little notes in the chart.

I later recorded my impressions of this hour of questioning:

She trusted nothing I had said and always seemed to be searching for something she just knew had to be there. I must be abusing some substance or hiding some medical problem; otherwise, I wouldn't be there. It's just a matter of tricking me into a confession. I felt very guilty though I had no reason to be. . . . Each reply I made was probed and analyzed. Similar to being cross-examined by a sharp lawyer who won't give up until hearing what she wants to hear.

With reluctant hesitation, the physician finally informed me that "I will allow you to donate today." Because I had not eaten breakfast, I was given a cup of water and a banana. I was instructed to eat this in her presence. I could not be trusted. My subjective reaction to the exam was one of mixed resentment and relief. The other new donors being processed along with me that day appeared to have similar reactions to the exam.

After spending 4.5 hours at the plasma center on my first visit, I still had not been called up for the final step of plasmapheresis. Other obligations at the university that afternoon forced me to leave before the process was completed, and because of the oppressiveness of the setting, I was in any event anxious to return to my more familiar world.

On inquiring, one of the staff told me it was all right to leave—just tell them next time that my chart was "in progress." On returning the next morning, a staff member looked at my chart and asked, "Why did you leave, Martin?" She was puzzled by this irregularity. I explained how long I had waited and vaguely referred to other matters I needed to attend that day. She simply looked at me, and her quizzical expression told me that this did not make much sense. How could a donor have anything else to do?

The screening process, with the exception of the exam, had to be repeated the next morning, and it took another 4 hours at the center until I had completed the entire process. After passing the screening for new donors, I was stood against the wall, photographed, and issued a donor identification card.

The plasmapheresis procedure itself was simple; I was called up, given my chart, and ordered to report to the "orange" station. There, a phlebotomist unwrapped a sterile "Y set" and connected me to the plasmapheresis machine. The area around the vein was brushed with an iodine paste, and the needle pushed in without warning. On this first donation, a blood sample was drawn from the tubing, and I was casually asked to hold a test tube filled with my still-warm blood while the phlebotomist made adjustments to the machine.

The machine completed the process of taking four "draws" of blood, separating a predetermined amount of plasma, and returning the red blood cells in roughly an hour. The phlebotomist later returned to remove the needle and wrapped my arm tightly with three or more turns of sticky white tape. I then returned my chart and bag of plasma to the front counter as I had observed the others do. After I initialed the chart, money was laid on the counter. As my field notes show, I was uncertain what to do after this:

> That seems to be it. I feel that there must be something else, but there doesn't appear to be. Perhaps someone should say thank you or something, or maybe check whether I feel well enough to leave. Nothing. I leave and it feels good to be out of there.

The entire process had left me feeling uncertain, angry, awkward, and anxious. As a novice donor, I had expected explanations and other ritual interactions from the staff to make the process easier. As Espeland (1984) put it, the initial experience of donating plasma can be "amorphous" (p. 136). In the urban setting, however, the experience was distressing. I felt as if I had been booked by the police after an arrest, then simply released without explanation.

THE MORAL EVALUATION AND CONTROL OF DONORS

The variety of people passing through the urban plasma center were automatically assigned the provisional status of bad blood regardless of their varied backgrounds. As a rule, the staff very rarely engaged in normal conversation with donors and donors did not ask questions or make requests. They were expected to do as they were told and no more. When donors did insist on carrying on a one-sided conversation with staff, they would listen with one ear while continuing to work or would simply ignore the donor.

This categoric identification of donors is an example of Lofland's (1973) thesis that strangers in urban settings are categorized primarily in terms of the location or setting in which they are encountered and less on the basis of visible demeanor or signs of status. It is more accurate, however, to say that donors were routinely devalued *unless* the bad blood stereotype was contradicted by the presented signs of a student. As discussed earlier, student donors are the preferred clientele of plasma centers. Because I did not present myself as a student, and in any event was too old for this to be readily believed, I was naturally regarded as bad blood by the staff.

When the "frat boys" came into the center one day, I observed a remarkable transformation in the normal demeanor of the staff. These were three young, White, male students wearing fraternity shirts who had evidently called in advance. Two of the staff members stopped what they were doing and carried on a friendly conversation with the boys. As I recorded in my fieldnotes,

> The fraternity boys came in today. Very remarkable. Three young White students wearing Greek letters. A staff member looked up and exclaimed, "It's the frat boys!" Two of the staff talked with them at length, and the interaction was very different from what I have observed to be the rule. . . . Quite a shock to see someone treated like a regular adult. . . .

This is obviously good blood and they have the right to march up to the counter and be attended to.

The staff smiled at the boys as they answered questions and offered advice about the best times to donate. Except for this one instance, I had not seen any donor engage in such civil and friendly relations with the staff.

In Roth's (1972) terms, donors were evaluated by staff as "deserving" or "undeserving" by applying "concepts of social worth common in the larger society" (pp. 840–43). For the majority of donors, their imputed identity as bad blood rendered them undeserving of the "service" provided by the staff. In the case of student donors, however, society provides a social status and worthiness that was readily accepted by the staff.

CONCLUSION

The urban plasma center is a prime example of an interaction order within which we can observe a devalued category of persons who must, as Goffman (1983) aptly put it, "constantly pay a very considerable price for their interactional existence" (p. 6). This rather obvious fact, according to Goffman, raises a basic problem because

> . . . over the short historic run at least, even the most disadvantaged categories continue to cooperate—a fact hidden by the manifest ill will their members may display in regard to a few norms while sustaining all the rest. Perhaps behind a willingness to accept the way things are ordered is the brutal fact of one's place in the social structure and the real or imagined cost of allowing oneself to be singled out as a malcontent. Whatever, there is no doubt that categories of individuals in every time and place have exhibited a disheartening capacity for overtly accepting miserable interactional arrangements. (p. 6)

The miserable interactional arrangements of the plasma center repelled the majority of new donors. For the regulars who remained, however, these conditions were endurable. It would be wrong to think that regular donors were no longer concerned with avowing a positive identity, however. From the interactionist standpoint, Snow and Anderson (1987) argued that the need for self-esteem is by no means secondary to survival needs. Based on observations of identity work among the homeless, they stated that

> the attempt to carve out and maintain a sense of meaning and self-worth seems especially critical for survival, perhaps because it is the thread that enables those situated on the margins or at the bottom to retain a sense of self and thus their humanity. (p. 1365)

Identifying the donor role with a job is a way to survive as a regular plasma donor. But this is a situated identity dictated by the needs and requirements of the staff. It is a demeaning and, as Goffman (1983) put it, disheartening way to survive.

REFERENCES

Ball, Donald W. 1967. An abortion clinic ethnography. *Social Problems* 14: 293–301.

Espeland, Wendy Nelson. 1984. Blood and money: Exploiting the embodied self. In *The existential self in society,* edited by Joseph Kotarba and Andrea Fontana, 131–55. Chicago: University of Chicago Press.

Espeland, Wendy Nelson, and Elizabeth Clemens. 1989. *Buying blood and selling truth: The cultural constitution of boundaries by emerging organizations.* Revised version of a paper presented at the annual meeting of the American Sociological Association, Atlanta, August 1988.

Goffman, Erving. 1974. *Frame analysis.* New York: Harper & Row.

———. 1983. The interaction order. *American Sociological Review* 48: 1–17.

Goodsell, Charles B. 1984. Welfare waiting rooms. *Urban Life* 12: 467–77.

Gross, Edward. 1986. Waiting at Mayo. *Urban Life* 15: 139–64.

Hall, E. T. 1964. *The hidden dimension.* New York: Doubleday.

Lofland, Lyn H. 1973. *A world of strangers: Order and action in urban public space.* New York: Basic Books.

Mann, Leon. 1969. Queue culture: The waiting line as a social system. *American Journal of Sociology* 75: 340–54.

Piliavin, Jane, Peter Callero, and Dorcas Evans. 1982. Addiction to altruism? Opponent-process theory and habitual blood donation. *Journal of Personality and Social Psychology* 43: 1200–13.

Roth, Julius. 1972. Some contingencies of the moral evaluation and control of clientele: The case of the hospital emergency service. *American Journal of Sociology* 77: 839–56.

Schwartz, Barry. 1975. *Queuing and waiting.* Chicago: University of Chicago Press.

Snow, David, and Leon Anderson. 1987. Identity work among the homeless: The verbal construction and avowal of personal identities. *American Journal of Sociology* 92: 1336–71.

Snow, David, Louis Zurcher, and Gideon Sjoberg. 1982. Interviewing by comment: An adjunct to the direct question. *Qualitative Sociology* 5: 285–311.

Titmuss, Richard M. 1971. *The gift relationship.* New York: Random House.

19

Discrediting Victims' Allegations of Sexual Assault

Prosecutorial Accounts of Case Rejections

LISA FROHMANN

C ase screening is the gateway to the criminal court system. Prosecutors, acting as gatekeepers, decide which instances of alleged victimization will be passed on for adjudication by the courts. A recent study by the Department of Justice (Boland et al. 1990) suggests that a significant percentage of felony cases never get beyond this point, with only cases characterized as "solid" or "convictable" being filed (Stanko 1981, 1982; Mather 1979). This paper will examine how prosecutors account for the decision to reject sexual assault cases for prosecution and looks at the centrality of discrediting victims' rape allegations in this justification.

A number of studies on sexual assault have found that victim credibility is important in police decisions to investigate and make arrests in sexual assault cases (LaFree 1981; Rose and Randall 1982; Kerstetter 1990; Kerstetter and Van Winkle 1990). Similarly, victim credibility has been shown to influence prosecutors' decisions at a number of stages in the handling of sexual assault cases (LaFree 1980, 1989; Chandler and Torney 1981; Kerstetter 1990).

Much of this prior research has assumed, to varying degrees, that victim credibility is a phenomenon that exists independently of prosecutors' interpretations and assessments of such credibility. Particularly when operationalized in terms of quantitative variables, victim credibility is treated statistically as a series of fixed, objective features of cases. Such approaches neglect the processes whereby prosecutors actively assess and negotiate victim credibility in actual, ongoing case processing.

An alternative view examines victim credibility as a phenomenon constructed and maintained through interaction (Stanko 1980). Several qualitative studies have begun to identify and analyze these processes. For example, Holmstrom and Burgess's (1983) analysis of a victim's experience with the

From "Discrediting Victims' Allegations of Sexual Assault," Lisa Frohmann, *Social Problems*, Vol. 38, No. 2, May 1991. © 1991 Society for the Study of Social Problems. Reprinted by permission of University of California Press Journal and the author.

institutional handling of sexual assault cases discusses the importance of victim credibility through the prosecutor's evaluation of a complainant as a "good witness." A "good witness" is someone who, through her appearance and demeanor, can convince a jury to accept her account of "what happened." Her testimony is "consistent," her behavior "sincere," and she cooperates in case preparation. Stanko's (1981, 1982) study of felony case filing decisions similarly emphasizes prosecutors' reliance on the notion of the "stand-up" witness—someone who can appear to the judge and jury as articulate and credible. Her work emphasizes the centrality of victim credibility in complaint-filing decisions.

In this article I extend these approaches by systematically analyzing the kinds of accounts prosecutors offer in sexual assault cases to support their complaint-filing decisions. Examining the justifications for decisions provides an understanding of how these decisions appear irrational, necessary, and appropriate to decision-makers as they do the work of a case screening. It allows us to uncover the inner, indigenous logic of prosecutors' decisions and the organizational structures in which those decisions are embedded (Garfinkel 1984).

I focus on prosecutorial accounting for case rejection for three reasons: First, since a significant percentage of cases are not filed, an important component of the case-screening process involves case rejection. Second, the organization of case filing requires prosecutors to justify case rejection, not case acceptance, to superiors and fellow deputies. By examining deputy district attorneys' (DDAs') reasons for case rejection, we can gain access to what they consider "solid" cases, providing further insight into the case-filing process. Third, in case screening, prosecutors orient to the rule—when in doubt, reject. Their behavior is organized more to avoiding the error of filing cases that are not likely to result in conviction than to avoiding the error of rejecting cases that will probably end in conviction (Scheff 1966). Thus, I suggest that prosecutors are actively looking for "holes" or problems that will make the victim's version of "what happened" unbelievable or not convincing beyond a reasonable doubt, hence unconvictable (see Miller [1970], Neubauer [1974], and Stanko [1980, 1981] for the importance of conviction in prosecutors' decisions to file cases). This bias is grounded within the organizational context of complaint filing.

DATA AND METHODS

The research was part of an ethnographic field study of the prosecution of sexual assault crimes by deputy district attorneys in the sexual assault units of two branch offices of the district attorney's offices in a metropolitan area on the West Coast.* Research was conducted on a full-time basis in 1989 for nine

---

*To protect the confidentiality of people and places studied, pseudonyms are used throughout this article.

months in Bay City and on a full-time basis in 1990 for eight months in Center Heights. Three prosecutors were assigned to the unit in Bay City, and four prosecutors to the unit in Center Heights. The data came from 17 months of observation of more than three hundred case screenings. These screenings involved the presentation and assessment of a police report by a sexual assault detective to a prosecutor, conversations between detectives and deputies regarding the "filability"/reject status of a police report, interviews of victims by deputies about the alleged sexual assault, and discussions between deputies regarding the file/reject status of a report. Since tape recordings were prohibited, I took extensive field notes and tried to record as accurately as possible conversation between the parties. In addition, I also conducted open-ended interviews with prosecutors in the sexual assault units and with investigating officers who handled these cases. The accounts presented in the data below include both those offered in the course of negotiating a decision to reject or file a case (usually to the investigating officer [IO] but sometimes with other prosecutors or to me as an insider), and the more or less fixed accounts offered for a decision already made (usually to me). Although I will indicate the context in which the account occurs, I will not emphasize the differences between accounts in the analysis.

The data were analyzed using the constant comparison method of grounded theory (Glaser and Strauss 1967). I collected all accounts of case rejection from both offices. Through constant comparison of the data, I developed coding schema which provide the analytic framework of the paper.

The two branches of the district attorney's office I studied cover two communities differing in socioeconomic and racial composition. Bay City is primarily a white middle-to-upper-class community, and Center Heights is primarily a black and Latino lower-class community. Center Heights has heavy gang-drug activity, and most of the cases brought to the district attorney were assumed to involve gang members (both the complainant and the assailant) or a sex-drug or sex-money transaction. Because of the activities that occur in this community, the prior relationships between the parties are often the result of gang affiliation. This tendency, in connection with the sex-drug and sex-money transactions, gives a twist to the "consent defense" in "acquaintance" rapes. In Bay City, in contrast, the gang activity is much more limited and the majority of acquaintance situations that came to the prosecutors' attention could be categorized as "date rape."

THE ORGANIZATIONAL CONTEXT
OF COMPLAINT FILING

Several features of the court setting that I studied provided the context for prosecutors' decisions. These features are prosecutorial concern with maintaining a high conviction rate to promote an image of the "community's legal protector," and prosecutorial and court procedures for processing sexual assault cases.

The promotion policy of the county district attorney's (DA) office encourages prosecutors to accept only "strong" or "winnable" cases for prosecution by using conviction rates as a measure of prosecutorial performance. In the DA's office, guilty verdicts carry more weight than a conviction by case settlement. The stronger the case, the greater likelihood of a guilty verdict, the better the "stats" for promotion considerations. The inducement to take risks—to take cases to court that might not result in conviction—is tempered in three ways: First, a pattern of not-guilty verdicts is used by the DA's office as an indicator of prosecutorial incompetency. Second, prosecutors are given credit for the number of cases they reject as a recognition of their commitment to the organizational concern of reducing the case load of an already overcrowded court system. Third, to continually pursue cases that should have been rejected outright may lead judges to question the prosecutor's competence as a member of the court.

Sexual assault cases are among those crimes that have been deemed by the state legislature to be priority prosecution cases. That is, in instances where both "sex" and "nonsex" cases are trailing (waiting for a court date to open), sexual assault cases are given priority for court time. Judges become annoyed when they feel that court time is being "wasted" with cases that "should" have been negotiated or rejected in the first place, especially when those cases have been given priority over other cases. Procedurally, the prosecutor's office handles sexual assault crimes differently from other felony crimes. Other felonies are handled by a referral system; they are handed from one DDA to another at each stage in the prosecution of the case. But sexual assault cases are vertically prosecuted; the deputy who files the case remains with it until its disposition, and therefore is closely connected with the case outcome.

ACCOUNTING FOR REJECTION
BECAUSE OF "DISCREPANCIES"

Within this organizational context, a central feature of prosecutorial accounts of case rejection is the discrediting of victims' allegations of sexual assault. Below I examine two techniques used by prosecutors to discredit victims' complaints: discrepant accounts and ulterior motives.

Using Official Reports and
Records to Detect Discrepancies

In the course of reporting a rape, victims recount their story to several criminal justice officials. Prosecutors treat consistent accounts of the incident over time as an indicator of a victim's credibility. In the first example two prosecutors are discussing a case brought in for filing the previous day.

DDA Tamara Jacobs: In the police report she said all three men were kissing the victim. Later in the interview she said that was wrong. It seems strange because there are things wrong on major events like oral copulation and intercourse . . . for example whether she had John's penis in her mouth. Another thing wrong is whether he forced her into the bedroom immediately after they got to his room or, as the police report said, they all sat on the couch and watched TV. This is something a cop isn't going to get wrong, how the report started. (Bay City)

The prosecutor questions the credibility of the victim's allegation by finding "inconsistencies" between the complainant's account given to the police and the account given to the prosecutor. The prosecutor formulates differences in these accounts as "discrepancies" by noting that they involve "major events"—events so significant no one would confuse them, forget them, or get them wrong. This is in contrast to some differences that may involve acceptable, "normal inconsistencies" in victims' accounts of sexual assault. By "normal inconsistencies," I mean those that are expected and explainable because the victim is confused, upset, or shaken after the assault.

The DDA also discredited the victim's account by referring to a typification of police work. She assumes that the inconsistencies in the accounts could not be attributed to the incorrect writing of the report by the police officer on the grounds that they "wouldn't get wrong how the report started." Similarly, in the following example, a typification of police work is invoked to discredit the victim's interview.

> **DDA Sabrina Johnson:** [T]he police report doesn't say anything about her face being swollen, only her hand. If they took pictures of her hand, wouldn't the police have taken a picture of her face if it was swollen? (Bay City)

The prosecutor calls the credibility of the victim's complaint into question by pointing to a discrepancy between her subsequent account of injuries received during the incident and the notation of injuries on the police reports taken at the time the incident was reported. Suspicion of the complainant's account is also expressed in the prosecutor's inference that if the police went to the trouble of photographing the victim's injured hand they would have taken pictures of her face had it also shown signs of injury.

In the next case the prosecutor cites two types of inconsistencies between accounts. The first set of inconsistencies is the victim's accounts to the prosecutor and to the police. The second set is between the account the victim gave to the prosecutor and the statements the defendants gave to the police. This excerpt was obtained during an interview.

> **DDA Tracy Timmerton:** The reason I did not believe her [the victim] was, I get the police report first and I'll read that, so I have read the police report which recounts her version of the facts but it also has the statement of both defendants. Both defendants were arrested at separate times and gave separate independent statements that were virtually the same. Her story when I had her recount it to me in the DA's office, the number of acts changed, the chronological order of how they happened has changed. (Bay City)

When the prosecutor compared the suspects' accounts with the victim's account, she interpreted the suspects' accounts as credible because both of their accounts, given separately to police, were similar. This rests on the assumption that if suspects give similar accounts when arrested together, they are presumed to have colluded on the story, but if they give similar accounts independent of the knowledge of the other's arrest, there is presumed to be a degree of truth to the story. This stands in contrast to the discrepant accounts the complainant gave to law enforcement officials and the prosecutor.

Using Official Typifications of
Rape-Relevant Behavior

In the routine handling of sexual assault cases prosecutors develop a repertoire of knowledge about the features of these crimes.* This knowledge includes how particular kinds of rape are committed, post-incident interaction between the parties in an acquaintance situation, and victims' emotional and psychological reactions to rape and their effects on victims' behavior. The typifications of rape-relevant behavior are another resource for discrediting a victim's account of "what happened."

Typifications of Rape Scenarios   Prosecutors distinguish between different types of sexual assault. They characterize these types by the sex acts that occur, the situation in which the incident occurred, and the relationship between the parties. In the following excerpt the prosecutor discredits the victim's version of events by focusing on incongruities between the victim's description of the sex acts and the prosecutor's knowledge of the typical features of kidnap-rape. During an interview a DDA described the following:

> **DDA Tracy Timmerton:** [T]he only acts she complained of was intercourse, and my experience has been that when a rapist has a victim cornered for a long period of time, they engage in multiple acts and different types of sexual acts and very rarely do just intercourse. (Bay City)

The victim's account is questioned by noting that she did not complain about or describe other sex acts considered "typical" of kidnap-rape situations. She only complained of intercourse. In the next example the DDA and IO are talking about a case involving the molestation of a teenage girl.

> **DDA William Nelson:** Something bothers me, all three acts are the same. She's on her stomach and has her clothes on and he has a "hard and long penis." All three times he is grinding his penis into her butt. It seems to me he should be trying to do more than that by the third time. (Center Heights)

---

*The use of practitioners' knowledge to inform decision making is not unique to prosecutors. For example, such practices are found among police (Bittner 1967; Rubinstein 1973), public defenders (Sudnow 1965), and juvenile court officials (Emerson 1969).

Here the prosecutor is challenging the credibility of the victim's account by comparing her version of "what happened" with his typification of the way these crimes usually occur. His experience suggests there should be an escalation of sex acts over time, not repetition of the same act.

Often the typification invoked by the prosecutor is highly situational and local. In discussing a drug-sex-related rape in Center Heights, for example, the prosecutor draws on his knowledge of street activity in the community and the types of rapes that occur there to question whether the victim's version of events is what "really" happened. The prosecutor is describing a case he received the day before to an investigating officer there on another matter.

> **DDA Kent Fernome:** I really feel guilty about this case I got yesterday. The girl is 20 going on 65. She is real skinny and gangly. Looks like a cluckhead [crack addict]—they cut off her hair. She went to her uncle's house, left her clothes there, drinks some beers and said she was going to visit a friend in Center Heights who she said she met at a drug rehab program. She is not sure where this friend Cathy lives. Why she went to Center Heights after midnight, God Knows? It isn't clear what she was doing between 12 and 4 A.M. Some gang bangers came by and offered her a ride. They picked her up on the corner of Main and Lincoln. I think she was turning a trick, or looking for a rock, but she wouldn't budge from her story. . . . There are lots of conflicts between what she told the police and what she told me. The sequence of events, the sex acts performed, who ejaculates. She doesn't say who is who. . . . She's beat up, bruises on face and a laceration on her neck. The cop and doctor say there is no trauma—she's done by six guys. That concerns me. There is no semen that they see. It looks like this to me—maybe she is a strawberry, she's hooking or looking for a rock, but somewhere along the line it is not consensual. . . . She is [a] real street-worn woman. She's not leveling with me—visiting a woman with an unknown address on a bus in Center Heights—I don't buy it. . . . (Center Heights)

The prosecutor questioned the complainant's reason for being in Center Heights because, based on his knowledge of the area, he found it unlikely that a woman would come to this community at midnight to visit a friend at an unknown address. The deputy proposed an alternative account of the victim's action based on his knowledge of activities in the community—specifically, prostitution and drug dealing—and questioned elements of the victim's account, particularly her insufficiently accounted for activity between 12 and 4 A.M., coming to Center Heights late at night to visit a friend at an unknown address, and "hanging out" on the corner.

The DDA uses "person-descriptions" (Maynard 1984) to construct part of the account, describing the complainant's appearance as a "cluckhead" and "street-worn." These descriptions suggested she was a drug user, did not have a "stable" residence or employment, and was probably in Center Heights in search of drugs. This description is filled in by her previous "participation in a drug rehab program," the description of her activity as "hanging out" and being "picked up" by gang bangers, and a medical report which states that no trauma or semen was found when she was "done by six guys." Each of these features of the account suggests that the complainant is a prostitute or "strawberry" who came to Center Heights to trade sex or money for drugs. This alternative scenario combined with "conflicts between what she told the police and what she told me" justify case rejection because it is unlikely that the prosecutor could get a conviction.

The prosecutor acknowledges the distinction between the violation of women's sexual/physical integrity—"somewhere along the line it wasn't consensual"—and prosecutable actions. The organizational concern with "downstream consequences" (Emerson and Paley, forthcoming) mitigate against the case being filed.

Typifications of Post-Incident Interaction   In an acquaintance rape, the interaction between the parties after the incident is a critical element in assessing the validity of a rape complaint. As implied below by the prosecutors, the typical interaction pattern between victim and suspect after a rape incident is not to see one another. In the following cases the prosecutor challenges the validity of the victims' allegations by suggesting that the complainants' behavior runs counter to a typical rape victim's behavior. In the first instance the parties involved in the incident had a previous relationship and were planning to live together. The DDA is talking to me about the case prior to her decision to reject.

> **DDA Sabrina Johnson:** I am going to reject the case. She is making it very difficult to try the case. She told me she let him into her apartment last night because she is easily influenced. The week before this happened [the alleged rape] she agreed to have sex with him. Also, first she says "he raped me" and then she lets him into her apartment. (Bay City)

Here the prosecutor raises doubt about the veracity of the victim's rape allegation by contrasting it to her willingness to allow the suspect into her apartment after the incident. This "atypical" behavior is used to discredit the complainant's allegation.

In the next excerpt the prosecutor was talking about two cases. In both instances the parties knew each other prior to the rape incident as well as having had sexual relations after the incident. As in the previous instance, the victims' allegations are discredited by referring to their atypical behavior.

> **DDA Sabrina Johnson:** I can't take either case because of the women's behavior after the fact. By seeing these guys again and having sex with them they are absolving them of their guilt. (Bay City)

In each instance the "downstream" concern with convictability is indicated in the prosecutor's talk—"She is making it very difficult to try the case" and "By seeing these guys again and having sex with them they are absolving them of their guilt." This concern is informed by a series of common-sense assumptions about normal heterosexual relations that the prosecutors assume judges and juries use to assess the believability of the victim: First, appropriate behavior within ongoing relationships is noncoercive and nonviolent. Second, sex that occurs within the context of ongoing relationships is consensual. Third, if coercion or violence occurs, the appropriate response is to sever the relationship, at least for a time. When complainants allege they have been raped by their partner within a continuing relationship, they challenge the taken-for-granted assumptions of normal heterosexual relationships. The prosecutors anticipate that this challenge will create problems for the successful prosecution of a case because they think that judges and jurors will use this typification to question the credibility of the victim's allegation. They assume that the triers of fact will assume that if there is "evidence" of ongoing normal heterosexual relations—she didn't leave and the sexual relationship continued—then there was no coercive sex. Thus the certitude that a crime originally occurred can be retrospectively undermined by the interaction between complainant and suspect after the alleged incident. Implicit in this is the assumed primacy of the normal heterosexual relations typification as the standard on which to assess the victim's credibility even though an allegation of rape has been made.

Typifications of Rape Reporting   An important feature of sexual assault cases is the timeliness in which they are reported to the police (see Torrey, forthcoming). Prosecutors expect rape victims to report the incident relatively promptly: "She didn't call the police until four hours later. That isn't consistent with someone who has been raped." If a woman reports "late," her motives for reporting and the sincerity of her allegation are questioned if they fall outside the typification of officially recognizable/explainable reasons for late reporting. The typification is characterized by the features that can be explained by Rape Trauma Syndrome (RTS). In the first excerpt the victim's credibility is not challenged as a result of her delayed reporting. The prosecutor describes her behavior and motives as characteristic of RTS. The DDA is describing a case to me that came in that morning.

> **DDA Tamara Jacobs:** Charlene was in the car with her three assailants after the rape. John (the driver) was pulled over by the CHP [California Highway Patrol] for erratic driving behavior. The victim did not tell the officers that she had just been raped by these three men. When she arrived home, she didn't tell anyone what happened for approximately 24 hours. When her best friend found out from the assailants (who were mutual friends) and confronted the victim, Charlene told her what happened. She then reported it to the police. When asked why she didn't report the crime earlier, she said that she was embarrassed and afraid they would hurt her more if she reported it to the police. The DDA went on to say that the victim's behavior and reasons

for delayed reporting were symptomatic of RTS. During the trial an expert in Rape Trauma Syndrome was called by the prosecution to explain the "normality" and commonness of the victim's reaction. (Bay City)

Other typical motives include "wanting to return home first and get family support" or "wanting to talk the decision to report over with family and friends." In all these examples, the victims sustained injuries of varying degrees in addition to the trauma of the rape itself, and they reported the crime within 24 hours. At the time the victims reported the incident, their injuries were still visible, providing corroboration for their accounts of what happened.

In the next excerpt we see the connection between atypical motives for delayed reporting and ulterior motives for reporting a rape allegation. At this point I focus on the prosecutors' use of typification as a resource for discrediting the victim's account. I will examine ulterior motives as a technique of discrediting in a later section. The deputy is telling me about a case she recently rejected.

DDA Sabrina Johnson: She doesn't tell anyone after the rape. Soon after this happened she met him in a public place to talk business. Her car doesn't start, he drives her home and starts to attack her. She jumps from the car and runs home. Again she doesn't tell anyone. She said she didn't tell anyone because she didn't want to lose his business. Then the check bounces, and she ends up with VD. She has to tell her fiance so he can be treated. He insists she tell the police. It is three weeks after the incident. I have to look at what the defense would say about the cases. Looks like she consented, and told only when she had to because of the infection and because he made a fool out of her by having the check bounce. (Bay City)

The victim's account is discredited because her motives for delayed reporting—not wanting to jeopardize a business deal—fall outside those considered officially recognizable and explicable.

Typifications of Victim's Demeanor   In the course of interviewing hundreds of victims, prosecutors develop a notion of a victim's comportment when she tells what happened. They distinguish between behavior that signifies "lying" versus "discomfort." In the first two exchanges the DDA and IO cite the victim's behavior as an indication of lying. Below, the deputy and IO are discussing the case immediately after the intake interview.

IO Nancy Fauteck: I think something happened. There was an exchange of body language that makes me question what she was doing. She was yawning, hedging, fudging something.

DDA Sabrina Johnson: Yawning is a sign of stress and nervousness.

IO Nancy Fauteck: She started yawning when I talked to her about her record earlier, and she stopped when we finished talking about it. (Bay City)

The prosecutor and the investigating officer collaboratively draw on their common-sense knowledge and practical work experience to interpret the yawns, nervousness, and demeanor of the complainant as running counter to behavior they expect from one who is "telling the whole truth." They interpret the victim's behavior as a continuum of interaction first with the investigating officer and then with the district attorney. The investigating officer refers to the victim's recurrent behavior (yawning) as an indication that something other than what the victim is reporting actually occurred.

In the next excerpt the prosecutor and IO discredit the victim's account by referencing two typifications—demeanor and appropriate rape-victim behavior. The IO and prosecutor are telling me about the case immediately after they finished the screening interview.

IO Dina Alvarez: One on one, no corroboration.

DDA William Nelson: She's a poor witness, though that doesn't mean she wasn't raped. I won't file a one-on-one case.

IO Dina Alvarez: I don't like her body language.

DDA William Nelson: She's timid, shy, naive, virginal, and she didn't do all the right things. I'm not convinced she is even telling the truth. She's not even angry about what happened to her. . . .

DDA William Nelson: Before a jury if we have a one on one, he denies it, no witnesses, no physical evidence or medical corroboration, they won't vote guilty.

IO Dina Alvarez: I agree, and I didn't believe her because of her body language. She looks down, mumbles, crosses her arms, and twists her hands.

DDA William Nelson: . . . She has the same mannerisms and demeanor as a person who is lying. A jury just won't believe her. She has low self-esteem and self-confidence. . . . (Center Heights)

The prosecutor and IO account for case rejection by characterizing the victim as unbelievable and the case as unconvictable. They establish their disbelief in the victim's account by citing the victim's actions that fall outside the typified notions of believable and expected behavior—"she has the same mannerisms and demeanor as a person who is lying," and "I'm not convinced she is even telling the truth. She isn't even angry about what happened." They assume that potential jurors will also find the victim's demeanor and post-incident behavior problematic. They demonstrate the unconvictability of the case by citing the "holes" in the case—a combination of a "poor witness" whom "the jury just won't believe" and "one on one, [with] no corroboration" and a defense in which the defendant denies anything happened or denies it was nonconsensual sex.

Prosecutors and investigating officers do not routinely provide explicit accounts of "expected/honest" demeanor. Explicit accounts of victim demeanor tend to occur when DDAs are providing grounds for discrediting a rape allegation. When as a researcher I pushed for an account of expected behavior, the following exchange occurred. The DDA had just concluded the interview and asked the victim to wait in the lobby.*

IO Nancy Fauteck: Don't you think he's credible?

DDA Sabrina Johnson: Yes.

LF: What seems funny to me is that someone who said he was so unwilling to do this talked about it pretty easily.

IO Nancy Fauteck: Didn't you see his eyes, they were like saucers.

DDA Sabrina Johnson: And [he] was shaking too. (Bay City)

This provides evidence that DDAs and IOs are orienting to victims' comportment and could provide accounts of "expected/honest" demeanor if necessary. Other behavior that might be included in this typification are the switch from looking at to looking away from the prosecutor when the victim begins to discuss the specific details of the rape itself; a stiffening of the body and tightening of the face as though to hold in tears when the victim begins to tell about the particulars of the incident; shaking of the body and crying when describing the details of the incident; and a lowering of the voice and long pauses when the victim tells the specifics of the sexual assault incident.

Prosecutors have a number of resources they call on to develop typification related to rape scenarios and reporting. These include how sexual assaults are committed, community residents and activities, interactions between suspect and defendants after a rape incident, and the way victims' emotional and psychological responses to rape influence their behavior. These typifications highlight discrepancies between prosecutors' knowledge and victims' accounts. They are used to discredit the victims' allegation of events, justifying case rejection.

As we have seen, one technique used by prosecutors to discredit a victim's allegations of rape as a justification of case rejection is the detection of discrepancies. The resources for this are official documents and records and typifications of rape scenarios and rape reporting. A second technique prosecutors use is the identification of ulterior motives for the victim's rape allegation.

ACCOUNTING FOR REJECTION BY "ULTERIOR MOTIVE"

Ulterior motives rest on the assumption that a woman consented to sexual activity and for some reason needed to deny it afterwards. These motives are drawn from the prosecutor's knowledge of the victim's personal history and the community in which the incident occurred. They are elaborated and supported by other techniques and knowledge prosecutors use in the accounting process.

---

*Unlike the majority of rape cases I observed, this case had a male victim. Due to lack of data, I am unable to tell if this made him more or less credible in the eyes of the prosecutor and police.

I identify two types of ulterior motives prosecutors use to justify rejection: The first type suggests the victim has a reason to file a false rape complaint. The second type acknowledges the legitimacy of the rape allegation, framing the motives as an organizational concern with convictability.

Knowledge of Victim's Current Circumstances

Prosecutors accumulate the details of the victims' lives from police interviews, official documents, and filing interviews. They may identify ulterior motives by drawing on this information. Note that unlike the court trial itself, where the rape incident is often taken out of the context of the victim's life, here the DDAs call on the texture of a victim's life to justify case rejection. In an excerpt previously discussed, the DDA uses her knowledge of the victim's personal relationship and business transactions as a resource for formulating ulterior motive. Drawing on the victim's current circumstances, the prosecutor suggests two ulterior motives for the rape allegation—disclosure to her fiance about the need to treat a sexually transmitted disease, and anger and embarrassment about the bounced check. Both of these are motives for making a false complaint. The ulterior motives are supported by the typification for case reporting. Twice unreported sexual assault incidents with the same suspect, a three-week delay in reporting, and reporting only after the fiance insisted she do so are not within the typified behavior and reasons for late reporting. Her atypical behavior provides plausibility to the alternative version of the events—the interaction was consensual and only reported as a rape because the victim needed to explain a potentially explosive matter (how she contracted venereal disease) to her fiance. In addition she felt duped on a business deal.

Resources for imputing ulterior motives also come from the specifics of the rape incident. Below, the prosecutor's knowledge of the residents and activities in Center Heights supply the reason: the type of activity the victim wanted to cover up from her boyfriend. The justification for rejection is strengthened by conflicting accounts between the victim and witness on the purpose for being in Center Heights. The DDA and IO are talking about the case before they interview the complainant.

> **DDA William Nelson:** A white girl from Addison comes to buy dope. She gets kidnapped and raped.
>
> **IO Brandon Palmer:** She tells her boyfriend and he beats her up for being so stupid for going to Center Heights. . . . The drug dealer positively ID'd the two suspects, but she's got a credibility problem because she said she wasn't selling dope, but the other two witnesses say they bought dope from her. . . .
>
> **LF:** I see you have a blue sheet [a sheet used to write up case rejections] already written up.
>
> **IO Brandon Palmer:** Oh yes. But there was no doubt in my mind that she was raped. But do you see the problems?

> **DDA William Nelson:** Too bad because these guys really messed her up. . . . She has a credibility problem. I don't think she is telling the truth about the drugs. It would be better if she said she did come to buy drugs. The defense is going to rip her up because of the drugs. He is going to say, isn't it true you had sex with these guys but didn't want to tell your boyfriend, so you lied about the rape like you did about the drugs, or that she had sex for drugs. . . . (Center Heights)

The prosecutor expresses doubt about the victim's account because it conflicts with his knowledge of the community. He uses this knowledge to formulate the ulterior motive for the victim's complaint—to hide from her boyfriend the "fact" that she traded sex for drugs. The victim, "a white woman from Addison," alleges she drove to Center Heights "in the middle of the night" as a favor to a friend. She asserted that she did not come to purchase drugs. The DDA "knows" that white people don't live in Center Heights. He assumes that whites who come to Center Heights, especially in the middle of the night, are there to buy drugs or trade sex for drugs. The prosecutor's scenario is strengthened by the statements of the victim's two friends who accompanied her to Center Heights, were present at the scene, and admitted buying drugs. The prosecutor frames the ulterior motives as an organizational concern with defense arguments and convictability. This concern is reinforced by citing conflicting accounts between witnesses and the victim. He does not suggest that the victim's allegation was false—"there is no doubt in my mind she was raped"; rather, the case isn't convictable—"she has a credibility problem" and "the defense is going to rip her up."

Criminal Connections

The presence of criminal connections can also be used as a resource for identifying ulterior motives. Knowledge of a victim's criminal activity enables prosecutors to "find" ulterior motives for her allegation. In the first excerpt the complainant's presence in an area known by police as "where prostitutes bring their clients" is used to formulate an ulterior motive for her rape complaint. This excerpt is from an exchange in which the DDA was telling me about a case he had just rejected.

> **DDA William Nelson:** Young female is raped under questionable circumstances. One on one. The guy states it is consensual sex. There is no corroboration, no medicals. We ran the woman's rap sheet, and she had a series of prostitution arrests. She's with this guy in the car in a dark alley having sex. The police know this is where prostitutes bring their customers, so she knew she had better do something fast unless she is going to be busted for prostitution, so, lo and behold, she comes running out of the car yelling "he's raped me." He says no. He picked her up on Long Beach Boulevard, paid her $25 and this is "where she brought me." He's real scared, he has no record. (Center Heights)

Above, the prosecutor, relying on police knowledge of a particular location, assumes the woman is a prostitute. Her presence in the location places her in a "suspicious" category, triggering a check on her criminal history. Her record of prostitution arrests is used as the resource for developing an ulterior motive for her complaint: To avoid being busted for prostitution again, she made a false allegation of rape. Here the woman's record of prostitution and the imminent possibility of arrest are used to provide the ulterior motive to discredit her account. The woman's account is further discredited by comparing her criminal history—"a series of prostitution arrests" with that of the suspect, who "has no record," thus suggesting that he is the more credible of the two parties.

Prosecutors and investigating officers often decide to run a rap sheet (a chronicle of a person's arrests and convictions) on a rape victim. These decisions are triggered when a victim falls into certain "suspicious" categories, categories that have a class/race bias. Rap sheets are not run on women who live in the wealthier parts of town (the majority of whom are white) or have professional careers. They are run on women who live in Center Heights (who are black and Latina), who are homeless, or who are involved in illegal activities that could be related to the incident.

In the next case the prosecutor's knowledge of the victim's criminal conviction for narcotics is the resource for formulating an ulterior motive. This excerpt was obtained during an interview.

> **DDA Tracy Timmerton:** I had one woman who had claimed that she had been kidnapped off the street after she had car trouble by these two gentlemen who locked her in a room all night and had repeated intercourse with her. Now she was on a cocaine diversion [a drug treatment program where the court places persons convicted of cocaine possession instead of prison], and these two guys' stories essentially were that the one guy picked her up, they went down and got some cocaine, had sex in exchange for the cocaine, and the other guy comes along and they are all having sex and all doing cocaine. She has real reason to lie, she was doing cocaine, and because she has then violated the terms of her diversion and is now subject to criminal prosecution for her possession of cocaine charge. She is also supposed to be in a drug program which she has really violated, so this is her excuse and her explanation to explain why she has fallen off her program. (Bay City)

The prosecutor used the victim's previous criminal conviction for cocaine and her probation conditions to provide ulterior motives for her rape allegation—the need to avoid being violated on probation for the possession of cocaine and her absence from a drug diversion program. She suggests that the allegation made by the victim was false.

Prosecutors develop the basis for ulterior motives from the knowledge they have of the victim's personal life and criminal connections. They create two types of ulterior motives, those that suggest the victim made a false rape complaint and those that acknowledge the legitimacy of the complaint but discredit the account because of its unconvictability. In the accounts prosecutors give, ulterior motives for case rejection are supported with discrepancies in victims' accounts and other practitioners' knowledge.

CONCLUSION

Case filing is a critical stage in the prosecutorial process. It is here that prosecutors decide which instances of alleged victimization will be forwarded for adjudication by the courts. A significant percentage of sexual assault cases are rejected at this stage. This research has examined prosecutorial accounts for case rejection and the centrality of victim discreditability in those accounts. I have elucidated the techniques of case rejection (discrepant accounts and ulterior motives), the resources prosecutors use to develop these techniques (official reports and records, typifications of rape-relevant behavior, criminal connections and knowledge of a victim's personal life), and how these resources are used to discredit victims' allegations of sexual assault.

This examination has also provided the beginnings of an investigation into the logic and organization of prosecutors' decisions to reject/accept cases for prosecution. The research suggests that prosecutors are orienting a "downstream" concern with convictability. They are constantly "in dialogue with" anticipated defense arguments and anticipated judge and juror responses to case testimony. These dialogues illustrate the intricacy of prosecutorial decision-making. They make visible how prosecutors rely on assumptions about relationships, gender, and sexuality (implicit in this analysis, but critical and requiring of specific and explicit attention) in complaint filing of sexual assault cases. They also make evident how the processes of distinguishing truths from untruths and the practical concerns of trying cases are central to these decisions. Each of these issues, in all its complexity, needs to be examined if we are to understand the logic and organization of filing sexual assault cases.

The organizational logic unveiled by these accounts has political implications for the prosecution of sexual assault crimes. These implications are particularly acute for acquaintance rape situations. As I have shown, the typification of normal heterosexual relations plays an important role in assessing these cases, and case conviction is key to filing cases. As noted by DDA William Nelson: "There is a difference between believing a woman was assaulted and being able to get a conviction in court." Unless we are able to challenge the assumptions on which these typifications are based, many cases of rape will never get beyond the filing process because of unconvictability.

REFERENCES

Bittner, Egon A. 1967. "The police on skid-row: A study of peace keeping." *American Sociological Review* 32: 699–715.

Boland, Barbara, Catherine H. Conly, Paul Mahanna, Lynn Warner, and Ronald Sones. 1990. *The Prosecution of Felony Arrests, 1987.* Washington, DC: Bureau of Justice Statistics, U.S. Department of Justice.

Chandler, Susan M., and Martha Torney. 1981. "The decision and the processing of rape victims through the criminal justice system." *California Sociologist* 4: 155–69.

Emerson, Robert M. 1969. *Judging Delinquents: Context and Process in Juvenile Court.* Chicago: Aldine.

Emerson, Robert M., and Blair Paley. Forthcoming. "Organizational horizons and complaint-filing." In *The Uses of Discretion,* ed. Keith Hawkins. Oxford: Oxford University Press.

Garfinkel, Harold. 1984. *Studies in Ethnomethodology.* Cambridge, G.B.: Polity Press.

Glaser, Barney, and Anselm Strauss. 1967. *The Discovery of Grounded Theory.* Chicago: Aldine.

Holmstrom, Lynda Lytle, and Ann Wolbert Burgess. 1983. *The Victim of Rape: Institutional Reactions.* New Brunswick, NJ: Transaction Books.

Kerstetter, Wayne A. 1990. "Gateway to justice: Police and prosecutorial response to sexual assaults against women." *Journal of Criminal Law and Criminology* 81: 267–313.

Kerstetter, Wayne A., and Barrik Van Winkle. 1990. "Who decides? A study of the complainant's decision to prosecute in rape cases." *Criminal Justice and Behavior* 17: 268–83.

LaFree, Gary D. 1980. "Variables affecting guilty pleas and convictions in rape cases: Toward a social theory of rape processing." *Social Forces* 58: 833–50.

————. 1981. "Official reactions to social problems: Police decisions in sexual assault cases." *Social Problems* 28: 582–94.

————. 1989. *Rape and Criminal Justice: The Social Construction of Sexual Assault.* Belmont, CA: Wadsworth.

Mather, Lynn M. 1979. *Plea Bargaining or Trial? The Process of Criminal-Case Disposition.* Lexington, MA: Lexington Books.

Maynard, Douglas W. 1984. *Inside Plea Bargaining: The Language of Negotiation.* New York: Plenum Press.

Miller, Frank. 1970. *Prosecution: The Decision to Charge a Suspect with a Crime.* Boston: Little, Brown.

Neubauer, David. 1974. *Criminal Justice in Middle America.* Morristown, NJ: General Learning Press.

Rose, Vicki M., and Susan C. Randall. 1982. "The impact of investigator perceptions of victim legitimacy on the processing of rape/sexual assault cases." *Symbolic Interaction* 5: 23–36.

Rubinstein, Jonathan. 1973. *City Police.* New York: Farrar, Straus & Giroux.

Scheff, Thomas. 1966. *Being Mentally Ill: A Sociological Theory.* Chicago: Aldine.

Stanko, Elizabeth A. 1980. "These are the cases that try themselves: An examination of extra-legal criteria in felony case processing." Presented at the Annual Meetings of the North Central Sociological Association, December. Buffalo, NY

————. 1981. "The impact of victim assessment on prosecutor's screening decisions: The case of the New York District Attorney's Office." *Law and Society Review* 16: 225–39.

————. 1982. "Would you believe this woman? Prosecutorial screening for 'credible' witnesses and a problem of justice." In *Judge, Lawyer, Victim, Thief,* ed. Nicole Hahn Rafter and Elizabeth A. Stanko (pp. 63–82). Boston: Northeastern University Press.

Sudnow, David. 1965. "Normal crimes: Sociological features of the penal code in a public defenders office." *Social Problems* 12: 255–76.

Torrey, Morrison. Forthcoming. "When will we be believed? Rape myths and the idea coming of a fair trial in rape prosecutions." *U.C. Davis Law Review.*

Waegel, William B. 1981. "Case routinization in investigative police work." *Social Problems* 28: 263–75.

Williams, Kristen M. 1978a. *The Role of the Victim in the Prosecution of Violent Crimes.* Washington, DC: Institute for Law and Social Research.

———. 1978b. *The Prosecution of Sexual Assaults.* Washington, DC: Institute for Law and Social Research.

PART V

Deviant Identity

Bureaucratic organizations are not the only contexts in which people are labeled as deviant; this labeling can also occur in interpersonal situations. Becoming deviant does not only entail having a definition of deviance and an environment in which it can occur; it also requires that people accept the identity and make it their own. In Part V we will examine this process: how the concept of deviance becomes applied to individuals and how it affects their self-conception.

IDENTITY DEVELOPMENT

We mentioned earlier that although many people engage in deviance, the label is applied to only a small percentage of them. Such labeling is tied to their formerly "secret deviance" (Becker 1963) becoming exposed, or to an abstract status coming to bear on their personal experience. Thus, Jews may not feel stigmatized unless they experience anti-Semitism, and embezzlers may not think of themselves as thieves until they are caught. When this happens, they enter the pathway to the deviant identity, a pathway that follows a certain trajectory. Howard Becker (1963) has suggested that we can think of this path as a **deviant career.**

Once people are publicly identified as deviant, their lives change in several ways. Others start to think of them differently. For example, suppose there has been a rash of thefts in a college dormitory, and Jessica, a freshman, is finally caught and identified as the culprit. She may or may not be reported to authorities and charged with theft; regardless, she will experience an informal labeling process. People will probably change their attitudes toward her as they find themselves talking about her behind her back. They may look back on her behavior and engage in **retrospective interpretation** (Kitsuse 1962) as they think about her differently in light of their new information. Where did she say she was when the last theft occurred? Where did she say she got the money to buy that new sweater? Jessica may develop what Goffman (1963) has called a **spoiled identity,** one with a damaged reputation.

Jessica's dormmates and former friends may then engage in what Lemert (1951) has called the **dynamics of exclusion,** deriding and ostracizing her from their social group. When she enters a room, she may notice that a sudden hush falls over the conversation. People may not feel comfortable leaving her alone in their room. They may exclude her from their supper plans and study groups. She may become progressively shut out from nondeviant activities and circles. At the same time, others may welcome or *include* her in their deviant circles or activities. She may find that she has developed a reputation that, though repelling to some groups, is attractive to others. They may welcome her as "cool" and invite her into their circles. The more people start to treat Jessica and interact with her as a deviant, the more she is likely to **internalize the label** and regard herself as a deviant.

Once people are labeled as deviant and accept that label into their self-conceptions, a variety of outcomes may ensue. Everett Hughes (1945) has suggested that we all have a series of statuses or identities through which we relate to people, including those of sibling, child, friend, student, neighbor, and customer. These identities also derive from some of our demographic or occupational features, such as race, gender, age, religion, or social class. Hughes asserts that some statuses are very dominant, overpowering others and coloring the way in which people are viewed. Having a deviant identity may become one of these **master statuses,** rising to the top of the hierarchy and infusing people's self-concept and others' reactions. People who are labeled as deviants, such as heroin addicts, cult members, or homosexuals, may be viewed by others through this lens no matter what the situation or setting. In addition, every master status has a set of **auxiliary traits** that accompany it, and once people label someone with a deviant master status they will expect to see the relevant auxiliary traits. A heroin addict may be suspected of being a prostitute or thief, and a homosexual may be suspected of being sexually promiscuous or AIDS-infected. This type of identification spreads the image of deviance to cover the person as a whole and not just one part of him or her.

The process of developing a deviant master status and auxiliary traits helps explain the move from **primary deviance** to **secondary deviance.** As Lemert (1967) defined it, primary deviance refers to the initial type of deviance in which people engage, one that has not necessarily yielded them the master status of deviance. Getting caught and labeled reinforces the deviant identity, spoils people's reputations, and leads to altered self-concepts. People accept the deviant view of themselves; then, stripped of their fear of losing their good reputations, they may go on to engage in more and different kinds of deviant behavior. Secondary deviance thus refers both to the diffusion of involvement in deviance and the change in self-conception, whereby people interact with others through the deviant master status.

Becker's notion of the identity career can thus be seen to encompass several stages, beginning with the commission of the deviance and leading to individuals' apprehension and public identification, the changing attitudes and expectations of others towards those individuals, shifting social acceptance or rejection by their friends and acquaintances, individuals' internalization of the deviant label and self-identity and interaction through it, movement into groups of differential associates, and commission of further acts of deviance.

Some of these processes are illustrated by the three readings in this first section of Part V. "The Identity Change Process: A Field Study of Obesity," by Douglas Degher and Gerald Hughes, treats the change process from a career focus by looking at the sequence whereby individuals use external cues to "recognize" that their self-image as thin or normally weighted is inappropriate, and to "place" themselves in a new, more appropriate status closer to the fat side of the continuum. Toward this end, they integrate external status cues with an evolving internal self-identity to locate themselves in society.

Naomi Gerstel's "Divorce and Stigma" highlights the dynamics of exclusion associated with divorcees' shift from the status of married to that of divorced. Within a short time after the divorce, individuals progressively find themselves excluded from the company of their former couple friends and cast into the company of other divorcees. Treated as outsiders and social misfits, they join with other outcasts in developing feelings of devaluation and demoralization.

In "Anorexia Nervosa and Bulimia: The Development of Deviant Identities," Penelope McLorg and Diane Taub describe and analyze women's progression from socially conforming eating behavior through primary deviance and on to secondary deviance. Along the way, the women they studied moved through stages of more common fixations about dieting, to frustration with dieting and movement toward more radical solutions such as binging, purging, compulsive exercising, and lack of eating (devoid of reconception of their selves—primary deviance), and to recognition of serious eating disorders accompanied by the internalization of these labels into their self-identities. By interacting with others through the vehicle of their eating disorders, these women reinforced their anorectic and bulimic behaviors.

ACCOUNTS AND NEUTRALIZATIONS

Marvin Scott and Stanford Lyman (1968) have suggested that we all engage in instances of deviant behavior but that we desire to maintain a positive self-image in both our own eyes and the eyes of others. To do so, we offer **accounts** intended to explain and **normalize** our deviant behavior. Three kinds of accounts predominate: excuses, justifications, and techniques of neutralization.

In offering **excuses,** individuals admit the wrongfulness of their actions but distance themselves from the blame. These excuses may precede or follow the acts of deviance. They are often fairly standard phrases or ideas designed to soften the deviance and relieve individuals of their accountability. These may include appeals to accidents ("my computer malfunctioned and lost my file"), appeals to defeasibility ("I thought my roommate turned my paper in"), appeals to biological drives ("I couldn't stop myself"), and scapegoating ("she borrowed my notes and I couldn't get them back in time to study for the test").

In offering **justifications,** individuals accept responsibility for their actions but seek to have specific instances excused. In so doing, they try to legitimate the acts or their consequences. In drawing on justifications, individuals may invoke sad tales ("I turn tricks because I was sexually abused as a child") or the need for self-fulfillment ("taking hallucinogenic drugs expands my consciousness").

Techniques of neutralization attempt to resolve the contradictions between what people say and what they do. Even people who engage in deviance want to feel good about themselves, so they offer neutralizations to allow a conception of themselves as "moral" (Sykes and Matza 1957). These techniques include the denial of responsibility ("it wasn't my fault"), denial of injury ("no harm done, it was just a prank"), denial of the victim ("nobody was hurt," "he deserved it"), condemnation of the condemners ("society is unfair"), and the appeal to higher loyalties ("some things are more important").

Diana Scully and Joseph Marolla's "Convicted Rapists' Vocabulary of Motive: Excuses and Justifications" offers a fascinating glimpse into the rationalizations offered by a group of incarcerated rapists. Representing the most hard-core segment of the rapist population, those sentenced to prison time, these men gave a variety of accounts to legitimate their violent crimes. In dividing these accounts into the categories of excuses and justifications, Scully and Marolla illustrate these disavowal techniques and shed light on the repertoire of culturally available excuses that men use to neutralize this behavior.

Donald McCabe then presents some rationalizations for students' deviant behavior in "The Influence of Situational Ethics on Cheating Among College Students." Based on a survey of 6,000 students, McCabe shows how Sykes and Matza's neutralization techniques can be used in a familiar situation. Readers will no doubt recognize some of these common rationales, as they tie their everyday life surroundings to deviance concepts.

STIGMA MANAGEMENT

The label of "deviant" marks people with a stigma in the eyes of society. As we have seen, this label may lead to devaluation and exclusion. Consequently, people with deviant features learn how to "manage" their stigma so that they are not shamed or ostracized. This effort requires considerable social skills.

Goffman (1963) has suggested that people with potential deviant stigma fall into two categories: **the discreditable** and **the discredited.** The former are those with concealable deviant traits (ex-convicts, secret homosexuals) who may manage themselves so as to avoid the deviant stigma. The latter are either members of the former category who have revealed their deviance or those who cannot hide their deviance (the obese, the physically handicapped). These people's lives are characterized by a constant focus on secrecy and information control. Goffman observed that most discreditables engage in **passing** as "normals" in their everyday lives, concealing their deviance and fitting in with regular people. They may do this by avoiding contacts with "stigma symbols," those objects or behaviors that would tip people off to their deviant condition (an unwed father avoiding his pregnant girlfriend, AIDS patients avoiding their medication). Another technique for passing includes using "disidentifiers" such as props, actions, or verbal expressions to distract and fool others into thinking that they do not have the deviant stigma (homosexuals bragging about heterosexual sexual conquests or taking a date to the company picnic). Finally, they may "lead a double life," maintaining two different lifestyles with two distinct groups of people, one that knows about their deviance and one that does not.

In this endeavor, people may employ the aid of others to help conceal their deviance by **covering** for them. In these team performances, friends and family members may assist the deviants by concealing their identities, their whereabouts, their deficiencies, or their pasts. They may even coach the deviants on how to construct stories designed to hide their deviance.

Another form of stigma management, sometimes adopted when concealment fails, involves disclosing the deviance. People may do this for cathartic reasons (alleviating their burden of secrecy), therapeutic reasons (casting it in a positive light), or preventive reasons (so others don't find out in negative ways later, so that efforts at developing relationships are not totally wasted). Although many people find that their disclosures lead to rejection, others are more fortunate.

Disclosures of deviance can follow two courses. In observing the interactions between deviants and normals, Fred Davis (1961) noted that some nondeviant people go through a process of normalizing their relationship with the deviant person largely through failing to acknowledge the deviant trait. Although this disavowal usually begins with a conspicuous and stilted ignoring of the individual's deviance **(deviance disavowal),** it progresses through a stage in which more relaxed interaction begins, directed at features of the person other than his or her deviant stigma, and finally gets to the point at which the deviant stigma is overlooked and almost forgotten.

In contrast, deviant people can strive to normalize their relationships with nondeviants through **deviance avowal** (Turner 1972), in which they openly acknowledge their stigma and try to present themselves in a positive light. This avowal often takes the form of humor, "breaking the ice" by joking about their deviant attribute. In this way, they show others that they can take the perspective of the normal and also see themselves as deviant, thus forming a bridge to others. This action further asserts that they have nondeviant aspects and that they can see the world as others do.

Thus far, we have considered individual modes of adaptation to deviant stigma. Yet these stigma can also be managed through a group or collective effort. Many voluntary associations of stigmatized individuals exist, from the early organizations of prostitutes (COYOTE—Call Off Your Old Tired Ethics), to more recent ones such as the Gay Liberation Front, the Little People of America, the National Stuttering Project, and the Gray Panthers. Most well known are the twelve-step programs modeled after the tremendous success of Alcoholics Anonymous (AA), including such groups as Overeaters Anonymous, Narcotics Anonymous, and Gamblers Anonymous.

These groups vary in character. Some are organized as what Stanford Lyman (1970) has called **expressive** groups, whose primary function is to provide support for their members. This support can take the form of organizing social and recreational activities, dispersing legal or medical information, or offering services such as shopping, meals, or transportation. Expressive groups tend to be apolitical, helping their members adapt to their social stigma rather than evade it. They also serve as places where members can come together in the company of other deviants, avoid the censure of nonstigmatized normals, and seek collective solutions to their common problems. It is here that they can make disclosures to others without fear of rejection.

Lyman has also described **instrumental** groups, in which members come together not only to accomplish the expressive functions but also to organize for political activism. Building on Lemert's (1967) terms, Kitsuse (1980) has called this activity **tertiary deviation,** in which individuals reject the societal conception and treatment of their stigma and organize to change social definitions. They fight to get others to modify their views of the status or behavior in question so that society, like them, will no longer regard it as deviant. Examples of such groups include ACT UP, an AIDS organization whose members have tried to change social attitudes toward AIDS patients, the National Organization for Women (NOW), and the Disabled in Action.

In "Women Athletes as Falsely Accused Deviants: Managing the Lesbian Stigma," Elaine Blinde and Diane Taub consider the double violation of the feminine gender role associated with women's participation in intercollegiate sports and outsiders' common assumption that they are lesbians. Blinde and Taub place the falsely accused deviant within Becker's (1963) typology of deviant roles, comparing it to other cases that are not perceived as deviant and that actually engage in rule-breaking behavior. They then portray the techniques that women athletes use to conceal, deflect, and normalize their stigma through the use of several strategies that include passing, salience reduction, and direct confrontation.

Nancy Herman's "Return to Sender: Reintegrative Stigma-Management Strategies of Ex-Psychiatric Patients" discusses individuals' efforts to manage their stigma by addressing some of the problems encountered by former mental patients after their release from treatment. She poignantly describes individuals' painful decisions about whether to conceal or disclose their stigmatizing past and some of the factors on which they base their decisions. She then explores the ways in which they try to manage their disclosures, including some of the difficulties they encounter. Whereas some of their confidants prove sympathetic, others react poorly, rejecting and re-stigmatizing the former patients. Finally, Herman discusses ways in which the individuals she studied collectively organize and fight to manage their deviant stigma.

David Karp's "Illness Ambiguity and the Search for Meaning: A Self-Help Group for Affective Disorders" offers us an excellent example of an expressive, collective stigma-management group. Based on the AA model, this group of manic depressives met weekly to discuss ways of coping with their illness and its accompanying stigma. Through concrete suggestion and emotional empathy, they helped one another navigate the complex and often painful pathway through diagnosis, forging a series of relationships with physicians and psychiatrists, and dealing with the positive and negative aspects of their experiences with therapeutic drugs.

20

The Identity
Change Process
A Field Study of Obesity

DOUGLAS DEGHER AND GERALD HUGHES

T he interactionist perspective has come to play an important part in contemporary criminological and deviance theory. Within this approach,
deviance is viewed as a subjectively problematic identity rather than an
objective condition of behavior. At the core is the emphasis on "process" rather
than on viewing deviance as a static entity. To paraphrase vintage Howard
Becker, ". . . social groups create deviance by making the rules whose infraction constitutes deviance. Consequently, deviance is not a quality of the act . . .
but rather a consequence of the application by others of rules and sanctions to
an offender" (Becker, 1963, p. 9). Attention is focused upon the *interaction* between those being labeled deviant and those promoting the deviant label. In
the interactionist literature, emphases are in two major areas: (a) the conditions
under which the label "deviant" comes to be applied to an individual and the
consequences for the individual of having adopted that label (Tannenbaum,
1939; Lemert, 1951; Kitsuse, 1961, p. 247; Goffman, 1963; Baum, 1987, p. 96;
Greenberg, 1989, p. 79), and/or (b) the role of social control agents[1] in contributing to the application of deviant labels (Becker, 1963; Piliavin & Briar,
1964, p. 206; Cicourel, 1968; Schur, 1971; Conrad, 1975, p. 12).

Much of this literature frequently assumes that once an individual has been
labeled, the promoted label and attendant identity is either internalized or rejected. As Lemert proposes, the shift from primary to secondary deviance is a
categorical one, and is primarily a response to problems created by the societal
reaction (Lemert, 1951, p. 40).

What is most often neglected is an examination of the mechanistic features
of this identity shift. Our focus is on this "identity change process," which is
what we have chosen to call this identity shift. Of interest is how individuals
come to make some personal sense out of proffered labels and their attendant
identities.

From "The Identity Change Process: A Field Study of Obesity," Douglas Degher and
Gerald Hughes, *Deviant Behavior,* V. 12, No. 4, Taylor & Francis, Inc., Washington, D.C.,
pp. 385–402, 1991. Reprinted with permission. All rights reserved.

METHODOLOGY

The primary methodological tool employed to construct our identity change process model comes from "grounded" analysis. . . . The model presented in this paper emerged from comments and codes appearing in interviews with obese members of a weight reduction organization that had weekly meetings. The frequency of attendance allowed us to consider the members typical, and allowed us to suggest that major issues of obesity are trans-situational and temporally durable. If obesity disappeared tomorrow, we would still be able to apply the generic concepts generated from our data to make statements about the process of "identity change." As suggested by Hadden, Degher, and Fernandez, our focus is on process rather than on unit characteristics of social phenomena (Hadden, Degher, & Fernandez, 1989, p. 9). This provides us with insights that have import for major issues in sociological theory.

SITE SELECTION

Because obese individuals suffer both internally (negative self-concepts) and externally (discrimination), they possess what Goffman refers to as a "spoiled identity" (Goffman, 1963). This seems to be the case particularly in contemporary America with what may be described as an almost pathological emphasis on fitness. The boom in health clubs, sales of videotapes on fitness, diet books, and so forth promote a definition of the "healthy" physical presence. As Kelly (1990) sees it, the boom in physical fitness in the mid–1980s is an attempt by many people to create a specific image of an ideal body. Thus, body build becomes a crucial element in self-appraisal. Consequently, fat people are an ideal strategic group within which to study the "identity change process."

Obese people are not only the subject of negative stereotypes, they are also actively discriminated against in college admissions (Canning & Mayer, 1966, p. 1172), pay more for goods and services (Petit, 1974), receive prejudicial medical treatment (Maddox, Back, & Liederman, 1968, p. 287; Maddox & Liederman, 1969, p. 214), are treated less promptly by salespersons (Pauley, 1989, p. 713), have higher rates of unemployment (Laslett & Warren, 1975, p. 69), and receive lower wages (Register, 1990, p. 130). The obese label is one that seems to clearly fit Becker's description of a "master status," that is

> Some statuses in our society, as in others, override all other statuses and have a certain priority . . . the status deviant (depending on the kind of deviance) is this kind of master status . . . one will be identified as a deviant first, before other identifications are made. (Becker, 1983, p. 33)

Obese people are "fat" first, and only secondarily are seen as possessing ancillary characteristics.

The site for the field observations had to meet two requirements: (a) it had to contain a high proportion of obese, or formerly obese individuals; and

(b) these individuals had to be identifiable by the observer. The existence of a large number of national weight control organizations (a) whose membership is composed of individuals who have internalized an obese identity, and (b) who emphasize a radical program of identity change, make these organizations an excellent choice as strategic sites for study and analysis. The local franchise chapter of one of these national weight loss organizations satisfied both of our requirements, and was selected as the site for our study.

Attendance at the weekly meetings of this national weight control group is restricted to individuals who are current members of the organization. Since one requirement for membership is that the individual be at least 10 pounds over the maximum weight for his or her sex and height (according to New York Life tables), all of the people attending the meetings are, or were, overweight, and a high proportion of them are, or were, sufficiently overweight to be classified as obese.[2]

During the period of the initial field observations, the weekly membership of the group varied from 30 to 100 members, with an average attendance of around 60 members. Although there was a considerable turnover in membership, the greatest part of this turnover consisted of "rejoins" (individuals who had been members previously, and were joining again).

Although we have no quantitative data from which to generalize, the group membership appeared to represent a cross section of the larger community. The group included both male and female members, although females did constitute about three-fourths of the membership. Although the membership was predominantly white, a range of ethnicities, notably Hispanic and Native American, existed within the group. The majority of the members appeared to fall within the 30 to 50 age range, although there was a member as young as 11, and one over 70.

DATA COLLECTION

Two types of data were gathered for this study: field observations and in-depth interviews. The field observations were performed while attending meetings of a local weight control organization. The insights gained from these observations were used primarily to develop interview guides. There were two major sources of observation during this period: pre-meeting conversations; and exchanges during the meeting itself.[3] The observations were recorded in note form and served to provide an orientation for the subsequent interviews. The goal during this period of observation was to gain insight into the basic processes of obesity and the obese career.

The in-depth interviews were carried out with 29 members from the local group. The interviews were solicited on a voluntary basis, and each individual was assured anonymity. The interviewees were representative of the group membership. Although most were middle-aged, middle-income, white females, various age groups, ethnicities, marital statuses, genders, and social classes were represented.

These interviews lasted in length from ½ to 2½ hours, with the average interview being about 1 hour and 15 minutes in duration. The interviews produced almost 40 hours of taped discussion, which yielded more than 600 pages of typed transcript for coding.

THE IDENTITY CHANGE PROCESS

In conceptualizing the "identity change" process, the concept of "career" was employed. As Goffman notes, "career" refers, ". . . to any social strand of any person's course through life" (Goffman, 1961, p. 127). In the present paper, our concern is the change process that takes place as individuals come to see definitions of self in light of specific transmitted information.

An important aspect of this career model is what Becker referred to as "career contingencies," or ". . . those factors on which mobility from one position to another depends. Career contingencies include both the objective facts of social structure, and changes in the perspectives, motivations, and desires of the individual" (Becker, 1963, p. 24).

Thus, the "identity change" process must be viewed on two levels: a public (external) and a private (internal) level. As Goffman has stated, "One value of the concept of career is its two-sidedness. One side is linked to internal matters held dearly and closely, such as image of self and felt identity; the other side concerns official position, jural relations, and style of life and is part of a publicly acceptable institutional complex" (Goffman, 1961, p. 127).

On the public level, social status exists as part of the public domain; social status is socially defined and promoted. The social environment not only contains definitions and attendant stereotypes for each status, it also contains information, in the form of *status cues,* about the applicability of that status for the individual.

On the internal level, two distinct cognitive processes must take place for the identity change process to occur: first, the individual must come to recognize that the current status is inappropriate; and second, the individual must locate a new, more appropriate status. Thus, in response to the external status cues, the individual comes to recognize internally that the initial status is inappropriate; and then he or she uses the cues to locate a new, more appropriate status. The identity change occurs in response to, and is mediated through, the status cues that exist in the social environment (see figure on p. 241).

STATUS CUES: THE EXTERNAL COMPONENT

Status cues make up the public or external component of the identity change process. A status cue is some feature of the social environment that contains information about a particular status or status dimension. Because this paper is about obesity, the cues of interest are about "fatness." Such status cues provide information about whether or not the individual is "fat," and if so, how "fat."

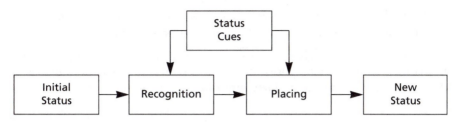

Visualization of the Identity Change Process (ICP)

"Recognizing" and "placing" comprise the internal component of the identity change process and occur in response to, and are mediated through, the status cues that exist in the social environment. In order to fully understand the identity change process, it is necessary to explain the interaction between outer and inner processes (Scheff, 1988, p. 396), or in our case, external and internal components of the process.

Status cues are transmitted in two ways: actively and passively. Active cues are communicated through interaction. For example, people are informed by peers, friends, spouses, etc., that they are overweight. The following are some typical comments that occurred repeatedly in the interviews in response to the question, "How did you know that you were fat?"

I was starting to be called chubby, and being teased in school.

When my mother would take me shopping, she'd get angry because the clothes that were supposed to be in my age group wouldn't fit me. She would yell at me.

Well, people would say, "When did you put on all your weight, Bob?" You know, something like that. You know, you kind of get the message, that, you know, I did put on weight.

A second category of cues might accurately be described as passive in form. The information in these cues exists within the environment, but the individual must in some way be sensitized to that information. For example, standing on a scale will provide an individual with information about weight. It is up to the individual to get on the scale, look at the numbers, and then make some sense out of them. Other passive cues might involve seeing one's reflection in a mirror, standing next to others, fitting in chairs, or, as frequently mentioned by respondents, the sizing of clothes. The comments below, all made in response to the question, "How did you know that you were fat?" are representative of passive cue statements.

I think that it was not being able to wear the clothes that the other kids wore.

How did I know? Because when we went to get weighed, I weighed more than my, uh, a girl my height should have weighed, according to the chart, according to all the charts that I used to read. That's when I first noticed that I was overweight.

I would see all these ladies come in and they could wear size 11 and 12, and I thought, Why can't I do that? I should be able to do that.

Both active and passive cues serve as mechanisms for communicating information about a specific status. As can be seen from the data, events occur that force the individual to evaluate his or her conceptions of self.

RECOGNIZING

The term "recognizing" refers to the cognitive process by which an individual becomes aware that a particular status is no longer appropriate. As shown in the figure, the process assumes the individual's acceptance of some initial status. For obese individuals, the initial status is that of "normal body build."[4] This assumption is based on the observation that none of our interviewees assumed that they were "always fat." Even those who were fat as children could identify when they became aware that they were "fat." Through the perception of discrepant status cues, the individual comes to recognize that the initial status is inappropriate. It is possible that the person will perceive the discrepant cues and will either ignore or reject them, in which case the initial status is retained. The factors regulating such a failure to recognize are important, but are not dealt with in this paper. Further research on this point is called for.

Status cues are the external mechanisms through which the recognizing takes place, but it is paying attention to the information contained in these cues that triggers the internal cognitive process of recognizing.

An important point is that the acceptance (or rejection) of a particular status does not occur simply because the individual possesses a set of objective characteristics. For example, two people may have similar body builds, but one may have a self-definition of "fat" whereas the other may not. There appears to be a rather tenuous connection between objective condition and subjective definition. The following comments are supportive of this disjunction.

I was really, as far as pounds go, very thin, but I had a feeling about myself that I was huge.

Well, I don't remember ever thinking about it until I was about in eighth grade. But I was looking back at pictures when I was little. I was always chunky, chubby.

This lack of necessary connection between objective condition and subjective definition points up an important and frequently overlooked feature of social statuses: the extent to which they are *self-evident*. Self-evidentiality refers to the degree to which a person who possesses certain objective status characteristics is *aware* that a particular status label applies to them.[5]

Some statuses possess a high degree of evidentiality: gender identification is one of these.[6] On the other hand, being beautiful or intelligent is somewhat non-self-evident. This is not to imply that individuals are either ignorant of these statuses or of the characteristics upon which they are assigned. People may know that other people are intelligent, but they may be unaware that the label is equally applicable to them.

One idea that emerged quite early from the interviews was that being "fat" is a relatively non–self-evident status. Individuals do not recognize that "fat" is a description that applies to them.[7] The objective condition of being overweight is not sufficient, in itself, to promote the adoption of a "fat" identity. This non–self-evidentiality is demonstrated in the following excerpts.

> I think that I just thought that it was a little bit here and there. I didn't think of it, and I didn't think of myself as looking bad. But you know, I must have.

> I have pictures of me right after the baby was born. I had no idea that I was that fat.

The self-evidentiality of a status is important in the discussion of the identity change process. The less self-evident a status, the more difficult the recognizing process becomes. Further, because recognition occurs in response to status cues, the self-evidentiality of a status will influence the type of cues that play the most prominent role in identity change.

A somewhat speculative observation should be made about status cues in the recognizing process. For our subjects, recognizing occurred primarily through active cues. When passive cues were involved, they typically were highly visible and unambiguous. In general, active cues appear to be more potent in forcing the individual's attention to the information that the current status is inappropriate. The predominance of these active cues is possibly a consequence of the relatively non–self-evident character of the "fat" status. It is probable that the less self-evident a status, the more likely that the recognizing process will occur through active rather than passive cues.

Once the individual comes to recognize the inappropriateness of the initial status, it becomes necessary to locate a new, more appropriate status. This search for a more appropriate status is referred to as the "placing" process."[8]

PLACING

Placing refers to a cognitive process whereby an individual comes to identify an appropriate status from among those available. The number of status categories along a status dimension influences the placing process. A status dimension may contain any number of status categories. If there are only two status dimensional categories, such as in the case of gender, the placement process is more or less automatic. When individuals recognize that they do not belong in one category, the remaining category becomes the obvious alternative. The greater the number of status categories, the more difficult the placing process becomes.

The body build dimension contains an extremely large number of categories. When an individual recognizes that he or she does not possess a "normal" body build, there are innumerable alternatives open. The knowledge that one's status lies toward the "fat" rather than the "thin" end of the continuum still presents a wide range of choices. In everyday conversation, we hear terms that describe these alternatives: chubby, porky, plump, hefty, full-figured, beer belly, etc. All are informal descriptions reflecting the myriad categories along the body build dimension.

> I wasn't real fat in my eyes. I don't think. I was just chunky.
>
> Not fat. I didn't exactly classify it as fat. I just thought, I'm, you know, I am a pudgy lady.
>
> I don't think that I have ever called myself fat. I have called myself heavy.

Even when individuals adopt a "fat" identity, they attempt to make distinctions about how fat they are. Because being fat is a devalued status, individuals attempt to escape the full weight of its negative attributes while still acknowledging the nonnormal status. The following responses exemplify this attempt to neutralize the pejorative connotations of having a "fat" status. The practice of differentiating one's status from others becomes vital in managing a fat identity.

> Q. How did you know that you weren't *that* fat?
>
> A. Well, comparing myself to others at the time, I didn't really feel that I was that fat. But I knew maybe because they didn't treat me the same way they treated people who were heavier than me. You know, I got teased lightly, but I was still liked by a lot of people, and the people that were heavy weren't.

As is apparent from this excerpt, the individual neutralized self-image by linking "fatness" with the level of teasing done by peers.

NEW STATUS

The final phase of the identity change process involves the acceptance of a new status. For our informants, it was the acceptance of a "fat" status, along with its previously mentioned pejorative characterizations.[9]

> I hate to look in mirrors. I hate that. It makes me feel so self-conscious. If I walk into a store, and I see my reflection in the glass, I just look away.
>
> We'd go somewhere and I would think, "I never look as good as everybody else." You know everybody always looks better. I'd cry before we'd go bowling because I'd think, "Oh, I just look awful."

As is clear, the final phase of the identity change process involves the internalization of a negative (deviant) definition of self. For many fat people accepting a new status means starting on the merry-go-round of weight reduction programs.[10] Many of these programs or organizations attempt to get members to accept a devalued status fully, and then work to change it. Consequently, individuals are forced to "admit" that they are fat and to "witness" in front of others.[11] The new identity becomes that of a "fat" person, which the weight reduction programs then attempt to transform. A further analysis of the impact of informal organizations on the identity change process will be attempted in another paper.

CONCLUSION

In this paper, we have attempted to fill a void within the interactionist literature by presenting an inductively generated model of the identity change process. The proposed model treats the change process from a career focus, and thus addresses both the external (public) and the internal (cognitive) features of the identity change.

We have suggested that the adoption of a new status takes place through two sequential cognitive processes, "recognizing" and "placing." First, the individual must come to recognize that a current status is no longer appropriate. Second, the individual must locate a new, more appropriate status from among those available. We have further suggested that these internal or cognitive processes are triggered by and mediated through status cues, which exist in the external environment. These cues can be either active or passive. Active cues are transmitted through interaction, whereas passive cues must be sought out by the individual.

We also found a relationship between the evidentiality of the status, that is, how obvious that status is to the individual, and the role of the different types of cues in the identity change process. Finally, we have suggested that the adoption of a new status is a trigger for further career changes.

Although the model presented in this paper was generated inductively from field data on obese individuals, we are confident that it may be fruitfully applied to the study of other deviant careers. It seems particularly appropriate where the identity involved has a low degree of self-evidentiality.

In addition, we feel that the focus upon the different types of status cues and their differing roles in the recognizing and placing processes can lead to a better understanding of how institutionally promoted identity changes occur.

NOTES

1. Included here is research on both rule creators and rule enforcers. We have not made an attempt to analytically separate the two types of investigation.

2. Some of the members had successfully lost their excess weight. When these people were present at a meeting, a leader was careful to introduce them to the other members of the class and to tell how much weight they had lost. This was done to uphold their claim to acceptance by the other group members.

3. Access to this information was gained from an "insider" perspective because one

of the researchers was well known among the membership, being an "off and on" member of the organization for 3 years. Thus, he was not confronted with the problem of gaining entry into a semi-closed social setting. Similarly, because the observer had "been an ongoing participant of the group," he did not have to desensitize the other members of the group to his presence.

4.   It is important to note that this process can operate generically. That is, it is not only applicable to the "identity change" from a "normal" to a "deviant" identity, but can encompass the reverse process as well. In a forthcoming project, we will use the process to analyze how various rehabilitation programs attempt to get individuals back to the initial status.

5.   This concept is different in an important way from what Goffman calls "visibility." He uses the term to refer to ". . . how well or how badly the stigma is adapted to provide means of communicating that the individual possesses it" (Goffman, 1963, p. 48). The focus of the concept is on how readily the social environment can identify that the individual possesses a stigmatized trait. The concept of self-evidentiality deals with how readily the individual can internalize possession of the stigmatized trait. The focus is upon the actor's perceptions, not on the audience.

6.   We are referring here to the physiological description of being male or female. We realize that sex roles are much less self-evident.

7.   Conversely, a number of individuals thought of themselves as "fat" or "obese," and were objectively "normal." In this case, the existence of objective indicators was insufficient to prevent the individual from adopting a "fat" identity.

8.   In some instances, recognizing and placing occur simultaneously. This is especially true when the cue involved is an active one, and contains information about both the initial and new statuses. For example, if peers call a child "fatty," this interaction informs the child that the "normal" status is inappropriate. At the same time, it informs the child that being "fat" is the appropriate status. Even here however, the individual must recognize before it is possible to place.

9.   This phase corresponds closely to that presented in much of the "subcultural" research. (See Schur, 1971; Becker, 1963; Sykes & Matza, 1957, p. 664.)

10. Weight Watchers, TOPS, Overeaters Anonymous, Diet Center, and OptiFast are typical examples of this type of program.

11. By witnessing, we are referring to the process whereby individuals come to renounce, in front of others, a former self and former behaviors associated with that self. Some religious groups, Synanon, Alcoholics Anonymous, etc., seem to encourage this type of degradation of self.

REFERENCES

Baum, L. (1987, August 3). Extra pounds can weigh down your career. *Business Week,* p. 96.

Becker, H. S. (1963). *Outsiders: Studies in the sociology of deviance.* New York: Free Press.

Canning, H., & Mayer, J. (1966). Obesity: Its possible effects on college acceptance. *New England Journal of Medicine,* 275(24); November, 1172–1174.

Cicourel, A. (1968). *The social organization of juvenile justice.* New York: Wiley.

Conrad, P. (1975). The discovery of hyperkinesis: Notes on the medicalization of deviant behavior. *Social Problems,* 23(1); October, 12–21.

Goffman, E. (1961). *Asylums.* Garden City, NY: Anchor.

———. (1963). *Stigma: Notes on the management of spoiled identity.* Englewood Cliffs, NJ: Prentice-Hall.

Greenberg, D. (1989). The antifat conspiracy. *New Scientist,* 22 (April 22): 79.

Hadden, S. C., Degher, D., & Fernandez, R. (1989). Sports as a strategic ethnographic arena. *Arena Review,* 13(1), 9–19.

Kelly, J. R. (1990). *Leisure* (2nd ed.). Englewood Cliffs, NJ: Prentice-Hall.

Kitsuse, J. (1962). Societal reactions to deviant behavior: Problems of theory and method. *Social Problems,* 9 (Winter): 247–256.

Laslett, B., & Warren, C. A. B. (1975). Losing weight: The organizational promotion of behavior change. *Social Problems,* 23(1), 69–80.

Lemert, E. (1951). *Social pathology.* New York: McGraw-Hill.

Maddox, G. L., Back, K. W., & Liederman, V. (1968). Overweight as social deviance and disability. *Journal of Health and Social Behavior,* 9(4): 287–298.

Maddox, G. L., & Liederman, V. (1969). Overweight as a social disability with medical implications. *Journal of Medical Education,* 9(4): 287–298.

Pauley, L. L. (1989). Customer weights as a variable in salespersons' response time. *Journal of Social Psychology,* 129: 713–714.

Petit, D. W. (1974). The ills of the obese. In G. A. Gray & J. E. Bethune (Eds.), *Treatment and management of obesity.* New York: Harper & Row.

Piliavin, I., & Briar, S. (1964). Police encounters with juveniles. *American Journal of Sociology,* (September): 206–214.

Register, C. A. (1990). Wage effects of obesity among young workers. *Social Science Quarterly,* 71 (March): 130–141.

Scheff, T. (1988). Shame and conformity: The deference emotion system. *American Journal of Sociology,* 53 (June): 395–406.

Schur, E. M. (1971). *Labeling deviant behavior: Its sociological implications.* New York: Harper & Row.

Sykes, G., & Matza, D. (1957). Techniques of neutralization: A theory of delinquency. *American Sociological Review,* (December): 664–670.

Tannenbaum, F. (1939). *Crime and the community.* New York: Columbia University Press.

21

Divorce and Stigma

NAOMI GERSTEL

By most accounts, tolerance of variation in family life has increased dramatically in the United States. Public opinion polls over the last two decades reveal declining disapproval of extended singlehood (Veroff et al., 1981), premarital sex and pregnancy (Gerstel, 1982), employment of mothers with young children (Cherlin, 1981), and voluntary childlessness (Huber and Spitze, 1983). Divorce resembles these other situations; in fact public tolerance of divorce appears to have increased especially dramatically over the last few decades (Veroff et al., 1981).

In the mid-1950's, Goode (1956: 10) could still observe: "We know that in our own society, divorce has been a possible, but disapproved, solution for marital conflict." However, comparing attitudes in 1958 and 1971, McRae (1978) found an increasing proportion of adults believing that divorce was only "sometimes wrong" while a decreasing proportion felt that it was "always wrong." These data, he claimed, indicated attitudes toward divorce had shifted "from moral absolutism to situational ethics" (1978: 228). In an analysis of panel data collected between 1960 and 1980, Thornton (1985) found that changes in attitudes toward divorce were not only large but pervasive: all subgroups—whether defined by age, class, or even religion—showed substantial declines in disapproval of marital separation.

What are the implications of declining disapproval of divorce? In historical perspective, it is clear that the divorced are no longer subject to the moral outrage they encountered centuries, or even decades, ago. Certainly, divorce is no longer treated as a sin calling for repressive punishment, as it was in theological doctrine and practice (be it Catholic or Protestant) until the beginning of the twentieth century (Halem, 1980; O'Neil, 1967). In electing a divorced president and many divorced senators and governors, U.S. citizens seem to have repudiated the idea that divorce is grounds for exclusion from public life. With the recent passage of no-fault divorce laws in every state, U.S. courts no longer insist on attributing wrongdoing to one party to a divorce (Weitzman, 1985).

Most recent commentators on divorce even argue that it is no longer stigmatized. For example, Spanier and Thompson (1984: 15) claim that "the social stigma associated with divorce has disappeared" and Weitzman (1981: 146)

From "Divorce and Stigma," Naomi Gerstel, Social Problems, Vol. 34, No. 2, April 1987.
© 1987 Society for the Study of Social Problems. Reprinted by permission of University of California Press Journals and the author.

suggests that "the decline in the social stigma traditionally attached to divorce is one of the most striking changes in the social climate surrounding divorce."

However, I argue in this paper that the stigma attached to divorce has disappeared in only two very limited senses. First, although other studies have shown a clear decline in disapproval of divorce as a general category, disapproval of divorced individuals persists contingent on the specific conditions of their divorce. Thus, as I show below, some divorced people experience disapproval and at least one party to a divorce often feels blamed.

Second, while many of the formal, institutional controls on divorce—imposed in the public realm of church or state—have weakened, the individual who divorces suffers informal, relational sanctions. These are the interpersonal controls that emerge more or less spontaneously in social life. I will present evidence indicating that the divorced believe the married often exclude them and that the divorced themselves frequently pull toward, yet devalue, others who divorce.

In these two senses, I argue that the divorced are still subject to the same social processes and evaluations associated with stigmatization more generally. As in Goffman's (1963: 3) classic formulation—which stresses both the conditional and relational aspects of stigma—my findings suggest that the divorced come to be seen (and to see themselves) as "of a less desired kind . . . reduced in our minds from a whole and usual person to a tainted, discounted one."

METHODS

My data come from interviews with 104 separated and divorced respondents: 52 women and 52 men. Based on a conception of marital dissolution as a process rather than a static life event, the research team sampled respondents in different stages of divorce: one-third of the respondents were separated less than one year; one-third separated one to two years; one-third separated two or more years. To obtain respondents, we could not rely on court records alone, for most couples who have filed for a divorce have already been living apart for at least a year. Thus, 61 percent of the respondents were selected from probate court records in two counties in the Northeast; the others came from referrals.[1] Comparisons between the court cases and referred respondents show no statistically significant differences on demographic characteristics.

Sample Characteristics

In contrast to the samples in most previous research on separation and divorce, the respondents are a heterogeneous group. They include people in the working class as well as in the middle class whose household incomes ranged from under $4,000 to over $50,000, with a median of $18,000 (with women's significantly lower than men's).[2] Levels of education varied widely: about one-fourth had less than a high school degree, and slightly less than one-fourth had four or more years of college. The sample also includes significant numbers whose primary source of income came from public assistance and from manual, clerical, and professional jobs. Only 11 percent were not currently employed while another 9 percent were working part-time. The median age of the respondents was 33 years, and the mean number of years married was nine. Finally, 30 percent of the sample had no children, 19 percent had one child, and 51 percent had more than one child.

FINDINGS

Social Exclusion: Rejection by the Married

Partners to a divorce not only split friends; they are often excluded from social interaction with the married more generally.[3] Many ex-husbands and ex-wives found they could not maintain friendships with married couples: about one-half of both men (43 percent) and women (58 percent) agreed with the statement: "Married couples don't want to see me now." Moreover, less than one-fourth (23 percent) of the women and men agreed: "I am as close to my married friends as when I was married."[4] By getting a divorce, then, they became marginal to at least part of the community on which they had previously relied.

One man summed up the views of many when he spoke of the "normal life" of the married.

> One of the things I recognized not long after I was separated is that this is a couple's world. People do things in couples, normally (C043, male).

Remembering his own marriage, he now recognized its impact:

> We mostly went out with couples. I now have little or no contact with them.

Discovering "they don't invite me anymore" or "they never call," many of the divorced felt rejected:

> The couples we shared our life with, uh, I'm an outsider now. They stay away. Not being invited to a lot of parties that we was always invited to. It's with males and females. It sucks (C030, male).

The divorced developed explanations for their exclusion. Finding themselves outsiders, some simply thought that their very presence destabilized the social life of couples: "I guess I threaten the balance" (N004, female). They found themselves social misfits in that world, using terms like "a third wheel" (N010, female; N027, male) and "odd person out" (N006, male) to describe their newly precarious relationship with the married.

Some went further, suggesting that those still in couples felt threatened by the divorce or were afraid it would harm their own marriages. "They say, 'My God, it's happening all over.' It scares them" (N019, male). Men and women expressed this form of rejection in terms of "contagion" (C027, male) and "a fear it's going to rub off on them" (N010, female).[5] Because the difficulties of marriage are often concealed, others found their divorce came as a surprise to married friends. That surprise reinforced the idea that "it can happen to anyone" and "so they tend to stay away" (C027, male).

A few turned the explanation around, suspecting that married couples rejected them out of jealousy rather than fear. One woman, speaking of a friend who no longer called, explained: "It was like me living out her fantasies" (N008, female). A salesman in his mid-40s, who had an affair before getting a divorce, believed:

> I get a kick out of it because . . . I am the envy of both men and women because, some of it is courage, others look upon it as freedom. Both words have been used a number of times. People become very envious, and a spouse of the envious person will feel extremely threatened (N043, male).

These few could turn an unpleasant experience into an enviable one. For a small minority, then, the experience of exclusion did not produce a sense of devaluation.

But more of the divorced, men as well as women, were troubled by the thought that old friends now defined them in terms of their sexual availability and, as a result, avoided them. One woman, a teacher's aide with a very young daughter and son at home, felt insulted that friends misconstrued her situation:

> Well, I now have no married friends. It's as if I all of a sudden became single and I'm going to chase after their husbands (N011, female).

And a plumbing contractor, unusual because he had custody of his five children, was particularly hurt by the image of sexual availability because he had resisted any sexual entanglements. Describing one woman who "couldn't understand why I didn't want to hop into bed with her," he noted "she told me there must be something wrong with me" (C047, male). He went on:

> I would say couples in general, there seems to be a, well, they are nice to me, but distant. I think the men don't want a single man around their wives.

And when asked, "Can you tell me about that?," he associated his seeming sexual availability with a threat to the cohesion of the community:

> (They) don't really want to involve me in things that are going on . . . neighborhood picnics . . . couples' things. . . . I have had men say that they figure that I'm out chasing women all over. So I'm considered somewhat of an unstable person.

He added that he was not the only person who had reached this conclusion:

> And this is quite common with divorced people. In group discussions, everyone seems to experience the same thing.

As his final comment indicates, the divorced talk to each other about this experience and generate a shared explanation for it—that they are viewed as somewhat "unstable." Thus, some divorced people come to believe that married acquaintances saw them as "misfits"—unstable individuals who could not maintain a stable marriage, a threat to the routines of a community made up of the "normal" married.[6]

The exclusion of the divorced from the social life they had enjoyed while married constitutes a negative sanction on divorce. This is not simply a functional process of friendship formation based on homogamy (cf. Lazarsfeld and Merton, 1964); it involves conflict, producing a sense of devaluation on the part of one group (the divorced) who feel rejected by another group still considered normal (the married).[7]

The divorced try to come to terms with their experience by talking to others who share it. Together they develop a shared understanding similar to what Goffman (1963: 5) calls a "stigma theory": the married feel uncomfortable, even threatened by them, and act as if divorce, as a "social disease," is contagious. Or divorce poses a threat because of the desired freedom and sexuality it (perhaps falsely) represents. Finally, divorced people mutually develop a broader explanation for the modern response to them: they are avoided because the dissolution of marriage is so common, so possible, that it becomes a real threat both to any given couple and to the social world built on, and routinized by, groups consisting of couples.

Colleagues and Demoralization

The separation of the divorced from the married is even more clearly apparent in the social life developed by the divorced themselves. The divorced pull away from the married and into the lives of others like them. Goffman (1963: 18) argues that the stigmatized turn to others like them in anticipation that "mixed social contact will make for anxious, unanchored interaction." Accordingly, many of the divorced said they felt "uncomfortable" (C042, female), "strained" (C003, male; N014 and C029, females), "strange," (C034, female), and "awkward" (C019, male) in a world composed of couples. And some abandoned the married: "I've been pulling away from my coupled friends" (C026, female).

Drawing together with other divorced, they develop as well as share their "sad tales" (Goffman, 1963: 19) and learn how to behave. In fact, over half of the people with whom these divorced men (52 percent) and women (62 percent) socialized were other divorced individuals, a far higher proportion than is found in the general population (U.S. Bureau of the Census, 1983). The divorced used many well-worn phrases to talk about others who shared their marital status: "birds of a feather flock together" (N021, male) and "likes attract likes" (N025, female).

Many discovered their interests and concerns, at least for a time, were based on their newfound marital status. When asked to respond to the statement, "I have more in common with singles now," over half of the men (55 percent) and women (58 percent) agreed. One 28-year-old working-class man sought out divorced people for the same reason many respondents avoided the married. He said of others who shared his marital status:

We have something in common. We have almost right off the bat something to talk about. It makes it easier to talk because you have gone through it (C018, male).

Equally important, respondents often felt they could turn to other divorced people as experienced "veterans" (Caplan, 1974) and "colleagues" (Best and Luckenbill, 1980) for their newfound marital state. These others served as role models, showing them ways to cope as spouseless adults and, in doing so, bolstered their new identities. A man referred to other divorced as providing an "experience bank" and elaborated by saying:

I solicited assistance, guidance from people who had gone through or who were going through similar experiences. So I might get a better understanding of how they reacted to it (C043, male).

So, too, other divorced people could help them do what Hochschild (1983: 254) has called "feeling work" or "the shaping, modulating, or inducing of feelings." That is, colleagues encouraged them to manage and change their feelings, and to realize how their marriage (or many marriages) were not as good as they had thought. That helped them disengage from their ex-spouses.

But these colleagues could do something more. They showed them that the life of a divorced adult was not all anguish and pain. One 30-year-old middle-class woman found she felt closest to her old friends who had been through divorce because:

I could talk to them and they helped me to talk about my problems. What went wrong. I felt they could understand. And it was interesting to hear what they had been through, you know (C016, female).

Importantly, she added:

I wanted to hear what it was like. It sorta helped me to think that they had made a go of their life again. They were happy afterwards.

The divorced needed reassurance that divorce did not imply a serious character flaw. They needed to find those who, after getting a divorce, had made a successful transition. A 32-year-old man, who in the first month "isolated" himself "for fear that people might think I'm doing something wrong," found a few months later that:

It's nice to hear people who have gone through the same experience, talking to me about it. It's nice to hear a lot of these things because you realize: "Hey, I'm not so bad. I'm not the only one this happens to." And looking at the person and seeing that they made it okay. And that I will, too (N009, male).

What Goffman (1963: 20) wrote more generally about the stigmatized, then, characterizes the modern divorced: "They can provide instruction for tricks of the trade and a circle of lament to which he can withdraw for moral support and comfort of feeling at home, at ease, accepted as a person who is really like any normal person." The divorced turn to others like themselves to get reassurance, advice, and encouragement, and to make sense of their often dislocated lives. Telling their story to those like themselves is therapeutic (Conrad and Schneider, 1980).

Yet, there are also pitfalls in this attraction to others like them. As time passes, the divorced may find themselves bored by constant discussion of divorce, that "the whole matter of focusing on atrocity tales . . . on the 'problem' is one of the largest penalties for having one" (Goffman, 1963: 21). As a man divorced close to two years put it:

> You see, I've been locked in too much with divorced people. You're relating to them relative to the separation. And what happened to you, when you did it, you know (N019, male).

The divorced, especially those who had been separated more than a year, spoke of how they were getting tired of "problems dominating the conversation" (C042, female, divorced three years) as they felt "dissipated" (C009, male, divorced two years) and "wanted to talk to people about something different" (N016, female, divorced a year-and-a-half) because "the less you talk about the problems, the less you think about them, and the less you feel about them" (N028, female, divorced a year-and-a-half). One man described how, in the first months of divorce, he had "learned a lot about divorce" from others who had the same experience because "they understood me." But then he went on to complain about the problems with this association: "You feel up and then someone drags you down with their problems" (C021, male). . . .

The Devaluation of Self

Perhaps the most striking evidence that the divorced devalue their own condition is found in their assessment of organizations established for the divorced. Only 10 percent of the respondents were in such groups.[8] In fact, most of the divorced—male as well as female—explicitly rejected such formal mechanisms of integration set up by and for others like them.

For the relatively small number of people who did join, such groups provided both a source of entertainment for their children as well as an opportunity to meet other adults. However, in explaining why they joined, the divorced typically stressed child care. Thus, children were not simply a reason for joining; they provided legitimation for membership. By explaining membership in instrumental, rather than expressive terms, and in terms of children rather than themselves, the divorced distanced themselves from the potentially damaging implications of membership for their own identity. In this sense, children provide a "face-saving device," much like those inventoried by Berk (1977) among people who attended singles dances.

The notion that groups for the divorced—and therefore those who join them—are stigmatized is substantiated still further by the comments of those who did not join. They gave a number of reasons for their reluctance. Some attributed their lack of participation to a lack of knowledge. Others simply felt they did not have the time or energy. When asked why she had not joined any divorce group, one 25-year-old saleswoman said:

> I've thought about it, but I have just never done anything about it. I know it is not getting me anywhere by not doing anything. Basically I am a lazy person (C024, female).

But while she first blamed herself for non-participation in these groups, she then went on to add a more critical note: "I think I would feel funny walking into a place like that." Her second thought reiterated a common theme—an attitude toward divorce and membership in organizations for them—which came through with compelling force. Many imagined that people who joined such groups were unacceptable in a variety of ways, or even that to join them was somehow a sign of weakness. For example:

> These people really don't have somebody to turn to. I guess that's the main reason for them belonging and I do have someone to turn to, matter of fact, more than one. They're really not sure of themselves; they're insecure (N013, female).

Such comments reveal that respondents saw divorce as a discredit, at least insofar as it became the axis of one's social life. Consequently, to join such groups was to reinforce the very devaluation they hoped to avoid. One welfare mother with a young child had been told by her social worker that joining a singles' group might alleviate the enormous loneliness she experienced. But she resisted:

> It's kind of degrading to me or something. Not that I'm putting these other people down. I could join something but I couldn't join something that was actually called a singles' group (N016, female).

Others reiterated the same theme. To them, groups of the divorced were "rejects looking, you know, going after rejects. They need a crutch" (N011, female). Or they asked rhetorically, "Is that for the very, very lonely?" (C016, female). These to them were "people with as many, if not worse, problems than I have" (C040, male) or "weirdos" (N029, female). As these comments show, the divorced were quick to put a pejorative label on groups consisting of other divorced.

Or, the divorced we spoke to felt that such groups were unacceptable because they were sexual marketplaces. In the words of a 28-year-old plant supervisor, who (like most others) had never actually been to any organization for the divorced:

> I refuse to go to a place where I'm looked at as a side of beef and women are looked at as sides of beef. It disgusts me (C040, male).

Association with such groups would reinforce the very view that so many of the divorced work so hard to dispel. While their rejection of such groups is a way to separate themselves from a stigmatized status (Berk, 1977: 542), the very character and strength of the rejection confirms that the status is stigmatized.

Thus, in distancing themselves, the divorced reveal that they share the belief that individuals who divorce, especially if they use that divorce to organize their social worlds, continue to be somehow tainted.

Despite their negative reaction to divorce groups, respondents did not reject all organized routes to friendship formation. The majority (82 percent) were members of at least one group—including, for example, sports, cultural, religious and service groups. Women participated in a median of 2.24 groups; men, a median of 3.56. In fact, many spoke of joining these other groups as a way to "make friends" and to cope with the loneliness they felt. Such groups may provide access to others who are divorced, but only coincidentally. These organizational memberships—and the relationships they allow—are legitimized by their *dissociation* from marital status. It is in this context that respondents' resistance to joining divorce groups becomes especially compelling as evidence for their devaluation of the status of divorce.

CONCLUSION

To argue that the divorced are no longer stigmatized is to misunderstand their experience. To be sure, divorce is now less deviant in a statistical sense than it was a decade ago. As a group, the divorced are not categorized as sinful, criminal, or even wrong. Moreover, even though the divorced lose married friends and have smaller networks than the married, they do not become completely ghettoized into subcultures of the divorced (Gerstel et al., 1985; Weiss, 1979). Finally, . . . the divorced themselves do not think that most of their kin and friends disapprove.

However, a decrease in statistical deviance, a relaxation of institutional controls by church or state, or a decline in categorical disapproval is not the same as the absence of stigmatization. Although a majority of Americans claim they are indifferent in principle to those who make a "personal decision" to leave a "bad" marriage, this indifference does not carry over into the social construction of private lives. The divorced believe they are the targets of informal relational sanctions—exclusion, blame, and devaluation. If we understand stigma as referring not simply to the realm of public sanctions but rather see it as emerging out of everyday experience, then we can see that the divorced continue to be stigmatized. . . .

NOTES

1.   A small number of these referrals were located through a "snowball" strategy: various people who heard about our study told us about individuals who had just separated. But the majority of referrals were located through respondents. At the end of each interview, we asked for the names of other people who had been separated less than a year and interviewed a maximum of one person named by each respondent.

2.   Women's mean household income was $14,000 while the men's mean income was $22,000.

3.   Using a variety of methods and samples, a number of other studies also find that the divorced lose married friends (e.g., Spanier and Thompson, 1984; Wallerstein and Kelly, 1980; Weiss, 1979).

4.   In contrast to my finding that divorced men and women were equally likely to experience a certain distance from the married, Hetherington et al. (1976: 422) found "dissociation from married friends was greater for women than for men." However, they studied only divorced parents of children in nursery school. Fischer's (1982) findings would lead us to believe that, of any group, mothers—married or divorced—of young children are most isolated. In fact, my data suggest that the gender differences may well characterize only this very special group. Among male respondents, the presence of children is not significantly associated with their belief that "married couples don't want to see them" or their feeling that they "are not as close to those couples." In contrast, for women, the presence of any children, especially young children, increases disassociation from the married. Among female respondents, the correlation between "married couples don't want to see me now" and the presence of any children is .25 (p < .05) and with having children less than 12-years-old is .39 (p < .01). This difference between women and men may well be a result of the fact that women obtain custody of the children far more often than do men. While Hetherington's findings are often cited as evidence for general gender differences in the social life of the divorced, this implication probably should be limited to this special group—the parents of young children.

5.   This belief that others see their "condition" as "contagious" is similar to the experience of others who are stigmatized, like the mentally ill. See Foucault (1967) for a discussion of the development of the belief that "unreason" is contagious, that anyone could "catch it," and the consequent movement to isolate the insane.

6.   Wallerstein and Kelly (1980: 33) hypothesize that still another factor may explain why the married move away from the divorced: the married "feel uncomfortable and inadequate in providing solace." While this is certainly a possible (and generous) explanation, the divorced nonetheless experience the loss of friendship as rejection and exclusion.

7.   To be sure, research on "single individuals"—be they widows (Lopata, 1979), never married (Stein, 1981), or divorced—suggests there is a general pattern of friendship based on homogeneity of marital status. Indeed, as Simmel (1950) points out in his classic work, the triad is a more unstable group than the dyad. Hence, a "third party" is likely to be excluded. Here, I am suggesting that such third parties, especially the divorced, are likely to interpret the separation of marital groups as exclusion and hence as devaluation.

8.   These findings suggest, of course, that those studies which draw entirely on members of singles' groups are seriously flawed: they represent a small and atypical population of the divorced.

REFERENCES

Berk, Bernard. 1977. "Face saving at the singles dance." *Social Problems* 24: 530–44.

Best, Joel, and David Luckenbill. 1980. "The social organization of deviants." *Social Problems* 28: 14–31.

Caplan, Gerald. 1974. *Social Supports and Community Mental Health*. New York: Behavioral Publications.

Cherlin, Andrew. 1981. *Marriage, Divorce and Remarriage*. Cambridge, MA: Harvard University Press.

Conrad, Peter, and Joseph W. Schneider. 1980. *Deviance and Medicalization: From Badness to Sickness*. St. Louis: C.V. Mosby.

Fischer, Claude. 1982. *To Dwell Among Friends*. Chicago: University of Chicago Press.

Foucault, Michel. 1967. *Madness and Civilization*. London: Tavistock.

Gerstel, Naomi. 1982. "The new right and the family." In Barbara Haber (ed.), *The Woman's Annual* (pp. 6–20). New York: G.K. Hall.

Gerstel, Naomi, Catherine Kohler Riessman, and Sarah Rosenfield. 1985. "Explaining the symptomatology of separated and divorced women and men: The role of material resources and social networks." *Social Forces* 64: 84–101.

Goffman, Erving. 1963. *Stigma*. Englewood Cliffs, NJ: Prentice-Hall.

Goode, William. 1956. *Women in Divorce*. New York: Free Press.

Halem, Lynne Carol. 1980. *Divorce Reform*. New York: Free Press.

Hetherington, E. M., M. Cox, and R. Cox. 1976. "Divorced fathers." *The Family Coordinator* 25: 417–28.

Hochschild, Arlie Russell. 1983. "Attending to, codifying and managing feelings: Sex differences in love." In Laurel Richardson and Verta Taylor (eds.), *Feminist Frontiers,* (pp. 250–62). Reading, MA: Addison–Wesley.

Huber, Joan, and Glenna Spitze. 1983. *Stratification: Children, Housework, and Jobs*. New York: Academic Press.

Lopata, Helena. 1979. *Women as Widows*. New York: Elsevier.

McRae, James A. 1978. "The secularization of divorce." In Beverly Duncan and Otis Dudley Duncan (eds.), *Sex Typing and Sex Roles* (pp. 227–42). New York: Academic Press.

O'Neil, William L. 1967. *Divorce in the Progressive Era*. New Haven: Yale University Press.

Simmel, Georg. 1950. *The Sociology of Georg Simmel*. Kurt H. Wolff, translator. New York: Free Press.

Spanier, Graham, and Linda Thompson. 1984. *Parting*. Beverly Hills, CA: Sage.

Stein, Peter (ed.). 1981. *Single Life*. Englewood Cliffs, NJ: Prentice–Hall.

Thornton, Arland. 1985. "Changing attitudes toward separation and divorce: Causes and consequences." *American Journal of Sociology* 90: 856–72.

U.S. Bureau of Census. 1983. "Marital status and living arrangements: March, 1983." Current Population Reports, Series P-20, #389. Washington, DC: U.S. Government Printing Office.

Veroff, Joseph, Elizabeth Douvan, and Richard A. Kulka. 1981. *The Inner American: A Self-Portrait from 1957–1976*. New York: Basic Books.

Wallerstein, Judith S., and Joan B. Kelly. 1980. *Surviving the Breakup*. New York: Basic Books.

Weiss, Robert. 1979. *Going It Alone*. New York: Basic Books.

Weitzman, Lenore. 1981. *The Marriage Contract*. New York: Free Press.

———. 1985. *The Divorce Revolution: The Unexpected Social and Economic Consequences for Women and Children in America*. New York: Free Press.

22

Anorexia Nervosa and Bulimia

The Development of
Deviant Identities

PENELOPE A. McLORG

AND DIANE E. TAUB

C urrent appearance norms stipulate thinness for women and muscularity for men; these expectations, like any norms, entail rewards for compliance and negative sanctions for violations. Fear of being overweight—of being visually deviant—has led to a striving for thinness, especially among women. In the extreme, this avoidance of overweight engenders eating disorders, which themselves constitute deviance. Anorexia nervosa, or purposeful starvation, embodies visual as well as behavioral deviation; bulimia, binge-eating followed by vomiting and/or laxative abuse, is primarily behaviorally deviant.

Besides a fear of fatness, anorexics and bulimics exhibit distorted body images. In anorexia nervosa, a 20-25 percent loss of initial body weight occurs, resulting from self-starvation alone or in combination with excessive exercising, occasional binge-eating, vomiting and/or laxative abuse. Bulimia denotes cyclical (daily, weekly, for example) binge-eating followed by vomiting or laxative abuse; weight is normal or close to normal (Humphries et al., 1982). Common physical manifestations of these eating disorders include menstrual cessation or irregularities and electrolyte imbalances; among behavioral traits are depression, obsessions/compulsions, and anxiety (Russell, 1979; Thompson and Schwartz, 1982).

Increasingly prevalent in the past two decades, anorexia nervosa and bulimia have emerged as major health and social problems. Termed an epidemic on college campuses (Brody, as quoted in Schur, 1984: 76), bulimia affects 13% of college students (Halmi et al., 1981). Less prevalent, anorexia nervosa was diagnosed in 0.6% of students utilizing a university health center (Stangler and Printz, 1980). However, the overall mortality rate of anorexia nervosa is 6% (Schwartz and Thompson, 1981) to 20% (Humphries et al., 1982); bulimia appears to be less life-threatening (Russell, 1979).

From "Anorexia Nervosa and Bulimia: The Development of Deviant Identities," Penelope A. McLorg and Diane E. Taub, *Deviant Behavior*, Vol. 8, 1987. Reprinted by permission of Taylor and Francis, Inc. All rights reserved.

Particularly affecting certain demographic groups, eating disorders are most prevalent among young, white, affluent (upper-middle to upper class) women in modern, industrialized countries (Crisp, 1977; Willi and Grossman, 1983). Combining all of these risk factors (female sex, youth, high socioeconomic status, and residence in an industrialized country), prevalence of anorexia nervosa in upper class English girls' schools is reported at 1 in 100 (Crisp et al., 1976). The age of onset for anorexia nervosa is bimodal at 14.5 and 18 years (Humphries et al., 1982); the most frequent age of onset for bulimia is 18 (Russell, 1979).

Eating disorders have primarily been studied from psychological and medical perspectives.[1] Theories of etiology have generally fallen into three categories: the ego psychological (involving an impaired child-maternal environment); the family systems (implicating enmeshed, rigid families); and the endocrinological (involving a precipitating hormonal defect). Although relatively ignored in previous studies, the sociocultural components of anorexia nervosa and bulimia (the slimness norm and its agents of reinforcement, such as role models) have been postulated as accounting for the recent, dramatic increases in these disorders (Schwartz et al., 1982; Boskind-White, 1985).[2]

Medical and psychological approaches to anorexia nervosa and bulimia obscure the social facets of the disorders and neglect the individuals' own definitions of their situations. Among the social processes involved in the development of an eating disorder is the sequence of conforming behavior, primary deviance, and secondary deviance. Societal reaction is the critical mediator affecting the movement through the deviant career (Becker, 1973). Within a framework of labeling theory, this study focuses on the emergence of anorexic and bulimic identities, as well as on the consequences of being career deviants.

METHODOLOGY

Sampling and Procedures

Most research on eating disorders has utilized clinical subjects or non-clinical respondents completing questionnaires. Such studies can be criticized for simply counting and describing behaviors and/or neglecting the social construction of the disorders. Moreover, the work of clinicians is often limited by therapeutic orientation. Previous research may also have included individuals who were not in therapy on their own volition and who resisted admitting that they had an eating disorder.

Past studies thus disregard the intersubjective meanings respondents attach to their behavior and emphasize researchers' criteria for definition as anorexic or bulimic. In order to supplement these sampling and procedural designs, the present study utilizes participant observation of a group of self-defined anorexics and bulimics.[3] As the individuals had acknowledged their eating disorders, frank discussion and disclosure were facilitated.

Data are derived from a self-help group, BANISH, Bulimics/Anorexics In Self-Help, which met at a university in an urban center of the mid-South. Founded by one of the researchers (D.E.T.), BANISH was advertised in local newspapers as offering a group experience for individuals who were anorexic or bulimic. Despite the local advertisements, the campus location of the meetings may have selectively encouraged university students to attend. Nonetheless, in view of the modal age of onset and socioeconomic status of individuals with eating disorders, college students have been considered target populations (Crisp et al., 1976; Halmi et al., 1981).

The group's weekly two-hour meetings were observed for two years. During the course of this study, thirty individuals attended at least one of the meetings. Attendance at meetings was varied: ten individuals came nearly every Sunday; five attended approximately twice a month; and the remaining fifteen participated once a month or less frequently, often when their eating problems were "more severe" or "bizarre." The modal number of members at meetings was twelve. The diversity in attendance was to be expected in self-help groups of anorexics and bulimics.

> . . . most people's involvement will not be forever or even a long time. Most people get the support they need and drop out. Some take the time to help others after they themselves have been helped but even they may withdraw after a time. It is a natural and in many cases *necessary* process (emphasis in original) (American Anorexia/Bulimia Association, 1983).

Modeled after Alcoholics Anonymous, BANISH allowed participants to discuss their backgrounds and experiences with others who empathized. For many members, the group constituted their only source of help; these respondents were reluctant to contact health professionals because of shame, embarrassment, or financial difficulties.

In addition to field notes from group meetings, records of other encounters with all members were maintained. Participants visited the office of one of the researchers (D.E.T.), called both researchers by phone, and invited them to their homes or out for a cup of coffee. Such interaction facilitated genuine communication and mutual trust. Even among the fifteen individuals who did not attend the meetings regularly, contact was maintained with ten members on a monthly basis.

Supplementing field notes were informal interviews with fifteen group members, lasting from two to four hours. Because they appeared to represent more extensive experience with eating disorders, these interviewees were chosen to amplify their comments about the labeling process, made during group meetings. Conducted near the end of the two-year observation period, the interviews focused on what the respondents thought antedated and maintained their eating disorders. In addition, participants described others' reactions to their behaviors as well as their own interpretations of these reactions. To protect the confidentiality of individuals quoted in the study, pseudonyms are employed.

Description of Members

The demographic composite of the sample typifies what has been found in other studies (Fox and James, 1976; Crisp, 1977; Herzog, 1982; Schlesier-Stropp, 1984). Group members' ages ranged from nineteen to thirty-six, with the modal age being twenty-one. The respondents were white, and all but one were female. The sole male and three of the females were anorexic; the remaining females were bulimic.[4]

Primarily composed of college students, the group included four non-students, three of whom had college degrees. Nearly all members derived from upper-middle or lower-upper class households. Eighteen students and two non-students were never-marrieds and uninvolved in serious relationships; two non-students were married (one with two children); two students were divorced (one with two children); and six students were involved in serious relationships. The duration of eating disorders ranged from three to fifteen years.

CONFORMING BEHAVIOR

In the backgrounds of most anorexics and bulimics, dieting figures prominently, beginning in the teen years (Crisp, 1977; Johnson et al., 1982; Lacey et al., 1986). As dieters, these individuals are conformist in their adherence to the cultural norms emphasizing thinness (Garner et al., 1980; Schwartz et al., 1982). In our society, slim bodies are regarded as the most worthy and attractive; overweight is viewed as physically and morally unhealthy—"obscene," "lazy," "slothful," and "gluttonous" (DeJong, 1980; Ritenbaugh, 1982; Schwartz et al., 1982).

Among the agents of socialization promoting the slimness norm is advertising. Female models in newspaper, magazine, and television advertisements are uniformly slender. In addition, product names and slogans exploit the thin orientation; examples include "Ultra Slim Lipstick," "Miller Lite," and "Virginia Slims." While retaining pressures toward thinness, an Ayds commercial attempts a compromise for those wanting to savor food: "Ayds . . . so you can taste, chew, and enjoy, while you lose weight." Appealing particularly to women, a nationwide fast-food restaurant chain offers low-calorie selections, so individuals can have a "license to eat." In the latter two examples, the notion of enjoying food is combined with the message to be slim. Food and restaurant advertisements overall convey the pleasures of eating, whereas advertisements for other products, such as fashions and diet aids, reinforce the idea that fatness is undesirable.

Emphasis on being slim affects everyone in our culture, but it influences women especially because of society's traditional emphasis on women's appearance. The slimness norm and its concomitant narrow beauty standards exacerbate the objectification of women (Schur, 1984). Women view themselves as visual entities and recognize that conforming to appearance expectations and "becoming attractive object[s] [are] role obligation[s]" (Laws, as quoted in Schur, 1984: 66). Demonstrating the beauty motivation behind dieting, a recent Nielsen survey indicated that of the 56 percent of all women aged 24 to 54 who dieted during the previous year, 76 percent did so for cosmetic, rather than health, reasons (Schwartz et al., 1982). For most female group members, dieting was viewed as a means of gaining attractiveness and appeal to the opposite sex. The male respondent, as well, indicated that "when I was fat, girls didn't look at me, but when I got thinner, I was suddenly popular."

In addition to responding to the specter of obesity, individuals who develop anorexia nervosa and bulimia are conformist in their strong commitment to other conventional norms and goals. They consistently excel at school and work (Russell, 1979; Bruch, 1981; Humphries et al., 1982), maintaining high aspirations in both areas (Theander, 1970; Lacey et al., 1986). Group members generally completed college-preparatory courses in high school, aware from an early age that they would strive for a college degree. Also, in college as well as high school, respondents joined honor societies and academic clubs.

Moreover, pre-anorexics and -bulimics display notable conventionality as "model children" (Humphries et al., 1982: 199), "the pride and joy" of their parents (Bruch, 1981: 215), accommodating themselves to the wishes of others. Parents of these individuals emphasize conformity and value achievement (Bruch, 1981). Respondents felt that perfect or near-perfect grades were expected of them; however, good grades were not rewarded by parents, because "A's" were common for these children. In addition, their parents suppressed conflicts, to preserve the image of the "all-American family" (Humphries et al., 1982). Group members reported that they seldom, if ever, heard their parents argue or raise their voices.

Also conformist in their affective ties, individuals who develop anorexia nervosa and bulimia are strongly, even excessively, attached to their parents. Respondents' families appeared close-knit, demonstrating palpable emotional ties. Several group members, for example, reported habitually calling home at prescribed times, whether or not they had any news. Such families have been termed "enmeshed" and "overprotective," displaying intense interaction and concern for members' welfare (Minuchin et al., 1978; Selvini-Palazzoli, 1978). These qualities could be viewed as marked conformity to the norm of familial closeness.[5]

Another element of notable conformity in the family milieu of pre-anorexics and -bulimics concerns eating, body weight/shape, and exercising (Kalucy et al., 1977; Humphries et al., 1982). Respondents reported their fathers' preoccupation with exercising and their mothers' engrossment in food preparation. When group members dieted and lost weight, they received an extraordinary amount of approval. Among the family, body size became a matter of "friendly rivalry." One bulimic informant recalled that she, her mother, and her coed sister all strived to wear a size 5, regardless of their heights and body frames. Subsequent to this study, the researchers learned that both the mother and sister had become bulimic.

As pre-anorexics and -bulimics, group members thus exhibited marked conformity to cultural norms of thinness, achievement, compliance, and parental attachment. Their families reinforced their conformity by adherence to norms of family closeness and weight/body shape consciousness.

PRIMARY DEVIANCE

Even with familial encouragement, respondents, like nearly all dieters (Chernin, 1981), failed to maintain their lowered weights. Many cited their lack of willpower to eat only restricted foods. For the emerging anorexics and bulimics, extremes such as purposeful starvation or binging accompanied by vomiting and/or laxative abuse appeared as "obvious solutions" to the problem of retaining weight loss. Associated with these behaviors was a regained feeling of control in lives that had been disrupted by a major crisis. Group members' extreme weight-loss efforts operated as coping mechanisms for entering college, leaving home, or feeling rejected by the opposite sex.

The primary inducement for both eating adaptations was the drive for slimness: with slimness came more self-respect and a feeling of superiority over "unsuccessful dieters." Brian, for example, experienced a "power trip" upon consistent weight loss through starvation. Binges allowed the purging respondents to cope with stress through eating while maintaining a slim appearance. As former strict dieters, Teresa and Jennifer used binging/purging as an alternative to the constant self-denial of starvation. Acknowledging their parents' desires for them to be slim, most respondents still felt it was a conscious choice on their part to continue extreme weight-loss efforts. Being thin became the "most important thing" in their lives—their "greatest ambition."

In explaining the development of an anorexic or bulimic identity, Lemert's (1951; 1967) concept of primary deviance is salient. Primary deviance refers to a transitory period of norm violations which do not affect an individual's self-concept or performance of social roles. Although respondents were exhibiting anorexic or bulimic behavior, they did not consider themselves to be anorexic or bulimic.

At first, anorexics' significant others complimented their weight loss, expounding on their new "sleekness" and "good looks." Branch and Eurman (1980: 631) also found anorexics' families and friends describing them as "well-groomed," "neat," "fashionable," and "victorious." Not until the respondents approached emaciation did some parents or friends become concerned and withdraw their praise. Significant others also became increasingly aware of the anorexics' compulsive exercising, preoccupation with food preparation (but not consumption), and ritualistic eating patterns (such as cutting food into minute pieces and eating only certain foods at prescribed times).

For bulimics, friends or family members began to question how the respondents could eat such large amounts of food (often in excess of 10,000 calories a day) and stay slim. Significant others also noticed calluses across the bulimics' hands, which were caused by repeated inducement of vomiting.

Several bulimics were "caught in the act," bent over commodes. Generally, friends and family required substantial evidence before believing that the respondents' binging or purging was no longer sporadic.

SECONDARY DEVIANCE

Heightened awareness of group members' eating behavior ultimately led others to label the respondents "anorexic" or "bulimic." Respondents differed in their histories of being labeled and accepting the labels. Generally first termed anorexic by friends, family, or medical personnel, the anorexics initially vigorously denied the label. They felt they were not "anorexic enough," not skinny enough; Robin did not regard herself as having the "skeletal" appearance she associated with anorexia nervosa. These group members found it difficult to differentiate between socially approved modes of weight loss—eating less and exercising more—and the extremes of those behaviors. In fact, many of their activities—cheerleading, modeling, gymnastics, aerobics—reinforced their pursuit of thinness. Like other anorexics, Chris felt she was being "ultra-healthy," with "total control" over her body.

For several respondents, admitting they were anorexic followed the realization that their lives were disrupted by their eating disorder. Anorexics' inflexible eating patterns unsettled family meals and holiday gatherings. Their regimented lifestyle of compulsively scheduled activities—exercising, school, and meals—precluded any spontaneous social interactions. Realization of their adverse behaviors preceded the anorexics' acknowledgment of their subnormal body weight and size.

Contrasting with anorexics, the binge/purgers, when confronted, more readily admitted that they were bulimic and that their means of weight loss was "abnormal." Teresa, for example, knew "very well" that her bulimic behavior was "wrong and unhealthy," although "worth the physical risks." While the bulimics initially maintained that their purging was only a temporary weight-loss method, they eventually realized that their disorder represented a "loss of control." Although these respondents regretted the self-indulgence, "shame," and "wasted time," they acknowledged their growing dependence on binging/purging for weight management and stress regulation.

The application of anorexic or bulimic labels precipitated secondary deviance, wherein group members internalized these identities. Secondary deviance refers to norm violations which are a response to society's labeling: "secondary deviation . . . becomes a means of social defense, attack or adaptation to the overt and covert problems created by the societal reaction to primary deviance" (Lemert, 1967: 17). In contrast to primary deviance, secondary deviance is generally prolonged, alters the individual's self-concept, and affects the performance of his/her social roles.

As secondary deviants, respondents felt that their disorders "gave a purpose" to their lives. Nicole resisted attaining a normal weight because it was not "her"—she accepted her anorexic weight as her "true" weight. For Teresa, bulimia became a "companion"; and Julie felt "every aspect of her life," including time management and social activities, was affected by her bulimia. Group members' eating disorders became the salient element of their self-concepts, so that they related to familiar people and new acquaintances as anorexics or bulimics. For example, respondents regularly compared their body shapes and sizes with those of others. They also became sensitized to comments about their appearance, whether or not the remarks were made by someone aware of their eating disorder.

With their behavior increasingly attuned to their eating disorders, group members exhibited role engulfment (Schur, 1971). Through accepting anorexic or bulimic identities, individuals centered activities around their deviant role, downgrading other social roles. Their obligations as students, family members, and friends became subordinate to their eating and exercising rituals. Socializing, for example, was gradually curtailed because it interfered with compulsive exercising, binging, or purging.

Labeled anorexic or bulimic, respondents were ascribed a new status with a different set of role expectations. Regardless of other positions the individuals occupied, their deviant status, or master status (Hughes, 1958; Becker, 1973), was identified before all others. Among group members, Nicole, who was known as the "school's brain," became known as the "school's anorexic." No longer viewed as conforming model individuals, some respondents were termed "starving waifs" or "pigs."

Because of their identities as deviants, anorexics' and bulimics' interactions with others were altered. Group members' eating habits were scrutinized by friends and family and used as a "catch-all" for everything negative that happened to them. Respondents felt self-conscious around individuals who knew of their disorders; for example, Robin imagined people "watching and whispering" behind her. In addition, group members believed others expected them to "act" anorexic or bulimic. Friends of some anorexic group members never offered them food or drink, assuming continued disinterest on the respondents' part. While being hospitalized, Denise felt she had to prove to others she was not still vomiting, by keeping her bathroom door open. Other bulimics, who lived in dormitories, were hesitant to use the restroom for normal purposes lest several friends be huddling at the door, listening for vomiting. In general, individuals interacted with the respondents largely on the basis of their eating disorder; in doing so, they reinforced anorexic and bulimic behaviors.

Bulimic respondents, whose weight-loss behavior was not generally detectable from their appearance, tried earnestly to hide their bulimia by binging and purging in secret. Their main purpose in concealment was to avoid the negative consequences of being known as a bulimic. For these individuals, bulimia connoted a "cop-out": like "weak anorexics," bulimics pursued thinness but yielded to urges to eat. Respondents felt other people regarded bulimia as "gross" and had little sympathy for the sufferer. To avoid these stigmas or "spoiled identities," the bulimics shrouded their behaviors.

Distinguishing types of stigma, Goffman (1963) describes discredited (visible) stigmas and discreditable (invisible) stigmas. Bulimics, whose weight was approximately normal or even slightly elevated, harbored discreditable stigmas.

Anorexics, on the other hand, suffered both discreditable and discredited stigmas—the latter due to their emaciated appearance. Certain anorexics were more reconciled than the bulimics to their stigmas: for Brian, the "stigma of anorexia was better than the stigma of being fat." Common to the stigmatized individuals was an inability to interact spontaneously with others. Respondents were constantly on guard against topics of eating and body size.

Both anorexics and bulimics were held responsible by others for their behavior and presumed able to "get out of it if they tried." Many anorexics reported being told to "just eat more," while bulimics were enjoined to simply "stop eating so much." Such appeals were made without regard for the complexities of the problem. Ostracized by certain friends and family members, anorexics and bulimics felt increasingly isolated. For respondents, the self-help group presented a non-threatening forum for discussing their disorders. Here, they found mutual understanding, empathy, and support. Many participants viewed BANISH as a haven from stigmatization by "others."

Group members, as secondary deviants, thus endured negative consequences, such as stigmatization, from being labeled. As they internalized the labels anorexic or bulimic, individuals' self-concepts were significantly influenced. When others interacted with the respondents on the basis of their eating disorders, anorexic or bulimic identities were encouraged. Moreover, group members' efforts to counteract the deviant labels were thwarted by their master statuses.

DISCUSSION

Previous research on eating disorders has dwelt almost exclusively on medical and psychological facets. Although necessary for a comprehensive understanding of anorexia nervosa and bulimia, these approaches neglect the social processes involved. The phenomena of eating disorders transcend concrete disease entities and clinical diagnoses. Multifaceted and complex, anorexia nervosa and bulimia require a holistic research design, in which sociological insights must be included.

A limitation of medical/psychiatric studies, in particular, is researchers' use of a priori criteria in establishing salient variables. Rather than utilizing predetermined standards of inclusion, the present study allows respondents to construct their own reality. Concomitant to this innovative approach to eating disorders is the selection of a sample of self-admitted anorexics and bulimics. Individuals' perceptions of what it means to become anorexic or bulimic are explored. Although based on a small sample, findings can be used to guide researchers in other settings.

With only five to ten percent of reported cases appearing in males (Crisp, 1977; Stangler and Printz, 1980), eating disorders are primarily a women's aberrance. The deviance of anorexia nervosa and bulimia is rooted in the visual objectification of women and attendant slimness norm. Indeed, purposeful starvation and binging/purging reinforce the notion that "a society gets the deviance it deserves" (Schur, 1979: 71). As recently noted (Schur, 1984), the sociology of deviance has generally bypassed systematic studies of women's norm violations. Like male deviants, females endure label applications, internalizations, and fulfillments.

The social processes involved in developing anorexic or bulimic identities comprise the sequence of conforming behavior, primary deviance, and secondary deviance. With a background of exceptional adherence to conventional norms, especially the striving for thinness, respondents subsequently exhibit the primary deviance of starving or binging/purging. Societal reaction to these behaviors leads to secondary deviance, wherein respondents' self-concepts and master statuses become anorexic or bulimic. Within this framework of labeling theory, the persistence of eating disorders, as well as the effects of stigmatization, is elucidated.

Although during the course of this research some respondents alleviated their symptoms through psychiatric help or hospital treatment programs, no one was labeled "cured." An anorexic is considered recovered when weight is normal for two years; a bulimic is termed recovered after being symptom-free for one and one-half years (American Anorexia/Bulimia Association Newsletter, 1985). Thus deviance disavowal (Schur, 1971), or efforts after normalization to counteract the deviant labels, remains a topic for future exploration.

NOTES

1. Although instructive, an integration of the medical, psychological, and sociocultural perspectives on eating disorders is beyond the scope of this paper.

2. Exceptions to the neglect of sociocultural factors are discussions of sex-role socialization in the development of eating disorders. Anorexics' girlish appearance has been interpreted as a rejection of femininity and womanhood (Orbach, 1979; Bruch, 1981; Orbach, 1985). In contrast, bulimics have been characterized as overconforming to traditional female sex roles (Boskind-Lodahl, 1976).

3. Although a group experience for self-defined bulimics has been reported (Boskind-Lodahl, 1976), the researcher, from the outset, focused on Gestalt and behaviorist techniques within a feminist orientation.

4. One explanation for fewer anorexics than bulimics in the sample is that, in the general population, anorexics are outnum-

bered by bulimics at 8 or 10 to 1 (Lawson, as reprinted in American Anorexia/Bulimia Association Newsletter, 1985: 1). The proportion of bulimics to anorexics in the sample is 6.5 to 1. In addition, compared to bulimics, anorexics may be less likely to attend a self-help group as they have a greater tendency to deny the existence of an eating problem (Humphries et al., 1982). However, the four anorexics in the present study were among the members who attended the meetings most often.

5. Interactions in the families of anorexics and bulimics might seem deviant in being inordinately close. However, in the larger societal context, the family members epitomize the norms of family cohesiveness. Perhaps unusual in their occurrence, these families are still within the realm of conformity. Humphries and colleagues (1982: 202) refer to the "highly enmeshed and protective" family as part of the "idealized family myth."

REFERENCES

American Anorexia/Bulimia Association. 1983. Correspondence. April.

American Anorexia/Bulimia Association Newsletter. 1985. 8(3).

Becker, Howard S. 1973. *Outsiders.* New York: Free Press.

Boskind-Lodahl, Marlene. 1976. "Cinderella's stepsisters: A feminist perspective on anorexia nervosa and bulimia." *Signs, Journal of Women in Culture and Society* 2: 342–56.

Boskind-White, Marlene. 1985. "Bulimarexia: A sociocultural perspective." In S. W. Emmett (ed.), *Theory and Treatment of Anorexia Nervosa and Bulimia: Biomedical, Sociocultural and Psychological Perspectives* (pp. 113–26). New York: Brunner/Mazel.

Branch, C. H. Hardin, and Linda J. Eurman. 1980. "Social attitudes toward patients with anorexia nervosa." *American Journal of Psychiatry* 137: 631–32.

Bruch, Hilde. 1981. "Developmental considerations of anorexia nervosa and obesity." *Canadian Journal of Psychiatry* 26: 212–16.

Chernin, Kim. 1981. *The Obsession: Reflections on the Tyranny of Slenderness.* New York: Harper and Row.

Crisp, A. H. 1977. "The prevalence of anorexia nervosa and some of its associations in the general population." *Advances in Psychosomatic Medicine* 9: 38–47.

Crisp, A. H., R. L. Palmer, and R. S. Kalucy. 1976. "How common is anorexia nervosa? A prevalence study." *British Journal of Psychiatry* 128: 549–54.

DeJong, William. 1980. "The stigma of obesity: The consequences of naive assumptions concerning the causes of physical deviance." *Journal of Health and Social Behavior* 21: 75–87.

Fox, K. C., and N. McI. James. 1976. "Anorexia nervosa: A study of 44 strictly defined cases." *New Zealand Medical Journal* 84: 309–12.

Garner, David M., Paul E. Garfinkel, Donald Schwartz, and Michael Thompson. 1980. "Cultural expectations of thinness in women." *Psychological Reports* 47: 483–91.

Goffman, Erving. 1963. *Stigma.* Englewood Cliffs, NJ: Prentice-Hall.

Halmi, Katherine A., James R. Falk, and Estelle Schwartz. 1981. "Binge-eating and vomiting: A survey of a college population." *Psychological Medicine* 11: 697–706.

Herzog, David B. 1982. "Bulimia: The secretive syndrome." *Psychosomatics* 23: 481–83.

Hughes, Everett C. 1958. *Men and Their Work.* New York: Free Press.

Humphries, Laurie L., Sylvia Wrobel, and H. Thomas Wiegert. 1982. "Anorexia nervosa." *American Family Physician* 26: 199–204.

Johnson, Craig L., Marilyn K. Stuckey, Linda D. Lewis, and Donald M. Schwartz. 1982. "Bulimia: A descriptive survey of 316 cases." *International Journal of Eating Disorders* 2(1): 3–16.

Kalucy, R. S., A. H. Crisp, and Britta Harding. 1977. "A study of 56 families with anorexia nervosa." *British Journal of Medical Psychology* 50: 381–95.

Lacey, Hubert J., Sian Coker, and S. A. Birtchnell. 1986. "Bulimia: Factors associated with its etiology and maintenance." *International Journal of Eating Disorders* 5: 475–87.

Lemert, Edwin M. 1951. *Social Pathology.* New York: McGraw-Hill.

———. 1967. *Human Deviance, Social Problems and Social Control.* Englewood Cliffs, NJ: Prentice-Hall.

Minuchin, Salvador, Bernice L. Rosman, and Lester Baker. 1978. *Psychosomatic Families: Anorexia Nervosa in Context.* Cambridge, MA: Harvard University Press.

Orbach, Susie. 1979. *Fat Is a Feminist Issue.* New York: Berkeley.

————. 1985. "Visibility/invisibility: Social considerations in anorexia nervosa—a feminist perspective." In S. W. Emmett (ed.), *Theory and Treatment of Anorexia Nervosa and Bulimia: Biomedical, Sociocultural, and Psychological Perspectives* (pp. 127–38). New York: Brunner/Mazel.

Ritenbaugh, Cheryl. 1982. "Obesity as a culture-bound syndrome." *Culture, Medicine and Psychiatry* 6: 347–61.

Russell, Gerald. 1979. "Bulimia nervosa: An ominous variant of anorexia nervosa." *Psychological Medicine* 9: 429–48.

Schlesier-Stropp, Barbara. 1984. "Bulimia: A review of the literature." *Psychological Bulletin* 95: 247–57.

Schur, Edwin M. 1971. *Labeling Deviant Behavior*. New York: Harper and Row.

————. 1979. *Interpreting Deviance: A Sociological Introduction*. New York: Harper and Row.

————. 1984. *Labeling Women Deviant: Gender, Stigma, and Social Control*. New York: Random House.

Schwartz, Donald M., and Michael G. Thompson. 1981. "Do anorectics get well? Current research and future needs." *American Journal of Psychiatry* 138: 319–23.

Schwartz, Donald M., Michael G. Thompson, and Craig L. Johnson. 1982. "Anorexia nervosa and bulimia: The socio-cultural context." *International Journal of Eating Disorders* 1(3): 20–36.

Selvini-Palazzoli, Mara. 1978. *Self-Starvation: From Individual to Family Therapy in the Treatment of Anorexia Nervosa*. New York: Jason Aronson.

Stangler, Ronnie S., and Adolph M. Printz. 1980. "DSM-III: Psychiatric diagnosis in a university population." *American Journal of Psychiatry* 137: 937–40.

Theander, Sten. 1970. "Anorexia nervosa." *Acta Psychiatrica Scandinavica Supplement* 214: 24–31.

Thompson, Michael G., and Donald M. Schwartz. 1982. "Life adjustment of women with anorexia nervosa and anorexic-like behavior." *International Journal of Eating Disorders* 1(2): 47–60.

Willi, Jurg, and Samuel Grossmann. 1983. "Epidemiology of anorexia nervosa in a defined region of Switzerland." *American Journal of Psychiatry* 140: 564–67.

23

Convicted Rapists' Vocabulary of Motive

Excuses and Justifications

DIANA SCULLY AND
JOSEPH MAROLLA

Psychiatry has dominated the literature on rapists since "irresistible impulse" (Glueck, 1925: 243) and "disease of the mind" (Glueck, 1925: 243) were introduced as the causes of rape. Research has been based on small samples of men, frequently the clinicians' own patient population. Not surprisingly, the medical model has predominated: rape is viewed as an individualistic, idiosyncratic symptom of a disordered personality. That is, rape is assumed to be a psychopathologic problem and individual rapists are assumed to be "sick." However, advocates of this model have been unable to isolate a typical or even predictable pattern of symptoms that are causally linked to rape. Additionally, research has demonstrated that fewer than 5 percent of rapists were psychotic at the time of their rape (Abel et al., 1980).

We view rape as behavior learned socially through interaction with others; convicted rapists have learned the attitudes and actions consistent with sexual aggression against women. Learning also includes the acquisition of culturally derived vocabularies of motive, which can be used to diminish responsibility and to negotiate a non-deviant identity.

Sociologists have long noted that people can, and do, commit acts they define as wrong and, having done so, engage various techniques to disavow deviance and present themselves as normal. Through the concept of "vocabulary of motive," Mills (1940: 904) was among the first to shed light on this seemingly perplexing contradiction. Wrong-doers attempt to reinterpret their actions through the use of a linguistic device by which norm-breaking conduct is socially interpreted. That is, anticipating the negative consequences of their behavior, wrong-doers attempt to present the act in terms that are both culturally appropriate and acceptable.

Following Mills, a number of sociologists have focused on the types of techniques employed by actors in problematic situations (Hall and Hewitt,

From "Convicted Rapists' Vocabulary of Motive: Excuses and Justifications," Diana Scully and Joseph Marolla, *Social Problems*, Vol. 31, No. 5, 1984. © 1984 Society for the Study of Social Problems. Reprinted by permission of University of California Press Journals and Diana Scully.

1970; Hewitt and Hall, 1973; Hewitt and Stokes, 1975; Sykes and Matza, 1957). Scott and Lyman (1968) describe excuses and justifications, linguistic "accounts" that explain and remove culpability for an untoward act after it has been committed. *Excuses* admit the act was bad or inappropriate but deny full responsibility, often through appeals to accident, or biological drive, or through scapegoating. In contrast, *justifications* accept responsibility for the act but deny that it was wrong—that is, they show in this situation the act was appropriate. *Accounts* are socially approved vocabularies that neutralize an act or its consequences and are always a manifestation of an underlying negotiation of identity.

Stokes and Hewitt (1976: 837) use the term "aligning actions" to refer to those tactics and techniques used by actors when some feature of a situation is problematic. Stated simply, the concept refers to an actor's attempt, through various means, to bring his or her conduct into alignment with culture. Culture in this sense is conceptualized as a "set of cognitive constraints—objects— to which people must relate as they form lines of conduct" (1976: 837), and includes physical constraints, expectations and definitions of others, and personal biography. Carrying out aligning actions implies both awareness of those elements of normative culture that are applicable to the deviant act and, in addition, an actual effort to bring the act into line with this awareness. The result is that deviant behavior is legitimized.

This paper presents an analysis of interviews we conducted with a sample of 114 convicted, incarcerated rapists. We use the concept of accounts (Scott and Lyman, 1968) as a tool to organize and analyze the vocabularies of motive which this group of rapists used to explain themselves and their actions. An analysis of their accounts demonstrates how it was possible for 83 percent (n = 114)[1] of these convicted rapists to view themselves as non-rapists.

When rapists' accounts are examined, a typology emerges that consists of admitters and deniers. Admitters (n = 47) acknowledged that they had forced sexual acts on their victims and defined the behavior as rape. In contrast, deniers[2] either eschewed sexual contact or all association with the victim (n = 35),[3] or admitted to sexual acts but did not define their behavior as rape (n = 32). . . . By and large, the deniers used justifications while the admitters used excuses. In some cases, both groups relied on the same themes, stereotypes, and images: some admitters, like most deniers, claimed that women enjoyed being raped. Some deniers excused their behavior by referring to alcohol or drug use, although they did so quite differently than admitters. Through these narrative accounts, we explore convicted rapists' own perceptions of their crimes. . . .

JUSTIFYING RAPE

Deniers attempted to justify their behavior by presenting the victim in a light that made her appear culpable, regardless of their own actions. Five themes run through attempts to justify their rapes: (1) women as seductresses; (2) women mean "yes" when they say "no"; (3) most women eventually relax and enjoy it; (4) nice girls don't get raped; and (5) guilty of a minor wrongdoing.

(1) Women as Seductresses

Men who rape need not search far for cultural language which supports the premise that women provoke or are responsible for rape. In addition to common cultural stereotypes, the fields of psychiatry and criminology (particularly the subfield of victimology) have traditionally provided justifications for rape, often by portraying raped women as the victims of their own seduction (Albin, 1977; Marolla and Scully, 1979). For example, Hollander (1924: 130) argues:

> Considering the amount of illicit intercourse, rape of women is very rare indeed. Flirtation and provocative conduct, i.e. tacit (if not actual) consent is generally the prelude to intercourse.

Since women are supposed to be coy about their sexual availability, refusal to comply with a man's sexual demands lacks meaning and rape appears normal. The fact that violence and, often, a weapon are used to accomplish the rape is not considered. As an example, Abrahamsen (1960: 61) writes:

> The conscious or unconscious biological or psychological attraction between man and woman does not exist only on the part of the offender toward the woman but, also, on her part toward him, which in many instances may, to some extent, be the impetus for his sexual attack. Often a women [sic] unconsciously wishes to be taken by force—consider the theft of the bride in Peer Gynt.

Like Peer Gynt, the deniers we interviewed tried to demonstrate that their victims were willing and, in some cases, enthusiastic participants. In these accounts, the rape became more dependent upon the victim's behavior than upon their own actions.

Thirty-one percent (n = 10) of the deniers presented an extreme view of the victim. Not only willing, she was the aggressor, a seductress who lured them, unsuspecting, into sexual action. Typical was a denier convicted of his first rape and accompanying crimes of burglary, sodomy, and abduction. According to the pre-sentence reports, he had broken into the victim's house and raped her at knife point. While he admitted to the breaking and entry, which he claimed was for altruistic purposes ("to pay for the prenatal care of a friend's girlfriend"), he also argued that when the victim discovered him, he had tried to leave but she had asked him to stay. Telling him that she cheated on her husband, she had voluntarily removed her clothes and seduced him. She was, according to him, an exemplary sex partner who "enjoyed it very much and asked for oral sex.[4] Can I have it now?" he reported her as saying. He claimed they had spent hours in bed, after which the victim had told him he was good looking and asked to see him again. "Who would believe I'd meet a fellow like this?" he reported her as saying.

In addition to this extreme group, 25 percent (n = 8) of the deniers said the victim was willing and had made some sexual advances. An additional 9 percent (n = 3) said the victim was willing to have sex for money or drugs. In two of these three cases, the victim had been either an acquaintance or picked up, which the rapists said led them to expect sex.

(2) Women Mean "Yes" When They Say "No"

Thirty-four percent (n = 11) of the deniers described their victim as unwilling, at least initially, indicating either that she had resisted or that she had said no. Despite this, and even though (according to pre-sentence reports) a weapon had been present in 64 percent (n = 7) of these 11 cases, the rapists justified their behavior by arguing that either the victim had not resisted enough or that her "no" had really meant "yes." For example, one denier who was serving time for a previous rape was subsequently convicted of attempting to rape a prison hospital nurse. He insisted he had actually completed the second rape, and said of his victim: "She semi-struggled but deep down inside I think she felt it was a fantasy come true." The nurse, according to him, had asked a question about his conviction for rape, which he interpreted as teasing. "It was like she was saying, 'rape me.'" Further, he stated that she had helped him along with oral sex and "from her actions, she was enjoying it." In another case, a 34-year-old man convicted of abducting and raping a 15-year-old teenager at knife point as she walked on the beach, claimed it was a pickup. This rapist said women like to be overpowered before sex, but to dominate after it begins.

> A man's body is like a Coke bottle, shake it up, put your thumb over the opening and feel the tension. When you take a woman out, woo her, then she says "no, I'm a nice girl," you have to use force. All men do this. She said "no" but it was a societal no, she wanted to be coaxed. All women say "no" when they mean "yes" but it's a societal no, so they won't have to feel responsible later.

Claims that the victim didn't resist or, if she did, didn't resist enough, were also used by 24 percent (n = 11) of admitters to explain why, during the incident, they believed the victim was willing and that they were not raping. These rapists didn't redefine their acts until some time after the crime. For example, an admitter who used a bayonet to threaten his victim, an employee of the store he had been robbing, stated:

> At the time I didn't think it was rape. I just asked her nicely and she didn't resist. I never considered prison. I just felt like I had met a friend. It took about five years of reading and going to school to change my mind about whether it was rape. I became familiar with the subtlety of violence. But at the time, I believed that as long as I didn't hurt anyone it wasn't wrong. At the time, I didn't think I would go to prison, I thought I would beat it.

Another typical case involved a gang rape in which the victim was abducted at knife point as she walked home about midnight. According to two of the rapists, both of whom were interviewed, at the time they had thought the victim had willingly accepted a ride from the third rapist (who was not interviewed). They claimed the victim didn't resist and one reported her as saying she would do anything if they would take her home. In this rapist's view, "She acted like she enjoyed it, but maybe she was just acting. She wasn't crying, she was engaging in it." He reported that she had been friendly to the rapist who abducted her and, claiming not to have a home phone, she gave him her office number—a tactic eventually used to catch the three. In retrospect, this young man had decided, "She was scared and just relaxed and enjoyed it to avoid getting hurt." Note, however, that while he had redefined the act as rape, he continued to believe she enjoyed it.

Men who claimed to have been unaware that they were raping viewed sexual aggression as a man's prerogative at the time of the rape. Thus they regarded their act as little more than a minor wrongdoing even though most possessed or used a weapon. As long as the victim survived without major physical injury, from their perspective, a rape had not taken place. Indeed, even U.S. courts have often taken the position that physical injury is a necessary ingredient for a rape conviction.

(3) Most Women Eventually Relax and Enjoy It

Many of the rapists expected us to accept the image, drawn from cultural stereotype, that once the rape began, the victim relaxed and enjoyed it.[5] Indeed, 69 percent (n = 22) of deniers justified their behavior by claiming not only that the victim was willing, but also that she enjoyed herself, in some cases to an immense degree. Several men suggested that they had fulfilled their victims' dreams. Additionally, while most admitters used adjectives such as "dirty," "humiliated," and "disgusted" to describe how they thought rape made women feel, 20 percent (n = 9) believed that their victim enjoyed herself. For example, one denier had posed as a salesman to gain entry to his victim's house. But he claimed he had had a previous sexual relationship with the victim, that she agreed to have sex for drugs, and that the opportunity to have sex with him produced "a glow, because she was really into oral stuff and fascinated by the idea of sex with a black man. She felt satisfied, fulfilled, wanted me to stay, but I didn't want her." In another case, a denier who had broken into his victim's house but who insisted the victim was his lover and let him in voluntarily, declared "She felt good, kept kissing me and wanted me to stay the night. She felt proud after sex with me." And another denier, who had hid in his victim's closet and later attacked her while she slept, argued that while she was scared at first, "once we got into it, she was ok." He continued to believe he hadn't committed rape because "she enjoyed it and it was like she consented."

(4) Nice Girls Don't Get Raped

The belief that "nice girls don't get raped" affects perception of fault. The victim's reputation, as well as characteristics or behavior which violate normative sex role expectations, are perceived as contributing to the commission of the crime. For example, Nelson and Amir (1975) defined hitchhike rape as a victim-precipitated offense.

In our study, 69 percent (n = 22) of deniers and 22 percent (n = 10) of admitters referred to their victims' sexual reputation, thereby evoking the stereotype that "nice girls don't get raped." They claimed that the victim was known to have been a prostitute, or a "loose" woman, or to have had a lot of affairs, or to have given birth to a child out of wedlock. For example, a denier who claimed he had picked up his victim while she was hitchhiking stated, "To be honest, we [his family] knew she was a damn whore and whether she screwed one or 50 guys didn't matter." According to pre-sentence reports this victim didn't know her attacker and he abducted her at knife point from the street. In another case, a denier who claimed to have known his victim by reputation stated:

> If you wanted drugs or a quick piece of ass, she would do it. In court she said she was a virgin, but I could tell during sex [rape] that she was very experienced.

When other types of discrediting biographical information were added to these sexual slurs, a total of 78 percent (n = 25) of the deniers used the victim's reputation to substantiate their accounts. Most frequently, they referred to the victim's emotional state or drug use. For example, one denier claimed his victim had been known to be loose and, additionally, had turned state's evidence against her husband to put him in prison and save herself from a burglary conviction. Further, he asserted that she had met her current boyfriend, who was himself in and out of prison, in a drug rehabilitation center where they were both clients.

Evoking the stereotype that women provoke rape by the way they dress, a description of the victim as seductively attired appeared in the accounts of 22 percent (n = 7) of deniers and 17 percent (n = 8) of admitters. Typically, these descriptions were used to substantiate their claims about the victim's reputation. Some men went to extremes to paint a tarnished picture of the victim, describing her as dressed in tight black clothes and without a bra; in one case, the victim was portrayed as sexually provocative in dress and carriage. Not only did she wear short skirts, but she was observed to "spread her legs while getting out of cars." Not all of the men attempted to assassinate their victim's reputation with equal vengeance. Numerous times they made subtle and off-hand remarks like, "She was a waitress and you know how they are."

The intent of these discrediting statements is clear. Deniers argued that the woman was a "legitimate" victim who got what she deserved. For example, one denier stated that all of his victims had been prostitutes; pre-sentence reports indicated they were not. Several times during his interview, he referred to them as "dirty sluts," and argued "anything I did to them was justified." Deniers also claimed their victim had wrongly accused them and was the type of woman who would perjure herself in court.

(5) Only a Minor Wrongdoing

The majority of deniers did not claim to be completely innocent and they also accepted some accountability for their actions. Only 16 percent (n = 5) of deniers argued that they were totally free of blame. Instead, the majority of deniers pleaded guilty to a lesser charge. That is, they obfuscated the rape by pleading guilty to a less serious, more acceptable charge. They accepted being over-sexed, accused of poor judgment or trickery, even some violence, or guilty of adultery or contributing to the delinquency of a minor, charges that are hardly the equivalent of rape.

Typical of this reasoning is a denier who met his victim in a bar when the bartender asked him if he would try to repair her stalled car. After attempting unsuccessfully, he claimed the victim drank with him and later accepted a ride. Out riding, he pulled into a deserted area "to see how my luck would go." When the victim resisted his advances, he beat her and he stated:

> I did something stupid. I pulled a knife on her and I hit her as hard as I would hit a man. But I shouldn't be in prison for what I did. I shouldn't have all this time [sentence] for going to bed with a broad.

This rapist continued to believe that while the knife was wrong, his sexual behavior was justified.

In another case, the denier claimed he picked up his under-age victim at a party and that she voluntarily went with him to a motel. According to pre-sentence reports, the victim had been abducted at knife point from a party. He explained:

> After I paid for a motel, she would have to have sex but I wouldn't use a weapon. I would have explained. I spent money and, if she still said no, I would have forced her. If it had happened that way, it would have been rape to some people but not to my way of thinking. I've done that kind of thing before. I'm guilty of sex and contributing to the delinquency of a minor, but not rape.

In sum, deniers argued that, while their behavior may not have been completely proper, it should not have been considered rape. To accomplish this, they attempted to discredit and blame the victim while presenting their own actions as justified in the context. Not surprisingly, none of the deniers thought of himself as a rapist. A minority of the admitters attempted to lessen the impact of their crime by claiming the victim enjoyed being raped. But despite this similarity, the nature and tone of admitters' and deniers' accounts were essentially different.

EXCUSING RAPE

In stark contrast to deniers, admitters regarded their behavior as morally wrong and beyond justification. They blamed themselves rather than the victim, although some continued to cling to the belief that the victim had contributed to the crime somewhat, for example, by not resisting enough.

Several of the admitters expressed the view that rape was an act of such moral outrage that it was unforgivable. Several admitters broke into tears at intervals during their interviews. A typical sentiment was,

> I equate rape with someone throwing you up against a wall and tearing your liver and guts out of you. . . . Rape is worse than murder . . . and I'm disgusting.

Another young admitter frequently referred to himself as repulsive and confided:

> I'm in here for rape and in my own mind, it's the most disgusting crime, sickening. When people see me and know, I get sick.

Admitters tried to explain their crime in a way that allowed them to retain a semblance of moral integrity. Thus, in contrast to deniers' justifications, admitters used excuses to explain how they were compelled to rape. These excuses appealed to the existence of forces outside of the rapists' control. Through the use of excuses, they attempted to demonstrate that either intent was absent or responsibility was diminished. This allowed them to admit rape while reducing the threat to their identity as a moral person. Excuses also permitted them to view their behavior as idiosyncratic rather than typical and, thus, to believe they were not "really" rapists. Three themes run through these accounts: (1) the use of alcohol and drugs; (2) emotional problems; and (3) nice guy image.

(1) The Use of Alcohol and Drugs

A number of studies have noted a high incidence of alcohol and drug consumption by convicted rapists prior to their crime (Groth, 1979; Queen's Bench Foundation, 1976). However, more recent research has tentatively concluded that the connection between substance use and crime is not as direct as previously thought (Ladouceur, 1983). Another facet of alcohol and drug use mentioned in the literature is its utility in disavowing deviance. McCaghy (1968) found that child molesters used alcohol as a technique for neutralizing their deviant identity. Marolla and Scully (1979), in a review of psychiatric literature, demonstrated how alcohol consumption is applied differently as a vocabulary of motive. Rapists can use alcohol both as an excuse for their behavior and to discredit the victim and make her more responsible. We found the former common among admitters and the latter common among deniers.

Alcohol and/or drugs were mentioned in the accounts of 77 percent (n = 30) of admitters and 84 percent (n = 21) of deniers and both groups were equally likely to have acknowledged consuming a substance—admitters, 77 percent (n = 30); deniers, 72 percent (n = 18). However, admitters said they had been affected by the substance; if not the cause of their behavior, it was at least a contributing factor. For example, an admitter who estimated his consumption to have been eight beers and four "hits of acid" reported:

> Straight, I don't have the guts to rape. I could fight a man but not that. To say, "I'm going to do it to a woman," knowing it will scare and hurt her, takes guts or you have to be sick.

Another admitter believed that his alcohol and drug use,

> . . . brought out what was already there but in such intensity it was uncontrollable. Feelings of being dominant, powerful, using someone for my own gratification, all rose to the surface.

**Rapists' Accounts of Own and Victims'
Alcohol and/or Drug (A/D) Use and Effect**

	Admitters n = 39 %	Deniers n = 25 %
Neither Self nor Victim Used A/D	23	16
Self Used A/D	77	72
Of Self Used, No Victim Use	51	12
Self Affected by A/D	69	40
Of Self Affected, No Victim Use or Effect	54	24
Self A/D Users Who Were Affected	90	56
Victim Used A/D	26	72
Of Victim Used, No Self Use	0	0
Victim Affected by A/D	15	56
Of Victim Affected, No Self Use or Effect	0	40
Victim A/D Users Who Were Affected	60	78
Both Self and Victim Used and Affected by A/D	15	16

In contrast, deniers' justifications required that they not be substantially impaired. To say that they had been drunk or high would cast doubt on their ability to control themselves or to remember events as they actually happened. Consistent with this, when we asked if the alcohol and/or drugs had had an effect on their behavior, 69 percent (n = 27) of admitters, but only 40 percent (n = 10) of deniers, said they had been affected.

Even more interesting were references to the victim's alcohol and/or drug use. Since admitters had already relieved themselves of responsibility through claims of being drunk or high, they had nothing to gain from the assertion that the victim had used or been affected by alcohol and/or drugs. On the other hand, it was very much in the interest of deniers to declare that their victim had been intoxicated or high: that fact lessened her credibility and made her more responsible for the act. Reflecting these observations, 72 percent (n = 18) of deniers and 26 percent (n = 10) of admitters maintained that alcohol or drugs had been consumed by the victim. Further, while 56 percent (n = 14) of deniers declared she had been affected by this use, only 15 percent (n = 6) of admitters made a similar claim. Typically deniers argued that the alcohol and drugs had sexually aroused their victim or rendered her out of control. For example, one denier insisted that his victim had become hysterical from drugs, not from being raped, and it was because of the drugs that she had reported him to the police. In addition, 40 percent (n = 10) of deniers argued that while the victim had been drunk or high, they themselves either hadn't ingested or weren't affected by alcohol and/or drugs. None of the admitters made this claim. In fact, in all of the 15 percent (n = 6) of cases where an admitter said the victim was drunk or high, he also admitted to being similarly affected.

These data strongly suggest that whatever role alcohol and drugs play in sexual and other types of violent crime, rapists have learned the advantage to be gained from using alcohol and drugs as an account. Our sample were aware that their victim would be discredited and their own behavior excused or justified by referring to alcohol and/or drugs.

(2) Emotional Problems

Admitters frequently attributed their acts to emotional problems. Forty percent (n = 19) of admitters said they believed an emotional problem had been at the root of their rape behavior, and 33 percent (n = 15) specifically related the problem to an unhappy, unstable childhood or a marital-domestic situation. Still others claimed to have been in a general state of unease. For example, one admitter said that at the time of the rape he had been depressed, feeling he couldn't do anything right, and that something had been missing from his life. But he also added, "being a rapist is not part of my personality." Even admitters who could locate no source for an emotional problem evoked the popular image of rapists as the product of disordered personalities to argue they also must have problems:

> The fact that I'm a rapist makes me different. Rapists aren't all there. They have problems. It was wrong so there must be a reason why I did it. I must have a problem.

Our data do indicate that a precipitating event, involving an upsetting problem of everyday living, appeared in the accounts of 80 percent (n = 38) of admitters and 25 percent (n = 8) of deniers. Of those experiencing a precipitating event, including deniers, 76 percent (n = 35) involved a wife or girlfriend. Over and over, these men described themselves as having been in a rage because of an incident involving a woman with whom they believed they were in love.

Frequently, the upsetting event was related to a rigid and unrealistic double standard for sexual conduct and virtue which they applied to "their" woman but which they didn't expect from men, didn't apply to themselves, and, obviously, didn't honor in other women. To discover that the "pedestal" didn't apply to their wife or girlfriend sent them into a fury. One especially articulate and typical admitter described his feeling as follows. After serving a short prison term for auto theft, he married his "childhood sweetheart" and secured a well-paying job. Between his job and the volunteer work he was doing with an ex-offender group, he was spending long hours away from home, a situation that had bothered his wife. In response to her request, he gave up his volunteer work, though it was clearly meaningful to him. Then, one day, he discovered his wife with her former boyfriend "and my life fell apart." During the next several days, he said his anger had made him withdraw into himself and, after three days of drinking in a motel room, he abducted and raped a stranger. He stated:

> My parents have been married for many years and I had high expectations about marriage. I put my wife on a pedestal. When I walked in on her, I felt like my life had been destroyed, it was such a shock. I was bitter and

angry about the fact that I hadn't done anything to my wife for cheating. I didn't want to hurt her [victim], only to scare and degrade her.

It is clear that many admitters, and a minority of deniers, were under stress at the time of their rapes. However, their problems were ordinary—the types of upsetting events that everyone experiences at some point in life. The overwhelming majority of the men were not clinically defined as mentally ill in court-ordered psychiatric examinations prior to their trials. Indeed, our sample is consistent with Abel et al. (1980) who found fewer than 5 percent of rapists were psychotic at the time of their offense.

As with alcohol and drug intoxication, a claim of emotional problems works differently depending upon whether the behavior in question is being justified or excused. It would have been counter-productive for deniers to have claimed to have had emotional problems at the time of the rape. Admitters used psychological explanations to portray themselves as having been temporarily "sick" at the time of the rape. Sick people are usually blamed for neither the cause of their illness nor for acts committed while in that state of diminished capacity. Thus, adopting the sick role removed responsibility by excusing the behavior as having been beyond the ability of the individual to control. Since the rapists were not "themselves," the rape was idiosyncratic rather than typical behavior. Admitters asserted a non-deviant identity despite their self-proclaimed disgust with what they had done. Although admitters were willing to assume the sick role, they did not view their problem as a chronic condition, nor did they believe themselves to be insane or permanently impaired. Said one admitter, who believed that he needed psychological counseling: "I have a mental disorder, but I'm not crazy." Instead, admitters viewed their "problem" as mild, transient, and curable. Indeed, part of the appeal of this excuse was that not only did it relieve responsibility, but, as with alcohol and drug addiction, it allowed the rapist to "recover." Thus, at the time of their interviews, only 31 percent (n = 14) of admitters indicated that "being a rapist" was part of their self-concept. Twenty-eight percent (n = 13) of admitters stated they had never thought of themselves as rapists, 8 percent (n = 4) said they were unsure, and 33 percent (n = 16) asserted they had been a rapist at one time but now were recovered. A multiple "ex-rapist," who believed his "problem" was due to "something buried in my subconscious" that was triggered when his girlfriend broke up with him, expressed a typical opinion:

I was a rapist, but not now. I've grown up, had to live with it. I've hit the bottom of the well and it can't get worse. I feel born again to deal with my problems.

(3) Nice Guy Image

Admitters attempted to further neutralize their crime and negotiate a non-rapist identity by painting an image of themselves as a "nice guy." Admitters projected the image of someone who had made a serious mistake but, in every other respect, was a decent person. Fifty-seven percent (n = 27) expressed regret and sorrow for their victim indicating that they wished there were a way to apologize for or amend their behavior. For example, a participant in a rape-murder, who insisted his partner did the murder, confided, "I wish there was something I could do besides saying 'I'm sorry, I'm sorry.' I live with it 24 hours a day and, sometimes, I wake up crying in the middle of the night because of it."

Schlenker and Darby (1981) explain the significance of apologies beyond the obvious expression of regret. An apology allows a person to admit guilt while at the same time seeking a pardon by signaling that the event should not be considered a fair representation of what the person is really like. An apology separates the bad self from the good self, and promises more acceptable behavior in the future. When apologizing, an individual is attempting to say: "I have repented and should be forgiven," thus making it appear that no further rehabilitation is required.

The "nice guy" statements of the admitters reflected an attempt to communicate a message consistent with Schlenker's and Darby's analysis of apologies. It was an attempt to convey that rape was not a representation of their "true" self. For example,

> It's different from anything else I've ever done. I feel more guilt about this. It's not consistent with me. When I talk about it, it's like being assaulted myself. I don't know why I did it, but once I started, I got into it. Armed robbery was a way of life for me, but not rape. I feel like I wasn't being myself.

Admitters also used "nice guy" statements to register their moral opposition to violence and harming women, even though, in some cases, they had seriously injured their victims. Such was the case of an admitter convicted of gang rape:

> I'm against hurting women. She should have resisted. None of us were the type of person that would use force on a woman. I never positioned myself on a woman unless she showed an interest in me. They would play to me, not me to them. My weakness is to follow. I never would have stopped, let alone pick her up without the others. I never would have let anyone beat her. I never bothered women who didn't want sex; never had a problem with sex or getting it. I loved her—like all women.

Finally, a number of admitters attempted to improve their self-image by demonstrating that, while they had raped, it could have been worse if they had not been a "nice guy." For example, one admitter professed to being especially gentle with his victim after she told him she had just had a baby. Others claimed to have given the victim money to get home or make a phone call, or to have made sure the victim's children were not in the room. A multiple rapist, whose pattern was to break in and attack sleeping victims in their homes, stated:

> I never beat any of my victims and I told them I wouldn't hurt them if they cooperated. I'm a professional thief. But I never robbed the women I raped because I felt so bad about what I had already done to them.

Even a young man, who raped his five victims at gun point and then stabbed them to death, attempted to improve his image by stating:

> Physically they enjoyed the sex [rape]. Once they got involved, it would be difficult to resist. I was always gentle and kind until I started to kill them. And the killing was always sudden, so they wouldn't know it was coming.

SUMMARY AND CONCLUSIONS

Convicted rapists' accounts of their crimes include both excuses and justifications. Those who deny what they did was rape justify their actions; those who admit it was rape attempt to excuse it or themselves. This study does not address why some men admit while others deny, but future research might address this question. This paper does provide insight on how men who are sexually aggressive or violent construct reality, describing the different strategies of admitters and deniers.

Admitters expressed the belief that rape was morally reprehensible. But they explained themselves and their acts by appealing to forces beyond their control, forces which reduced their capacity to act rationally and thus compelled them to rape. Two types of excuses predominated: alcohol/drug intoxication and emotional problems. Admitters used these excuses to negotiate a moral identity for themselves by viewing rape as idiosyncratic rather than typical behavior. This allowed them to reconceptualize themselves as recovered or "ex-rapists," someone who had made a serious mistake which did not represent their "true" self.

In contrast, deniers' accounts indicate that these men raped because their value system provided no compelling reason not to do so. When sex is viewed as a male entitlement, rape is no longer seen as criminal. However, the deniers had been convicted of rape, and like the admitters, they attempted to negotiate an identity. Through justifications, they constructed a "controversial" rape and attempted to demonstrate how their behavior, even if not quite right, was appropriate in the situation. Their denials, drawn from common cultural rape stereotypes, took two forms, both of which ultimately denied the existence of a victim.

The first form of denial was buttressed by the cultural view of men as sexually masterful and women as coy but seductive. Injury was denied by portraying the victim as willing, even enthusiastic, or as politely resistant at first but eventually yielding to "relax and enjoy it." In these accounts, force appeared merely as a seductive technique. Rape was disclaimed: rather than harm the woman, the rapist had fulfilled her dreams. In the second form of denial, the victim was portrayed as the type of woman who "got what she deserved." Through attacks on the victim's sexual reputation and, to a lesser degree, her emotional state, deniers attempted to demonstrate that since the victim wasn't a "nice girl," they were not rapists. Consistent with both forms of denial was the self-interested use of alcohol and drugs as a justification. Thus, in contrast to admitters, who accentuated their own use as an excuse, deniers emphasized the victim's consumption in an effort to both discredit her and make her appear more responsible for the rape. It is important to remember that deniers did not invent these justifications. Rather, they reflect a belief system which has historically victimized women by promulgating the myth that women both enjoy and are responsible for their own rape.

While admitters and deniers present an essentially contrasting view of men who rape, there were some shared characteristics. Justifications particularly, but also excuses, are buttressed by the cultural view of women as sexual commodities, dehumanized and devoid of autonomy and dignity. In this sense, the sexual objectification of women must be understood as an important factor contributing to an environment that trivializes, neutralizes, and, perhaps, facilitates rape.

Finally, we must comment on the consequences of allowing one perspective to dominate thought on a social problem. Rape, like any complex continuum of behavior, has multiple causes and is influenced by a number of social factors. Yet, dominated by psychiatry and the medical model, the underlying assumption that rapists are "sick" has pervaded research. Although methodologically unsound, conclusions have been based almost exclusively on small clinical populations of rapists—that extreme group of rapists who seek counseling in prison and are the most likely to exhibit psychopathology. From this small, atypical group of men, psychiatric findings have been generalized to all men who rape. Our research, however, based on volunteers from the entire prison population, indicates that some rapists, like deniers, viewed and understood their behavior from a popular cultural perspective. This strongly suggests that cultural perspectives, and not an idiosyncratic illness, motivated their behavior. Indeed, we can argue that the psychiatric perspective has contributed to the vocabulary of motive that rapists use to excuse and justify their behavior (Scully and Marolla, 1984).

Efforts to arrive at a general explanation for rape have been retarded by the narrow focus of the medical model and the preoccupation with clinical populations. The continued reduction of such complex behavior to a singular cause hinders, rather than enhances, our understanding of rape.

NOTES

1.   These numbers include pretest interviews. When the analysis involves either questions that were not asked in the pretest or that were changed, they are excluded and thus the number changes.

2.   There is, of course, the possibility that some of these men really were innocent of rape. However, while the U.S. criminal justice system is not without flaw, we assume that it is highly unlikely that this many men could have been unjustly convicted of rape, especially since rape is a crime with traditionally low conviction rates. Instead, for purposes of this research, we assume that these men were guilty as charged and that their attempt to maintain an image of non-rapist springs from some psychologically or sociologically interpretable mechanism.

3.   Because of their outright denial, interviews with this group of rapists did not contain the data being analyzed here and, consequently, they are not included in this paper.

4.   It is worth noting that a number of deniers specifically mentioned the victim's alleged interest in oral sex. Since our interview questions about sexual history indicated that the rapists themselves found oral sex marginally acceptable, the frequent mention is probably another attempt to discredit the victim. However, since a tape recorder could not be used for the interviews and the importance of these claims didn't emerge until the data was being coded and analyzed, it is possible that it was mentioned even more frequently but not recorded.

5.   Research shows clearly that women do not enjoy rape. Holmstrom and Burgess (1978) asked 93 adult rape victims, "How did it feel sexually?" Not one said they enjoyed it. Further, the trauma of rape is so great that it disrupts sexual functioning (both frequency and satisfaction) for the overwhelming majority of victims, at least during the period immediately following the rape and, in fewer cases, for an extended period of time (Burgess and Holmstrom, 1979; Feldman-Summers et al., 1979). In addition, a number of studies have shown that rape victims experience adverse consequences prompting some to move, change jobs, or drop out of school (Burgess and Holmstrom, 1974; Kilpatrick et al., 1979; Ruch et al., 1980; Shore, 1979).

REFERENCES

Abel, Gene, Judith Becker, and Linda Skinner. 1980. "Aggressive behavior and sex." *Psychiatric Clinics of North America* 3(2): 133–151.

Abrahamsen, David. 1960. *The Psychology of Crime.* New York: John Wiley.

Albin, Rochelle. 1977. "Psychological studies of rape." *Signs* 3(2): 423–435.

Burgess, Ann Wolbert, and Lynda Lytle Holmstrom. 1974. *Rape: Victims of Crisis.* Bowie: Robert J. Brady.

———. 1979. "Rape: Sexual disruption and recovery." *American Journal of Orthopsychiatry* 49(4): 648–657.

Feldman-Summers, Shirley, Patricia E. Gordon, and Jeanette R. Meagher. 1979. "The impact of rape on sexual satisfaction." *Journal of Abnormal Psychology* 88(1): 101–105.

Glueck, Sheldon. 1925. *Mental Disorders and the Criminal Law.* New York: Little, Brown.

Groth, Nicholas A. 1979. *Men Who Rape.* New York: Plenum Press.

Hall, Peter M., and John P. Hewitt. 1970. "The quasi-theory of communication and the management of dissent." *Social Problems* 18(1): 17–27.

Hewitt, John P., and Peter M. Hall. 1973. "Social problems, problematic situations, and quasi-theories." *American Journal of Sociology* 38(3): 367–374.

Hewitt, John P., and Randall Stokes. 1975. "Disclaimers." *American Sociological Review* 40(1): 1–11.

Hollander, Bernard. 1924. *The Psychology of Misconduct, Vice and Crime.* New York: Macmillan.

Holmstrom, Lynda Lytle, and Ann Wolbert Burgess. 1978. "Sexual behavior of assailant and victim during rape." Paper presented at the annual meetings of the American Sociological Association, San Francisco, September 2–8.

Kilpatrick, Dean G., Lois Veronen, and Patricia A. Resnick. 1979. "The aftermath of rape: Recent empirical findings." *American Journal of Orthopsychiatry* 49(4): 658–669.

Ladouceur, Patricia. 1983. "The relative impact of drugs and alcohol on serious felons." Paper presented at the annual meetings of the American Society of Criminology, Denver, November 9–12.

McCaghy, Charles. 1968. "Drinking and deviance disavowal: The case of child molesters." *Social Problems* 16(1): 43–49.

Marolla, Joseph, and Diana Scully. 1979. "Rape and psychiatric vocabularies of motive." In Edith S. Gomberg and Violet Franks (eds.), *Gender and Disordered Behavior: Sex Differences in Psychopathology* (pp. 301-318). New York: Brunner/Mazel.

Mills, C. Wright. 1940. "Situated actions and vocabularies of motive." *American Sociological Review* 5(6): 904–913.

Nelson, Steve, and Amir Menachem. 1975. "The hitchhike victim of rape: A research report." In Israel Drapkin and Emilio Viano (eds.), *Victimology: A New Focus* (pp. 47–65). Lexington, KY: Lexington Books.

Queen's Bench Foundation. 1976. *Rape: Prevention and Resistance*. San Francisco: Queen's Bench Foundation.

Ruch, Libby O., Susan Meyers Chandler, and Richard A. Harter. 1980. "Life change and rape impact." *Journal of Health and Social Behavior* 21(3): 248–260.

Schlenker, Barry R., and Bruce W. Darby. 1981. "The use of apologies in social predicaments." *Social Psychology Quarterly* 44(3): 271–278.

Scott, Marvin, and Stanford Lyman. 1968. "Accounts." *American Sociological Review* 33(1): 46–62.

Scully, Diana, and Joseph Marolla. 1984. "Rape and psychiatric vocabularies of motive: Alternative perspectives." In Ann Wolbert Burgess (ed.), *Handbook on Rape and Sexual Assault*. New York: Garland Publishing.

Shore, Barbara K. 1979. "An Examination of Critical Process and Outcome Factors in Rape." Rockville, MD: National Institute of Mental Health.

Stokes, Randall, and John P. Hewitt. 1976. "Aligning actions." *American Sociological Review* 41(5): 837–849.

Sykes, Gresham M., and David Matza. 1957. "Techniques of neutralization." *American Sociological Review* 22(6): 664–670.

24

The Influence of Situational Ethics on Cheating Among College Students

DONALD L. McCABE

Numerous studies have demonstrated the pervasive nature of cheating among college students (Baird 1980; Haines, Diekhoff, LaBeff, and Clark 1986; Michaels and Miethe 1989; Davis et al. 1992). This research has examined a variety of factors that help explain cheating behavior, but the strength of the relationships between individual factors and cheating has varied considerably from study to study (Tittle and Rowe 1973; Baird 1980; Eisenberger and Shank 1985; Haines et al. 1986; Ward 1986; Michaels and Miethe 1989; Perry, Kane, Bernesser, and Spicker 1990; Ward and Beck 1990).

Although the factors examined in these studies (for example, personal work ethic, gender, self-esteem, rational choice, social learning, deterrence) are clearly important, the work of LaBeff, Clark, Haines, and Diekhoff (1990) suggests that the concept of situational ethics may be particularly helpful in understanding student rationalizations for cheating. Extending the arguments of Norris and Dodder (1979), LaBeff et al. conclude

> that students hold qualified guidelines for behavior which are situationally determined. As such, the concept of situational ethics might well describe
> . . . college cheating [as having] rules for behavior [that] may not be considered rigid but depend on the circumstances involved (1990, p. 191).

LaBeff et al. believe a utilitarian calculus of "the end justifies the means" underlies this reasoning process and "what is wrong in most situations might be considered right or acceptable if the end is defined as appropriate" (1990, p. 191). As argued by Edwards (1967), the situation determines what is right or wrong in this decision-making calculus and also dictates the appropriate principles to be used in guiding and judging behavior.

Sykes and Matza (1957) hypothesize that such rationalizations, that is, "justifications for deviance that are seen as valid by the delinquent but not by the

From "The Influence of Situational Ethics on Cheating Among College Students,"
Donald L. McCabe, *Sociological Inquiry*, Vol. 62:3 (Summer 1992). Reprinted by permission of the University of Texas Press and the author.

legal system or society at large" (p. 666), are common. However, they challenge conventional wisdom that such rationalizations typically follow deviant behavior as a means of protecting "the individual from self-blame and the blame of others after the act" (p. 666). They develop convincing arguments that these rationalizations may logically precede the deviant behavior and "[d]isapproval from internalized norms and conforming others in the social environment is neutralized, turned back, or deflated in advance. Social controls that serve to check or inhibit deviant motivational patterns are rendered inoperative, and the individual is freed to engage in delinquency without serious damage to his self-image" (pp. 666–667).

Using a sample of 380 undergraduate students at a small southwestern university, LaBeff et al. (1990) attempted to classify techniques employed by students in the neutralization of cheating behavior into the five categories of neutralization proposed by Sykes and Matza (1957): (1) denial of responsibility, (2) condemnation of condemners, (3) appeal to higher loyalties, (4) denial of victim, and (5) denial of injury. Although student responses could easily be classified into three of these techniques, denial of responsibility, appeal to higher loyalties, and condemnation of condemners, LaBeff et al. conclude that "[i]t is unlikely that students will either deny injury or deny the victim since there are no real targets in cheating" (1990, p. 196).

The research described here responds to LaBeff et al. in two ways; first, it answers their call to "test the salience of neutralization . . . in more diverse university environments" (p. 197) and second, it challenges their dismissal of denial of injury and denial of victim as neutralization techniques employed by students in their justification of cheating behavior.

METHODOLOGY

The data discussed here were gathered as part of a study of college cheating conducted during the 1990–1991 academic year. A seventy-two-item questionnaire concerning cheating behavior was administered to students at thirty-one highly selective colleges across the country. Surveys were mailed to a minimum of five hundred students at each school and a total of 6,096 completed surveys were returned (38.3 percent response rate). Eighty-eight percent of the respondents were seniors, nine percent were juniors, and the remaining three percent could not be classified. Survey administration emphasized voluntary participation and assurances of anonymity to help combat issues of non-response bias and the need to accept responses without the chance to question or contest them.

The final sample included 61.2 percent females (which reflects the inclusion of five all-female schools in the sample and a slightly higher return rate among female students) and 95.4 percent U.S. citizens. The sample paralleled the ethnic diversity of the participating schools (85.5 percent Anglo, 7.2 percent Asian, 2.6 percent African American, 2.2 percent Hispanic and 2.5 percent other); their religious diversity (including a large percentage of students who claimed no religious preference, 27.1 percent); and their mix of undergraduate majors (36.0 percent humanities, 28.8 percent social sciences, 26.8 percent natural sciences and engineering, 4.5 percent business, and 3.9 percent other).

RESULTS

Of the 6,096 students participating in this research, over two-thirds (67.4 percent) indicated that they had cheated on a test or major assignment at least once while an undergraduate. This cheating took a variety of different forms, but among the most popular (listed in decreasing order of mention) were: (1) a failure to footnote sources in written work, (2) collaboration on assignments when the instructor specifically asked for individual work, (3) copying from other students on tests and examinations, (4) fabrication of bibliographies, (5) helping someone else cheat on a test, and (6) using unfair methods to learn the content of a test ahead of time. Almost one in five students (19.1 percent) could be classified as active cheaters (five or more self-reported incidents of cheating). This is double the rate reported by LaBeff et al. (1990), but they asked students to report only cheating incidents that had taken place in the last six months. Students in this research were asked to report all cheating in which they had engaged while an undergraduate—a period of three years for most respondents at the time of this survey.

Students admitting to any cheating activity were asked to rate the importance of several specific factors that might have influenced their decisions to cheat. These data establish the importance of denial of responsibility and condemnation of condemners as neutralization techniques. For example, 52.4 percent of the respondents who admitted to cheating rated the pressure to get good grades as an important influence in their decision to cheat, with parental pressures and competition to gain admission into professional schools singled out as the primary grade pressures. Forty-six percent of those who had engaged in cheating cited excessive workloads and an inability to keep up with assignments as important factors in their decisions to cheat.

In addition to rating the importance of such preselected factors, 426 respondents (11.0 percent of the admitted cheaters) offered their own justifications for cheating in response to an open-ended question on motivations for cheating. These responses confirm the importance of denial of responsibility and condemnation of condemners as neutralization techniques. They also support LaBeff et al.'s (1990) claim that appeal to higher loyalties is an important neutralization technique. However, these responses also suggest that LaBeff et al.'s dismissal of denial of injury as a justification for student cheating is arguable.

As shown in the table, denial of responsibility was the technique most frequently cited (216 responses, 61.0 percent of the total) in the 354 responses classified into one of Sykes and Matza's five categories of neutralization. The most common responses in this category were mind block, no understanding

**Neutralization Strategies:
Self-Admitted Cheaters**

Strategy	Number	Percent
Denial of responsibility	216	61.0
Mind block	90	25.4
No understanding of material	31	8.8
Other	95	26.8
Condemnation of condemners	99	28.0
Pointless assignment	35	9.9
No respect for professor	28	7.9
Other	36	10.2
Appeal to higher loyalties	24	6.8
Help a friend	10	2.8
Peer pressure	9	2.5
Other	5	1.5
Denial of injury	15	4.2
Cheating is harmless	9	2.5
Does not matter	6	1.7

of the material, a fear of failing, and unclear explanations of assignments. (Although it is possible that some instances of mind block and a fear of failing included in this summary would be more accurately classified as rationalization, the wording of all responses included here suggests that rationalization preceded the cheating incident. Responses that seem to involve post hoc rationalizations were excluded from this summary.) Condemnation of condemners was the second most popular neutralization technique observed (99 responses, 28.0 percent) and included such explanations as pointless assignments, lack of respect for individual professors, unfair tests, parents' expectations, and unfair professors. Twenty-four respondents (6.8 percent) appealed to higher loyalties to explain their behavior. In particular, helping a friend and responding to peer pressures were influences some students could not ignore. Finally fifteen students (4.2 percent) provided responses that clearly fit into the category of denial of injury. These students dismissed their cheating as harmless since it did not hurt anyone or they felt cheating did not matter in some cases (for example, where an assignment counted for a small percentage of the total course grade).

Detailed examination of selected student responses provides additional insight into the neutralization strategies they employ.

Denial of Responsibility

Denial of responsibility invokes the claim that the act was "due to forces outside of the individual and beyond his control such as unloving parents" (Sykes and Matza 1957, p. 667). For example, many students cite an unreasonable workload and the difficulty of keeping up as ample justification for cheating:

Here at . . . , you must cheat to stay alive. There's so much work and the quality of materials from which to learn, books, professors, is so bad that there's no other choice.

It's the only way to keep up.

I couldn't do the work myself.

The following descriptions of student cheating confirm fear of failure is also an important form of denial of responsibility:

. . . a take-home exam in a class I was failing.

. . . was near failing.

Some justified their cheating by citing the behavior of peers:

Everyone has test files in fraternities, etc. If you don't, you're at a great disadvantage.

When most of the class is cheating on a difficult exam and they will ruin the curve, it influences you to cheat so your grade won't be affected.

All of these responses contain the essence of denial of responsibility: the cheater has deflected blame to others or to a specific situational context.

Denial of Injury

As noted in the table, denial of injury was identified as a neutralization technique employed by some respondents. A key element in denial of injury is whether one feels "anyone has clearly been hurt by [the] deviance." In invoking this defense, a cheater would argue "that his behavior does not really cause any great harm despite the fact that it runs counter to the law" (Sykes and Matza 1957, pp. 667–668). For example, a number of students argued that the assignment or test on which they cheated was so trivial that no one was really hurt by their cheating.

These grades aren't worth much therefore my copying doesn't mean very much. I am ashamed, but I'd probably do it the same way again.

If I extend the time on a take-home it is because I feel everyone does and the teacher kind of expects it. No one gets hurt.

As suggested earlier, these responses suggest the conclusion of LaBeff et al. that "[i]t is unlikely that students will . . . deny injury" (1990, p. 196) must be re-evaluated.

The Denial of the Victim

LaBeff et al. failed to find any evidence of denial of the victim in their student accounts. Although the student motivations for cheating summarized in the table support this conclusion, at least four students (0.1% of the self-admitted cheaters in this study) provided comments elsewhere on the survey instrument which involved denial of the victim. The common element in these responses was a victim deserving of the consequences of the cheating behavior and cheating was viewed as "a form of rightful retaliation or punishment" (Sykes and Matza 1957, p. 668).

This feeling was extreme in one case, as suggested by the following student who felt her cheating was justified by the

> realization that this school is a manifestation of the bureaucratic capitalist system that systematically keeps the lower classes down, and that adhering to their rules was simply perpetuating the institution.

This "we" versus "they" mentality was raised by many students, but typically in comments about the policing of academic honesty rather than as justification for one's own cheating behavior. When used to justify cheating, the target was almost always an individual teacher rather than the institution and could be more accurately classified as a strategy of condemnation of condemners rather than denial of the victim.

The Condemnation of Condemners

Sykes and Matza describe the condemnation of condemners as an attempt to shift "the focus of attention from [one's] own deviant acts to the motives and behavior of those who disapprove of [the] violations. [B]y attacking others, the wrongfulness of [one's] own behavior is more easily repressed or lost to view" (1957, p. 668). The logic of this strategy for student cheaters focused on issues of favoritism and fairness. Students invoking this rationale describe "uncaring, unprofessional instructors with negative attitudes who were negligent in their behavior" (LaBeff et al. 1990, p. 195). For example:

> In one instance, nothing was done by a professor because the student was a hockey player.

> The TAs who graded essays were unduly harsh.

> It is known by students that certain professors are more lenient to certain types, e.g., blondes or hockey players.

> I would guess that 90% of the students here have seen athletes and/or fraternity members cheating on an exam or papers. If you turn in one of these culprits, and I have, the penalty is a five-minute lecture from a coach and/or administrator. All these add up to a "who cares, they'll never do anything to you anyway" attitude here about cheating.

Concerns about the larger society were an important issue for some students:

> When community frowns upon dishonesty, then people will change.

> If our leaders can commit heinous acts and then lie before Senate committees about their total ignorance and innocence, *then why can't I cheat a little?*

In today's world you do anything to be above the competition.

In general, students found ready targets on which to blame their behavior and condemnation of the condemners was a popular neutralization strategy.

The Appeal to Higher Loyalties

The appeal to higher loyalties involves neutralizing "internal and external controls . . . by sacrificing the demands of the larger society for the demands of the smaller social groups to which the [offender] belongs. [D]eviation from certain norms may occur not because the norms are rejected but because other norms, held to be more pressing or involving a higher loyalty, are accorded precedence" (Sykes and Matza 1957, p. 669). For example, a difficult conflict for some students is balancing the desire to help a friend against the institution's rules on cheating. The student may not challenge the rules, but rather views the need to help a friend, fellow fraternity/sorority member, or roommate to be a greater obligation which justifies the cheating behavior.

Fraternities and sororities were singled out as a network where such behavior occurs with some frequency. For example, a female student at a small university in New England observed:

There's a lot of cheating within the Greek system. Of all the cheating I've seen, it's often been men and women in fraternities and sororities who exchange information or cheat.

The appeal to higher loyalties was particularly evident in student reactions concerning the reporting of cheating violations. Although fourteen of the thirty-one schools participating in this research had explicit honor codes that generally require students to report cheating violations they observe, less than one-third (32.3 percent) indicated that they were likely to do so. When asked if they would report a friend, only 4 percent said they would and most students felt that they should not be expected to do so. Typical student comments included:

Students should not be sitting in judgment of their own peers.

The university is not a police state.

For some this decision was very practical:

A lot of students, 50 percent, wouldn't because they know they will probably cheat at some time themselves.

For others, the decision would depend on the severity of the violation they observed and many would not report what they consider to be minor violations, even those explicitly covered by the school's honor code or policies on academic honesty. Explicit examination or test cheating was one of the few violations where students exhibited any consensus concerning the need to report violations. Yet even in this case many students felt other factors must be considered. For example, a senior at a woman's college in the Northeast commented:

It would depend on the circumstances. If someone was hurt, *very likely.* If there was no single victim in the case, if the victim was [the] institution . . . , then *very unlikely.*

Additional evidence of the strength of the appeal to higher loyalties as a neutralization technique is found in the fact that almost one in five respondents (17.8 percent) reported that they had helped someone cheat on an examination or major test. The percentage who have helped others cheat on papers and other assignments is likely much higher. Twenty-six percent of those students who helped someone else cheat on a test reported that they had never cheated on a test themselves, adding support to the argument that peer pressure to help friends is quite strong.

CONCLUSIONS

From this research it is clear that college students use a variety of neutralization techniques to rationalize their cheating behavior, deflecting blame to others and/or the situational context, and the framework of Sykes and Matza (1957) seems well-supported when student explanations of cheating behavior are analyzed. Unlike prior research (LaBeff et al. 1990), however, the present findings suggest that students employ all of the techniques described by Sykes and Matza, including denial of injury and denial of victim. Although there was very limited evidence of the use of denial of victim, denial of injury was not uncommon. Many students felt that some forms of cheating were victimless crimes, particularly on assignments that accounted for a small percentage of the total course grade. The present research does affirm LaBeff et al.'s finding that denial of responsibility and condemnation of condemners are the neutralization techniques most frequently utilized by college students. Appeal to higher loyalties is particularly evident in neutralizing institutional expectations that students report cheating violations they observe.

The present results clearly extend the findings of LaBeff et al. into a much wider range of contexts as this research ultimately involved 6,096 students at thirty-one geographically dispersed institutions ranging from small liberal arts colleges in the Northeast to nationally prominent research universities in the South and West. Fourteen of the thirty-one institutions have long-standing honor-code traditions. The code tradition at five of these schools dates to the late 1800s and all fourteen have codes that survived the student unrest of the 1960s. In such a context, the strength of the appeal to higher loyalties and the denial of responsibility as justifications for cheating is a very persuasive argument that neutralization techniques are salient to today's college student. More importantly, it may suggest fruitful areas of future discourse between faculty, administrators, and students on the question of academic honesty.*

*The author would like to acknowledge the support of the Rutgers Graduate School of Management Research Resources Committee, Exxon Corporation, and First Fidelity Bancorporation.

REFERENCES

Baird, John S. 1980. "Current Trends in College Cheating." *Psychology in Schools* 17: 512–522.

Davis, Stephen F., Cathy A. Grover, Angela H. Becker, and Loretta N. McGregor. 1992. "Academic Dishonesty: Prevalence, Determinants, Techniques, and Punishments." *Teaching of Psychology*. In press.

Edwards, Paul. 1967. *The Encyclopedia of Philosophy,* no. 3, Paul Edwards (ed.). New York: Macmillan Company and Free Press.

Eisenberger, Robert, and Dolores M. Shank. 1985. "Personal Work Ethic and Effort Training Affect Cheating." *Journal of Personality and Social Psychology* 49: 520–528.

Haines, Valerie J., George Diekhoff, Emily LaBeff, and Robert Clark. 1986. "College Cheating: Immaturity, Lack of Commitment, and the Neutralizing Attitude." *Research in Higher Education* 25: 342–354.

LaBeff, Emily E., Robert E. Clark, Valerie J. Haines, and George M. Diekhoff. 1990. "Situational Ethics and College Student Cheating." *Sociological Inquiry* 60: 190–198.

Michaels, James W., and Terance Miethe. 1989. "Applying Theories of Deviance to Academic Cheating." *Social Science Quarterly* 70: 870–885.

Norris, Terry D., and Richard A. Dodder. 1979. "A Behavioral Continuum Synthesizing Neutralization Theory, Situational Ethics and Juvenile Delinquency." *Adolescence* 55: 545–555.

Perry, Anthony R., Kevin M. Kane, Kevin J. Bernesser, and Paul T. Spicker. 1990. "Type A Behavior, Competitive Achievement-Striving, and Cheating Among College Students." *Psychological Reports* 66: 459–465.

Sykes, Gresham M., and David Matza. 1957. "Techniques of Neutralization: A Theory of Delinquency." *American Sociological Review* 22: 664–670.

Tittle, Charles, and Alan Rowe. 1973. "Moral Appeal, Sanction Threat, and Deviance: An Experimental Test." *Social Problems* 20: 488–498.

Ward, David. 1986. "Self-Esteem and Dishonest Behavior Revisited." *Journal of Social Psychology* 123: 709–713.

Ward, David, and Wendy L. Beck. 1990. "Gender and Dishonesty." *Journal of Social Psychology* 130: 333–339.

25

Women Athletes as
Falsely Accused Deviants
Managing the Lesbian Stigma

ELAINE M. BLINDE AND
DIANE E. TAUB

Gender represents a powerful normative system that both evaluates and controls the behavior of men and women (Schur 1984). This system entails socially constructed conceptualizations of behavior intricately tied to societal perceptions of "masculinity" and "femininity" (Keller 1978).

Women occupy a unique position in this system of gender norms. Not only does it define appropriate and inappropriate behaviors, but the behaviors sanctioned for women are generally devalued in the broader society and often contribute to their subordination (Smith 1978). Women who engage in non-traditional gender behaviors capable of enhancing opportunities and self-actu-alization are subjected to various forms of social stigmatization (Schur 1984).

Widespread violation of gender norms by women constitutes a serious threat to the entire gender system. It is thus not surprising that women are stig-matized and labeled as deviant when their behaviors challenge traditional gen-der norms (Anderson 1988). As Schur (1984) suggests, stigmatization of behaviors not conforming to rigid gender-role expectations reinforces women's overall subordination as it restricts women's roles and potential. Moreover, op-erating from a relatively disadvantaged position of power, women are more vulnerable to stigmatization (Anderson 1988). Stigmatization thus represents a means of social control as it preserves the traditional gender system. Fear of being labeled deviant keeps women "in their place" and reduces challenges to prevailing gender norms (Schur 1984). Not surprisingly, individuals who oc-cupy positions of power or privilege, the "deviance-definers," benefit from continued subordination or suppression of the less powerful (Becker 1963).

Although all women experience devaluation and stigmatization by virtue of being female (Schur 1984), some women occupy roles or engage in behav-iors that make them even more susceptible to deviant labeling. This labeling is particularly indicative of women who violate multiple categories of gender

From "Women Athletes as Falsely Accused Deviants," Elaine M. Blinde and Diane E. Taub, *The Sociological Quarterly*, V. 33, No. 4, pp. 521–533. Copyright © 1992 by Mid-west Sociology Society. Reprinted by permission of University of California Press and Diane E. Taub.

norms, including (1) presentation of self (e.g., emotions, nonverbal communication, appearance, speech), (2) marriage and maternity, (3) sexuality (e.g., sexual behavior/orientation), and (4) occupational choice (Schur 1984).

One group of women judged to violate multiple categories of gender norms and thus subjected to various forms of deviance labeling and stigmatization is athletes. Women athletes are frequently perceived to cross or extend the boundaries of socially constructed definitions of "femininity" (Theberge 1985; Willis 1982). In a culture that traditionally equates athleticism with masculinity, women who participate in sport are often viewed as masculine, unladylike, or manly (Willis 1982). Such descriptors imply that women athletes violate presentation of self and occupational gender norms (Schur 1984). Moreover, both the popular press and research literature have examined the assumed conflict between athleticism and femininity (Colker and Widom 1980; Griffin 1987; Hall 1988). This presumed incompatibility, along with equating sport and masculinity, results in a belief system linking women athletes with lesbianism (Lenskyj 1991). The lesbian label, representing a violation of sexuality norms (Schur 1984), is based on the idea that women who challenge traditional gender-role behavior cannot be "real" women (Lenskyj 1991).

The lesbian label as applied to women athletes is particularly significant given the assumed threat of lesbianism (as well as male homosexuality) to the prevailing gender system (Goodman 1977; Lenskyj 1991; Schur 1984). Stigmatization of lesbians is common as the phenomena of "heterosexual assumption" and "heterosexual privilege" ensure that lesbians realize they are indeed norm violators, deviants, and subjects of oppression (Schur 1984; Wolfe 1988). Such stigmatization reflects contempt for those who stretch the boundaries of culturally defined gender roles.

Since the behavior of women athletes is often interpreted to challenge gender norms, they may be subject to various forms of devaluation and stigmatization. Although athleticism represents the initial discrediting attribute, its linkage with lesbianism magnifies the devaluation and stigmatization associated with female athletes. The present study examines the manner in which such forms of devaluation and stigmatization impact on women athletes and how they manage the lesbian label (and accompanying masculine image) attached to their sport participation. Not only has there been a paucity of research investigating the specific topic of the stigmatization of women athletes as lesbians, but the entire issue of the perceived presence of lesbianism in women's sport has been neglected (Griffin 1987).

METHODOLOGY

Sample

Athletic directors at seven large Division I universities were contacted to obtain names and addresses of all varsity women athletes for the purpose of telephone interviews. These universities represent a wide range of athletic conference affiliations and geographical regions. A larger number of universities than needed was contacted because of the projected low response rates; constraints both on time for compiling lists of athletes and utilization of athletes as research respondents were anticipated.

Three universities (two midwestern and one southern) provided names and addresses of all women athletes with at least one year of completed athletic eligibility who were participating in the school's sport programs during the 1990–1991 school year. All sport programs offered by the universities were represented. Varsity women athletes, rather than intramural or informal sport participants, were the focus of this study because of the salience of their athletic identities and their greater sport commitment. Four universities declined to participate due to time demands and institutional restrictions on athletes serving as research participants.

A final sample size between 20 and 30 was desired given its manageability for the projected interview length, yet sufficiently broad-based to depict a wide range of athletic experiences. An initial random sample of 48, equally selected from the three universities, was drawn to allow for potential nonrespondents.

Each athlete was mailed a letter explaining the importance and general purpose of the study. Because the interview schedule focused on topics other than lesbianism and since the lesbian issue was lodged in the broader area of stereotypes, the cover letter did not specifically mention the topic of lesbianism. Moreover, identifying the lesbian issue as a component of the interview could have influenced an athlete's decision to participate, thus leading to sample bias. Athletes were encouraged to return an informed consent form indicating their willingness to participate in a tape-recorded telephone interview. Based on this initial contact, 16 athletes returned informed consent forms.

To increase the sample to the originally desired size (20–30), 30 additional individuals were randomly drawn and equally selected from the three lists. An additional eight athletes agreed to participate, thus providing for a final sample of 24 (representing a distribution of 10–5–9 among the three universities).

These 24 athletes were currently participating in a variety of intercollegiate sports—basketball (5), track and field (4), volleyball (3), swimming (3), softball (3), tennis (2), diving (2), and gymnastics (2). With an average age of 20.2 years and predominantly white (92%), the sample consisted of 2 freshmen, 9 sophomores, 5 juniors, and 8 seniors. The majority (22) were receiving athletic scholarships.

Procedures

Two trained interviewers conducted in-depth and semi-structured tape-recorded telephone interviews that lasted 50–90 minutes. The interview schedule focused on several aspects of the intercollegiate sport experience, one of which related to societal perceptions of women's sport and female athletes. The lesbian topic usually surfaced in responses to questions regarding (1) positive and negative connotations associated with women's sport and female athletes, and (2) stereotypes associated with women athletes. In response to these two questions, 17 of the 24 athletes initiated discussion of the lesbian topic. When asked whether they were aware of the lesbian label being associated with women athletes, 6 of the 7 remaining athletes indicated strong familiarity with this association.

Further, several questions explored how athletes dealt with the lesbian label, for example, (1) why they felt women athletes are labeled lesbian, (2) which athletes or teams are most likely to be labeled, (3) who applies this label, (4) how this label affects athletes, and (5) the degree to which athletes discuss this topic among themselves. Questions were open-ended to allow athletes to discuss the most relevant aspects of their own experiences.

RESULTS AND DISCUSSION

Since athletes were highly aware that women in sport are labeled lesbian, the manner in which these athletes dealt with the lesbian label was examined. Although the study was not developed within a preexisting framework, data analysis revealed that the processes underlying athletes' responses paralleled concepts established by Becker (1963) and Goffman (1963).

This article focuses on Becker's (1963) construction of the "falsely accused deviant" and Goffman's (1963) conceptualization of "stigma management." Not only do these two concepts enhance our knowledge of the experiences of women athletes, but the athletes' experiences add further understanding and validation to the conceptualizations of the "falsely accused deviant" and "stigma management."

Falsely Accused Deviant

The vast majority of respondents thought that labeling women athletes lesbian (or at least questioning sexual preference) was quite common. One athlete referred to it as a "societal blanket covering all of collegiate female athletes."

Despite athletes' perception that the lesbian label is prevalent in women's sport and often indiscriminately applied to women athletes, it would be unreasonable to assume all women athletes are lesbian. Although our intent was not to ascertain the sexual orientation of respondents, it was clear that a large majority responded to questions as to suggest that they were not lesbians. For example, phrases like "lesbians don't bother me as long as they don't hit on me," "I have a boyfriend," "the label does not pertain to me," "I'm not really the person to ask about lesbianism because I don't really see enough of it," "people know how I am" (suggesting heterosexuality), or "it's not a problem I'm faced with" were common.

Another way athletes implied distance from lesbianism was through out-group (e.g., "they" or "them") and in-group (e.g., "those of us who aren't" or "we joke about them") terminology. Even though women athletes as a whole frequently are viewed as the out-group in terms of social deviance, a secondary level of deviance exists within the athletic group. That is, within this deviant group, athletes use in-group and out-group terminology to distinguish between the nonlesbian and lesbian athlete.

	Obedient Behavior	Rule-Breaking Behavior
Perceived as Deviant	Falsely accused	Pure deviant
Not Perceived as Deviant	Conforming	Secret deviant

Becker's Typology of Deviants

Becker 1963, p. 20.

Given this widespread disassociation from lesbianism, it was assumed that the majority of athletes interviewed were not lesbians. However, since the label is indiscriminately applied to women athletes, many are incorrectly labeled lesbian, typifying what Becker (1963) terms the "falsely accused deviant." One difficulty noted by Klemke and Tiedeman (1990) in the study of false accusations is documenting that accusations are indeed false. Relative to our research, the need of respondents to mislead interviewers should have been substantially reduced as the interviews were anonymous and conducted by strangers over the telephone.

The concept of the "falsely accused deviant" derives from Becker's (1963) original classification scheme of deviant behavior. Basic to Becker's typology are two factors: whether an individual (1) is a rule violator, and (2) has been labeled deviant. In contrast to the "pure deviant," correctly labeled a rule violator; the "conformist," correctly labeled a rule abider; and the "secret deviant," a rule violator not labeled as such, the "falsely accused deviant" is incorrectly labeled deviant (see figure above).

Unfortunately, Becker (1963) does not systematically examine or discuss the "falsely accused deviant." In fact, this category of deviant is underresearched compared to the other three types. Neglect of the category exists despite its potential to enhance understanding of labeling, especially in terms of conditions and social processes involved in creating false conceptions of social reality (Klemke and Tiedeman 1990). Conceptualizations surrounding the "falsely accused deviant" are particularly useful in understanding why the lesbian label permeates women's sport.

Given the factors that lead to the indiscriminate stigmatization of women athletes, both lesbian and nonlesbian athletes adopt strategies to manage the lesbian stigma. As discussed in the following section, a variety of stigma management mechanisms are utilized by athletes to manage these labels.

Stigma Management

According to Goffman (1963), stigmas represent discrediting attributes that reflect a discrepancy between individuals' virtual (assumed) and actual (real) social identities. These attributes fall outside the range of what is considered ordinary or natural and generally "spoil" the social identity of the stigma possessor. Since the stigma taints and discredits the individual, attempts are generally undertaken to control and manage the discrediting attribute. The need for such management mechanisms is not necessarily limited to individuals who possess the discrediting attribute; stigma management is arguably the domain of the "falsely accused" as well.

As noted previously, sport participation (especially elite competition) is often considered outside the range of what is ordinary or normal for women. Moreover, given the widespread labeling of women athletes as lesbian and the corresponding stigmatization, women athletes must manage these discrediting attributes.

The particular strategy utilized to manage a stigma largely depends on the degree to which the attribute is visible or perceivable to others. Goffman (1963) distinguishes between individuals possessing attributes easily identifiable (i.e., discredited individuals) and not immediately perceivable (i.e., discreditable individuals). Techniques designed to manage the impact of these two types of attributes differ; the discredited individual manages tension while the discreditable manages information. Although the degree to which others can identify women as athletes varies, athletic status (and the accompanying lesbian label) is generally not immediately perceivable, thus resulting in a discreditable stigma. As Goffman (1963) suggests, such a stigma is associated with the control and management of information.

Concealment One basic stigma management mechanism noted by athletes was concealment. As an information management technique to prevent association with the lesbian label, women athletes sometimes simply conceal information about their athleticism. Elliott, Ziegler, Altman, and Scott (1990) and Goffman (1963) identify several forms of concealment, including self-segregation, passing, and use of disidentifiers.

Common in the responses of athletes was *self-segregation,* a condition where the stigmatized interact primarily with those sharing a similar affliction or those considered "wise." Wise individuals are familiar and comfortable with the stigmatized but do not have the stigma (Goffman 1963). Athletes frequently indicated that most of their friends are athletes and interaction with the general student body is limited. This compressed social network reduces athletes' interactions with those most likely to react negatively toward their athletic participation. Such avoidance limits disruption in the pursuit of athletic goals (Elliott et al. 1990). Although the reason for self-segregation is not exclusively stigma management (e.g., some indicated they relate better to other athletes or that time prohibits developing friendships outside of sport), the comfortable, nonthreatening environment other athletes provide may enhance the prevalence of self-segregation. As one athlete indicated, those outside the women's sport community most commonly initiate the lesbian labeling.

Self-segregation, however, is not without its complications. As Adler and Adler (1987) suggest, self-segregation often isolates athletes both socially and culturally from the university student body. The protective environment of self-segregation is offset by the sometimes narrow and limited athletic subculture.

Since self-segregation is not always practical, women athletes also conceal their athletic identity by *passing*. As information management, passing involves controlling the disclosure of discrediting information about oneself to conceal the stigma entirely (Elliott et al. 1990). Several women described techniques they (or other athletes) use so others will not know they are athletes. For example, some simply withhold information about their athletic status in conversations with outsiders. Alternatively, one individual stated athletes sometimes "over-talk" the lesbian topic to disassociate the label from themselves.

Other athletes relied on what Goffman (1963) terms *disidentifiers* to separate themselves from things most stereotypically associated with lesbianism. For example, to contradict the masculine image identified with lesbianism, athletes sometimes consciously accentuated their femininity by wearing dresses, skirts, makeup, and earrings or by letting their hair grow. Others indicated that being seen with men or having a boyfriend is a common technique to reaffirm their heterosexuality. In a few instances this need to identify with men fosters more extreme behaviors including "hanging on men" in public or establishing a reputation as promiscuous with men. Some women athletes take care not to constantly be seen in public with groups of women; others were cautious about the nature of physical contact publicly displayed with women. In all cases, the athlete's presentation of self is consciously monitored or altered to conform to appropriate gender roles.

Although effective in controlling information, passing behaviors and disidentifiers can negatively affect individuals' psychological well-being, as well as alienate them from their personal identities (Goffman 1963). Goffman's work suggests that athletes who engage in passing may experience feelings of disloyalty and self-contempt as they disassociate themselves not only from their athletic identities but from qualities representing their personal essence.

In another form of disidentification, some athletes distanced themselves from athletes more seriously "contaminated" or stigmatized. This mechanism is accomplished through techniques such as making "rude comments" about a suspected lesbian teammate, establishing team cliques on the basis of sexual orientation, and being "very critical and mean" to lesbian teammates behind their backs.

Since the lesbian label is not equally applied to all athletes or teams, association with "certain" athletes increases the chance of stigmatization. In response to the question why women athletes are labeled lesbian, athletes identified three categories—appearance, personality characteristics, and nature of the sport activity.

Relative to appearance or externally identifiable characteristics, factors such as dress, hair style, body build, body posture/carriage, muscularity, and mannerisms were mentioned as underlying usage of the labels "dyke," "butch," or "lesbian." Athletes with personality characteristics such as being assertive, outgoing, strong, independent, aggressive, and hard-nosed are most likely to be labeled lesbian.

Regarding the nature of the sport activity, participants in team sports such as softball, basketball, and field hockey are more often recipients of the lesbian label. When asked why team sports are a more likely target, athletes indicated that such activities require more athleticism and strength, involve more physical contact, and are more commonly viewed as sports played by men. These qualities associated with team sports reflect traits socially defined as masculine (Bem 1974). Stigmatization of these team sports may discourage women from both engaging in "masculine sports" and participating in activities with potential to develop cooperation, teamwork, and solidarity among women (Lenskyj 1986). Self-distancing from those more seriously "contaminated," although effective for controlling information, may at the same time prevent women athletes from bonding together as a collective.

In a few instances athletes disassociate themselves from the lesbian label by publicly making fun of lesbian athletes or criticizing homosexuality. This disparaging behavior frequently occurs in the course of conversations with "outsiders" when the topic of women athletes is discussed. As Goffman (1963) suggests, a discreditable individual often finds it dangerous to refrain from joining in the vilification of one's own group. Such a strategy may protect the social identity of the individual athlete but impede the development and maintenance of the collective identity of women athletes.

When concealment is not possible, a stigmatized individual may adopt other stigma management strategies. Included are salience reduction through deflection and direct confrontation through normalization (Elliott et al. 1990).

Deflection Individuals may use deflection to reduce the importance or salience of a discrediting attribute. Representing the original source of stigmatization, the athlete role is publicly downplayed by athletes. Not only do women athletes use disidentifiers to distance themselves from this role, but they attempt to accentuate the significance of nonsport roles and attributes. It is thus important to many to be viewed as more than an athlete; efforts to demonstrate mastery in other areas (e.g., student role, social role) are common. For example, several athletes mentioned they want to do well in the classroom so that others will not identify them exclusively as athletes. Moreover, some athletes highlighted their social role by mentioning dating and party activities.

Normalization In some situations it is difficult to reduce the visibility of the discrediting attribute (i.e., athletic status) when it is salient and relevant. This difficulty is particularly true of athletes who are visible on campus or the focus of media and where athletic status frequently impacts on the dynamics of social interaction. Rather than conceal or deflect, the individual has no choice but to directly confront the stigma. The ideal outcome for the athlete is a state of normalization where the discrediting attribute loses its stigmatizing capability. To accomplish normalization, strategies attempt to redefine the stigma and re-educate "normals" (Elliott et al. 1990). Although not widespread, redefinition efforts generally emphasize the positive characteristics or contributions of female athletes while re-education consists of athletes informing outsiders that lesbianism is not prevalent in women's sport or that sexual preference should be a non-issue.

Even though most respondents thought the lesbian label is unfairly applied to women athletes, attempts at normalization are difficult given their assumed violation of multiple gender norms. Although some athletes indicated the label did not bother them and that people could "think what they wanted," little evidence existed to suggest that the discrediting attribute of lesbian has lost its stigmatizing capability. Efforts to redefine or reeducate are generally futile since the vast majority of women athletes do not directly confront these labels, preferring to engage in other stigma coping strategies.

This refusal to confront the lesbian label resembles what Elliott and colleagues (1990) term capitulation. Even if false or inaccurate, the lesbian label assumes a master status for the individual (Becker 1963; Schur 1979) and becomes central to her identity and interactions. The stigmatized accept both the stereotypes placed upon them and the accompanying stigmatization. Athletes were generally not proactive in fighting the stereotypes; resistance to these stereotypes and labels usually emerges after a direct confrontation or challenge from an outsider. For example, one respondent stated "I am not an activist," while another indicated she "would not bring up the topic unless something negative was said."

Acceptance of the normative definition of deviance is often true of socially marginalized groups as they (1) rely on others for self-definition, (2) engage in self-hate, and (3) identify with the aggressor (Kitzinger 1987; Nobles 1973; Sarnoff 1951; Thomas 1971). Interestingly, several remarks from respondents demonstrate that women athletes accept and internalize societal stereotypes about themselves and incorporate these images into their identity accounts and personal interactions. For example, internalization of societal beliefs about appearance norms was evident in such comments as "I don't look like an athlete so I rarely am labeled a lesbian" and "I've never seen somebody that looks like a girl called a dyke or lesbian." One athlete dismissed lesbianism as a problem on her team since her "teammates were pretty." This same athlete implied, however, that lesbianism might be more prevalent on other teams since "most of the athletes on these teams looked like guys."

Self-hate was occasionally noted in responses as well. Although generally indicating the lesbian label angers them, athletes also mentioned that it makes them feel "unattractive" and "less desirable to men" and leads some to "always worry about how I look." Identification with aggressors was also evident in that some nonlesbian athletes engage in negative conversations that criticize or mock lesbian athletes.

Nevertheless, the presence of the deviant label does not necessarily garner negative outcomes (Elliott et al. 1990; Goffman 1963). As a few athletes' responses indicated, being labeled lesbian makes them stronger individuals and less dependent on what outsiders think of them. For example, learning to cope with the lesbian label helped one athlete "find confidence" in herself and encouraged another to learn more about homosexuality through reading and coursework. Moreover, some athletes claimed exposure to the issue of homosexuality makes them less judgmental and more accepting/respectful of dissimilar others.

Women athletes thus utilize a variety of stigma management techniques to control and manage information about their athleticism. This effort is somewhat paradoxical since most respondents were generally proud to be athletes and viewed their athletic experience as positive. The prevailing negative societal linkage of female athleticism with masculinity and lesbianism often overrides positive self-definitions of their athleticism.

CONCLUSION

This research examines ways the devaluation and stigmatization of women athletes impact on their social interactions and how they manage the lesbian label attached to their sport participation. Drawing upon Becker's (1963) construction of the "falsely accused deviant" and Goffman's (1963) conceptualization of "stigma management," the present analysis explicates both the social conditions and processes that facilitate false labeling, and stigma management strategies of women athletes. Although all respondents are deviant in that society views female athleticism in opposition to traditional gender norms (Schur 1984), indiscriminate application of the lesbian label to women athletes denotes many as falsely accused deviants.

In general, Klemke and Tiedeman's (1990) typology of error-types underlying false accusations (i.e., pure, intentional, legitimatized, and victim-based) and discussion of societal conditions fostering false labeling (i.e., perceived threat, lack of power and outsider status, and stereotype adoption) assist in understanding why false accusations are often directed at women athletes. Such factors indicate that labeling is not a random process and that the study of false accusation should focus on not only the falsely accused but the false accusers as well (Klemke and Tiedeman 1990). As Klemke and Tiedeman (1990) argue, the "falsely accused deviant" is a conceptually rich category for analysis since it represents the quintessential example of labeling and offers insight into the processes that generate false conceptions of social reality.

Responding to labels directed at women in sport, the athletes in this sample utilize a variety of stigma management techniques. These methods consist of concealment, including self-segregation, passing, and use of disidentifiers; deflection; and normalization. These strategies suggest that the falsely accused deviant reacts in a manner similar to the "pure deviant" (Becker 1963; Goffman 1963; Lemert 1972). Thus, stigma management is the domain of both the lesbian and nonlesbian athlete. Coping with a "spoiled identity," women athletes adopt various techniques to control and manage information about their athletic identities.

The role of labeling is undeniably central to understanding the experiences of women athletes forced to resort to stigma management strategies in the absence of deviant behavior. Female athletes acquire the deviant label because more powerful groups impose their definition of morality on the athletic act (Erikson 1962; Kitsuse 1962). Critical in this labeling are the stigma-laden meanings female athleticism evokes and the processes by which these ideas are perceived and applied. Such meanings are relayed to women, reminding them of "their place" in the gender system (Henley and Freeman 1979).

Overall lack of power confounds the ability of women athletes to actively challenge deviant labels. Not only do women in general lack power, but the discrediting attributes of athlete and lesbian diminish the power position of women athletes even further. Generally lacking organizational backing and viewed as outsiders, women athletes, rather than actively challenge the stigma, rely on stigma management. This passivity not only has negative social and psychological outcomes for the athlete, but enhances the overall power of the label itself. Since athletes resort to actions where silence and denial are central and internalization of deviant labels is frequent, the likelihood of resistance to these labels is significantly reduced.

ACKNOWLEDGMENTS

The authors thank two anonymous reviewers for their insightful and thorough comments.

REFERENCES

Adler, Peter, and Patricia A. Adler. 1987. "Role Conflict and Identity Salience: College Athletics and the Academic Role." *Social Science Journal* 24: 443–455.

Anderson, Margaret L. 1988. *Thinking About Women: Sociological Perspectives on Sex and Gender.* 2nd ed. New York: Macmillan.

Becker, Howard S. 1963. *Outsiders: Studies in the Sociology of Deviance.* New York: Free Press.

Bem, Sandra L. 1974. "The Measurement of Psychological Androgyny." *Journal of Consulting and Clinical Psychology* 42: 155–162.

Birenbaum, Arnold. 1970. "On Managing a Courtesy Stigma." *Journal of Health and Social Behavior* 11: 196–206.

Blinde, Elaine M. 1987. "Contrasting Models of Sport and the Intercollegiate Sport Experience of Female Athletes." Ph.D. dissertation, Department of Physical Education, University of Illinois, Urbana-Champaign.

Colker, Ruth, and Cathy S. Widom. 1980. "Correlates of Female Athletic Participation: Masculinity, Femininity, Self-Esteem, and Attitudes Toward Women." *Sex Roles* 6: 47–58.

Elliott, Gregory, C., Herbert L. Ziegler, Barbara M. Altman, and Deborah R. Scott. 1990. "Understanding Stigma: Dimensions of Deviance and Coping." In *Deviant Behavior,* edited by Clifton D. Bryant (pp. 423–443). New York: Hemisphere.

Erikson, Kai T. 1962. "Notes on the Sociology of Deviance." *Social Problems* 9: 307–314.

Goffman, Erving. 1963. *Stigma: Notes on the Management of Spoiled Identity.* Englewood Cliffs, NJ: Prentice-Hall.

Goodman, Bernice. 1977. *The Lesbian: A Celebration of Difference.* New York: Out and Out Books.

Griffin, Patricia S. 1987. "Homophobia, Lesbians, and Women's Sports: An Exploratory Analysis." Paper presented at the annual meetings of the American Psychological Association, New York.

Hall, M. Ann. 1988. "The Discourse of Gender and Sport: From Femininity to Feminism." *Sociology of Sport Journal* 5: 330–340.

Henley, Nancy, and Jo Freeman. 1979. "The Sexual Politics of Interpersonal Behavior." In *Women: A Feminist Perspective,* edited by Jo Freeman (pp. 391–401). Palo Alto: Mayfield.

Keller, Evelyn F. 1978. "Gender and Science." *Psychoanalysis and Contemporary Thought* 1: 409–433.

Kitsuse, John I. 1962. "Societal Reaction to Deviant Behavior: Problems of Theory and Method." *Social Problems* 9: 247–256.

Kitzinger, Celia. 1987. *The Social Construction of Lesbianism.* London: Sage.

Klemke, Lloyd W., and Gary H. Tiedeman. 1990. "Toward an Understanding of False Accusation: The Pure Case of Deviant Labeling." In *Deviant Behavior,* edited by Clifton D. Bryant (pp. 266–286). New York: Hemisphere.

Lemert, Edwin M. 1972. *Human Deviance, Social Problems, and Social Control.* 2nd ed. Englewood Cliffs, NJ: Prentice-Hall.

Lenskyj, Helen. 1986. *Out of Bounds: Women, Sport and Sexuality.* Toronto: Women's Press.

———. 1991. "Combating Homophobia in Sport and Physical Education." *Sociology of Sport Journal* 8: 61–69.

Nobles, Wade W. 1973. "Psychological Research and the Black Self-Concept: A Critical Review." *Journal of Social Issues* 29: 11–31.

Pollak, Otto. 1952. "The Errors of Justice." *The Annals of the American Academy of Political and Social Science* 284: 115–123.

Sarnoff, Irving. 1951. "Identification with the Aggressor: Some Personality Correlates of Anti-Semitism Among Jews." *Journal of Personality* 20: 199–218.

Scheff, Thomas J. 1964. "The Societal Reaction to Deviance: Ascriptive Elements in the Psychiatric Screening of Mental Patients in a Midwestern State." *Social Problems* 11: 401–413.

Schur, Edwin M. 1979. *Interpreting Deviance: A Sociological Introduction.* New York: Harper and Row.

———. 1984. *Labeling Women Deviant: Gender, Stigma, and Social Control.* New York: Random House.

Smith, Dorothy E. 1978. "A Peculiar Eclipsing: Women's Exclusion from Man's Culture." *Women's Studies International Quarterly* 1: 281–295.

Theberge, Nancy. 1985. "Toward a Feminist Alternative to Sport as a Male Preserve." *Quest* 37: 193–202.

Thomas, Charles W. 1971. *Boys No More.* Beverly Hills: Glencoe.

Willis, Paul. 1982. "Women in Sport in Ideology." In *Sport, Culture and Ideology,* edited by Jennifer Hargreaves (pp. 117–135). London: Routledge and Kegan Paul.

Wolfe, Susan J. 1988. "The Rhetoric of Heterosexism." In *Gender and Discourse: The Power of Talk,* vol. 30, edited by Alexandra Dundas Todd and Sue Fisher (pp. 199–224). Norwood, NJ: Ablex.

26

Return to Sender

Reintegrative Stigma-Management Strategies of Ex-Psychiatric Patients

NANCY J. HERMAN

Although scholars have addressed the exit phase of deviant careers (cf. Adler and Adler 1983; Faupel 1991; Frazier 1976; Glassner et al. 1983; Harris 1973; Inciardi 1970; Irwin 1970; Luckenbill and Best 1981; Meisenhelder 1977; and Ray 1961), the issue of reintegrating deviants into society has received little sociological attention. So too has little attention been given to the wide array of factors affecting role exit and reintegration. . . .

Despite the preponderance of sociological research on the mentally ill, there is a dearth of ethnographically-based studies dealing with the post-hospital lives of ex-psychiatric patients. Such studies, as they do exist, deal largely with chronic ex-patients living in halfway houses or boarding homes, or involved in specific aftercare treatment programs (Cheadle et al. 1978; Estroff 1981; Lamb and Goertzel 1977; Reynolds and Farberow 1977). Little systematic attention has been given to ex-patients' perceptions of mental illness as a stigmatizable/stigmatizing attribute,[1] the numerous problems they face on the outside, the ways such persons manage discreditable information about themselves in the context of social interaction with others, and the consequences of employing these strategies for altering their deviant identities and social reintegration. In this paper I address these deficits in the sociological literature by presenting ethnographic evidence from a study of 146 non-chronic,[2] ex-psychiatric patients. First, I begin with a discussion of the setting and methods used in this study. Second, I illustrate how ex-patients came to perceive their attribute as potentially stigmatizing. Third, I analyze the five strategies these persons developed and employed in their "management work." Finally, I address the implications of adopting such stratagems for identity transformation and social reintegration. In this paper I hope to contribute to the existing literature on stigma, deviant career exit, management work, and the reintegration of deviants.

From "Return to Sender: Reintegrative Stigma-Management Strategies of Ex-Psychiatric Patients," Nancy J. Herman, *Journal of Contemporary Ethnography*, Vol. 22, No. 3, October 1993. Reprinted by permission of Sage Publications, Inc.

SETTING AND METHODS

My interest in mental illness and psychiatric patients has been a long-standing one. My father was employed as an occupational therapist for twenty-five years at a large psychiatric institution in a metropolitan city in Ontario, Canada. Throughout my childhood and adolescent years, I made frequent trips to the institution, interacting with many of the patients on the "admission" and "back" wards, in the hospital canteen, in the occupational therapy workshop, and on the grounds. Moreover, during those years, my father often brought a number of his patients into our home (those, in particular, whose families had abandoned them or who did not have any relatives) to spend Easter, Thanksgiving, and Christmas holidays with our family.

My childhood interest in mental patients sparked my initial involvement in this topic, and I began to study mental patients in 1980 as a graduate student at McMaster University. Having a father who was highly esteemed by the institutional gatekeepers greatly facilitated my access. . . . I spent eight months studying the institutionalization of psychiatric patients ethnographically (Herman 1981). After completing this study on the pre- and in-patient phases of the patients' careers, I became interested in learning about the post-patient phase of their careers. As a result of the movement toward deinstitutionalization, chronic patients, once institutionalized for periods of years, were being released into the community. Moreover, newly diagnosed patients were being hospitalized for brief periods and shortly released. I became interested in examining the post-hospital social worlds and experiences of these ex-psychiatric clients.

I began a four-year research project on this topic in January 1981 (see Herman 1986). In contrast to my earlier study on institutionalized mental patients, where I encountered numerous problems with institutional gatekeepers, this time I encountered relatively few such problems. Rather than dealing with the same provincial institution to obtain a list of discharged patients, I contacted the Director of Psychiatric Services and Professor of Psychiatry at the Medical School affiliated with my university. After hearing about my research proposal, he granted his support for the project and served as my sponsor to the Ethics Committee of the hospital. Approximately one month later, I was given access to a listing of discharged (chronic[3] and non-chronic) ex-psychiatric patients who had been released from the psychiatric wards of seven general hospitals and from two psychiatric institutions in the Southern Ontario area between 1975 and including 1981. In order to protect the identities of those ex-patients not wishing to participate in my study, and to avoid litigation against the hospital for violating rights of confidentiality, I agreed to only be given access to the names of those individuals who agreed to participate. A stratified random sample of 300 ex-patients was formed from the overall discharge list. Upon drawing this sample, the hospitals sent out a letter to each potential subject on my behalf, outlining the general nature of the study, my identity, and my affiliation. Two weeks later, hospital officials made follow-up telephone calls asking potential subjects if they were willing to participate in the study. I was subsequently given a list containing the names of 285 willing participants, 146 non-chronic and 139 chronic ex-patients.[4]

I initially conducted informal interviews with each of the ex-patients in coffee shops, in their homes, in malls, or at their places of work. The interviews lasted from three to five and one-half hours. These interviews not only provided me with a wealth of information about the social worlds of ex-patients, but many subjects invited me to subsequently attend and participate with them in various social settings, including self-help group meetings, activist group meetings and protest marches, and therapy sessions. In addition, I frequently met ex-patients where they worked during coffee and lunch breaks and was able to observe them interacting with co-workers. I ate lunches and dinners in their homes (as they did in mine) and watched them interacting with family members, friends, and neighbors. Each Wednesday afternoon, I met a group of six ex-patients at a local doughnut shop where they would discuss the problems they were facing "on the outside" and collectively search for possible remedies.

PERCEPTIONS OF MENTAL ILLNESS AS STIGMA

In his classic work *Stigma*, Goffman (1963: 4) distinguishes between stigmata that are "discrediting" and those that are "discreditable"; the former refer to attributes that are immediately apparent to others, such as obesity, physical abnormalities, and blindness; the latter refer to attributes that are not visible or readily apparent to others, such as being a secret homosexual or ex-prisoner. Mental illness is conceptualized, for the most part, as a discreditable or potentially stigmatizing attribute in that it is not readily apparent to others. It is equally important to note, however, that for some ex-psychiatric patients, mental illness is a deeply discrediting attribute. Many ex-patients, especially chronic ones, were rendered discredited by inappropriate patterns of social interaction and the side effects of medication, which took the forms of twitching, swaying, jerking, and other bizarre mannerisms (see Herman 1986).[5] When ex-patients made voluntary disclosures of personal information about their psychiatric histories or when other people somehow became aware of their histories, ex-patients were categorized in negative terms, as representing some sort of personal failure to "measure up" to the rest of society. Tom, age 45, summed it up for most of the ex-patients:

> Having been diagnosed as a psychiatric patient with psychotic tendencies is the worst thing that has ever happened to me. It's shitty to be mentally ill; it's not something to be proud of. It makes you realize just how different you are from everybody else—they're normal and you're not. Things are easy for them; things are hard for you. Life's a ball for them; life's a bitch for you! I'm like a mental cripple! I'm a failure at life! . . .

MENTAL ILLNESS AND STRATEGIES OF
INFORMATION MANAGEMENT

Some studies on the discreditable (Edgerton 1967; Humphreys 1972; Ponse 1976) have suggested that individuals either disclose their attribute to others *or* make attempts to actively conceal such information about their selves. Other studies (Bell and Weinberg 1978; Miall 1986; Schneider and Conrad 1980; Veevers 1980) have suggested that being a "secret deviant" is far more complex than either choosing to disclose or not disclose one's "failing." These studies suggest that individuals *selectively* conceal such information about themselves at certain times, in certain situations, and with certain individuals and freely disclose the same information at other times, in other situations, and with other individuals. Concealment and disclosure, then, are contingent upon a "complex interaction of one's learned perceptions of the stigma [of their attribute], actual 'test' experiences with others before/or after disclosure, and the nature of the particular relationship involved" (Schneider and Conrad 1980: 39).

The complex reality of how individuals selectively conceal and disclose information was evident in the case of non-chronic ex-psychiatric patients. Examination of their post-hospital worlds revealed not only that many ex-patients faced economic hardships, had problems coping in the community, and experienced adverse side effects from their "meds"[6] but also that their perception of mental illness as a potentially stigmatizing attribute presented severe problems in their lives. Many lived their lives in states of emotional turmoil, afraid and frustrated—deciding who to tell or not tell, when to tell and when not to tell, and how to tell. Joan, a 56-year-old waitress, aptly summed it up for most non-chronics:

> It's a very difficult thing. It's not easy to distinguish the good ones from the bad ones. . . . You've gotta figure out who you can tell about your illness and who you better not tell. It is a tremendous stress and strain that you have to live with 24 hours a day!

Ex-psychiatric patients learned how, with whom, and under which circumstances to disclose or conceal their discreditable aspects of self, largely through a process of trial and error, committing numerous *faux pas* along the way. Frank, a 60-year-old factory worker, spoke of the number of mistakes he made in his "management work":

> I was released over two years now. And since then, I've developed an ulcer trying to figure out how to deal with my "sickness"—that is, how or whether others could handle it or not. I screwed up things a few times when I told a couple of guys on the bowling team. I made a mistake and thought that they were my buddies and would accept it.

In fact, even if no *faux pas* were committed, there was no guarantee that others would accept preferred meanings and definitions of self. As Charlie, a 29-year-old graduate student, hospitalized on three occasions, remarked:

I'm not a stupid person. I learned how to handle effectively the negative aspects of my sickness—I mean how others view it. I've been doing OK now since my discharge, but still, each time I'm entering a new situation, I get anxious; I'm not always a hundred percent sure of whether to tell or not to—especially in the case of dating relationships. Even if you've had success in telling certain types of people, there's always the chance—and it happens more than you think, that people will just not "buy" what you're trying so desperately to "sell" them.

Nearly 80 percent of the non-chronics in this study engaged in some form of information control about their illnesses and past hospitalizations. Specifically, the stratagems adopted and employed by the ex-patients, resembling those observed in other deviant groups (cf. Davis 1961; Hewitt and Stokes 1975; Levitin 1975; Miall 1986; Schneider and Conrad 1980), included (1) selective concealment, (2) therapeutic disclosure, (3) preventive disclosure, and (4) political activism—stratagems adopted by ex-patients in their effort to lessen or avoid the stigma potential of mental illness, elevate self-esteem, renegotiate societal conceptions of mental illness as a discreditable attribute, and alter deviant identities.

Selective Concealment

Selective concealment can be defined as the selective withholding or disclosure of information about self perceived as discreditable in cases where secrecy is the major stratagem for handling information about an attribute. Especially during the period directly following their psychiatric treatment, the majority of non-chronics had a marked desire to conceal such information about their selves from all others. Decisions about disclosure and concealment were made on the basis of their perceptions of others—that is, whether they were "safe others" or "risky others." So too were decisions based on prior, negative experiences with "certain types" of others. Speaking of her classification of others into "trustworthy" and "untrustworthy" others, Dawina, a 46-year-old secretary, institutionalized on seven occasions, said:

It's like this. There are two types of people out there, "trustworthy" ones—the people who will be understanding and supportive—and "untrustworthy" ones. Out of all of my friends and relations and even the people I work with at the company, I only decided to tell my friend Sue.

Moreover, there was a hierarchical pattern of selective disclosure based upon the individual's perceived degree of closeness and the ex-patient's revealing his or her discreditable attribute. In general, such information was most frequently revealed to family members, followed by close friends, and then acquaintances, a pattern also reflected in the literature on epileptics (Schneider and Conrad 1980) and involuntary childless women (Miall 1986). As Sarah, a 36-year-old mother of two, put it:

When I was discharged, I didn't automatically hide from everyone the fact that I was hospitalized for a nervous breakdown again. But I didn't go and tell everyone either. I phoned and told my relatives in "Logenport," and I confided in two of my close, good friends here in town.

Further, selective disclosures to normal others were frequently made to test reactions. Similar to Schneider and Conrad's (1980) epileptics, the ex-patients continued to disclose mental illness contingent upon responses they had received to previous disclosures. Rudy, a 39-year-old man hospitalized on ten occasions, stated:

> You learn through trial and error. When I was let out back in 1976, I was still naive, you know. I decided to tell a few people. Boy, was that a mistake. They acted as if I had AIDS. Nobody wanted anything further to do with me. . . . Since then, I've pretty much dummied up and not told anyone!

In those cases where concealment was the dominant strategy of information management, ex-patients usually disclosed only to one or two individuals. As Simon, a 25-year-old ex-patient, aptly expressed:

> I decided from the moment that my treatment ended, I would tell as few people as possible about my stay in the psychiatric hospital. I figured that it would be for the best to "keep it under a lid" for the most part. So, to this day, I've only confided in my friend Paul and a neighbor who had a similar illness a while back.

The employment of concealment as a stratagem of information management took the following forms: avoidance of selected "normals," redirection of conversations, withdrawal, the use of disidentifiers, and the avoidance of stigma symbols. Speaking on his efforts to redirect conversations, Mark, a 34-year-old non-chronic, explained:

> Look, you've got to remain on your toes at all times. More often than not, somebody brings up the topic about my past and starts probing around. Sometimes these people won't let up. . . . I use the tactic where I change the subject, answer their question with a question. . . . I try to manipulate the conversation so it works out in my favor.

For still others, concealment of their discreditable attribute was achieved through withdrawal. Over two-thirds of the ex-patients in this study engaged in withdrawal as a form of concealment, especially during the early months following discharge. Speaking of his use of this technique, Harry, a college junior, remarked:

> Sometimes when I'm at a party or some type of gathering with a number of people, I just remain pretty reticent. I don't participate too much in the conversations. . . . I'm really unsure how much to tell other people. For the most part, I just keep pretty quiet and remain a wallflower. People may think I'm shy or stuck-up, but I'd rather deal with that than with the consequences of others finding out that I'm a mental patient.

A third technique, employed by over one-third of the ex-patients to conceal their discreditable aspects from others, involved the use of disidentifiers

(Goffman 1963: 44). That is, ex-patients utilized misleading physical or verbal symbols in an effort to prevent normal others from discovering their "failing." Similar to homosexuals (Carrier 1976; Delph 1978), unwed parents (Christensen 1953; Pfuhl 1978), and lesbians (Ponse 1978), who frequently made use of disidentifiers in their management work, non-chronic ex-patients also employed such techniques. Specifically, disidentifiers took the form of making jokes about psychiatric patients while in the presence of "normal" others and participating in protests *against* the integration of ex-patients into the community. Mike, a 26-year-old ex-patient, recently released after three hospitalizations, remarked (with some remorse) on his use of this tactic:

> They wanted to use this house down the street for a group home for discharged patients. All the neighbors on the street were up in arms over it. It didn't upset me personally, but the neighbors made up this petition, and to protect myself, I not only signed it, but I also went door-to-door convincing other neighbors to sign it and "keep those mentals out". . . . I felt sort of bad afterwards, but what else could I do?

In a similar vein, Morgan, a 49-year-old history professor, explained how his joke-telling aided in the concealment of his attribute:

> I conceal this information about myself—my psychiatric past—by frequently telling jokes about mental patients to my colleagues in the elevator, and sometimes even my lectures, and in everyday conversation. It's really a great ploy to use. People may think the jokes are in bad taste, but at the very least, it helps me to keep secret my illness.

A final form of concealing information on the part of ex-patients was through the avoidance of stigma symbols (cf. Goffman 1963: 43)—signs that would bring into the forefront or disclose their discreditable attribute. It is interesting to note that the data presented here on non-chronics and their avoidance of stigma symbols support observations made of other deviant groups, for example, transsexuals (Bogdan 1974; Kando 1973) and unwed fathers (Pfuhl 1978). . . . Among the 146 ex-patients studied, over two-thirds avoided contact with such stigma symbols as other ex-mental patients with whom they had become friends while institutionalized and self-help groups for ex-patients. So, too, did they avoid frequenting drop-in centers, attending dances and bingo games for ex-patients, and, in general, placing themselves where other "patients and ex-patients hung out." For still others, avoidance of stigma symbols entailed not attending post-hospital therapy sessions. Margarette, a stocky, middle-aged woman of German descent, explained her avoidance of post-discharge therapy sessions in the following manner:

> After I was released, my psychiatrist asked that I make appointments and see him every two weeks for follow-up maintenance treatments. But I never did go because I didn't want someone to see me going into the psychiatric department of "Meadowbrook Hospital" and sitting in the wait-

ing room of the "Nut Wing." Two of my nosy neighbors are employed at that hospital, and I just couldn't take the chance of them seeing me there one day.

In sum, as a strategy of information management, selective concealment of their attribute and past hospitalizations was done to protect themselves from the perceived negative consequences that might result from the revelation of their illness—an "offensive tactical maneuver" through which ex-patients attempted (although often unsuccessfully) to mitigate the stigma potential of mental illness on their daily lives. Notably, employing concealment as a strategy of information management was a *temporal* process. The majority of ex-patients employed this strategy primarily during the first eight months following their discharge. During this time, in particular, they expressed feelings of anxiety, fear, and trepidation. As time passed, however, ex-patients began to test reactions. They encountered both positive and negative responses from certain "normals," and their strong initial desires for secrecy were replaced by alternative strategies.

Therapeutic Disclosures

Therapeutic disclosure can be defined as the selective disclosure of a discreditable attribute to certain "trusted," "empathetic" supportive others in an effort to renegotiate personal perceptions of the stigma of "failing." . . . Thirty-six percent of the ex-patients felt that discussing their mental illnesses and past hospitalizations, getting it off their chests in a cathartic fashion, functioned to alleviate much of the burden of their loads. Attesting to the cathartic function disclosing served, Vincent, a 29-year-old ex-patient, remarked:

> Finally, letting it all out, after so many secrets, lies, it was so therapeutic for me. Keeping something like this all bottled up inside is self-destructive. When I came clean, this great burden was lifted from me!

Therapeutic disclosure was most often carried out with family members, close friends, and with other ex-psychiatric patients—individuals "sharing the same fate." Ida, age 52, discussing the circumstances surrounding her disclosure to a neighbor who had also been hospitalized in a psychiatric facility at one time, said:

> At first, I was apprehensive to talk about it. But keeping it inside of you all bottled up is no good either. One day, I walked down the street to a neighbor of mine and she invited me in to have tea. I knew what had happened to her years ago (her deceased husband confided in my husband). I let out all my anxieties and fears to her that afternoon. . . .
> I told her everything and she was so sympathetic. . . . She knew exactly what I was going through. Once I let it all out, I felt so much better.

Even in cases where ex-patients disclosed to individuals who turned out to be unsympathetic and unsupportive, some considered this therapeutic:

> When I came out of hiding and told people about my sickness, not every-
> one embraced me. A lot of people are shocked and just tense up. Some
> just stare. . . . A few never call you after that time or make up excuses not
> to meet with you. . . . But I don't care, because overall, telling made me
> feel better.

Just as therapeutic disclosure functioned to relieve ex-patients' anxieties
and frustrations, it also allowed for the renegotiation of personal perceptions
of mental illness as a discreditable attribute. Speaking of the manner by which
she came to redefine mental illness in her own mind as a less stigmatizing attribute, Edith explained:

> When I finally opened up and started talking about it, it really wasn't so
> bad after all. My Uncle John was very supportive and helped me to put
> my mind at rest, to realize that having mental illness isn't so bad; it's not
> like having cancer. He told me that thousands of people go into the hos-
> pital each year for psychiatric treatment and probably every third person I
> meet has had treatment. . . . After much talking, I no longer think of my-
> self as less human, but more normally. . . . Having mental illness isn't the
> blight I thought it was.

In short, then, ex-patients employed therapeutic disclosure in order to relieve
feelings of frustration and anxiety, to elevate their self-esteem, and to renegotiate (in their own minds) personal perceptions of mental illness as stigmatizing.

Preventive Disclosure

Preventive disclosure can be described as the selective disclosure to "normals"
of a discreditable attribute in an effort to influence others' actions or perceptions about the ex-patient or about mental illness in general (cf. Miall 1986;
Schneider and Conrad 1980). Preventive disclosure of their mental illness and
past hospitalizations occurred in situations where ex-patients anticipated future rejection by "normal" others. In order to minimize the pain of subsequent rejection, 34 percent of the sample decided that the best strategy to
employ with certain people was preventive disclosure *early* in their relationships. As Hector, a 40-year-old janitor, said:

> I figured out that, for me, it is best to inform people right off the bat
> about my mental illness. Why? Because you don't waste a lot of time de-
> veloping relationships and then are rejected later. That hurts too much.
> Tell them early and if they can't deal with it, and run away, you don't get
> hurt as much!

Preventive disclosure, then, represented a way ex-patients attempted to prevent a drop in their status at a later date, or a way of testing acquaintances in
an effort to establish friendship boundaries.

Just as non-chronics used preventive disclosure to avoid future stigma and
rejection, so too did they employ this strategy to influence normals' attitudes
about themselves and about mental illness in general. Specifically, ex-patients

used the following devices: (1) medical disclaimers (cf. Hewitt and Stokes 1975; Miall 1986; Schneider and Conrad 1980); (2) deception/coaching (cf. Goffman 1963; Miall 1986; Schneider and Conrad 1980); (3) education (cf. Schneider and Conrad 1980); and (4) normalization (cf. Cogswell 1967; Davis 1961, 1963; Levitin 1975; McCaghy 1968; Scott 1969).

Medical Disclaimers

Fifty-two percent of the ex-patients frequently used medical disclaimers in their management work—"blameless, beyond-my-control medical interpretation(s)" developed in order to "reduce the risk that more morally disreputable interpretations might be applied by naive others" discovering their failing (Schneider and Conrad 1980: 41). Such interpretations were often used by ex-patients to evoke sympathy from others and to ensure that they would be treated in a charitable manner. As Dick, an unemployed laborer, put it:

> When I tell people about my hospitalization in a psychiatric hospital, I immediately emphasize that the problem isn't anything I did, it's a biological one. I didn't ask to get sick; it was just plain biology; or my genes that fucked me up. I try to tell people in a nice way so that they see mental illness just like other diseases—you know, cancer or the mumps. It's not my parents' fault or my own. . . . I just tell them, "Don't blame me, blame my genes!"

In a similar vein, Anna, a 29-year-old waitress, explained her use of medical disclaimers:

> Talking about it is quite tricky. When I tell them about it, I'm careful to emphasize that the three times I was admitted, was due to a biochemical imbalance—something that millions of people get. I couldn't do anything to help myself—I ate properly, didn't drink or screw around. It's not something I deserved. When you give people the facts and do it in a clinical fashion, you can sway many of them to sympathize with you.

Further, eleven ex-patients revealed their mental illness and past hospitalizations as a side effect of another medical problem or disease, such as childbirth, stroke, or heart disease, thereby legitimizing what otherwise might be considered a potentially stigmatizing condition. As Rebecca, age 36, confessed:

> I have had heart problems since birth. I was a very sick baby. I've had four operations since that time and I've been on all kinds of medications. The stress of dealing with such an illness led to my depression and subsequent breakdowns. . . . When my friends hear about mental illness in this light, they are very empathetic.

While Sue spoke of her successes in influencing others' perceptions about her attribute and mental illness in general, Lenny lamented about his failure with the same strategy:

> Life's not easy for ex-nuts, you know. I tried telling two of my drinking
> buddies about my schizophrenia problem one night at the bar. I thought if
> I told 'em that it's a "disease" like having a heart problem that they would
> understand and pat me on the butt and say it didn't matter to them and
> that I was OK. Shit, it didn't work out like I planned; they flipped out on
> me. Sid couldn't handle it at all and just let out of there in a hurry; Jack
> stayed around me for about twenty minutes and then made some excuse
> and left.

In sum, through the use of medical disclaimers, ex-patients hoped to elevate
their self-esteem and to renegotiate personal perceptions of mental illness as a
non-stigmatizing attribute.

Deception/Coaching

Deception differed from strategies of concealment in that with the former,
ex-patients readily disclosed their illness and past hospitalizations but explicitly
distorted the conditions or circumstances surrounding it. Similar to Miall's
(1986) involuntary childless women and Schneider and Conrad's (1980)
epileptics, about one-third of the ex-patients employed deceptive practices
developed with the assistance of coaches. Coaches included parents, close
friends, spouses, and other ex-patients sharing the same stigma. Coaches actively provided ex-patients with practical suggestions on how to disclose their
attribute in the least stigmatizing manner and present themselves in a favorable
light. Maureen, age 32, explained of her "coaching sessions" with relatives:

> My parents and grandma really helped me out in terms of what I should
> say or tell others. They were so afraid I'd be hurt that they advised me
> what to tell my school mates, the manager at Wooldo where I got hired.
> We had numerous practice exercises where we'd role-play and I'd rehearse
> what I would say to others. . . . After a while I became quite convincing.

Moreover, it is interesting to note that about one quarter of the ex-patients
employed deceptive practices together with medical disclaimers. As Benjamin,
age 62, aptly expressed:

> To survive in this cruel, cold world, you've got to be sneaky. I mean, that
> you've got to try to win people over to your side. Whoever you decide to
> tell about your illness, you've got to make it clear that you had nothing to
> do with getting sick; nobody can place blame on anyone. . . . And you've
> got to color the truth about how you ended up in the hospital by telling
> heart-sob stories to get people sympathetic to you. You never tell them
> the whole truth or they'll shun you like the plague!

Education

A third form of preventive disclosure used by ex-patients to influence others'
perceptions of them and their ideas about mental illness was education. . . .
Twenty-eight percent of the ex-patients revealed their attribute in an effort to

educate others. Marge, age 39, speaking on her efforts to educate friends and neighbors, said:

> I have this urge inside of me to teach people out there, to let them know that they've been misinformed about mental illness and mental patients. We're not the way the media has portrayed us. That's why people are afraid of us. I feel very strongly that someone has to tell people the truth . . . give them the facts. . . . And when they hear it, they're amazed sometimes and begin to treat me without apprehension. . . . Each time I make a breakthrough, I think more highly of myself too.

Ex-patients did not automatically attempt to educate everyone they encountered but, rather, based on subjective typification of normals, made value judgments about whom to "educate." Brenda, speaking on this matter, explained:

> You just can't go ahead and tell everyone. You ponder who it is, what are the circumstances, and whether you think that they can be educated about it. There are some people that these efforts would be fatal and fruitless. Others, however, you deem as a potential. And these are the people you work with.

While education proved successful for some ex-patients in their management activities with certain individuals, others found it less successful. Jim, recalling one disastrous experience with a former poker buddy, said:

> I really thought he would learn something from my discussion of the facts. I really misjudged Fred. I thought him to be an open-minded kind of guy but perhaps just naive, so I sat him down one afternoon and made him my personal "mission." I laid out my past and then talked to him about all the kinds of mental illnesses that are out there. He reacted terrible. All his biases came out, and he told me that all those people should be locked up and the key thrown away—that they were a danger to society. He was probably thinking the same thing about me too!

Following Goffman (1963: 101), medical disclaimers, deception/coaching, and education are forms of "disclosure etiquette"—they are formulas for revealing a stigmatizing attribute "in a matter of fact way, supporting the assumption that those present are above such concerns while preventing them from trapping themselves into showing that they are not."

Normalization

A final form of preventive disclosure employed by ex-psychiatric patients to manage stigma was normalization. This concept is drawn from Davis's (1963) study on children with polio and is akin to deviance disavowal (cf. Davis 1961). Normalization is a strategy individuals use to deny that their behavior or attribute is deviant. It "seeks to render normal and morally acceptable that which has heretofore been regarded as abnormal and immoral" (Pfuhl 1986: 163). Similar to observations made on pedophiles (McCaghy 1968), the obese (Millman 1980), the visibly handicapped (Levitin 1975), and paraplegics

(Cogswell 1967), about one quarter of the ex-psychiatric patients I studied also employed this same strategy. Such persons were firmly committed to societal conceptions of normalcy and were aware that according to these standards, they were disqualified—they would never "measure up." Yet ex-patients made active attempts at rationalizing and downplaying the stigma attached to their failing. So, for example, they participated in a full round of normal activities and aspired to normal attainments. They participated in amateur theater groups, played competitive sports such as hockey and tennis, enrolled in college, and the like. Ex-patients whose stigma could be considered "discreditable," that is, not readily or visibly apparent to others, would disclose such information for preventive reasons, thereby rendering them "discredited" in the eyes of others. They would then attempt to negotiate with normals for preferred images, attitudes, roles, and non-deviant conceptions of self and definitions of mental illness as less stigmatizing. Discussing his utilization of this technique, "Weird Old Larry," age 59, said:

> The third time I got out [of the hospital], I tried to fit right in. I told some of my buddies and a couple of others about my sickness. It was easier to get it out in the open. But what I tried to show 'em was that I could do the same things they could, some of them, even better. I beat them at pool, at darts; I could outdrink them, I was holding down two jobs—one at the gas station and at K-mart. I tried to show them I was normal. I was cured! The key to success is being up-front and making them believe you're just as normal as them. . . . You can really change how they see and treat you.

If successfully carried out, this avowal normalized relations between ex-patients and others.

This is not to imply, however, that the strategy of normalization worked for all patients in all situations. Similar to Millman's (1980: 78) overweight females who were accepted in certain roles but treated as deviant in others, many ex-patients expressed similar problems. Frederick, speaking on this problem with respect to co-workers, said:

> It's really tragic, you know. When I told the other people at work that I was a manic-depressive but was treated and released, I emphasized that I was completely normal in every way . . . but they only accepted me normally part of the time, like when we were in the office. . . . But they never really accepted me as their friend, as one of "the boys"; and they never invited me over to dinner with their wife and family—they still saw me as an ex-crazy, not as an equal to be worthy being invited to dinner, or playing with their kids.

It is interesting to note that just as ex-patients whose attribute was discreditable employed the strategy of normalization, so, too, did other ex-patients with discrediting attributes (conditions *visibly apparent* or *known* to others) employ this same technique. Explaining how medication side effects rendered him discredited and how he attempted to reduce the stigma of mental illness through normalization, Ross said:

Taking all that dope the shrinks dish out makes my hands tremble. Look at my shaking legs too. I never used to have these twitches in my face either, but that's just the side effects, a bonus you get. It really fucks things up though. If I wanted to hide my illness, I couldn't; everyone just looks at me and knows. . . . So, what I do is to try to get people's attention and get them to see my positive side—that I can be quite normal, you know. I emphasize all the things that I can do!

In short, by presenting themselves as normals, ex-patients hoped to elicit positive responses from others whose reactions were deemed to be important. From a social-psychological perspective, others accepted and reinforced a non-deviant image of self through this process of negotiation, allowing ex-patients to achieve more positive, non-deviant identities.

In many cases, ex-psychiatric patients progressed from one strategy to another as they managed information about themselves. Specifically, they moved from a strategy of initial selective concealment to disclosure for therapeutic and preventive reasons. According to the ex-patients, such a progression was linked to their increased adjustment to their attribute as well as the result of positive responses from others to the revelation of their mental illness.

Political Activism

Just as ex-psychiatric patients developed and employed a number of individualized forms of information management to deal with the stigma potential of mental illness, enhance self-images, and alter deviant identities, they also employed one collective management strategy[7] to achieve the same ends, namely, joining and participating in ex-mental patient activist groups (cf. Anspach, 1979). Such groups, with their goal of self-affirmation, represent what Kitsuse (1980: 9) terms "tertiary deviation"—referring to the deviant's confirmation, assessment, and rejection of the negative identity embedded in secondary deviation, and the transformation of that identity into a positive and viable self-conception."

Political activism served a three-fold function for ex-patients: (1) it repudiated standards of normalcy (standards to which they couldn't measure up) and the deviant labels placed upon these individuals; (2) it provided them with a new, positive, non-deviant identity, enhanced their self-respect, and afforded them a new sense of purpose; and (3) it served to propagate this new, positive image of ex-mental patient to individuals, groups, and organizations in society. The payoff from political activism was then, personal as well as social.

Similar to such activist groups as the Gay Liberation Front, the Disabled in Action, the Gray Panthers, or the Radical Feminist Movement, ex-mental patient activists rejected prevailing societal values of normalcy through participation in their groups. They repudiated the deviant identities, roles, and statuses placed on them. Moreover, these individuals flatly rejected the stigma associated with their identities. Steve, a 51-year-old electrician, aptly summed it up for most ex-patient activists:

The whole way society had conceived of right and wrong, normal and abnormal is all wrong. They somehow have made us believe that to be mentally ill is to be ashamed of something; that these people are to be feared, that they are to blame for their sickness. Well I don't accept this vein anymore.

Upon repudiating prevailing cultural values and deviant identities, ex-patient activists collectively redefined themselves in a more positive, non-deviant light according to their *own* newly constructed set of standards. Speaking of her embracing a new non-deviant identity, Susan, age 39, who recently returned to teaching school, said:

I no longer agree to accept what society says is normal and what is not. It's been so unfair to psychiatric patients. Who are they to say, just because we don't conform, that we're rejects of humanity. . . . The labels they've given us are degrading and make us feel sick. . . . [The labels] have a negative connotation to them. . . . So, we've gotten together and liberated ourselves. We've thrown away the old labels and negative images of self-worth, and we give ourselves new labels and images of self-worth—as human beings who should be treated with decency and respect.

In contrast to other ex-patients who employed various individual management strategies to deal with what they perceived to be their *own* problems—personal failings—ex-patient activists saw their problems not as personal failings or potentially stigmatizing attributes but as *societal* problems. To the extent that ex-patients viewed their situations in this manner, it allowed them to develop more positive self-images. Speaking of this process as one of "stigma conversion," Humphreys (1972: 142) writes:

In converting his stigma, the oppressed person does not merely exchange his social marginality for political marginality. . . . Rather, he emerges from a stigmatized cocoon as a transformed creature, one characterized by the spreading of political wings. At some point in the process, the politicized "deviant" gains a new identity, an heroic self-image as crusader in a political cause.

Sally, a neophyte activist, placed the "blame" on society for her deviant self-image:

It's not any of our faults that we ended up the way we did. I felt guilty for a long time. . . . I crouched away feeling that I had something that made me "different" from everyone else, a pock on my life. . . . But I learned at the activist meetings that none of it was my fault. It was all society's fault—they're the one who can't deal with anything that is different. Now I realize that having mental illness is nothing to be ashamed of; it's nothing to hide. I'm now proud of who . . . and what I am!

Just as political activism, as contrasted with other adaptive responses to stigma, sought, in repudiating the dominant value system, to provide ex-patients with positive, non-deviant statuses, so too did it attempt to propagate this new positive, normal image of ex-psychiatric patient to others in society. Thus, through such activities as rallies, demonstrations, protest marches, attendance at conferences on human rights for patients, lobbyist activities directed toward politicians and the medical profession, and the production of newsletters, ex-patient activists sought to promote social change. Specifically, they sought to counter or remove the stigma associated with their "differentness" and present society with an image of former psychiatric patients as "human beings" capable of self-determination and political action. Abe, the president of the activist group, aptly summed up the aim of political activism during a speech to selected political figures, media personnel, and "upstanding" citizens:

> Simply put, we're tired of being pushed around. We reject everything society says about us, because it's just not accurate. We reject the type of treatment we get . . . both in the hospital . . . and out. We don't like the meaning of the words [people] use to describe us—"mentals" and "nuts." We see ourselves differently, just as good and worthy as everybody out there. In our newsletter, we're trying to get across the idea that we're not the stereotypical mental patient you see in the movies. We're real people who want to be treated equally under the Charter of Rights. We're not sitting back now, we're fighting back!

In sum, then, through participation in political activist groups, many ex-patients internalized an ideology that repudiated societal values and conventional standards of normalcy, rejected their deviant identities and statuses, adopted more positive, non-deviant identities, and attempted to alter society's stereotypical perceptions about mental patients and mental illness in general. . . .

NOTES

1. The stigmatized status of individuals and the information management strategies they employ have been well-documented for other groups such as the retarded (Edgerton 1967), epileptics (Schneider and Conrad 1980), secret homosexuals (Humphreys 1975), involuntary childless women (Miall 1986), swingers (Bartell 1971), and lesbians (Ponse 1976), among others.

2. Chronicity, for the purposes of this study, was defined *not* in diagnostic terms, that is, "chronic schizophrenic"; rather, it was defined in terms of duration, continuity, and frequency of hospitalizations. Specifically, the term "non-chronic" refers to those individuals hospitalized for periods of less than two years, those institutionalized on a discontinuous basis, those hospitalized on fewer than five occasions,

or those treated in psychiatric wards of general hospitals.

3. The term "chronic" ex-psychiatric patient refers to those institutionalized in psychiatric hospitals for periods of two years or more, institutionalized on a continual basis, or hospitalized on five or more occasions.

4. The decision to stratify the sample by chronicity was based upon my interests and prior fieldwork activities. My earlier study (Herman 1981) indicated that when we speak of "deinstitutionalized" or "discharged" patients, we cannot merely assume that they are one homogeneous grouping of individuals with like characteristics, and similar post-hospital social situations, experiences, and perceptions of reality. Rather, prior research led me to believe that there might be distinct

subgroups of individuals with varying perceptions of reality and experiences (see Herman 1986).

5.   In Estroff's (1981) ethnography on chronic ex-mental patients, she points out the catch-22 situation in which they find themselves. Ex-patients need to take their medications regularly in order to remain on the outside. Ironically, however, in an effort to become more like others, they take "meds" that make them "different." The various side effects reinforce their deviant identities.

6.   See Herman (1986) for a detailed discussion of such other post-hospital problems.

7.   Following Lyman's (1970) typology of deviant voluntary associations, ex-mental patient political activist groups represent an "instrumental-alienative" type of association. It is interesting to note that chronic ex-patients also employed one collective form of stigma management; specifically, they formed and participated in deviant subcultures (see Herman 1987).

REFERENCES

Adler, Patricia A. 1992. "The 'post' phase of deviant careers: Reintegrating drug traffickers." *Deviant Behavior* 13: 103–126.

Adler, Patricia A., and Peter Adler. 1983. Shifts and oscillations in deviant careers: The case of upper-level drug dealers and smugglers. *Social Problems* 31: 195–207.

Anspach, Renee. 1979. From stigma to identity politics: Political activism among the physically disabled and former mental patients. *Social Science and Medicine* 13A: 765–773.

Bartell, Gilbert D. 1971. *Group Sex: A Scientist's Eyewitness Report on the American Way of Swinging*. New York: Wyden.

Bell, Alan, and Martin S. Weinberg. 1978. *Homosexualities: A Study of Diversity Among Men and Women*. New York: Simon and Schuster.

Bogdan, Robert. 1974. *Being Different: The Autobiography of Jane Fry*. New York: Wiley.

Carrier, J. M. 1976. Family attitudes and Mexican male homosexuality. *Urban Life* 50: 359–375.

Cheadle, A. J., H. Freeman, and J. Korer. Chronic schizophrenic patients in the community. *British Journal of Psychiatry* 132: 221–227.

Christensen, Harold T. 1953. Studies in child spacing: Premarital pregnancy as measured by the spacing of the first

birth from marriage. *American Sociological Review* 18: 53–59.

Cogswell, B. 1967. Rehabilitation of the paraplegic: Processes of socialization. *Sociological Inquiry* 37: 11–26.

Davis, Fred. 1961. Deviance disavowal: The management of strained interaction by the visibly handicapped. *Social Problems* 9: 120–132.

———. 1963. *Passage Through Crisis: Polio Victims and Their Families*. Indianapolis: Bobbs-Merrill.

Delph, E. 1978. *The Silent Community: Public Homosexual Encounters*. Beverly Hills, CA: Sage.

Edgerton, Robert. 1967. *The Cloak of Competence: Stigma in the Lives of the Mentally Retarded*. Berkeley: University of California Press.

Estroff, Sue E. 1981. *Making It Crazy: An Ethnography of Psychiatric Clients in an American Community*. Berkeley: University of California Press.

Faupel, Charles E. 1991. *Shooting Dope: Career Patterns of Hard-Core Heroin Users*. Gainesville: University of Florida Press.

Frazier, Charles. 1976. *Theoretical Approaches to Deviance*. Columbus, OH: Merrill.

Glassner, Barry, Margret Ksander, Bruce Berg, and Bruce D. Johnson. 1983. A note on the deterrent effect of juvenile vs. adult jurisdiction. *Social Problems* 31: 219–221.

Goffman, Erving. 1961. *Asylums*. New York: Doubleday.

———. 1963. *Stigma*. Englewood Cliffs, NJ: Prentice-Hall.

Harris, Mervyn. 1973. *The Dilly Boys*. Rockville, MD: New Perspectives.

Herman, Nancy J. 1981. *The Making of a Mental Patient: An Ethnographic Study of the Processes and Consequences of Institutionalization upon Self-Images and Identities*. Unpublished master's thesis, McMaster University, Hamilton, Ontario.

———. 1986. *Crazies in the Community: An Ethnographic Study of Ex-Psychiatric Clients in Canadian Society—Stigma, Management Strategies and Identity Transformation*. Unpublished Ph.D. dissertation, McMaster University, Hamilton, Ontario.

———. 1987. "Mixed nutters" and "looney tuners": The emergence, development, nature, and functions of two informal, deviant subcultures of chronic, ex-psychiatric patients. *Deviant Behavior* 8: 235–258.

Hewitt, J., and R. Stokes. 1975. Disclaimers. *American Sociological Review* 40: 1–11.

Humphreys, Laud. 1972. *Out of the Closets: The Sociology of Homosexual Liberation*. Englewood Cliffs, NJ: Prentice-Hall.

———. 1975. *Tearoom Trade*. New York: Aldine de Gruyter.

Inciardi, James. 1975. *Careers in Crime*. Chicago: Rand McNally.

Irwin, John. 1970. *The Felon*. Englewood Cliffs, NJ: Prentice-Hall.

Kando, T. 1973. *Sex Change: The Achievement of Gender Identity Among Feminized Transsexuals*. Springfield, IL: Charles C. Thomas.

Kitsuse, John. 1980. Presidential address. *Society for the Study of Social Problems* 9: 1–13.

Lamb, J., and V. Goertzel. 1977. The long-term patient in the era of community treatment. *Archives of General Psychiatry* 34: 679–682.

Levitin, T. 1975. Deviants as active participants in the labelling process: The case of the visibly handicapped. *Social Problems* 22: 548–557.

Luckenbill, David F., and Joel Best. 1981. Careers in deviance and respectability: The analogy's limitations. *Social Problems* 29: 197–206.

Lyman, Stanford M. 1970. *The Asian in the West*. Reno and Las Vegas, NV: Western Studies Center, Desert Research Institute.

McCaghy, Charles H. 1968. Drinking and deviance disavowal: The case of child molesters. *Social Problems* 16: 43–49.

Meisenhelder, Thomas. 1977. An exploratory study of exiting from criminal careers. *Criminology* 15: 319–334.

Miall, Charlene E. 1986. The stigma of involuntary childlessness. *Social Problems* 33(4): 268–282.

Millman, Marcia. 1980. *Such a Pretty Face*. New York: Norton.

Pfuhl, Erdwin H., Jr. 1978. The unwed father: A "non-deviant" rule breaker. *Sociological Quarterly* 19 (Winter): 113–128.

Ponse, Barbara. 1976. Secrecy in the lesbian world. *Urban Life* 5: 313–338.

Ray, Marsh. 1961. The cycle of abstinence and relapse among heroin addicts. *Social Problems* 9: 132–140.

Reynolds, David K., and Norman Farberow. 1977. *Endangered Hope: Experiences in Psychiatric Aftercare Facilities*. Berkeley: University of California Press.

Schneider, J., and P. Conrad. 1980. In the closet with illness: Epilepsy, stigma potential and information control. *Social Problems* 28(1): 32–44.

Scott, Robert. 1969. The socialization of blind children. In *Handbook of Socialization Theory and Research*, edited by D. Goslin. Chicago: Rand McNally.

Veevers, Jean. 1980. *Childless by Choice*. Toronto: Butterworths.

Illness Ambiguity and the Search for Meaning

A Self-Help Group for Affective Disorders

DAVID A. KARP

This article reports on a group of people suffering from depression or manic depression who met weekly in "sharing and caring" groups to talk about the nature of their "illness" and the various life contingencies associated with it. It shortly became apparent that one central function of this support group was to provide a forum for making more coherent an elusive and uncertain condition. Much of people's discussion in these groups centered on the very difficult problems that depression and mania created in their lives. Most of those attending the meetings had experienced multiple hospitalizations, and many described severely dysfunctional family and work lives. The material discussed in the meetings was often poignant, and one could not help but be struck by the courage these people displayed as they daily grappled with extraordinary and debilitating pain.

The group talk also illustrated the uncertainty, ambiguity, and lack of clarity attached to the experience of having an "affective disorder." Much of the discussion centered on the confusions surrounding the nature and causes of depression, the labels that most accurately described their condition, the problems in the use of psychotropic drugs, the utility of psychotherapy, the biochemical basis for their condition, and the extent to which they were victims of a disease beyond their control. The frequency with which these and related issues came up in group discussion revealed their problematic character. Group members could not easily achieve consensus on these matters, and they were often the basis for debate concerning the meaning of their common experience. How they collectively created explanations, understandings, agreements, and common "illness ideologies" in order to impose some order onto a hazy and ill-understood life condition is the subject matter for this article. . . .

Although group discussion covered a range of topics, certain questions and themes kept recurring, and it is around these issues that participants tried to work toward consensus about the meaning of their shared condition. Each of

From "Illness Ambiguity and the Search for Meaning: A Case Study of a Self-Help Group for Affective Disorders," David A. Karp, *Journal of Contemporary Ethnography*, Vol. 21, No. 2, 1992. Reprinted by permission of Sage Publications, Inc.

these questions refers to an interpretive dilemma that had to be resolved by individuals as they tried to comprehend their illness. From the array of issues discussed, four interconnected questions constantly reasserted themselves:

1. What is involved in getting a proper diagnosis, and what are the appropriate labels to attach to our common difficulty?
2. To what extent are we responsible for causing the problem, and what is our personal responsibility for remedying it?
3. How much ought we, or must we, rely on medical experts to provide relief for our situation?
4. Are drugs the ultimate salvation for those with affective disorders?

. . . In the following section, I describe in greater detail the setting of this research and the problems faced in gathering the data. I then discuss, in turn, how the subjects thought about the diagnoses of depression and manic depression, assessed cause and responsibility for their condition, evaluated their relationship with therapeutic experts, and interpreted the experience of taking psychotropic medications. Taken together, these features of group talk illuminate how persons collaboratively constructed a *version* of what affective disorders are all about and what one ought to do about them.

SETTING AND METHOD

The organization studied had been in existence for 5 years. I became a participant and observer in August 1988 and attended meetings regularly through November 1990. When I first attended group meetings, they were being held at a local mental hospital that had allocated to the organization three small rooms in an out-of-the-way building on its grounds. At the time, approximately 35 to 50 people attended the Wednesday night meetings. By the end of the observation period, about 120 people attended the meetings each week, and the hospital had made available to the organization a number of rooms in its relatively new cafeteria building. On a typical evening, after the president of the group greeted those present and various announcements of general interest were made, the groups for that evening were indicated. Several "open" groups were available, but along with them were groups on such specific topics as "coping with mania," "coping with depression," "spiritual healing," and "Why do I live?" In addition there were groups each week for newcomers, for family and friends, and for young people.

Significantly, as the group increased in size and, by that criterion, became more successful, the running of the meetings became somewhat more formalized. When I first attended, decisions about which members would "run" the groups were made very informally. Normally, two or three of the group's officers volunteered to do this, but, as often, people who had been coming for a couple of months or more were approached on an ad hoc basis to run a group. Within the last 6 months of my participation, the decision was made to require potential group leaders to take a 12-week leadership training course given by a consultant on small group dynamics. Consequently, only those persons who had successfully completed the training could be group leaders. As a result, a relatively small cadre of individuals repeatedly served as group leaders.

At the outset of each group session, the leader read the specific rules that were expected to govern discussion. Then, the 15 to 20 people in each group introduced themselves, described the central dimensions of their problem, and indicated what they hoped to get out of the discussion. Particular attention was given to individuals who described themselves as having an immediate crisis. If no one had a special problem to raise, group leaders suggested possible foci for the discussion. Usually, conversation happened quite naturally and group leaders needed to do little more than ensure orderliness to the discussion.

The membership in the group was fluid. Some persons attended only once or twice without returning. Others came periodically, attending only when they were in the middle of an "episode" and stopped when they felt better. Others came regularly, even when they were feeling well. Of the approximately 120 persons who attended each week, I would estimate that between 20 and 30 constituted a core of "regulars" whom I knew by name. Because the population of persons attending shifted from week to week, it was difficult to establish with confidence the demographic characteristics of those who came. Observation, however, allows the generalizations that the group was composed about equally of men and women; that, beginning with the late teen years, all age categories were represented; and that the members were overwhelmingly White. Based on casual conversation with regulars and on individuals' self-identification during meetings, attendees were drawn from points all along the class structure and from a range of religious groups. An estimate, based on group discussion, is that about 80% of the members had been hospitalized at one time or another and that about 50% were unemployed for reasons largely related to their mental health.

Because this research involved entering into a group of people that met expressly to discuss personal identity information about a mental condition that is often deeply stigmatizing, questions about my research identity and matters of confidentiality were critical considerations from the outset. Although there is no consensus among social scientists about the appropriateness of disclosing one's research role, especially when it may compromise the study itself (see Douglas 1976; Humphreys 1970; Punch 1985), my own stance is closest to the view that researchers should not deliberately misrepresent their identities in order to gain access to a group (Bulmer 1982; Erikson 1970). My decision from the beginning was to present myself as an individual who has grappled periodically with depression but also as a social scientist who might eventually "be interested in writing about how people live with depression in their daily lives." This article, therefore, is truly an instance of a researcher attempting, as Mills (1959) put it, to "transform private troubles into public issues." . . .

It would have been illegitimate for someone to attend the meetings of this group without having experienced an affective disorder. This was partially confirmed at a later meeting when I approached Alex (pseudonyms used throughout), another of the organization's officers, who routinely tape-recorded lectures delivered to the group. I explained that I would like to get a copy of an earlier talk as I was considering the possibility of writing about depression. At this point, he said, "Well, you *are* someone who experiences this illness, right?" When I indicated that this was so, Alex seemed more than willing to provide me a copy of the tape. Moreover, at one time or another, nearly everyone participated in group discussion. People described their episodes of depression or mania, talked extensively about their experience with medications, their encounters with therapists of all sorts, and the negative impact that their problem has had on family and work life. To be a member in good standing, one needed an experiential basis for discussing such issues. It would have been hard to participate without having a "story" to tell. Of course, a researcher could invent stories, but to do so would be, in my view, an unacceptably cynical misrepresentation of identity. . . .

What follows is a rendering of the repeating themes in the talk of group members and an analysis of how that talk was directed toward the creation of a coherent reality about the meaning of having an affective disorder.

THE SOCIAL CONSTRUCTION OF AFFECTIVE DISORDERS

The data collected in the groups suggest that the persons studied faced a number of interpretive tasks as they endeavored to make sense out of their situation. Without exception persons had to define what was wrong with them and to attach a label to their experience. The process of acquiring an acceptable diagnosis was often arduous. In an ongoing way, everyone sought to understand the causes of their condition. Because no one definitively knows the causes of affective disorders, efforts to understand whether nature, nurture, or some combination of the two explained the "illness" were always incomplete. Although people eventually arrived at "theories" about causation, an inevitable dilemma was the inherent untestability of such theories. . . . All of the people in this group, once diagnosed, began an extensive "therapeutic career" involving relations with psychiatric experts who prescribed a variety of psychotropic drugs. Everyone, therefore, necessarily evaluated his or her relationship with these experts and the effects on them of mind-altering medications.

Talking About Diagnosis

Just as people introduce themselves at AA meetings, group sessions often began with personal introductions that took the form "My name is Dick and I'm a [manic] depressive." Such an introduction often included as well the year in which they were first diagnosed. Indeed, members of the group dated themselves from the time of first diagnosis. The point of diagnosis was an important symbolic benchmark in a difficult illness career. The importance of having a diagnosis was also evidenced by the frequency with which people making an initial acquaintance asked, "And when were you diagnosed?" For purposes of self-identification in the context of this group, description of one's diagnosis (or diagnoses) was a key piece of information. Self-identification as having a unipolar or bipolar disorder, as being obsessive-compulsive, as suffering from agoraphobia, and so on often set the pattern of subsequent conversation.

Acquiring a clear conception of what one has and having a label to attach to confounding feelings and behaviors was significant because most individuals in the group, particularly older people, often went years without being able to give a name to their situation. People often remarked on the increase in knowledge about depression in the past 15 or 20 years and then told stories of years of absolute bewilderment during which psychiatric help was not widely available.

Although some individuals continued to doubt the veracity of the diagnosis given them, most persons were thankful for being able to name what they experienced. To name something objectified it and gave it a tangible reality. The creation of an illness reality was, then, a first and critical step in the effort to make sense of things. On several occasions, persons, while complaining about their own current hardships, were reminded by group members, "At least you're diagnosed." Such a comment referred to the millions of people who cannot begin to take steps to help themselves because they do not know what is wrong with them. Illustrating the idea that people begin to understand what troubles them through a series of successive approximations, one man, probably in his 60s, reported,

> At first I thought I needed more sleep than other people. Then I realized that I had mood swings. Then I learned that I had depressive periods. Then I learned that I had bipolar depression. Then I learned from the doctors that I inherited this from my grandmother. This was a learning process that took several years.

Just as social scientists (Brown 1987; Rosenberg 1984) have been critical of the meaning of psychiatric diagnoses and the process through which they are established, group members were often critical of doctors who misdiagnosed them for years. . . . The sense of consternation and anger at doctors, whom individuals believe gave incorrect diagnoses, is evident in the comments of a middle-aged woman who offered this history, starting with a major episode experienced while living abroad:

> I flew home. I don't know how I made it. Just all my symptoms came back. I didn't know what was happening to me, and my parents are saying it's my fault and I don't know what's up. I think to myself, "I've been totally fine, so why is this happening to me?". . . I tried to pull myself together. I still didn't have a diagnosis. I even decided to get remarried to someone who was willing to put up with me, but I have to say that I didn't have a strong base of confidence. . . . I didn't think I knew any answers. . . . I was 26 at the time and in the middle of what I now see as my

second depressive episode in 10 years. . . . So I walk into my old therapist's office and say, "I've got this problem. I can't sleep and I feel like I'm
walking on water." And he said, "Oh, what I'd like you to do, I'd like you
to go back on the medication (taken years earlier)." I said, "I'd like to get
another opinion." . . . He never heard me say that before. I never thought
to say that. I said, "What is my diagnosis?" He told me, "You are psycho-
bizarre" but I couldn't find that anywhere in the lexicon. . . . I finally did
see a consultant at [names hospital]. Well, in a 1-hour appointment . . . he
said, "I think you're a manic depressive." Nowhere had I been told that.
"And you have a disease." Nowhere had I been told that. I was not
pleased, but he gave me some literature . . . and recommended a specialist.

. . . Group members, however, felt ambivalent about the labels associated
with their diagnosis. For many, receiving a diagnosis was a double-edged thing.
On one hand, knowing that you "have" something that doctors see as a specific illness imposes definitional boundaries onto an array of behaviors and
feelings that previously had no name. To be diagnosed suggests the possibility
that the condition can be treated and that one's suffering can be diminished.
However, many individuals were unsettled by the "mental illness" label. At
one meeting, for example, the question of labels came up as a topic of discussion. People around the room expressed their preferences, with some wishing
to avoid the word "illness" altogether. They spoke about having a "chemical
imbalance" or "emotional disorder." Some found the words "disease" and
"illness" appropriate but tried to avoid the word "mental," referring instead to
their "emotional" illness or disease. The discussion indicated that these people
recognized and had thought a great deal about the power of words in influencing personal identity. Several agreed with one person's sentiment that "you
do not want to think that you have a mental illness." . . .

Evaluating Cause and Responsibility

Among the stated goals of this organization, as described in one of its publications, was "to educate the patient and the public on the biochemical nature of
mania and depression." Although other documents explicitly stated that the
organization "has not and will not form any opinions or comments on politics, religion or other controversial matters or on the merits of various forms
of medicine or therapy in the treatment of illness," it had plainly adopted the
position that the root cause of affective disorders was biochemical. Only on
the matter of cause had the organization taken an "official" position, one that
also reflected the beliefs of most of the members of the group. Although there
was no unanimity on the point, belief in the biochemical nature of the illness
was widely shared in the group. . . .

Often, people voiced a fatalism surrounding the onset of episodes, suggesting the view that they were deterministically and involuntarily buffeted
about by faulty brain chemistry. One woman, for example, said, "I need the
doctor to tell me it's not just my attitude, that there is something chemical

going on. You can't just snap out of a chemical depression." A related and common theme was that people fall into states of depression or mania even when there is no precipitating event. People often said things like "You just never know when it's [a period of depression] going to hit" or "I've had a few good days recently, but you never know when it's going to fall apart." One person described working as manageable "until disaster comes." Another described her life as "iffy and tentative" and elicited head nods from others in the groups when she said, "It just comes on. You can't explain it sometimes. It seems to happen sort of randomly." When they voiced the language of bio-chemical cause, direct analogies were often made to organic illnesses, as in the case of the woman who said, "The illness comes upon me, sort of like a bout of pneumonia out of the blue." . . .

To manage their identities in a way that preserved a positive sense of self, people effectively split themselves in two. It was quite common to hear people talk about the fact that when they were overly aggressive, acting out, irritable, and so on, "it was the illness talking," not them. One woman whose husband was a manic depressive said that sometimes "it is unclear whether the illness is talking or whether [he] is talking." In a different meeting, a woman, who described a history of beginning to take college courses only to drop out, offered the following analysis of her behavior: "I would always start courses and then drop out. I used to think I was easily bored. Now I don't blame myself or the class. It's not me or the class, it's the illness." Yet another woman who was undecided about returning to work as a preschool teacher wondered out loud: "Do I not want to go back to it, or is it because of the illness that I don't want to return?" These and similar comments were meant to reaffirm the view that people suffering from depression are actually intelligent, proper people whose interactional competence is sometimes compromised by a disease. . . .

Looking for Dr. Right

As several cultural observers (Derber, Schwartz, and Magrass 1990; Gross 1978; Lasch 1980) have noted, the behavior of persons in today's society is dominated by "experts." Experts advise us on virtually every aspect of our lives. Today, experts follow us through the life course. They are there when we are born and follow us each step along the way, eventually to our graves. Many have come to feel reliant on experts to tell them how to maintain their health, how to become educated, and how to raise their children. Freidson (1970) indicated the enormously expanded role of professionals in our daily lives with his comment that "the relation of the expert to modern society seems in fact to be one of the central problems of our time, for at its heart lie the issues of democracy and freedom and the degree to which ordinary men can shape the character of their own lives" (p. 336). Most critical to the present inquiry, however, is the fact that medical experts now tell us when our bodies and our "selves" need repair and the proper procedures for doing it. The medical profession has played a key role in the rise of what might be termed a "therapeutic culture" (Rieff 1966) in which ever greater numbers of behaviors fall under the aegis of a "medical model" (Conrad and Schneider 1980). According to some critics, especially of psychiatry (Goffman 1961; Laing 1967; Szasz 1974), the medical model is used to support an essentially political reality.

Like social scientists, those who had long histories as patients were uncertain about the proper role of psychiatrists in defining reality and the adequacy of their conceptual knowledge for curing the soul. The typical person in the group had years of experience with psychiatrists, and much conversation concerned their relationships with particular psychiatrists, their sense of dependency on them, and questions about the value of psychotherapy in combating affective disorders. In a word, the underlying feeling about psychiatry was "ambivalence." On one hand, these individuals saw psychiatrists as the professional experts who held out the possibility of helping them. However, for every person who talked of having found a warm, sensitive, and sensible psychiatrist, there were those who told psychiatric horror stories involving, as they saw it, insensitivity and incompetence. In one group meeting, when asked about possible topics for discussion, one person jokingly suggested "therapists I have known." Several people described themselves as "veterans of the system" and told of years searching for the "right" therapist. In a newsletter, one group member used an interesting analogy in talking about illness treatment and the search for good psychiatric help:

> The solution to the problem? Preventive maintenance. Any mechanic or fleet operator will tell you that it's far cheaper and more effective to change the oil regularly than it is to do major overhauls. And when trying to solve a particularly confounding problem with your car, it may require going to a couple of different mechanics until you find someone who knows how to do the job right. If you aren't satisfied with your current mechanic, don't go back. The same is true of affective disorders. Regular check-ups and medication maintenance add immeasurably to our ability to lead normal lives. When you find a competent therapist, you'll know it. Being in the care of someone you trust is a wonderfully reassuring feeling.

The phrase "When you find a competent therapist, you'll know it" is noteworthy because it parallels comments often heard in everyday conversation and popular music about falling in love. People's relations with a series of therapists might be seen as similar to a pattern of "serial monogamy"; finding someone you believe is right for you, becoming dependent on that person for meeting certain needs, making a commitment, eventually realizing that you may have made the wrong choice, leaving the relationship, and searching once more for the person who is *really* right for you. Stories interchangeable with the following were frequently heard:

> My father died in March of 1977 and again I had another bad depression. My second therapist was having some of his own problems and deep down inside I realized that he could never help me. . . . I hired a new consultant and we both agreed that a new therapist was necessary at this time. This therapist, who is now still my therapist . . . is very caring. He is very insightful and he made me feel that he wanted to work with me, and he saw a lot of hope for me that I had not for myself. Despite the fact that I had

no hope and was totally discouraged, he saw that I wanted to get better, and he did everything possible to show how much he believed in me. That was something I never had in my life, a truly caring person.

Although some persons claimed to have found the right psychiatrist, the more typical pattern appeared to be disillusionment about the efficacy of psychiatry. In the course of a discussion where two people were talking about their respective psychiatrists, one indicated that he was switching from Dr. X to Dr. Y. The second responded by saying, "They send you to Y when they don't know what to do with you." Others were heard to comment on different occasions: "Psychiatrists don't really understand, and often they're not helping the situation." "There are some good, some bad, but mostly bad." "I can't stand therapy. It never seems relevant to my disease." "I'd like to see a shrink with the disorder." If some were disillusioned with psychiatrists, others were downright angry:

> For all the research that's gone on since 1940 about drug research, psychiatrists still know jack-squat about it. This is a really frustrating thing for me. They are rip-off artists, but I need them to dispense my drugs.

Even those who spoke favorably about their psychiatrists ("My present therapist and therapy is very positive"; "With the work and encouragement of my current therapist, I am able to avoid regression and even step ahead") came to recognize that therapy could not relieve their uncertainty. One frequently heard the sentiment that after years of dealing with depression, the patient knows more about it than the doctor ("We really know ourselves inside out and they don't"). . . . Despite their physicians' best efforts, these patients eventually realized that their therapists could not clear away their confusion about depression. . . . What emerged from the group discussion was the definition of affective disorders as troubles that must ultimately be remedied by the individuals who suffered from them. Implicit in such a definition is an antipsychiatry ideology that demands, at the very least, a greater democracy between doctor and patient in efforts to treat the problem.

Interpreting the Drug Experience

Every person I met in this group had taken psychotropic medications at one point or another. Although some people periodically stopped taking drugs, the vast majority were currently taking medication(s). Many in the group saw two doctors: psychotherapists with whom they talked and psychopharmacologists who prescribed and monitored their medications. Over the years, most people had taken multiple drugs in what seemed to be a trial and error process. Of all the topics raised in the groups, shared experience about drug use was the one that came up at nearly all the meetings. Because drugs literally altered people's feeling and perceptions, their use required especially constant evaluation and interpretation.

Group discussion about the effectiveness of psychotropic medications affirmed the view that compliance or noncompliance with physicians' directives

about drug use must be understood in terms of the daily lives of patients themselves. Rather than being passive recipients of medical advice, group members
created "theories" about drug use in terms of their practical experience with
illness. In this regard, persons taking psychotropic medications responded in
much the same way as the epilepsy patients described by Conrad (1985). For
both groups, medications rarely solved their problem, and in the face of uncertainty about the value of medications, patients constructed their own drug
management strategies. Conrad's argument that "regulating medication represents an attempt to assert some degree of control over a condition that appears
at times to be beyond control" (p. 36) fits well with the perspectives voiced
and constructed in the group.

For many people in the group, there was a great deal of confusion surrounding the use of drugs. Conversation revealed that the vast majority had been taking medications for years, and many were on multiple medications
simultaneously. The number of permutations and combinations of drugs and
dosage (lithium, tricyclic antidepressants, monoamine oxidase [MAO] inhibitors,
and tranquilizers) taken by individuals in the group was bewilderingly large, and
arriving at the "right" combination of drugs was a puzzling and frustrating affair. Persons talked regularly about stopping one drug, beginning another,
changing doses of their current medications, and sometimes making the decision to stop drugs altogether. Just as individuals became expert about different
forms of psychotherapy, the majority of persons in the group had taken a wide
variety of drugs over the years and had definite opinions about them.

Discussion in the group often centered on an individual who had just
begun a new drug and did not quite know what to expect. Such a person
could count on there being others in the group who were either currently
taking the same drug or had taken it in the past. It was common for persons
to trade information about dosage levels, side effects, and interactions among
medications. The problem of making sense of the drug experience was compounded by the fact that the same medication that helped one individual could
precipitate a terrible experience for another. Nevertheless, group members relied on others to determine whether a medication was working for them,
when they could expect it to "kick in," and when the side effects were sufficiently noxious that they should either get off the drug or alter its dosage. It
appears that the information supplied by this subculture of fellow drug users
was at least as important as the advice offered by psychopharmacologists in
determining a drug's effectiveness.

Although some individuals told of finding a drug that "saved my life," the
more common response was the recognition that the drugs, like psychotherapy, constituted, at best, a partial answer to their problem. Just as there was a
disillusionment that often set in with their therapy, many expressed disillusionment about medications. Stories suggested that individuals began with
high hopes that a new medication would finally be the one to help them.
Sometimes, it did dramatically change things for the better, but more often, it
either modified their symptoms only slightly or exacerbated them. Sometimes,
individuals reported feeling well for periods of time on certain medications,

only to then have a depressive episode that obviously called into question the drug's effectiveness. Several times, I heard individuals describe having "broken through" their lithium (experiencing mania or depression in spite of the drug), and casual conversations often went like this:

"How are you?"

"I've been feeling really terrible for the last couple of weeks."

"But I thought you were one of Prosac's miracles [said somewhat tongue in cheek to imply that there are no miracles]."

. . . Despite confusion about the efficacy of drugs, the consensus was that one should not try "to go it alone" (without drugs). At one meeting, for example, a young man announced his plan to stop taking his lithium because, as he put it, "I don't like feeling drugged. I'm like a whirling dervish and that's just the way I am." In response, one individual offered the idea that wanting to get off medications "is a stage through which all of us go." Another concurred with the observation that "we are all struggling with what to do with medications." Not everyone, however, agreed that the best course is to keep taking medications. I periodically heard opinions equivalent to the following: "There is no reason to take pills all the time when your episodes are infrequent" or "Pills are not the absolute God." Such contrasts in view indicate that group discussion could only provide the broad interpretive frame within which individuals tried to understand the value of medications in helping them to maximize the quality of their lives. It was, however, a frame reasserting the view that patients are the ultimate experts on what it means to have an affective disorder and, consequently, that group members should, whenever possible, avoid giving doctors autonomy over their lives. In this regard, the group provided a nascent ideology posing a potential threat to medical dominance in the treatment of affective disorders.

CONCLUSION

This article proceeded from a cornerstone of sociological thought. A key idea of the symbolic interaction perspective (Blumer 1969; Hewitt 1986; Karp and Yoels 1986) is that persons' feelings and behaviors arise from their "definition of the situation": that to appreciate people's thoughts, feelings, and actions, we need to inquire into how they arrive at definitions of the situations in their lives. However, we need to acknowledge that some life situations require more extensive definitional efforts than others. This article has centered on the role of conversation in a support group to aid individuals in better comprehending a life situation—having an affective disorder—that was intrinsically ambiguous and problematic. Although groups of the sort described in this article could not absolutely resolve questions about the correctness of particular diagnoses, the extent to which persons ought to feel responsible for their condition, or how to deal with psychiatrists and the medications they prescribe, there was something very powerful for individuals in learning that others shared their confusions. Beyond this, however, participation in the group was compelling because members' talk provided a "vocabulary" through which to see what affective disorders were all about. . . .

REFERENCES

Blumer, H. 1969. *Symbolic interaction: Perspective and method*. Englewood Cliffs, NJ: Prentice-Hall.

Brown, P. 1987. Diagnostic conflict and contradiction in psychiatry. *Journal of Health and Social Behavior* 28: 37–50.

Bulmer, M. 1982. *Social research ethics*. London: Macmillan.

Conrad, P. 1985. The meaning of medications: Another look at compliance. *Social Science and Medicine* 20: 29–37.

Conrad, P., and J. Schneider. 1980. *Deviance and medicalization*. St. Louis: Mosby.

Derber, C., W. Schwartz, and Y. Magrass. 1990. *Power in the highest degree: Professionals and the rise of a new mandarin class*. New York: Oxford University Press.

Douglas, J. 1976. *Investigative social research*. Beverly Hills, CA: Sage.

Erikson, K. 1970. A comment on disguised observation in sociology. In *Qualitative sociology,* edited by W. Filstead, pp. 252–60. Chicago: Markham.

Freidson, E. 1970. *Profession of medicine*. New York: Harper & Row.

Goffman, E. 1961. *Asylums*. New York: Anchor Books.

————. 1963. *Behavior in public places*. New York: Free Press.

Gross, M. 1978. *The psychological society*. New York: Random House.

Hewitt, J. 1986. *Self and society*. Boston: Allyn & Bacon.

Humphreys, L. 1970. *Tearoom trade: Impersonal sex in public places*. Chicago: Aldine.

Karp, D., and W. Yoels. 1986. *Sociology and everyday life*. Itasca, IL: Peacock.

Laing, R. 1967. *The politics of experience*. New York: Ballantine.

Lasch, C. 1980. Life in the therapeutic state. *New York Review of Books,* 12 June, 24–31.

Mills, C. W. 1959. *The sociological imagination*. New York: Oxford University Press.

Punch, M. 1985. *The politics and ethics of fieldwork*. Beverly Hills, CA: Sage.

Rieff, P. 1966. *Triumph of the therapeutic*. New York: Harper & Row.

Rosenberg, M. 1984. A symbolic interactionist view of psychosis. *Journal of Health and Social Behavior* 25: 289–302.

Szasz, T. 1974. *The myth of mental illness*. New York: Harper & Row.

PART VI

Relations
Among Deviants

We begin our turn to a closer examination of the lives and activities of deviants. Once they get past dealing with outsiders, they must deal with other members of their deviant communities and with the specifics of accomplishing their deviance. There are several ways of looking at how deviants organize their lives. We start here by looking at the relationships among groups of deviants, focusing on the character, structure, and consequences of different organizational forms. These encompass the structure or patterns of relationships in which individuals entwine when they engage in the pursuit of deviance.

As Best and Luckenbill (1980) have noted in their analysis of the social organization of deviants, relationships among deviants can follow many models. These vary along a dimension of sophistication, involving complexity, coordination, and purposiveness. Deviant associations vary in their numbers of members, the task specialization among members, the stratification within the group, and the amount of authority concentrated in the hands of a leader or leaders. Some groups of deviants are loose and flexible, with members entering or leaving at their own will, uncounted or monitored by no one. Others maintain more rigid boundaries, with access granted only by the consent of one or more insiders. Membership rituals may vary from none to highly specific acts that must be performed by prospective inductees, thereby granting them not only membership but also a place in the pecking order once they are inside. In some ways, rigidity inside deviant groups is related to their insulation from conventional society: the more that a group's members withdraw into a social and economic world of their own, the more that they will develop norms and rules to guide them, replacing those of the outside order.

Groups of deviants also vary in their organizational sophistication, with the more organized groups capable of more complex activities. Such organized groups provide greater resources and services to their members—they pass on the norms, values, and lore of their deviant subculture; they teach novices specific skills and techniques when necessary; and they help one another out when they get in trouble. As a result, individuals who join more tightly knit deviant scenes tend to be better protected from the efforts of social control agents and more deeply committed to a deviant identity.

The three selections in Part VI look at **subcultures, gangs,** and **formal organizations** of deviants. These readings are organized along a continuum that rises progressively in organizational sophistication. Kathryn Fox's article, "Real Punks and Pretenders: The Social Organization of a Counterculture," examines the stratification within a Midwestern city's punk scene, or subculture. Starting at the center, she looks at the hardcore punks and then makes her way outward to the softcore and preppie punks. She examines the immersion of each group in the punk lifestyle and their commitment to punk ideals, activities, style of dress, and mode of survival. Membership in the hardcore inner group required a more serious dedication than did participation in the transitory outer fringe.

Columbus Hopper and Johnny Moore's "Women in Outlaw Motorcycle Gangs" offers a journey into a dangerous and isolated deviant world. Gang members are more serious and involved with one another than are members of the nihilistic punk scene, with strong barriers to achieving entry and status. Stratification marks insiders as well, as women hold a decidedly inferior position within this society. Hopper and Moore's essay is at once intriguing and chilling, for it offers a glimpse into a hidden subculture that inspires intense loyalty in some and fear in others. This portrait reveals the kinds of bonds that exist among members of gangs.

If scene participants have loose relationships characterized by socializing together and if gang participants have tighter relationships based on living, partying, and fighting together, individuals who belong to deviant formal organizations share the bond of membership in a large-scale, highly structured deviant association that extends over time and space. Deviant formal organizations entail more highly sophisticated relationships than the less-structured or committed crews, or rings, where we first see an emerging division of labor so that individual members learn specific skills and pursue these specialized tasks within the group. Members of deviant formal organizations work together in enterprises that may be specialized or diverse, tightly connected from top to bottom or portioned out in authority to separate sub-spheres, ethnically and/or familially homogeneous, and capable of earning vast amounts of money. Roy Godson and William Olson offer a rare depiction of the modern breed of cartels in "International Organized Crime," an essay that contains a description and analysis of this most sophisticated form of deviant group's nature, scope, type of organization, history, and international consequences for crime, law enforcement, and the erosion of legitimate government. In so doing, they contrast this level of deviant relationships to the less organizationally sophisticated gangs that operate domestically in our major urban areas.

28

Real Punks and Pretenders

The Social Organization
of a Counterculture

KATHRYN JOAN FOX

Modern Western society has been characterized by a variety of antiestablishment style countercultures following in succession (i.e., the Teddy Boys of 1953–1957, the Mods and Rockers of 1964–1966, the Skinheads of 1967–1970, and the Punk Rockers of the late 1970s; Taylor, 1982). The punk culture is but the latest in this series. Since most studies of youth- and style-oriented groups are British (Frith, 1982), very little has been written from a sociological perspective about punks in the United States. This study is an attempt to fill that void.

Whether or not the punk scene in the United States could be legitimately classified as a social movement is debatable. Most writers on this contemporary phenomenon agree that American punks have a more amorphous, less articulated ideological agenda than punks elsewhere (Brake, 1985; Street, 1986). While the punk scene in England responded to youth unemployment and working-class problems, the phenomenon in the United States was more closely connected to style than to politics. Street (1986: 175) notes that even for English punks, "politics was part of the style." I would argue that this was even more the case for American punks. The consciousness of the youth in the United States did not parallel the identification with the plight of youth found in Europe. Nonetheless, the "style" code for punks in the United States contained an insistent element of conflict with the dominant value system. The consensual values among the punks, as ambiguous as they were, could best be understood by their contradictory quality with reference to mainstream society. In this respect, punk in America fit the definition of a "counterculture" offered by Yinger (1982: 22-23). According to this definition, the salient feature of a counterculture is its contrariness. Further, as opposed to individual deviant behavior, punks constituted a counterculture in that they shared a specific normative system. Certain behaviors were considered punk, while others were not. Indeed, style was the message and the means of expression. Observation of behaviors that were consistent with punk sensibilities

From "Real Punks and Pretenders: The Social Organization of a Counterculture," Kathryn J. Fox, *Journal of Contemporary Ethnography,* Vol. 16, No. 3, 1987. Reprinted by permission of Sage Publications, Inc.

were viewed as indicative of punk "beliefs." These behaviors, along with verbal pronouncements, verified commitment. Within the groups of punks I studied, the degree of commitment to the counterculture lifestyle was the variable that determined placement within the hierarchy of the local scene.

Previous portrayals of youthful, antiestablishment style cultures have discussed their norms and values (Berger, 1967; Davis, 1970; Hebdige, 1981; Yablonsky, 1968), their relationship to conventional society (Cohen, 1972; Douglas, 1970; Flacks, 1967), their focal concerns and ideology (Flacks, 1967; Miller, 1958), and their relationship to social class (Brake, 1980; Hall and Jefferson, 1975; Mungham and Pearson, 1976). With the exception of Davis and Munoz (1968), Kinsey (1982), and Yablonsky (1959), few of the studies of antiestablishment, countercultural groups discuss their implicit stratification. In this essay I will describe and analyze the various categories of membership in the punk scene and show how members of these strata differ with regard to their ideology, appearance, taste, lifestyle, and commitment.

I begin by discussing how I became interested in the topic and the methods I employed to gain access to the group and to gather data. I then offer a description of the setting and the people who frequented this scene. Next I offer a structural portrayal of the social organization of this punk scene, showing how the layers of membership form. I then examine each of the three membership categories (hardcore punks, softcore punks, and preppie punks), as well as the spectator category, focusing on the differences in their attitudes, behavior, and involvement with this antiestablishment style culture. I conclude by outlining the contributions each of these types of members makes to the continuing existence of the punk movement and, more broadly, by describing the relation between the punk counterculture and conventional society.

METHODS

My interest in the punk movement dates back to 1978. At that time, I attended a local punk bar fairly regularly and wore my hair and clothing in "punk" style, albeit not the radical version. I also visited a major northeastern city at about that same time, when the punk scene was in full flower, and spent several nights visiting what are now famous punk hangouts. My early interest and involvement in this scene laid the groundwork for this study, as it permitted me to gain knowledge of the punk vernacular, styles, and motives. The research continued, with active, weekly participation, through the middle of 1986. I have continued to keep a close watch on national trends and developments in punk culture. Further, I continually frequent local punk bars in an effort to deepen my understanding and to observe the decline of this counterculture. However, I conducted the bulk of my interviews in the fall of 1983 over a period of about two months. During that time, I attended "punk night" at a local bar once a week. In addition, I was invited to other punk functions, such as parties, midnight jam sessions, and public property destruction events. I thereby observed approximately 30 members of this movement with some degree of regularity. I used mainly observational techniques, along with some participation. Although I frequented numerous punk gatherings, my participatory role was constrained by the limited time I spent there. I also had to tread a line between covert and overt roles. While some people knew I was researching this setting, I could not reveal this to others because they might have denied me further access to the group. This created a problem, much the same as that experienced by Henslin (1972) and Adler (1985), in that I had to be careful about what I said and to whom I confided my research interests. This "tightrope effect" severely limited my active participation in the scene.

Nevertheless, by following the investigative research techniques advocated by Douglas (1976), I was able to gain the trust of some key members. I tried to establish friendly relations by running errands for them, buying them drinks and food, and driving them to pick up their welfare checks and food stamps. After several weeks, when I began to be recognized, I was able to broach the topic of doing interviews with several people. I formally interviewed nine people at locations outside the bar. These tape-recorded interviews were unstructured and open-ended. Additionally, I conducted 15 informal interviews at the bar. Finally, I had countless conversations with members, nonmembers, interested bystanders, and social scientists who had an interest in the punk counterculture. In all, my somewhat punk appearance, similar age, regularity at the scene, and apparent acceptance of their lifestyle allowed me to move within the scene freely and easily.

SETTING

The research took place in a small cowboy bar, "The Glass Gun," which was transformed into a punk bar one night a week. The bar was situated in a southwestern city with a population of about 500,000. The city itself is located in the "Bible Belt," characterized by conservative religious and political views. The bar was a small, dark, and dilapidated place. There was a stage area where the bands played, surrounded by a wooden rail. Wobbly tables and torn chairs formed a U-shape around the stage. There was a pool table in the corner, which the punks rarely used. Most of the patrons of the club stood at or near the bar.

For the punks in this city, the Glass Gun was the only place to congregate regularly at that time. On these designated nights, local punk bands played to an audience of about 20 people; some were punks, others were not. The typical audience ranged in age from about 16 to 30, although a few were younger or older. Basically, the punk counterculture was a youth phenomenon. It seemed to attract young, single, mobile people. Snow et al. (1980) have suggested that these characteristics make a person more "structurally available" for movement recruitment. Within the punk scene the number of men and women was fairly equal.

The punk style codes were somewhat diverse. Different styles existed for different kinds of punk. Pfohl (1985: 381) has referred to Hebdige's description of punk style as "the outrageous disfigurement of commonsensical images of aesthetics and beauty and the abrasive, destructive codes of punk style. These are aesthetic inversions of the normal, or consensus-producing, rituals of the dominant culture's style." The basic identifiable element was a subculturally accepted punk hairstyle. These ranged from a very short, uneven haircut, sticking straight up in front, to an American Indian mohawk style, to a shaven head. Along with the haircut, a punk fashion prevailed. The two were inextricably associated. The fashion ranged from torn, faded jeans, T-shirts, and army boots to expensive leather outfits.

The punk dress code was also fairly androgynous. There was no real distinction between male and female fashions. Both men and women wore faded jeans, although leather pants and miniskirts were also quite common among the women. The middle-class punk women, who tended to be students, dressed in a more traditionally feminine manner, glorifying and exaggerating the "glamour girl" image reminiscent of the sixties. This included tight skirts, teased hair, and dark, heavy makeup. The other punk women identified with a more masculine, working-class image, deemphasizing their feminine attributes. Both sexes also wore and admired leather jackets. It was also quite common to see both men and women with multiple pierced earrings all the way around the outside of their ears. Men sometimes wore eye makeup as well. (One man wore miniskirts, makeup, and rhinestones. However, this type of behavior occurred infrequently.) Basically, punk style ran counter to what the dominant culture would deem aesthetically pleasing. One major reason punks dressed as they did was to set themselves apart and to make themselves recognizable. The image consisted of dark, drab clothing, short, spiky, "homemade" haircuts, and blank, bored, expressionless faces reminiscent of those of concentration camp prisoners.[1] The punks created a new aesthetic that revealed their lack of hope, cynicism, and rejection of societal norms.

THE SOCIAL ORGANIZATION
OF THE PUNK SCENE

Like the youth gangs that Yablonsky (1962) studied, members of this local punk scene constituted a "near-group." The membership was impermanent and shifting, members' expectations were not always clearly defined, consensus within the group was problematic, and the leadership was vague. Yet out of this uncertainty surfaced an apparent consensus about the stratification of the local community and the roles of the three types of members and peripheral hangers-on who participated in this scene. These four typologies can be hierarchically arranged by the presence (or absence) and intensity of their commitment to the punk counterculture and their consequent display of the punk affectations and belief system. They thus formed a series of outwardly expanding concentric circles, with the most committed members occupying the core, inner roles, and the least involved participants falling around the periphery.

Starting from the center, the number of members occupying each stratum progressively increased as the commitment level of the participants diminished. The categories to which I refer come from the terms used by the participants themselves.[2] The *hardcore punks* were the most involved in the scene, and derived the greatest amount of prestige from their association with it. They set the trends and standards for the rest of the members. The *softcore punks* were less dedicated to the antiestablishment lifestyle and to a permanent association with this counterculture, yet their degree of involvement was still high. They were greater in number and, while highly respected by the less committed participants, did not occupy the same social status within the group as the hardcores. Their roles were, in a sense, dictated by the hardcores, whom they admired, and who defined the acceptable norms and values. The *preppie punks* were only minimally committed, constituting the largest portion of the actual membership. They were held in low esteem by the two core groups, following their lead but lacking the inner conviction and degree of participation necessary to be considered socially desirable within the scene. Finally, the *spectators* made up the largest part of the crowd at any public setting where a punk event transpired. They were not truly members of the group, and therefore did not necessarily revere the actions and dedication of the hardcores as did the two intermediary groups. They did not attempt to follow the standards of those committed to this near-group. They were merely outsiders with an interest in the punk scene.

These four groups constituted the range of participants who attended and were involved with, to varying degrees, the punk counterculture. I will now examine in greater detail their styles, beliefs, practices, intentions, and roles in the scene.

PUNKS AND COMMITMENT

At the time of this study, the group of punks was small and disorganized. The number of people at any given punk event had steadily declined since my first encounters with the scene in 1978. Punk was no longer a new phenomenon, and this particularly conservative community did not provide a very conducive atmosphere for a large countercultural group to flourish. Every member of the group expressed dissatisfaction and boredom with the events (or rather, lack of events) within the scene. Even within the limits necessitated by the relatively small size of the group, there was a great deal of variation in terms of punk roles and characteristics. The qualities attributed to the different roles were based upon commitment to the scene. The punks' perceptions of levels of commitment were based principally on their evaluation of physical appearances and lifestyles. The punks categorized members of the scene on these bases and invented terms to describe them. The four types of participants, described above, varied according to their level of commitment to the scene.

Hardcore Punks

Hardcore punks made up the smallest portion of the scene's membership. In the eyes of the other punks, though, they were the essence of the local movement. The hardcores expressed the greatest loyalty to the punk scene as a whole. Although the hardcores embodied punk fashion and lifestyle codes to the highest degree, their commitment to the counterculture went much deeper than that. As one hardcore punk said:

> There's been so much pure bullshit written about punks. Everyone is shown with a safety pin in their ear or blue hair. The public image is too locked into the fashion. That has nothin' to do with punk, really. . . . For me, it is just my way of life.

The feature that distinguished hardcore punks from other punks was their belief in, and concern for, the punk counterculture. In this sense, the hardcore punks had gone beyond commitment; they had undergone the process of conversion (Snow et al., 1986). In other words, not only did they have membership status, but they believed in and espoused the virtues and ideology of the counterculture. Although many hardcores differed on what the counterculture's core values were, they all expressed some concern with punk ideology. These values were ambiguous at best, but included a distinctly antiestablishment, anarchistic sentiment. Street (1986: 175) has described punk as celebrating chaos and "a life lived only for the moment." The associated value system of punk was understood by the incorporation of cynicism and a distrust of authority. In keeping with other subcultures that intentionally distinguish themselves from the dominant culture, the punk aesthetic, lifestyle, and worldview directly confronted those of the larger society and its traditions. While the other types of punks made no reference to group beliefs or values, the hardcores revered the counterculture. For them, being punk had a profound effect on all aspects of their lives. As one hardcore said:

> There are a lot of punks around, even real punks, who don't mean it. At least, not all the way, like I do. Sometimes I feel so good about punk that I cry. And when I see people getting into some band with real punk lyrics, it's like a religious experience.

It was precisely this belief in "punk" as an external reality, like a higher good, that set the hardcores apart. Similar to Sykes and Matza's (1957) "appeal to higher loyalties," hardcore punks based their rejection of conventional society on their commitment to their antiestablishment lifestyles and beliefs. This imbued their self-identity with a sense of seriousness and purpose. Unlike other punks, they did not view their punk identity as a temporary role or a transitory fashion, but as a permanent way of life. As one hardcore member said:

> Punk didn't influence me to be the way I am much. I was always this way inside. When I came into punk, it was what I needed all my life. I could finally be myself.

Without exception, the hardcores reported having always held the values or qualities associated with the punk counterculture. The local scene, in fact, was just a convenient way of expressing these ideas collectively. Perhaps the most essential value professed by the punks was a genuine disdain for the conventional system. Their use of the term *system* here referred to a general concept of the way the material world works: bureaucracies, power structures, and competition for scarce goods. This "system" further referred to the ethic of deferred gratification, conventional hard work for profit, and the concept of private property. While this bears some similarity to Flacks's (1967) discussion of the student movement and Davis's (1970) portrayal of hippies, hardcore punks generally had a disdain for these earlier youth subcultures. There was a general attitude among punks of the need to create and maintain their own distinctive style. Kinsey found this same feature in the antibourgeois "killum and eatum" subculture. According to Kinsey (1982: 316), "K and E offered an attractive setting as its ideology presented an excellent vehicle for expressing hostility toward conventional society." This contempt for authority and the conventional culture was, in fact, such an essential value for the punks that if one expressed prosystem sentiments or support for the present administration, one could not be considered a committed member, no matter how well one looked the part. Overt behavioral and physical attributes, though, were major ways hardcore punks showed disdain for the system. Particular characteristics were essential for consideration as hardcores. Most fundamentally, a verbal commitment to punk values and the punk scene, in general, was required. For example, John, the epitome of a hardcore punk, claimed to hate the system. He talked about the inequality of the system quite often. In John's words:

> Punk set me free. It let me out of the system. I can walk the streets now and do what I want and not live by the demands of the system. When I walk the streets, I am a punk, not a bum.

However, this verbal pronouncement had to be backed by a certain lifestyle that further indicated commitment to the group. This lifestyle consisted of escaping the system in some way. Almost all of the hardcores were unemployed and lived in old, abandoned houses or moved into the homes of friends for periods of time. Some survived from the charity of sycophantic, less committed punks. Others worked in jobs that they considered to be outside the system, such as musicians in rock bands or artists.

Another central feature of the lifestyle was the hardcores' use of dangerous drugs. Many hardcores indulged heavily in sniffing glue. Glue was inexpensive and readily available to the punks. Its use also symbolized the self-destructive, nihilistic attitude of hardcores and their desire to live outside of society's norms. As one member said:

> It is kind of like a competition, a show-off thing. . . . See who has the most guts by seeing who can burn his brain up first. It is like a total lack of care about anything, really.

This closely corresponds to Davis and Munoz's (1968) description of "freaks." Both punks and freaks were "in search of drug kicks as such, especially if [their] craving carries [them] to the point of drug abuse where [their] health, sanity and relations with intimates are jeopardized" (Davis and Munoz, 1968: 306). Again, here we see a rejection of anything the larger society sees as "sensible."

However, the most salient feature of the hardcore lifestyle was the radical physical appearance. In every case, people who were labeled as hardcore had drastically altered some aspect of their bodies. For example, in addition to the hairstyles discussed earlier, they often had tattoos, such as swastikas, on their arms or faces. Brake (1985: 78) has referred to the use of the swastika as a symbol for punks that was actually devoid of any political significance. Rather, the swastika was a "symbol of contempt" employed as a means of offending the traditional culture. The hardcore punks did their best to alienate themselves from the larger society.

According to Kanter (1972), the first requisite in the principle of a gestalt sociology is that a group forms maximum commitment to this higher ideal by sharply differentiating itself from the larger society. The hardcore members did this by going through the initiation rite of passage: semipermanently altering their appearance. As one hardcore said:

> Did you see Russell's mohawk? I'm so glad for him. He finally decided to go for it. Now he is a punk everywhere . . . no way he can hide it now.

This was similar to certain religious cults, such as the Hare Krishnas, where a drastic change in appearance was required for consideration as a total convert (Rochford, 1985). The punk counterculture informally imposed the same prerequisite. By doing something so out of the ordinary to their appearance, the punks voluntarily deprived themselves of some of the larger society's coveted goods. For example, many of the hardcores were desperately poor. They said that they knew all they would have to do to obtain a job would be to grow their hair into a conventional style; yet they refused. This kind of action based on commitment was what Becker (1960) has called "side bets," where committed people act in such ways that affect their other interests separate and apart from their commitment interests. By making specific choices, people who are committed sacrifice the possible benefits of their other roles. An important characteristic of Becker's notion of side bets is that people are fully aware of the potential ramifications of their actions. This point was illustrated by one punk:

> Some of my friends that aren't punk say, "Why don't you get a job? All you'd have to do is grow your hair out or get a wig and you could get a job." I mean, I know I could. Don't they think I knew that when I did it? It was a big step when I finally cut my hair in a mohawk.

For the hardcore punk, being punk was worth the sacrifices; it was perceived as an inherently good quality. In this respect, the hardcores differed from other types of punks. They held the larger punk scene in esteem. As one loyal member said:

> It really pisses me off when people act like ours is the only punk group in the world. They don't even care what bigger and better groups exist. If this whole thing ended tomorrow, they wouldn't care what happened to the whole punk scene.

The hardcores continually expressed their disgust with the local scene. Much like the hippies studied by Davis (1970), "the scene," in itself, was the message. While not enough people joined the group to satisfy the punks, they were, nonetheless, grateful that they had any kind of group environment to which they could attach themselves. The Glass Gun, with its regularly scheduled "punk night," nonhostile attitude toward them, and coterie of interested bystanders, at least gave them a place to express their values collectively. It was essential in maintaining the group's solidarity and social organization.

Softcore Punks

The softcore punks made up a larger portion of the local scene than the hardcores. There were around fifteen softcore punks. There was one fundamental difference between the hardcore and softcore punks. For the hardcores, it was not sufficient just to be antiestablishment or to wear one's hair in a certain way. Rather, one had to embody the punk lifestyle and ideology in all possible ways. As one hardcore punk put it:

> Everybody thinks she is hardcore because she looks so hardcore. I mean, yeah, she has a mohawk, and she won't get a job and she says she's for anarchy, but she doesn't care that much about being punk. She likes all these different kinds of music and stuff. She seems sometimes like she is just in it for fun. She even says she'll be whatever's in when punk goes out!

The softcore punks lived similar lifestyles to the hardcores. However, the element of "seriousness" about the scene, so pervasive among the hardcores, was absent among the softcores. Visually, the two types were basically indistinguishable. They were different only in their level of commitment. The commitment for softcores was to the lifestyle and the image only, not to "punk" as an ideology or an intrinsically valuable good. The softcores made no pretense of concern for either the larger counterculture or the feeling of permanence about their punk roles. As one softcore, Beth, said:

> Everyone thinks I am so serious about it because I have a mohawk. Some people just can't get past it. Sometimes I get tired of it. Other times, I like to play jokes on people; like another friend of mine who has a mohawk, we'll walk down the street and point at someone with regular hair and say, "Wow! Look at him, he's weird, he doesn't have a mohawk." The fact is, if everyone did have one, I'd do something different to my hair.

The softcores identified with the punk image only temporarily. This distinguished their level of commitment from the *conversion* of the hardcores. The softcores' interest in the scene had only to do with what it could offer them at the present time. While participating, they did what was considered a good job of being "punk." However, if a new cultural trend surfaced, it would be just as likely that they would use their energy effectively to create that particular image. As one softcore punk said:

> I've spent time identifying myself as a hippie, then as a women's libber, then an ecologist, and now as a punk. I'm punk now, but I am in the process of changing into something else. I don't know what. I'm getting bored with this scene. But for now, if I'm gonna do it, I'll do it right.

The softcore punks were somewhat committed in that they participated in some of the more drastic elements of punk lifestyle. For example, softcores had their hair cut in severe ways, just like the hardcores. They were, at least temporarily, committed to being punk (or playing punk) in that they "cut" themselves off from some of society's goods as well. However, the softcores did not share the self-destructive bent of the hardcores. The drugs that they consumed, such as marijuana, alcohol, and amphetamines, were not so potentially dangerous. Yet, because of their apparent visual commitment, and because of the lip service they gave to punk values, softcores were viewed as members in good standing. The hardcores liked and respected the softcores; the two groups associated freely. Some hardcores considered softcores to be simply members in transition. Lofland and Stark (1965) have suggested that movements themselves play a role in promoting the ideology in the new members, rather than the members coming to the movement because its ideology coincides with their own established beliefs. This was the case for the softcore punks. They did not claim to have held punk values before becoming punk. As Anne, a softcore, recalled:

> It was scary to me at first. The hype from the magazines and stuff—all this weird shit, y'know. Then I went there and just hung out. The reason it was frightening is that a lot of people had different ideas about life than me. And I had to change myself to be with them. I had to be more intense, be an outcast. It was exciting because there seemed to be an element of danger in it—like living on the edge.

A process of simply happening onto the scene was typical of softcore members. Many recounted the feelings of purposelessness that preceded the drift into the punk scene. This drift is similar to the drifts that occur in other deviant lifestyles (Matza, 1964). What Matza called the "mood of desperation" often caused people to drift into delinquency or deviant lifestyles. As Joanie said:

> I was really doing nothing with my life and I just kinda accidentally came into the punk scene. I gradually got involved in it that way. The music, and the people to an extent, really raised my consciousness about the system.

Softcores' verbal recognition of punk attitudes, such as awareness of the system, helped to validate their punk performances. Hardcores felt that verbal commitment was an essential first step to further commitment. For this reason, the hardcores accepted the softcores and considered them to be genuine and authentic in their punk identity. Such identification with punk values, along with a typical punk lifestyle, made the distinction between hardcores and softcores difficult. Again, the distinction became clear only with regard to the level of commitment, or seriousness, of the two types. One softcore made this qualification more apparent:

They get mad at me and think I'm insincere or whatever 'cause I like to
have fun. I take my politics serious, too, but I feel if you are here, you
might as well enjoy it. They think being punk is so serious, they are depressed or stoned all the time.

This statement indicates that the hardcores defined the situation for the
local scene. The hardcores decided what differentiated real punks (or committed punks) from pretenders. The hardcores considered only themselves and
the softcores to be real. The "realness" of a punk was based on the level of
commitment. The level was judged on the basis of willingness to sacrifice
other identities for the punk identity. To prove this, a member would have to
make permanent his or her punk image. What the punk identity offered was
status within its own subculture for those who could not or would not achieve
it in conventional society (Cohen, 1955). However, commitment to the deviant identity did not stem from a forced label. On the contrary, commitment
to the punk identity was a "self-enhancing attachment" (Goffman, in Stebbins, 1971). The punks' self-esteem was enhanced by the approval they received. It would follow, then, that the more consistent one's behavior was with
the superficial signs of commitment, the more prestige one would be able to
obtain. Doug, a softcore, commented on this aspect of subcultural prestige
among the punks:

> In their own way, they're elitist. It's kind of like because they're not part of
> the general run of things, because they've actually *chosen* to be rejected in
> a lot of cases, they've kind of set up their own little social order. It seems
> to me like it's based on, like a contest, who can be more cool than who.
> With the really hardcore punks, it's who can self-destruct first; in the
> name of punk, I guess.

The hardcores and the softcores used the same criteria to judge commitment. Both types agreed that the difference between them was their levels of
commitment. Both types fully realized that the softcores did not share the
same loyalty to, and identification with, the punk counterculture as a whole.
Although both types expressed some commitment to the punk identity and
lifestyle, they both realized that the hardcores viewed their own identities as
permanent and the softcores' as temporary.

Preppie Punks

The preppie punks made up an even larger portion of the crowd at punk
events. The preppies frequented the scene, but approached it similarly to a
costume party. They were concerned with the novelty and the fashion. The
preppies bore some resemblance to Yablonsky's (1968) "plastic hippies" in that
they were drawn to the excitement of the scene. Whereas the core members
acted nonchalant and natural about being punk, the preppies could not hide
their enthusiasm about being part of the scene. This feature contributed to the
core members' perceptions of preppies as "not real" punks. As one core member said:

It really kills me when these preppie girls come up to me and say "Oh, wow, you're so punk; you're so new wave," like I'm really trying or something.

Preppie punks did not lead the lifestyle of the core members. The preppie punks tended to be from middle-class families, whereas the core punks were generally from lower- or working-class backgrounds.[3] Preppie punks often lived with their parents; they tended to be younger, and were often in school or in respectable, system-sanctioned jobs. This quasi-commitment meant that preppies had to be able to turn the punk image on and off at will. For example, a preppie punk hairstyle, although short, was styled in such a versatile way that it could be manipulated to look punk sometimes and conventional at other times. Preppie fashion was much the same way. Mary, a typical preppie punk, put her regular clothes together in a way she thought would look punk. She ripped up her sorority T-shirt. She bought outfits that were advertised as having the "punk look." Her traditional bangs transformed into "punk" bangs, standing straight up using hair spray or setting gel. The distinguishing feature of preppie punks was the manufactured quality of their punk look. This obvious ability to change roles kept the preppies from being considered real or committed. The preppies were not willing to give anything up for a punk identity. As one softcore said:

> They come in with their little punk outfits from Ms. Jordan's [an exclusive clothing store] and it's written all over 'em: money. They think they can have their nice little jobs and their semipunk hairdo and live with mom and dad and be a real punk, too. Well, they can't.

Another said of preppies:

> It's a little hard to take when you have nothing and they try to have everything. Having all that goes against punk. They gotta choose to not have it. Otherwise, they're just playing a game.

The preppies liked to disavow their punk association in situations that would sanction them negatively for such associations. This state of "dual commitments," in which they never had to reject the conventional world in order to be marginally a part of the group, was characteristic of preppie punks (Cohen, 1955). Kanter (1972) has described a process of conversion and commitment that is commonly found in communes. The first step in the process was the renunciation of previous identities. According to this model, the preppie punks would not be considered committed at all. Thus they could not have been categorized as punks in any meaningful sense. As one core member said, "Being 'punk' to them is like playing cowboys and Indians."

Criticizing and joking about the preppies made up a large portion of core members' conversations. Some truly disliked the preppies and others were flattered by their feeble attempts at imitation of core behavior. For example, when a preppie punk approached one core member, he rolled his eyes and said, "Here comes my fan club," with a half-embarrassed smile and a distinct look of pleasure on his face.

Also, the financial function that the preppies served to core members made them more tolerable. Preppies almost always had jobs or survived by their parents' support. Many of the core members subsisted on the continued generosity of their devout fans. The preppies were more than willing to help the other punks. Preppies sometimes offered hardcores financial help in the form of buying them groceries, driving them places, and providing them with cigarettes, alcohol, and other drugs. Because of this, many punks felt that they could not afford to reject outwardly those who were less committed. As Anne said, "One of these days, this kindness is going to dry up."

Yet the joking and poking fun at preppies was a constant activity. It served to separate, for the committed punks, "us" from "them." It reinforced their sense of being the only real punks. Again the distinction made by core members was grounded in the preppies' attempts to play numerous roles. Haircut and clothing were the decisive clues. The real punks could spot a preppie from a distance; they never had to say a word. As one core member said,

> Oh look, she's punked out her hair. Yes, we're impressed. Tomorrow she'll look just like a Barbie doll again.

Perhaps the most definitive statement separating the real punks from the preppies referred to lifestyle:

> All I know is that I live this seven days a week, and they just do it on weekends.

The preppies, though, while definitely removed from core members, still played an important role in the scene.

Spectators

The category of spectators referred to everyone who observed the scene fairly regularly, but were not punks themselves. This type consisted of, literally, "everyone else." They made up the largest portion of the crowd at the Glass Gun on any given night. They were different from the preppie punks in that they did not try to look punk. They made no pretense of commitment to the scene at all. They did not identify themselves as punks; they had no stake in the scene. Spectators consisted of all different types of people and varied in their occupations, clothes, and reasons for being there. The only common denominator this group shared was the desire to stand back and watch, rather than to participate actively in punk activities. One spectator said of his involvement in this scene:

> People on the fringe are usually voyeurs of a sort. They like to be on the receiver's end of what's happening. Maybe punk is really their alter-ego. And maybe that need is satisfied just by watching and pretending. That's how it is with me, anyway.

The spectators liked to observe the fashion, to listen to the music, and to be "in the know" about the scene. They were, in other words, punk appreciators.

For the most part, spectators on the fringe were ignored by the core members. They never received the attention that preppies did because they made no attempts to "play punk." However, if a spectator appeared on the scene looking completely antithetical to punk, core members would simply laugh or say something derogatory about them and drop the subject. For example, one time a hippie-looking character came in and one punk said, "Oh my God, I think we're in a time warp," to which another punk responded, "Maybe we should tell him that Woodstock's over and that it is 1983."

Following such statements, the punks would watch the spectator's reaction to the scene. For the most part, except as a diversion, the punks were uninterested in the spectators. They did not generally associate with them or talk about them much. Presumably this was the case because of the tremendous turnover in spectator membership.

Most spectators either slowly began to identify with the group (most core members started out as spectators) or stopped frequenting the punk events. There were, however, some loyal spectators. They would frequent the club. They knew most of the punks at least slightly. The punks generally liked this sort of spectator because they provided the punks with an audience. Every type of punk thrived on an audience. The punks needed people to shock. The spectator served that function. The attitude that the members had toward the spectators was one of tolerance and indifference. As one core member said of them:

> They're into it for the novelty. It's like going to the circus for them, to be a part of something new and exciting. But that's okay. I like going to the circus, too; I just like being in it better.

Thus though spectators were only peripheral to the scene they provided an alternative set of norms that functioned to delineate the social boundaries of the counterculture. . . .

NOTES

1. I am indebted to David Matza and John Torpey for this analogy.

2. With the exception of the term soft-core, all of the distinctions between categories came directly from the participants. The members did make a distinction between hardcore and what I am calling soft-core punks. However, the softcores were referred to simply as "punks" by the hard-cores, in an effort to distinguish the "hard-core" quality they attributed to themselves. I chose to refrain from using the term punk to apply to one specific category so that I can use the term more freely and generally, and to avoid confusion.

3. Very little information is provided in this text about the class, race, and ethnicity of these participants. The community from which these data come is relatively homogeneous. The few references to class are more impressionistic; that is, based upon knowledge of family occupations, school districts, and so on. However, the dearth of this kind of data stems from the fact that I was more interested in the features the members had in common than in the distinctions between them, with the exception of their differing levels of commitment and their styles.

REFERENCES

Adler, P. A. (1985). *Wheeling and Dealing*. New York: Columbia Univ. Press.

Becker, H. S. (1960). "Notes on the concept of commitment." *Amer. J. of Sociology* 66: 32–40.

Berger B. (1967). "Hippie morality—more old than new." *Transaction* 5: 19–27.

Brake, M. (1980). *The Sociology of Youth Culture and Youth Subcultures*. London: Routledge.

———. (1985). *Comparative Youth Culture: The Sociology of Youth Culture and Subcultures in America, Britain, and Canada*. London: Routledge.

Cohen, A. (1955). *Delinquent Boys*. Glencoe, IL: Free Press.

Cohen, S. (1972). *Folk Devils and Moral Panics*. New York: St. Martin's.

Davis, F. (1970). "Focus on the flower children: Why all of us may be hippies some day." In J. Douglas (ed.), *Observations of Deviance* (pp. 327–340). New York: Random House.

Davis, F., and L. Munoz. (1968). "Heads and freaks: Patterns and meanings of drug use among hippies." *J. of Health and Social Behavior* 9: 156–164.

Douglas, J. (1970). *Youth in Turmoil*. Washington, DC: National Institute of Mental Health.

———. (1976). *Investigative Social Research*. Newbury Park, CA: Sage.

Flacks, R. (1967). "The liberated generation: An exploration of the roots of student protest." *J. of Social Issues* 23: 52–75.

———. (1971). *Youth and Social Change*. Chicago: Markham.

Frith, S. (1982). *Sound Effects*. New York: Pantheon.

Hall, S., and T. Jefferson (eds.). (1975). *Resistance Through Rituals*. London: Hutchinson.

Hebdige, D. (1981). *Subcultures: The Meaning of Style*. New York: Methuen.

Henslin, J. (1972). "Studying deviance in four settings: Research experiences with cabbies, suicides, drug users, and abortionees." In J. Douglas (ed.), *Research on Deviance* (pp. 35–70). New York: Random House.

Kanter, R. M. (1972). *Commitment and Community: Communes and Utopia in Sociological Perspective*. Cambridge, MA: Harvard Univ. Press.

Kinsey, B. A. (1982). "Killum and eatum: Identity consolidation in a middle class poly-drug abuse subculture." *Symbolic Interaction* 5: 311–324.

Lofland, J., and R. Stark. (1965). "Becoming a world saver: A theory of conversion to a deviant perspective." *Amer. Soc. Rev.* 30: 862–875.

Matza, D. (1964). *Delinquency and Drift*. New York: John Wiley.

Miller, W. (1958). "Lower class culture as a generating milieu of gang delinquency." *J. of Social Issues* 14: 5–19.

Mungham, G., and G. Pearson (eds.). (1976). *Working Class Youth Culture*. London: Routledge.

Pfohl, S. (1985). *Images of Deviance and Social Control*. New York: McGraw-Hill.

Rochford, E. B., Jr., (1985). *Hare Krishnas in America*. New Brunswick, NJ: Rutgers Univ. Press.

Snow, D. A., E. B. Rochford, Jr., S. K. Worden, and R. D. Benford. (1986). "Frame alignment and mobilization." *Amer. Soc. Rev.* 51: 464–481.

Snow, D. A., L. Zurcher, Jr., and S. Ekland-Olson. (1980). "Social networks and social movements: A micro-structural approach to differential recruitment." *Amer. Soc. Rev.* 45: 787–801.

Stebbins, R. A. (1971). *Commitment to Deviance*. Westport, CT: Greenwood.

Street, J. (1986). *Rebel Rock: The Politics of Popular Music*. Oxford: Basil Blackwell.

Sykes, G., and D. Matza. (1957). "Techniques of neutralization." *Amer. Soc. Rev.* 22: 664–670.

Taylor, I. (1982). "Moral enterprise, moral panic, and law-and-order campaigns." In M. M. Rosenberg et al. (eds.), *A Sociology of Deviance* (pp. 123–149). New York: St. Martin's.

Yablonsky, L. (1959). "The delinquent gang as a near-group." *Social Problems* 7: 108–117.

———. (1962). *The Violent Gang.* New York: Macmillan.

———. (1968). *The Hippie Trip.* New York: Pegasus.

Yinger, J. M. (1982). *Countercultures: The Promise and Peril of a World Turned Upside Down.* New York: Free Press.

Young, J. (1973). "The hippie solution: An essay in the politics of leisure." In I. Taylor and L. Taylor (eds.), *Politics and Deviance* (pp. 182–208). Harmondsworth, England: Penguin.

29

Women in Outlaw Motorcycle Gangs

COLUMBUS B. HOPPER
AND JOHNNY MOORE

This article is about the place of women in gangs in general and in outlaw
motorcycle gangs in particular. Street gangs have been observed in New
York dating back as early as 1825 (Asbury, 1928). The earliest gangs origi-
nated in the Five Points district of lower Manhattan and were composed mostly
of Irishmen. Even then, there is evidence that girls or young women participated
in the organizations as arms and ammunition bearers during gang fights. . . .

Studies have shown that girls have participated in street gangs as auxiliaries,
as independent groups, and as members in mixed-gender organizations. While
gangs have varied in age and ethnicity, girls have had little success in gaining
status in the gang world. As reported by Bowker (1978, Bowker and Klein,
1983), however, female street gang activities were increasing in most respects;
he thought that independent gangs and mixed groups were increasing more
than were female auxiliary units.

Unlike street gangs that go back for many years, motorcycle gangs are relatively new. They first came to public attention in 1947 when the Booze Fighters, Galloping Gooses, and other groups raided Hollister, California (Morgan,
1978). This incident, often mistakenly attributed to the Hell's Angels, made
headlines across the country and established the motorcycle gangs' image. It
also inspired *The Wild Ones,* the first of the biker movies released in 1953,
starring Marlon Brando and Lee Marvin.

Everything written on outlaw motorcycle gangs has focused on the men in
the groups. Many of the major accounts (Eisen, 1970; Harris, 1985; Montegomery, 1976; Reynolds, 1967; Saxon, 1972; Thompson, 1967; Watson, 1980;
Wilde, 1977; Willis, 1978; Wolfe, 1968) included a few tantalizing tidbits of
information about women in biker culture but in none were there more than a
few paragraphs, which underscored the masculine style of motorcycle gangs
and their chauvinistic attitudes toward women.

Although the published works on outlaw cyclists revealed the fact that
gang members enjoyed active sex lives and had wild parties with women,
the women have been faceless; they have not been given specific attention as

functional participants in outlaw culture. Indeed, the studies have been so one-sided that it has been difficult to think of biker organizations in anything other than a masculine light. We have learned that the men were accompanied by women but we have not been told anything about the women's backgrounds, their motivations for getting into the groups or their interpretations of their experiences as biker women.

From the standpoint of the extant literature, biker women have simply existed; they have not had personalities or voices. They have been described only in the contemptuous terms of male bikers as "cunts," "sluts," "whores," and "bitches." Readers have been given the impression that women were necessary nuisances for outlaw motorcyclists. A biker Watson (1980: 118) quoted, for example, summed up his attitude toward women as follows: "Hell," he said, "if I could find a man with a pussy, I wouldn't fuck with women. I don't like 'em. They're nothing but trouble."

In this article, we do four things. First, we provide more details on the place of women in arcane biker subculture, we describe the rituals they engage in, and we illustrate their roles as money-makers. Second, we give examples of the motivations and backgrounds of women affiliated with outlaws. Third, . . . we show how the place of biker women has changed over the years of our study and we suggest a reason for the change. We conclude by noting the impact of sex role socialization on biker women.

METHODS

The data we present were gathered through participant observation and interviews with outlaw bikers and their female associates over the course of 17 years. Although most of the research was done in Mississippi, Tennessee, Louisiana, and Arkansas, we have occasionally interviewed bikers throughout the nation, including Hawaii.[1] The trends and patterns we present, however, came from our study in the four states listed.

During the course of our research, we have attended biker parties, weddings, funerals, and other functions in which outlaw clubs were involved. In addition, we have visited in gang clubhouses, gone on "runs" and enjoyed cookouts with several outlaw organizations.

It is difficult to enumerate the total amount of time or the number of respondents we have studied because of the necessity of informal research procedures. Bikers would not fill out questionnaires or allow ordinary research methods such as tape recorders or note taking. The total number of outlaw motorcyclists we studied over the years was certainly several hundred. In addition to motorcycle gangs in open society, we also interviewed and corresponded with male and female bikers in state and federal prisons.

The main reason we were able to make contacts with bikers was the background of Johnny Moore, who was once a biker himself. During the 1960s, "Big John" was president of Satan's Dead, an outlaw club on the Mississippi Gulf Coast. He participated in the rituals we describe, and his own experience and observations provided the details of initiation ceremonies that we relate. As a former club president, Moore was able to get permission for us to visit biker clubhouses, a rare privilege for outsiders.[2]

Most of our research was done on weekends because of our work schedules and because the gangs were more active at this time. The bikers usually had a large party one weekend a month, or more often when the weather was nice, and we were invited to many of these.

At some parties, such as the "Big Blowout" each spring in Gulfport, there were a variety of nonmembers present to observe the motorcycle shows and "old lady" contests as well as to enjoy the party atmosphere. These occasions were especially helpful in our study because bikers were "loose" and easier to approach while partying. We spent more time with three particular "clubs," as outlaw gangs refer to themselves, because of their proximity.

In addition to studying outlaw bikers themselves, we obtained police reports, copies of Congressional hearings that deal with motorcycle gangs, and indictments that were brought against prominent outlaw cyclists. Our attempt was to study biker women and men in as many ways as possible. We were honest in explaining the purpose of our research to our respondents. They were told that our goal was only to learn more about outlaw motorcycle clubs as social organizations. . . .

Problems in Studying Biker Women

Although it was difficult to do research on outlaw motorcycle gangs generally, it was even harder to study the women in them. In many gangs, the women were reluctant to speak to outsiders when the men were present. We did not hear male bikers tell the women to refrain from talking to us. Rather, we often had a man point to a woman and say, "Ask her," when we posed a question that concerned female associates. Usually, the woman's answer was, "I don't know." Consequently, it took longer to establish rapport with female bikers than it did with the men.

Surprisingly, male bikers did not object to our being alone with the women. Occasionally, we talked to a female biker by ourselves and this is when we were able to get most of our information and quotations from them. In one interview with a biker and his woman in their home, the woman would not express an opinion about anything. When her man left to help a fellow biker whose motorcycle had broken down on the road, the woman turned into an articulate and intelligent individual. Upon the return of the man, however, she resumed the role of a person without opinions.

THE PLACE OF WOMEN IN
OUTLAW MOTORCYCLE GANGS

Although national,[3] outlaw motorcycle clubs of the 1980s had restricted their membership to adult males (Quinn, 1983), women were important in the outlaw life-style we observed. We rarely saw a gang without female associates sporting colors similar to those the men wore.

To the casual observer, all motorcycle gang women might have appeared the same. There were, however, two important categories of women in the biker world: "mamas" and "old ladies." A mama belonged to the entire gang. She had to be available for sex with any member and she was subject to the authority of any brother. Mamas wore jackets that showed they were the "property" of the club as a whole.

An old lady belonged to an individual man; the jacket she wore indicated whose woman she was. Her colors said, for example, "Property of Frog." Such a woman was commonly referred to as a "patched old lady." In general terms, old ladies were regarded as wives. Some were in fact married to the members whose patches they wore. In most instances, a male biker and his old lady were married only in the eyes of the club. Consequently, a man could terminate his relationship with an old lady at any time he chose, and some men had more than one old lady.

A man could require his old lady to prostitute herself for him. He could also order her to have sex with anyone he designated. Under no circumstances, however, could an old lady have sex with anyone else unless she had her old man's permission.

If he wished to, a biker could sell his old lady to the highest bidder, and we saw this happen. When a woman was auctioned off, it was usually because a biker needed money in a hurry, such as when he wanted a part for his motorcycle or because his old lady had disappointed him. The buyer in such transactions was usually another outlaw.

RITUALS INVOLVING WOMEN

Outlaw motorcycle gangs, as we perceived them, formed a subculture that involved rituals and symbols. Although each group varied in its specific ceremonies, all of the clubs we studied had several. There were rites among bikers that had nothing to do with women and sex but a surprising number involved both.

The first ritual many outlaws were exposed to, and one they understandably never forgot, was the initiation into a club. Along with other requirements, in some gangs, the initiate had to bring a "sheep" when he was presented for membership. A sheep was a woman who had sex with each member of the gang during an initiation. In effect, the sheep was the new man's gift to the old members.

Group sex, known as "pulling a train," also occurred at other times. Although some mamas or other biker groupies (sometimes called "sweetbutts") occasionally volunteered to pull a train, most instances of train pulling were punitive in nature. Typically, women were being penalized for some breach of biker conduct when they pulled a train.

An old lady could be forced to pull a train if she did not do something her old man told her to do, or if she embarrassed him by talking back to him in front of another member. We never observed anyone pulling a train but we were shown clubhouse rooms that were designated "train rooms." And two women told us they had been punished in this manner.

One of the old ladies who admitted having pulled a train said her offense was failing to keep her man's motorcycle clean. The other had not noticed that her biker was holding an empty bottle at a party. (A good old lady watched her man as he drank beer and got him another one when he needed it without having to be told to do so.) We learned that trains were pulled in vaginal, oral, or anal sex. The last was considered to be the harshest punishment.

Another biker ritual involving women was the earning of "wings," a patch similar to the emblem a pilot wears. There were different types of wings that showed that the wearer had performed oral sex on a woman in front of his club. Although the practice did not exist widely, several members of some groups we studied wore wings.

A biker's wings demonstrated unlimited commitment to his club. One man told us he earned his wings by having oral sex with a woman immediately after she had pulled a train; he indicated that the brothers were impressed with his abandon and indifference to hygiene. Bikers honored a member who laughed at danger by doing shocking things.[4]

The sex rituals were important in many biker groups because they served at least one function other than status striving among members. The acts ensured that it was difficult for law enforcement officials, male or female, to infiltrate a gang.

BIKER WOMEN AS MONEY-MAKERS

Among most of the groups we studied, biker women were expected to be engaged in economic pursuits for their individual men and sometimes for the entire club. Many of the old ladies and mamas were employed in nightclubs as topless and nude dancers. Although we were not able to get exact figures on the proportion of "table dancers" who were biker women, in two or three cities almost all of them were working for outlaw clubs.

A lot of the dancers were proud of their bodies and their dancing abilities. We saw them perform their routines in bars and at parties. At the "Big Blowout" in Gulfport, which is held in an open field outside of the city, in 1987 and 1988 there was a stage with a sound system set up for the dancers. The great majority of the 2,000 people in attendance were bikers from around the country so the performances were free.

Motorcycle women who danced in the nightclubs we observed remained under the close scrutiny of the biker men. The men watched over them for two reasons. First, they wanted to make sure that the women were not keeping money on the side; second, the cyclists did not want their women to be exploited by the bar owners. Some bikers in one gang we knew beat up a nightclub owner because they thought he was "ripping off" the dancers. The man was beaten so severely with axe handles that he had to be hospitalized for several months.

While some of the biker women limited their nightclub activities to danc-ing, a number of them also let the customers whose tables they danced on know they were available for "personal" sessions in a private place. As long as they were making good money regularly, the bikers let the old ladies choose their own level of nightclub participation. Thus some women danced nude only on stage; others performed on stage and did table dances as well. A smaller number did both types of dances and also served as prostitutes.

Not all of the money-making biker women we encountered were em-ployed in such "sleazy" occupations. A few had "square" jobs as secretaries, factory workers, and sales persons. One biker woman had a job in a bank. A friend and fellow biker lady described her as follows: "Karen is a chameleon. When she goes to work, she is a fashion plate; when she is at home, she looks like a whore. She is every man's dream!" Like the others employed in less pres-tigious labor, however, Karen turned her salary over to her old man on payday.

A few individuals toiled only intermittently when their bikers wanted a new motorcycle or something else that required more money than they usu-ally needed. The majority of motorcycle women we studied, however, were regularly engaged in work of some sort.

MOTIVATIONS AND
BACKGROUNDS OF BIKER WOMEN

In view of the ill treatment the women received from outlaws, it was surpris-ing that so many women wanted to be with them. Bikers told us there was never a shortage of women who wanted to join them and we observed this to be true. Although it was unwise for men to draw conclusions about the rea-sons mamas and old ladies chose their life-styles, we surmised three interre-lated factors from conversations with them.

First, some women, like the male bikers, truly loved and were excited by motorcycles. Cathy was an old lady who exhibited this trait. "Motorcycles have always turned me on," she said. "There's nothing like feeling the wind on your titties. Nothing's as exciting as riding a motorcycle. You feel as free as the wind."

Cathy did not love motorcycles indiscriminately, however. She was im-bued with the outlaw's love for the Harley Davidson. "If you don't ride a Hog," she stated, "you don't ride nothing. I wouldn't be seen dead on a rice burner" (Japanese model). Actually, she loved only a customized bike or "chopper." Anything else she called a "garbage wagon."

When we asked her why she wanted to be part of a gang if she simply loved motorcycles, Cathy answered:

> There's always someone there. You don't agree with society so you find
> someone you like who agrees with you. The true meaning for me is to
> express my individuality as part of a group.

Cathy started "putting" (riding a motorcycle) when she was 15 years old and she dropped out of school shortly thereafter. Even with a limited education, she gave the impression that she was a person who thought seriously. She had a butterfly tattoo that she said was an emblem of the freedom she felt on a bike. When we talked to her, she was 26 and had a daughter. She had ridden with several gangs but she was proud that she had always been an old lady rather than a mama.

The love for motorcycles had not dimmed for Cathy over the years. She still found excitement in riding and even in polishing a chopper. "I don't feel like I'm being used. I'm having fun," she insisted. She told us that she would like to change some things if she had her life to live over, but not biking. "I feel sorry for other people; I'm doing exactly what I want to do," she concluded.

A mama named Pamela said motorcycles thrilled her more than anything else she had encountered in life. Although she had been involved with four biker clubs in different sections of the country, she was originally from Mississippi and she was with a Mississippi gang when we talked to her. Pamela said she graduated from high school only because the teachers wanted to get rid of her. "I tried not to give any trouble, but my mind just wasn't on school."

She was 24 when we saw her. Her family background was a lot like most of the women we knew. "I got beat a lot," she remarked. "My daddy and my mom both drank and ran around on each other. They split up for good my last year in school. I ain't seen either of them for a long time."

Cathy described her feelings about motorcycles as follows:

> I can't remember when I first saw one. It seems like I dreamed about them even when I was a kid. It's hard to describe why I like bikes. But I know this for sure. The sound a motorcycle makes is really exciting—it turns me on, no joke. I mean really! I feel great when I'm on one. There's no past, no future, no trouble. I wish I could ride one and never get off.

The second thing we thought drew women to motorcycle gangs was a preference for macho men. "All real men ride Harleys," a mama explained to us. Generally, biker women had contempt for men who wore suits and ties. We believed it was the disarming boldness of bikers that attracted many women.

Barbara, who was a biker woman for several years, was employed as a secretary in a university when we talked to her in 1988. Although Barbara gradually withdrew from biker life because she had a daughter she wanted reared in a more conventional way, she thought the university men she associated with were wimps. She said:

> Compared to bikers, the guys around here (her university) have no balls at all. They hem and haw, they whine and complain. They try to impress you with their intelligence and sensitivity. They are game players. Bikers come at you head on. If they want to fuck you, they just say so. They don't care what you think of them. I'm attracted to strong men who know what they want. Bikers are authentic. With them, what you see is what you get.

Barbara was an unusual biker lady who came from an affluent family. She was the daughter of a highly successful man who owned a manufacturing and distributing company. Barbara was 39 when we interviewed her. She had gotten into a motorcycle gang at the age of 23. She described her early years to us:

I was rebellious as long as I can remember. It's not that I hated my folks. Maybe it was the times (1960s) or something. But I just never could be the way I was expected to be. I dated "greasers," I made bad grades; I never applied myself. I've always liked my men rough. I don't mean I like to be beat up, but a real man. Bikers are like cowboys; I classify them together. Freedom and strength I guess are what it takes for me.

Barbara did not have anything bad to say about bikers. She still kept in touch with a few of her friends in her old club. "It was like a family to me," she said. "You could always depend on somebody if anything happened. I still trust bikers more than any other people I know." She also had become somewhat reconciled with her parents, largely because of her daughter. "I don't want anything my parents have personally, but my daughter is another person. I don't want to make her be just like me if she doesn't want to," she concluded.

A third factor that we thought made women associate with biker gangs was low self-esteem. Many we studied believed they deserved to be treated as people of little worth. Their family backgrounds had prepared them for subservience.

Jeanette, an Arkansas biker woman, related her experience as follows:

My mother spanked me frequently. My father beat me. There was no sexual abuse but a lot of violence. My parents were both alcoholics. They really hated me. I never got a kind word from either of them. They told me a thousand times I was nothing but a pain in the ass.

Jeanette began hanging out with bikers when she left home at the age of 15. She was 25 when we talked to her in 1985. Although he was dominating and abusive, her old man represented security and stability for Jeanette. She said he had broken her jaw with a punch. "He straightened me out that time," she said. "I started to talk back to him but I didn't get three words out of my mouth." Her old man's name was tattooed over her heart.

In Jeanette's opinion, she had a duty to obey and honor her man. They had been married by another biker who was a Universal Life minister. "The Bible tells me to be obedient to my husband," she seriously remarked to us. Jeanette also told us she hated lesbians. "I go in lesbian bars and kick ass," she said. She admitted she had performed lesbian acts but she said she did so only when her old man made her do them. The time her man broke her jaw was when she objected to being ordered to sleep with a woman who was dirty. Jeanette believed her biker had really grown to love her. "I can express my opinion once and then he decides what I am going to do," she concluded.

In the opinions of the women we talked to, a strong man kept a woman in line. Most old ladies had the lowly task of cleaning and polishing a motorcycle every day. They did so without thanks and they did not expect or want any praise. To them, consideration for others was a sign of weakness in a man. They wanted a man to let them know who was boss. . . .

THE CHANGING ROLE OF
BIKER WOMEN

During the 17 years of our study, we noticed a change in the position of women in motorcycle gangs. In the groups we observed in the 1960s, the female participants were more spontaneous in their sexual encounters and they interacted more completely in club activities of all kinds. To be sure, female associates of outlaw motorcycle gangs have never been on a par with the men. Biker women have worn "property" jackets for a long time, but in the outlaw scene of 1989, the label had almost literally become fact.

Bikers have traditionally been notoriously active sexually with the women in the clubs. When we began hanging out with bikers, however, the men and the women were more nearly equal in their search for gratification. Sex was initiated as much by the women as it was by the men. By the end of our study, the men had taken total control of sexual behavior, as far as we could observe, at parties and outings. As the male bikers gained control of sex, it became more ceremonial.

While the biker men we studied in the late 1980s did not have much understanding of sex rituals, their erotic activities seemed to be a means to an end rather than an end in themselves, as they were in the early years of our study. That is to say, biker sex became more concerned with achieving status and brotherhood than with "fun" and physical gratification. We used to hear biker women telling jokes about sex but even this had stopped.

The shift in the position of biker women was not only due to the increasing ritualism in sex; it was also a consequence of the changes in the organizational goals of motorcycle gangs as evidenced by their evolving activities. As we have noted, many motorcycle gangs developed an interest in money; in doing so, they became complex organizations with both legal and illegal sources of income (McGuire, 1986).

When bikers became more involved in illegal behavior, they followed the principles of sex segregation and sex typing in the underworld generally. The low place of women has been well documented in the studies of criminal organizations (Steffensmeier, 1983). The bikers did not have much choice in the matter. When they got involved in financial dealings with other groups in the rackets, motorcycle gangs had to adopt a code that had prevailed for many years; they had to keep women out of "the business."

Early motorcycle gangs were organized for excitement and adventure; money-making was not important. Their illegal experiences were limited to individual members rather than to the gang as a whole. In the original gangs, most male participants had regular jobs, and the gang was a part-time organization that met about once a week. At the weekly gatherings, the emphasis was on swilling beer, soaking each other in suds, and having sex with the willing female associates who were enthusiastic revelers themselves. The only money the old bikers wanted was just enough to keep the beer flowing. They did not regard biker women as sources of income; they thought of them simply as fellow hedonists.

Most of the gangs we studied in the 1980s required practically all of the members' time. They were led by intelligent presidents who had organizational ability. One gang president had been a military officer for several years. He worked out in a gym regularly and did not smoke or drink excessively. In his presence, we got the impression that he was in control, that he led a disciplined life. In contrast, when we began our study, the bikers, including the leaders, always seemed on the verge of personal disaster.

A few motorcycle gangs we encountered were prosperous. They owned land and businesses that had to be managed. In the biker transition from hedonistic to economic interests, women became defined as money-makers rather than companions. Whereas bikers used to like for their women to be tattooed, many we met in 1988 and 1989 did not want their old ladies to have tattoos because they reduced their market value as nude dancers and prostitutes. We also heard a lot of talk about biker women not being allowed to use drugs for the same reason. Even for the men, some said drug usage was not good because a person hooked on drugs would be loyal to the drug, not to the gang.

When we asked bikers if women had lost status in the clubs over the years, their answers were usually negative. "How can you lose something you never had?" a Florida biker replied when we queried him. The fact is, however, that most bikers in 1989 did not know much about the gangs of 20 years earlier. Furthermore, the change was not so much in treatment as it was in power. It was a sociological change rather than a physical one. In some respects, women were treated better physically after the transition than they were in the old days. The new breed did not want to damage the "merchandise."

An old lady's status in a gang of the 1960s was an individual thing, depending on her relationship with her man. If her old man wanted to, he could share his position to a limited extent with his woman. Thus the place of women within a gang was variable. While all women were considered inferior to all men, individual females often gained access to some power, or at least they knew details of what was happening.

By 1989, the position of women had solidified. A woman's position was no longer influenced by idiosyncratic factors. Women had been formally defined as inferior. In many biker club weddings, for example, the following became part of the ceremony:

> You are an inferior woman being married to a superior man. Neither you nor any of your female children can ever hold membership in this club or own any of its property.

Although the bikers would not admit that their attitudes toward women had shifted over the years, we noticed the change. Biker women were completely dominated and controlled as our study moved into the late 1980s. When we were talking to a biker after a club funeral in North Carolina in 1988, he turned to his woman and said, "Bitch, if you don't take my dick out, I'm going to piss in my pants." Without hesitation, the woman unzipped his trousers and helped him relieve himself. To us, this symbolized the lowly place of women in the modern motorcycle gang.

CONCLUSION

Biker women seemed to represent another version of what Romenesko and Miller (1989) have referred to as a "double jeopardy" among female street hustlers. Like the street prostitutes, most biker women came from backgrounds in which they had limited opportunities in the licit or conventional world, and they faced even more exploitation and subjugation in the illicit or deviant settings they had entered in search of freedom.

It is ironic that biker women considered themselves free while they were under the domination of biker men. They had the illusion of freedom because they lived with men who were bold and unrestrained. Unlike truly liberated women, however, the old ladies and mamas did not compete with men; instead, they emulated and glorified male bikers. Biker women thus illustrated the pervasive power of socialization and the difficulty of changing deeply ingrained views of the relations between the sexes inculcated in their family life. They believed that they should be submissive to men because they were taught that males were dominant. While they adamantly stated that they were living the life they chose, it was evident that their choices were guided by values that they had acquired in childhood. Although they had rebelled against the strictures of straight society, their orientation in gender roles made them align with outlaw bikers, the epitome of macho men.

NOTES

1. We briefly observed the Alii ("Chiefs"), a native Hawaiian gang, located on the island of Hawaii, the "Big Island." Motorcycle gangs on the island of Oahu may have developed before the California clubs. Lord (1978) described a club that volunteered its services during the attack on Pearl Harbor in 1941. The bikers, wearing their colors, carried messengers and officers on motorcycles from one place to another because automobiles could not move efficiently during the traffic jams caused by the battle. This display of patriotism was not the only instance. Sonny Barger, the president of the Oakland chapter, sent President Lyndon Johnson a letter offering to send the Hell's Angels to Vietnam. Some outlaws were active in "Toys for Tots" drives and blood drives. These activities were seldom noted; that led bikers to the slogan "When we do right, nobody remembers; when we do wrong, nobody forgets."

2. Bikers presented "courtesy cards" to people that they believed deserved the privilege of biker acceptance. "Big John," as Moore was called by outlaws, had cards from outlaw motorcycle clubs throughout the country.

3. A national club had chapters in different regions of the country. For more details on gang vocabulary and argot and the distribution of specific gangs, see Hopper and Moore (1983, 1984).

4. Although the initiation ceremony was the culmination of a biker's efforts to become a member of a club, it usually required a period of a year or more before a man would be made a member in full standing. During this time, a person was a "probate." He rode with the gang, and to the public he appeared to be a regular member, but a probate was not trusted until he proved himself worthy of complete membership. He did this by showing his courage and disregard for danger.

REFERENCES

Asbury, H. (1928). *The Gangs of New York*. New York: Alfred A. Knopf.

Bowker, L. (1978). *Women, Crime, and the Criminal Justice System*. Lexington, MA: D. C. Heath.

Bowker, L., and M. Klein. (1983). "The etiology of female juvenile delinquency and gang membership: A test of psychological and social structural explanations." *Adolescence* 8: 731–751.

Eisen, J. (1970). *Altamont*. New York: Avon Books.

Harris, M. (1985). *Bikers*. London: Faber & Faber.

Hopper, C., and J. Moore. (1983). "Hell on wheels: The outlaw motorcycle gangs." *J. of Amer. Culture* 6: 58–64.

———. (1984). "Gang slang." *Harpers* 261: 34.

Lord, W. (1978). *Day of Infamy*. New York: Bantam.

McGuire, P. (1986). "Outlaw motorcycle gangs: Organized crime on wheels." *National Sheriff* 38: 68–75.

Montegomery, R. (1976). "The outlaw motorcycle subculture." *Canadian J. of Criminology and Corrections* 18: 332–342.

Morgan, R. (1978). *The Angels Do Not Forget*. San Diego: Law and Justice.

Quinn, J. (1983). Outlaw Motorcycle Clubs: A Sociological Analysis. M.A. thesis: University of Miami.

Reynolds, F. (1967). *Freewheeling Frank*. New York: Grove Press.

Rice, R. (1963). "A reporter at large: The Persian queens." *New Yorker* 39: 153.

Romenesko, K., and E. Miller. (1989). "The second step in double jeopardy: appropriating the labor of female street hustlers." *Crime and Delinquency* 35: 109–135.

Saxon, K. (1972). *Wheels of Rage*. (privately published).

Steffensmeier, D. (1983). "Organization properties and sex-segregation in the underworld: Building a sociology theory of sex differences in crime." *Social Forces* 61: 1010–1032.

Thompson, H. (1967). *Hell's Angels*. New York: Random House.

Watson, J. (1980). "Outlaw motorcyclists as an outgrowth of lower class values." *Deviant Behavior* 4: 31–48.

Wilde, S. (1977). *Barbarians on Wheels*. Secaucus, NJ: Chartwell Books.

Willis, P. (1978). *Profane Culture*. London: Routledge & Kegan Paul.

Wolfe, T. (1968). *The Electric Kool-Aid Acid Test*. New York: Farrar, Straus & Giroux.

30

International Organized Crime

ROY GODSON
AND WILLIAM J. OLSON

O n August 18, 1989, in the middle of an adoring crowd, a gunman murdered Luis Carlos Galan, the leading candidate for the presidency of
Colombia. The act shocked the nation. Political murder was not a
new phenomenon in Columbia. What was new were the perpetrators. The
assassin was not an individual acting alone, nor was he a member of a guerrilla
group, of which Colombia has several violent examples. Instead, the assassin
was a hireling, a *sicario,* acting on the orders of the Medellín Cartel, one of the
world's major international criminal organizations. The murder was only
the beginning. The Medellín Cartel launched a full-scale terrorist assault on
the country. Public facilities and newspaper offices were bombed. Members of
leading families were kidnapped. Hundreds of policemen were murdered. A
national airline flight was blown up in mid-flight. The cartel waged war against
the Colombian state. Its aim was to force the government to come to terms
with the cartels, in effect, to share power with the drug traffickers.

The Colombian government, which had been trying to control not only
Medellín groups but also the less prominent Cali group, now faced a far more
violent and direct threat to its institutions. It called for help. The United States,
which was already trying to support Colombian efforts, came forward with an
emergency aid package and promises of more aid to follow. The United States
provided upwards of $400 million in police, military, and advisory assistance
over five years. These funds were intended to eliminate the major drug trafficking organizations in Colombia, and many millions more were spent to
help attack the overseas operations responsible for producing and transship-
ping almost all the cocaine in the world.

The level of support and cooperation was unprecedented. And the threat?
A drug-trafficking organization with the wealth and power to challenge the
internal stability of one country while it defied the power and authority of the
world's remaining superpower. Hubris? Perhaps, but the nature of the confrontation and the fact that it is not over says something fundamental about
the modern world, about the nature of state power, international relations,
and the stability of governments. And the situation in Colombia is far from
the whole picture.

From "International Organized Crime," Roy Godson and William J. Olson, *Society,*
Jan./Feb. 1995, pp. 18–29. Copyright © 1995 Transaction Publishers. Reprinted by permission of Transaction Publishers; all rights reserved.

In 1991 and 1992 the disruptive influence of organized criminal activity rocked the foundations of political stability in Italy. Since World War II organized crime has carried out a wide range of domestic activities, using violence, extortion, bribery, and murder to advance its interests. More recently the Sicilian Mafia specifically has been locked in a murderous struggle with the government, assassinating judges, policemen, and those seen as interfering in its operations, and, more ominously, using its economic power to try to corrupt the political process itself. The scandals surrounding official corruption linked to the Sicilian, Neapolitan, and Calabrian Mafias have touched the very heart of the Italian government, undermining its credibility and effectiveness.

In Burma, large parts of the country are under the control of separatist movements who finance their activities by selling opium on the international market. The government itself engages in or countenances the production and trafficking of opium to help finance its operations. All over Asia, organized criminal groups, Pakistani, Thai, Chinese, or Japanese, operate vast international organizations trafficking in drugs or engaging in a wide variety of other criminal activities, in many cases with the complicity of local government and military officials.

In the former Soviet Union and in the struggling states of Eastern Europe, criminal organizations, long held in check, are beginning to grow. They have developed international links to improve their own organizational abilities and marketing contacts. Of more concern, these criminal groups are penetrating local governments (which are often struggling for cohesion and lacking resources) by using bribery and violence to win protection for their expanding operations. Governmental resources, strained to cope with a wide range of social and economic problems, are completely inadequate to respond; in fact, governments are unable to assess the extent of the problem accurately. Political paralysis and economic hardship have combined to give various criminal organizations considerable freedom to operate, even when local governments are not cooperating with criminal elements. Governments, however, are not the only institutions vulnerable to criminal penetration.

In a New York courtroom in 1992, Clark Clifford, an adviser to presidents and one of the most respected men in the United States, was called as a defendant in a case involving a vast illegal international financial enterprise, the Bank of Credit and Commerce International (BCCI). So far, BCCI is the biggest such case, but it illustrates only too graphically the extent to which banking and financial systems are vulnerable to penetration, manipulation, and fraud by criminal groups. The mechanisms whereby incredible sums of illegal proceeds—perhaps $300 billion in drug money alone—are laundered and massive frauds are perpetrated through the world's financial markets are still only dimly understood, but the realities of the process underscore the permeability of the system. This permeability of governments and private business and the growth of major international organized crime raise concerns for the future.

Whether in the developed or in the developing world, criminal organizations' scope of action and range of capabilities are undergoing a profound change. Decline in political order, deteriorating economic circumstances, a growing underground economy that habituates people to working outside the legal framework, easy access to arms, the massive flow of emigrants and refugees, and the normal difficulties involved in engendering meaningful state-to-state cooperation are working to the advantage of criminal organizations. The rise of better-organized, internationally based criminal groups with vast financial resources is creating a new threat to the stability and security of the international system. As Senator John Kerry noted, "this is new. This is something that none of us has ever experienced before. It is not ideological. It has nothing to do with right or left, but it is money-oriented, greed-based criminal enterprise that has decided to take on the lawful institutions and civilized society." The growth of these organizations presents a major challenge to the quality of life in the United States and to U.S. interests.

ORGANIZED CRIME AND BUSINESS

Although major criminal organizations pose a new and compelling challenge to national and international interests, the extent of the threat should not be exaggerated. It is clear that the wealth and power of individual organizations has grown and there are increasing signs of international links between various criminal organizations. This does not mean, however, that there is an integrated, centrally directed criminal conspiracy. The first business of criminal organizations is usually business, its promotion and protection. In this sense criminal organizations are similar to legitimate enterprises. Like the activities of their legal twins, the activities of separate "corporations" can be cooperative or competitive by turns. The long-term threat from these organizations is subtle and more insidious than images of criminal masterminds seeking to dominate the globe in some vast, shared, and centrally coordinated enterprise. It is important and difficult to measure and to understand the true nature and composition of the threat.

Criminal organizations, of course, are not new. Oliver Twist's Fagin is only one of the many memorials to the possibilities of an organized criminal underground. Nor is there anything particularly new about the ethnic composition of such organizations; the Sicilian Mafia and the Chinese Triads, in particular, are of venerable lineage, and many have had international dimensions. Yet there is something fundamentally different in the threat that such organizations now pose to organized society that must be understood so that the United States and its friends and allies around the world can undertake a reasonable, timely, and effective response. First, however, it is useful to define the nature and contours of the emerging problem.

It is essential to come to terms with what is meant by criminal organization. What distinguishes it from regular criminal activity, from a legitimate business enterprise, or from economic activity in underground or informal economies?

In many parts of the world, there are forms of entrepreneurial activity that are classified as illegitimate if they do not meet the state's test for permits, licenses, and so on. These informal economic activities now account for a significant proportion of all meaningful economic activity in many parts of the Third World. In Peru, for example, upwards of 50 percent of the economically active population are engaged in the informal economy, generating as much as 40 percent of the gross domestic product (GDP), all of it illegal under existing law. (This does not include illegal coca cultivation and cocaine processing.) There is considerable organization in this effort. This does not mean, however, that this activity is indicative of organized crime, although there are some important parallels. It is important to keep this distinction in mind. It was not that long ago, for instance, that all nonstate economic activity was illegal in the former Soviet Union. As the economy of Russia struggles to make the transition from statist control to free market and many of the people formerly engaged in illegal activities begin to emerge into the new economy, it will be important to understand and to be able to draw the distinction between true criminals and those who were criminals by definition. What is involved? There are a variety of definitions and definitional approaches, but several elements are essential.

First, the activities involved must be criminal. They must violate laws for which there is a punishment prescribed by a legal authority capable of enforcing the laws. This raises a problem, as the trouble with informal economies indicates. Clearly, many societies, including our own, make the informal economy illegal and provide punishment. By definition, then, these activities are criminal, and when they become a major component of the GDP outside the formal economy and the tax base these activities can adversely affect the growth of the formal economy. However, the informal economy can also make a major contribution to the quality of life, as in the Soviet case, where the average citizen would have been far worse off far sooner if it had not been for private entrepreneurs providing a wide range of goods and services. The dividing line is blurred. Nevertheless, most societies, past and present, have defined a number of behaviors—such as murder, drug trafficking, prostitution, extortion, kidnapping, and theft—as not only outside the law but fundamentally wrong. By contrast, the informal economy essentially encompasses "normal" economic activity that is legitimate in the formal sector. But the primary concern here is not just with crime, per se, but with major crime, and there are additional features to help sharpen the picture.

A second central characteristic of the crime, which does present a significant threat to established societies, is that the criminals engaged in particular activities must be organized. While trite, it is important to understand that the individuals involved are not acting alone, and the activities in which they are engaged are not random. Further, the economic behavior in mature criminal organizations, unlike that of groups in an illegal, informal economy in which groups may loosely cooperate, is intentional. The activities are usually directed by identifiable leaders. Mature criminal organizations operate with varying organizational structures for a common purpose, which is outside the law.

Organization is key, yet there is no standardized organizational chart for a developed criminal organization. There are varying types of organizational patterns. Like a legitimate business enterprise, a criminal organization may employ various features best designed to carry out its purposes. Thus it can be vertically organized and fairly tightly controlled, as are the Colombian cocaine cartels; or it may be regionally organized, often around functions, as is the American Mafia; or it may be even more loosely organized, as are the Jamaican Yardies in England. It may have a quasi-religious character, as do the Chinese Triads, or a semi-political/military organization, as does the Shan United Army in Burma. The organizational nature, then, while a key component, is not the sole defining feature.

The purposes to which the organization is dedicated ultimately define its legality or criminality. The activities of criminal organizations are equivalent to many of the efforts of legitimate business: export-import, trade in various articles, wholesale and retail sales, services. The members of criminal organizations, however, seek to operate in areas outside legal guidelines and generally trade in items also defined as illegal—such as drugs. However, as in the case of the Chinese Triads, they may use their organization for both legal and illegal activities.

Interestingly, there are indications that many of the present major criminal organizations began not as deliberately criminal groups but as protective-benevolent or secret societies for the welfare of persecuted ethnic or political groups. The Chinese Triads and the Sicilian Mafia began not as bodies to carry out illegal economic activities but as organizations to provide a type of community for their members. They provided aid and comfort to members in tough times. They shielded them against turmoil, anarchy, and arbitrary government. They served as resistance movements. Robin Hood and Jesse James of legend and the movies were the leaders of such groups. In the government's eyes they were thugs. In legend they were heroes. This shadowy area between heroic resistance to injustice and criminal activity will continue to pose difficulties in sorting out criminals from patriots.

Over time, however, these groups, acting outside the law and in opposition to existing authority, typically began to focus increasingly on illegal (and legal) acts for profit rather than for a cause. As the profit motive took precedence, resistance took a back seat. Not all criminal organizations, however, began as failed resistance efforts. Many evolved from groups of thieves or gangs that always sought profit outside the law. Some of these groups claimed political motives. But a gang, while organized, is generally a very small and localized affair with very limited means. In general, major criminal organizations of the type that are of greatest concern here have grown into substantial enterprises, often with transnational connections, involve hundreds if not thousands of "employees," and are no longer confined to localized areas. The third point to note about criminal organizations, then, is that their underlying purpose is to make a profit from illegal acts and that they employ a large number of people whose activities are coordinated over time. The organization usually is neither ephemeral nor temporary.

A fourth characteristic of major criminal organizations is their willingness to use violence to promote and protect their interests. Violence and crime often go together. But criminal organizations use violence deliberately. They control its use and direct it in specific cases to achieve defined goals. Violence is a "business" tool and not a random or individual act. There is sometimes a lack of discipline or acts of individual cruelty inside organized crime, but generally violence is used for business purposes. It is directed outward to intimidate or eliminate rivals and threats, and it is directed inward to enforce discipline within the organization. The level of violence may vary—the Medellín Cartel, for example, is far more prone to resort to violence than is the Cali Cartel—but it remains a consciously controlled instrument.

A corollary to the use of violence is the purposeful use of bribery. It should be clear that criminal organizations are in direct conflict with police and other governmental agencies. In some cases, the criminal organization may have more firepower than the state—as with the Shan United Army—but generally they cannot sustain direct violence against official bodies. Instead, they use bribery in order to corrupt the legal system and evade prosecution. The availability of large amounts of ready cash allow the criminal organizations considerable flexibility in using the power of money to suborn government officials on a large scale, sometimes including government ministers. In many parts of the world bribery and corruption are endemic. In these cases, organized crime is just another business group that uses bribery to go about pursuing its business.

FAMILY TIES

Finally, although it is not true in every case, most major criminal organizations have a family or ethnic base. They operate out of a small, tight-knit community. Family ties, especially in circumstances where a member's family can be held accountable for the member's disloyalty, present a practical solution to the problem of ensuring fidelity and obedience. Codes of allegiance, rituals, ethnic bonds, and quasi-religious ceremonies, whether linked to familial ties or not, also help to engage the compliance and loyalty of individuals within organizations. In addition, they create distinguishing codes of recognition that reduced the chances of hostile penetration. The fact that involvement in such an organization places an individual outside the law and subject to arrest also helps to reinforce the ties that bind, and when all else fails, murder is the final sanction to ensure loyal silence.

This is not hard to understand. Members of criminal organizations need to trust each other. The illegal and dangerous nature of the work plus the fact that law enforcement will seek to penetrate and disrupt the organization make some form of security discrimination vital to survival. It is easier to enforce operations security within a group known to one another and bound by ties of kinship or race. It is also easier to control membership and to spot outsiders. One of the advantages that ethnic-based criminal organizations have when they operate outside of their place of origin is that language and culture offer added barriers to fend off or identify outsiders. It is essential for such organizations to protect themselves from a hostile environment, to guard access to information, to ensure the success of operations, to minimize losses, and to guarantee loyalty.

Characteristically, then, organized crime is defined by a more or less formal structure that endures over time, is directed toward a common purpose by a recognizable leadership operating outside the law, is quite often based on family or ethnic identity, and is prepared to use violence or other means to promote and protect common interests and objectives. As noted at the outset, however, there is nothing particularly new about such organizations; they have existed for centuries. A number of factors are now at work that argue that the nature and role of these organizations are raising to a new level the threat that they pose to social order and the stability of nations.

There are three major new characteristics of organized crime at the end of the twentieth century. First is the broader, global canvas of the traditional criminal activity. Second is the growth of transnational links between criminal organizations and between criminal organizations and other groups. Third is the growing ability and power of international criminal organizations (ICOs) to threaten the stability of states, to undermine democratic institutions, to hinder economic development, to undermine alliance relationships, and to challenge even a superpower.

There are a number of factors that aid the ICOs' growing ability to challenge individual states. These factors are not so much characteristics of ICOs as of the environment in which they exist. They benefit, for example, from weak governments without the will or the resources to cope with rich and powerful ICOs. They enjoy fantastic sums of money generated by illegal activities, especially drug production, which gives them maximum flexibility to employ bribery or violence to achieve their ends. They are able to take advantage of the movement of large numbers of people internationally, which gives various organizations a recruitment base around the world. They are also able to capitalize on a decline in economic and political order, especially thriving on the growth of parallel or informal economies that subvert loyalty from the nation–state and government and habituate people to operating outside the legal framework. The ready availability of sophisticated arms and other technologies gives ICOs better means to protect and promote their interests. And systemic limits on the ability of individual states and international organizations to coordinate effective transnational anticriminal programs provide ICOs with maneuvering room to adjust to enforcement threats.

These features, taken separately or together, mean that today's international criminal syndicates are powerful enough to challenge, sometimes to destabilize, and so far, rarely, to control small, weak states. As a recent U.S. Senate report noted, these "new international criminals" represent a threat to international security that no single state can control alone.

These three major features, global operations, transnational links, and the ability to challenge national authorities, will be considered in turn.

THE NEW INTERNATIONAL CRIMINAL

The major ICOs have all the same characteristics and features of more traditional, nationally based criminal organizations. What distinguishes their ability to conduct global operations on the order of a major multinational corporation is their transnational scale and their ability to challenge national and international authority. Disposing of large quantities of ready cash, diversified into a wide range of activities, and employing a workforce spread around the globe, the ICOs represent a different order of magnitude in criminal operations.

The ICOs differ from traditional criminal organizations in the *global scope of their operations.* Traditional organized crime groups have their roots within individual countries, and although they may have overseas connections they do not operate on an international scale. Their organizations and operations are confined to nations or regions, and cities within nations. The America Mafia, also commonly known as La Cosa Nostra (LCN), is a well-known example of this older type. LCN emerged in the 1930s from conflict among gangs of Sicilian immigrants in U.S. cities. Although its members drew on the traditions of their Sicilian origins, LCN was never a subsidiary or arm of the Sicilian Mafia; it is a distinctly American organization. While LCN has had international connections, these have been largely related to being buyers of alcohol and heroin from foreign groups, not as being part of those groups. LCN's organization and operations are essentially domestic and regional.

Many other organized crime groups operating in the United States are primarily of this type. Black and Hispanic groups, for example, are generally localized gangs. They often dominate criminal activity in their respective ethnic neighborhoods. Typically, they control much street-level drug dealing and some distribution activity above the street level. They often interface with international traffickers who control wholesaling and regional distribution. Motorcycle gangs, such as the Hell's Angels, have also become well organized and engage in a wide variety of criminal activity, sometimes developing international suppliers.

The new international criminal groups differ sharply from these domestic groups. They are organized for and engage in large-scale criminal activity across international boundaries. Perhaps the greatest such organizations are Colombian. The Colombian cartels are vertically integrated global businesses. They have tens of thousands of specialized employees and associated individuals or businesses worldwide. Similarly, the Chinese Triads, although more loosely structured, have also developed extensive overseas operations, often in the wake of Chinese emigration. Such global networks provide mobility, an effective communications infrastructure, and international connections for criminal enterprise and sometimes for noncriminal groups who want to use their services. These structures also enhance the criminal groups' ability to create whole new markets for goods and services. This can be done either by creating new products, like "crack," which revolutionized the U.S. cocaine market in the mid-1980s, or by opening up new market areas, as the cocaine cartels have attempted to do in Europe, establishing links to the Mafia or other European criminal organizations.

International networks also provide flexibility to adapt quickly and creatively to enforcement efforts. For example, after U.S. enforcement efforts shut off the flow of heroin originating from Turkey, new sources developed in Southeast and Southwest Asia and Mexico. U.S. law enforcement agencies noted in 1991 the "inherent flexibility" of traffickers in shifting routes and modes of transport in response to enforcement efforts. When enforcement was stepped up in Florida and the Caribbean, cocaine was routed increasingly through Central America and Mexico. There is another aspect to this flexibility: by operating in the international arena, crossing national boundaries at will, the ICOs are often able to thwart localized law enforcement efforts. Diversification of operations and locale and diffusion of risk greatly enhance the ICOs' ability to recover from losses. They are also better able to adjust to changing situations in one country and exploit gaps in international law enforcement cooperation.

One of the emerging characteristics of the major ICOs is the extent of their *transnational links:* their growing interconnectivity with other transnational, nonstate actors. Although traditional criminal organizations have links to similar groups outside their own countries, they generally do not conduct extensive overseas operations and have only limited contact with other criminal groups. The major ICOs, however, have an international focus and operate vast transnational business empires. Their activities go beyond establishing subsidiaries. In addition to operating across international boundaries, the new ICOs have also begun to establish links to other nonstate groups, such as insurgents and terrorists, and similar criminal organizations. These linkages are diverse and in some cases tenuous and tense, but the trend that is emerging is toward closer cooperative relationships. The nature of the relationship can be quite complex.

Traditional, nationally based criminal organizations pose a variety of threats to public order and legitimate business. The U.S. experience with LCN is typical: corruption of officials, penetration of unions, money laundering, prostitution, street crime, gambling, violent internecine power struggles, drug trafficking; in short, the whole range of criminal activity that can be organized for profit. As rich and powerful as these organizations have been, however, they are generally in no position to challenge political order directly. Indeed, in the case of the LCN, recent law enforcement successes have decimated its ranks and may have dealt it a crippling blow. The major ICOs, however, pose a more dangerous threat, one that is again defined by its scope and sheer audacity: a *challenge to authority.*

The Medellín Cartel felt powerful enough to challenge the sovereignty and integrity of the Colombian state, and along with it the United States. The last five years of struggle have damaged the Medellín Cartel, at considerable loss of life and capital, but the Colombian cocaine entrepreneurs remain in business, not only evading annihilation but continuing to prosper and diversify their operations. Alone or in conjunction with various guerrilla groups in Colombia and Peru, they are able to control large areas of the Andes in defiance of government authority. They are also able to transship immense quantities of drugs through the Caribbean, Central America, and Mexico to the

United States. In many cases, even in totalitarian Cuba, the Medellín Cartel has been able to corrupt officials throughout the hierarchy, including ranking military officers and ministers of state. Their ability to penetrate governments has even led to major disruptions in the relations between nations, as the Enrique Camarena case in Mexico showed only too clearly.

There is also mounting evidence to indicate that the major ICOs can have a profound effect on local economies. In the Andes, for example, the diversion of labor into illegal activities, the destruction of land and its use for nonproductive crops, and the generation of inflationary pressures all work to undermine the viability of local economies, already none too strong in Bolivia and Peru. Added to this is the penetration of financial markets and the international banking system. Corruption of financial institutions is now a major and growing concern, as the ICOs use international financial networks to launder money and in some cases to provide cover to further illegal activities. Even major developed countries are not immune. The French government, for example, has become concerned with the penetration of the Italian Mafia in southern France. This is also one of the major reasons the British government established the National Criminal Intelligence Service.

The major ICOs enjoy a number of advantages denied to smaller groups. The scale of their operations, their growing international connectivity, and the power and influence that they have garnered at the expense of governments make them a formidable challenge. They enjoy the advantages of economy of scale in their operations. Their enormous financial resources and diversification operations give them maneuvering room in responding to law enforcement encroachments. Their ability to operate across many legal jurisdictions reinforces this inherent flexibility. Moreover, their contacts with other nonstate actors, such as insurgent groups and terrorists, means that they can call increasingly upon allies to help distract the government or even challenge its authority. The cumulative effect of these advantages puts them in a position of great strength.

THE FUTURE

International criminal groups in the United States and worldwide are likely to expand. A number of factors are likely to aid their growth.

Economics of Production For small farmers in many countries, choosing to grow drug-related crops makes the most sense economically. Markets for other commodities are less profitable and less stable. In many cases, even where the necessary marketing infrastructure and expertise exist, government controls make entry into those markets difficult or impossible for peasants. At the same time, drug entrepreneurs are expanding into markets where drugs have not been a major problem in the past. Without dramatic and unlikely changes, raw materials for drug production will continue to be readily available.

Furthermore, the United States is one of the world's most lucrative markets for illegal as well as legal enterprise. It will continue to attract trade in illegal products. Europe, too, is a large market, and the creation of the Common Market has significantly reduced the extent of border controls that might have helped to restrain drug trafficking and other transnational criminal activities.

International Ungovernability The growth of international crime parallels a global trend toward ungovernability, that is, the declining ability of governments to govern, to manage a modern state, and to provide adequate or effective services. In some cases criminal organizations have been able to capitalize on the fact that large areas, such as the Andes and the Amazon regions of Latin America, were never under much central government control. They have moved into these remote regions and have begun to provide the major form of authority in much of the region. In other cases, they have begun to contest local control of areas with the government, as have groups in Burma. There are dozens of places, such as Peru and Burma, where state authority has lapsed in whole or in part. It is not limited to small states, either. There are indications that areas of some of the Central Asian Republics have been given over to drug cultivation. Similarly, areas in Mexico, Pakistan, and southern China appear to be largely beyond government control. This situation provides favorable conditions for criminal groups and bases of operations and safe havens in areas key to drug trafficking and alien smuggling. Experts in political geography predict continuing global fragmentation. Criminal organizations thrive where governments are weak.

Immigration Streams Ethnic criminal organizations are likely to follow immigration patterns. They do not always do so, but there has often been a strong correlation between the two. In the 1990s economic pressures and widespread ethnic turmoil are likely to generate refugees and immigrants from regions where international criminal groups are based. Between 1980 and 1990 the Asian population in the United States alone grew by 108 percent, from 3.5 million to 7.3 million. The Chinese population grew by 104 percent, from 806,000 to 1.6 million. While the vast majority of immigrants are law-abiding, criminal organizations tend to exploit immigrant communities in a variety of ways. They provide cover and concealment. Immigrant pools also provide a pool of recruits. In addition, the immigrants are usually fearful of law enforcement. Their recent experience in their country of origin makes them reluctant to cooperate with the police in their new countries. The police, moreover, historically do not provide the same degree of service to immigrants. The immigrants do not have important political connections, and the police find it difficult to cooperate with them because of their strange cultures and languages. Hence many experts anticipate increased international organized criminal activity accompanying the immigration of Russians, East Europeans, Asians, Middle Easterners, Kurds, and others.

Border Porosity The United States' long open borders with Mexico and Canada provide ready access for criminals and illegal goods, and tens of thousands of miles of U.S. coastline are virtually uncontrollable. The opening of free-trade areas, such as the North American Free Trade Agreement and the EC, will lower many existing safeguards and customs inspections as well.

Trends in Technology Continued advances in technology and international transportation will facilitate growth in international organized crime. The ease of modern communications makes contact among international criminal organizations easy, fast, and more secure. For example, new digital technologies make it more difficult for law enforcement bodies to intercept their communications. The movement of trillions of dollars in wire transfers each day makes it possible for many actors to evade state monitoring.

Relative Disorganization of Law Enforcement Preventing, disrupting, and successfully prosecuting organized crime in most parts of the world is difficult enough in the best of times. Many traditional organized criminal organizations have survived the onslaught of law enforcement organizations for decades.

Now, however, the U.S. and other states are faced with international criminal groups. As was described earlier, they are bigger and more powerful than most of their predecessors. They operate globally, making it impossible for law enforcement in any one jurisdiction to neutralize major parts of their activities. While some degree of cooperation exists among law enforcement agencies, and new initiatives are getting under way, many observers believe that it is inadequate to the task. In 1992, for example, a U.S. Senate report noted that there is little evidence to suggest that either U.S. or foreign law enforcement entities are currently equipped to meet the challenge of this new breed of international criminal.

PART VII

Deviant Acts

lthough the structure of deviant associations is revealing, in Part VII we investigate the characteristics of the acts of deviance themselves. Deviant acts involve one or more people aiming to accomplish a particular deviant goal. These vary widely in character from those enduring over a period of months to the more fleeting encounters that last only a few minutes, from those conducted alone to those requiring the participation of several or many people, and from those where the participants are face-to-face to those where they are physically separated. At the same time, acts of deviance can be looked at in terms of what they have in common. All deviant acts consist of purposeful behavior intended to accomplish a gratifying end, require the coordination of participants (if there are more than one), and depend on individuals reacting flexibly to unexpected events that may arise in this relatively unstructured and unregulated arena. Like the relations among deviants, deviant acts fall along a continuum of sophistication and organizational complexity. Following Best and Luckenbill's (1981) typology, we arrange them here according to the minimum number of their participants and the intricacy of the relations among these participants.

Some deviant acts can be accomplished by a lone **individual,** without recourse to the assistance or presence of other people. This does not mean that others cannot accompany the deviant, either before or during the deviant act, or even that two deviants cannot commit acts of individual deviance together. Rather, the

383

defining characteristic of individual deviance is that it can be committed by one person, to that person, on that person, alone. A teenager's suicide, a drug addiction, a skid row transient's alcoholism, and a self-induced abortion are all examples of individual behavioral deviance. Nonbehavioral forms of individual deviance include obesity, minority group status, a physical handicap, and a deviant belief system (such as alternative religious or political beliefs).

For this section we include an article by Shearon Lowery and Charles Wetli on "Sexual Asphyxia: A Neglected Area of Study." Sexual asphyxia, otherwise known as the "ultimate orgasm," is a solitary practice pursued by individuals, primarily young men, who desire to raise their level of orgasmic pleasure through self-strangulation. So deviant that practitioners rarely reveal their interest in this act, even to friends or lovers, it is accomplished outside of any deviant support group or subculture. Lowery and Wetli outline some of the features of this highly secretive form of individual deviance, the characteristics of individuals who engage in it, and the driving force behind their involvement.

A second type of deviant act involves the **cooperation** of at least two voluntary participants. This cooperation usually involves the transfer of illicit goods, such as pornography, arms, or drugs, or the provision of deviant services, such as those in the sexual or medical realm. Cooperative deviant acts may involve the exchange of money. When this is absent, participants usually trade reciprocal acts. They both come to the interaction wanting to give and get something. In deviant sales, one participant supplies an illicit good or service in exchange for money. One or more of the participants in such acts may be earning a living through this means.

Laud Humphreys' "Tearoom Trade: Homosexual Behavior in Public Restrooms" offers an excellent illustration of a type of cooperative deviant act that does not involve the transfer of money. He studied impersonal homosexual behavior that occurred in public restrooms between anonymous consenting adults. In a fascinating description of this hidden subculture, he lays out the progression of the deviant act, tracing the way that participants located and identified each other, how they signaled their interest in engaging the deviant act and their particular role, and how they culminated the act. Individuals engaging in this behavior often lived their lives as covert homosexuals, passing as straights to avoid the stigma of deviance.

A second example of cooperative deviance can be found in "Turn-Ons for Money: Interactional Strategies of the Table Dancer," by Carol Rambo Ronai and Carolyn Ellis. They describe a relatively recent form of strip-dancing practiced in some nightclubs. Dancers not only (partially) remove their clothing on the stage, but they also contract for semi-private performances in specially designed pits where greater contact with the customer is possible. This article discusses how these women "work" the customers to sell their table dances, from contact through culmination, how individuals among them construct certain presentational styles to appeal to specific types of clients, and how they cultivate relationships with "regular" customers who form the basis for their steady income.

The final type of deviant act is one of **conflict** between the involved parties. A perpetrator forces the interaction on the unwilling other, or an act entered into through cooperation turns out with one party "setting up" the other. In either case, the core relationship between the interactants is one of hostility, with one person getting the more favorable outcome. Conflictual acts may be carried out through secrecy, trickery, or physical force, but they end up with one person giving up goods or services to the other, involuntarily or without adequate compensation. Conflictual acts may be highly volatile in character, for victims may complain to the authorities or enlist the aid of outside parties if they have the chance. To be successful, therefore, perpetrators must control not only their victims' activities but also the victims' perception of what is going on. Such acts can range from kidnapping and blackmail to theft, fraud, arson, and assault.

The reading by Patricia Martin and Robert Hummer considers the situation of "Fraternities and Rape on Campus." This compelling article examines the culture of masculinity that flourishes within college fraternities, and how men engage in bonding and status building by degrading women and making them objects of sexual conquest. This article offers insight into the culture of masculinity that binds male participants to potentially abusive behavior through its hierarchical status system, as well as into the way that women accept the values of this culture and passively follow them to their own exploitation. It also specifically describes the strategies that fraternity men used to accomplish their rapes, from using alcohol as a weapon to employing violence. It concludes that basic features of both society and fraternities foster the continuation of such rampant behavior. This article offers an interesting comparison to Gilbert's critique of the sexual abuse surveys in Part II.

Next, Robert Prus and C. R. D. Sharper look at a more organized and systematically planned type of conflictual deviant acts in their discussion of "Pool and Golf Hustling." This article takes us into the illicit subculture of traveling hustlers who make their living by setting up unsuspecting "marks" and using deception and misinformation to win their money. Performing alone or in the type of deviant relationships known as rings or crews, hustlers operate intricate deviant capers designed to provide their primary, often sole source of income. Prus and Sharper's subjects, traveling deviants who live and work on the road for the majority of each year, are tightly bound to one another and to the deviant subculture that supports and facilitates their scams. Practitioners have a highly specialized set of skills, equipment, roles, and rules, with clear codes that must not be violated. This selection offers us a glimpse into the ideology, planning, and practice of the professional hustlers who match their physical talents in the competitive arena of sport with their interpersonal talents that are designed to mislead potential targets into unwitting exploitation.

31

Sexual Asphyxia
A Neglected Area of Study

SHEARON A. LOWERY
AND CHARLES V. WETLI

T he sociological study of sexual deviance has traditionally been limited to the examination and explication of prostitution and homosexuality, although some scholars have also considered premarital and extramarital sexual behavior to be deviant. Even when such a "broad" definition of sexual deviance has been employed, however, other types of sexual behavior such as sadomasochism and bondage have been typically excluded. Consequently, such practices have remained virtually ignored in the literature of sociology. In addition, judging from an examination of that literature, more unusual sexual practices such as sexual asphyxia are apparently unknown to sociologists. Consider, for example, the following case:

> The sheriff's office received a call from the frantic wife stating she had found her husband bound, gagged, and murdered. The Police and Medical Investigator rushed to the scene to find a 27-year-old male deceased on the bed. The deceased was clad in brassiere, woman's panties, red negligee, and panty hose, all of which the wife identified as hers. He was facedown with his legs flexed at the knees. The ankles were bound with four loops of clothesline, loosely knotted. The clothesline was attached to an elastic strap with metal hooks on both ends. The strap was looped about a dog collar encircling the neck. A bath towel was between the dog collar and the skin of the anterior neck. A handkerchief was tied about the shaft and the end of the penis was stained with seminal fluid. A handkerchief was stuffed in the mouth, and a bandana was tied about the face at mouth level, knotted in the back. A large bathroom mirror had been removed from the door and rested against the dresser where it could be visualized by the deceased. (Lewman, 1978: 11)

Although this scene suggests homicide or suicide, the actual manner of death is accidental. This case is in fact a rather typical example of the sexual asphyxia syndrome familiar to medical examiners and to a lesser extent to some police investigators and psychiatrists.

From "Sexual Asphyxia: A Neglected Area of Study," Shearon A. Lowery and Charles V. Wetli, *Deviant Behavior*, V. 3, No. 1, Taylor & Francis, Inc., Washington, D.C., pp. 19–39, 1982. Reprinted with permission. All rights reserved.

Sexual asphyxia is an autoerotic activity practiced almost exclusively by males. Lethal cases are almost always characterized by an individual male engaged in solitary sexual activity while simultaneously creating a self-induced mechanical or chemical asphyxia. The purpose of the asphyxia is to heighten the sexual pleasure. Cases of lethal sexual asphyxia among females are extremely rare; to our knowledge, only three instances have been reported (Sass, 1974; Danto, 1980; Sullivan and Wray, 1981).

Among males transvestism is frequently concomitant with the practice and there may be erotic literature or mirrors in view of the deceased. Sometimes the deceased's hands are tied behind his back, but close inspection invariably reveals that the victim himself was responsible for this. In addition, it is done in such a way as to permit a quick escape should that be necessary. Finally, the practitioner will have devised some mechanism to impede the flow of oxygen to the brain. This will usually be a ligature (often with a soft material or padding about the neck to prevent abrasion), a plastic bag, or an inhalent (gas, chemical aerosol, etc.). The subjective result is a giddiness and exhilaration; but the objective consequence is asphyxia resulting from mechanical restriction of the airway or chemical replacement of oxygen. If the process continues unabated the individual loses consciousness as a result of the lack of oxygen and the increased retention of carbon dioxide. Death will ensue unless the individual spontaneously breaks away from the asphyxiating device. If that device is a ligature, there is little likelihood of this happening once consciousness has been lost. Instead, the airway becomes further obstructed, as does the venous return from the brain, and finally the arterial blood supply to the brain is disrupted.

The most frequent method used to produce the asphyxia, and the one most discussed, involves ligatures applied to the neck. Ropes, scarves, and the like are tied about the neck in such a manner that they may be manipulated to control the flow of blood and oxygen to the brain. Such choking creates feelings of pleasure and may induce erections and even orgasms in some individuals (Resnik, 1972). Various descriptions of sexual asphyxia appear in literary sources; a particularly graphic passage can be found in DeSade's *Justine:*

> He got upon the stool, the rope around his neck. . . . He got ready and beckoned her to pull away the stool. Hanging by his neck for a while, his tongue lolling way out, his eyes bulging; but soon, beginning to swoon away, he motioned feebly to Justine to set him loose. On being revived he said, "Oh Therese! One has no idea of such sensations, what a feeling! It surpasses anything I know." (p. 101)

In addition, anthropologists have pointed out that certain ethnic and cultural groups (e.g., Eskimos) are known to choke each other as part of their sexual activity. In such cultures it is common for the children to suspend themselves by the neck during play (Walsh, Stahl, Unger, Lilienstern, and Stephens, 1977).

The limited research that has been done indicates that it is likely that medical authorities underestimate the frequency of death resulting from sexual asphyxia. Some investigators (Sass, 1974; Litman and Swearingen, 1972; Resnik, 1972) have argued that sexual asphyxial practices are relatively common in the

United States. Rosenblum and Faber (1979) think that it is reasonable to estimate that at least 250 deaths occur in the United States each year. This estimate is probably very conservative because it is likely that many sexual
asphyxial deaths are judged to be either homicide or suicide because of a lack
of awareness of the practice in many localities,[1] the sensitivity of the sexual aspects of the practice, or the social status of the victims.[2] For example, it is possible that some deaths among youths that have been attributed to glue sniffing
could actually have been the result of the practice of sexual asphyxia. More research is necessary before we can adequately address the issue of the reliability
of these national estimates; however, it is important that the issues be raised
and the subject investigated.

Although it is difficult to estimate the number of annual deaths from this
cause it is even more difficult to estimate the number of individuals who engage in this sexual practice (Resnik, 1972; Rosenblum and Faber, 1979). Because we know that those practitioners for whom sexual asphyxia proved to
be fatal took elaborate precautions not to die, we must take this factor into account in any attempt to estimate the number of practitioners. Thus, if we accept the view that those who engage in the behavior do not intend to die,[3] it
becomes evident that the actual number of practitioners of sexual asphyxia
may be considerable. Because of the precautions taken by these individuals, it
is likely that the number of fatalities is small compared to the total number of
incidents. In fact, researchers of sexual asphyxial practices posit that "such
practices are common" (Litman and Swearingen, 1972: 11) and certainly not
the "oddity or rarity" that many people assume (Enos, 1975: 134). But whatever the number of practitioners or the frequency of performance, sexual asphyxia is always potentially lethal and thus deserves careful sociological and
psychological examination as well as legal and medical scrutiny.

SOCIAL CHARACTERISTICS OF THE
VICTIMS AND THE DEATH SCENE

Almost all of the handful of scientific studies examining sexual asphyxia deal
with cases that were discovered because the practitioners were unfortunate
enough to die in the process. These postmortem studies do, however, delineate certain patterns. In contrast to most other forms of sexual deviance, sexual asphyxia appears to be an activity confined to middle- and upper-class
white males. These males tend to be young; most of the recorded fatalities
have occurred among teenage or young adult males who were unmarried at
the time of their death. One review of 43 such deaths (Walsh et al., 1977)
found the majority of the victims to have been younger than 25; and only 13
were married at the time of death.

Such behavior usually begins in adolescence, when it is a solitary act
(Resnik, 1972). However, as the practitioner matures into adulthood the syndrome may become less lethal. This is because he may be able to involve partners in the process to protect him. Such partners are, however, used solely for

the purpose of protection of the practitioner and take no part in the act itself. In all other aspects sexual asphyxia remains a solitary act.

The characteristics of the scene of discovery of the victims of sexual asphyxia lead to the conclusion that the practitioner requires solitude and privacy. The body is usually discovered in the victim's residence; many times it is in a bedroom, bathroom, attic, closet, or basement—somewhere that a door might be locked. If the act is performed outdoors, the most frequent setting is a secluded wooded area or an abandoned or little-used structure. There is an obvious need on the part of the participant to avoid any intrusion or interruption during the act (Sass, 1974).

Danto (1980) reported that the majority of the bodies are found either naked or partially clothed and that there is frequently evidence of transvestism (articles of female clothing such as dresses, brassieres, or panty hose). Moreover, there is usually evidence of penile erection and ejaculation. However, evidence of ejaculation must be carefully evaluated because emission of seminal fluid is a frequent consequence of rigor mortis.

It is estimated that in from one-third to one-half of the cases the hands, body, and feet are bound in some manner. And although some of these binding mechanisms are extremely complex, it can usually be demonstrated that the ligatures and ropes could readily be tied and released by the victim himself. In addition, in some cases the scrotal sac may also be bound with string, thread, or rope. Mirrors are frequently present and it is thought that as many as half of the victims may have been viewing themselves in a mirror and/or using erotic materials found near the body. Instances associated with self-inflicted pain or infibulation have also been reported (Sass, 1974).

INTERVIEWS WITH LIVING PRACTITIONERS

Among the studies that report aspects of the practice of sexual asphyxia, only one has involved the interviewing of current practitioners. One additional study (Rosenblum and Faber, 1979) reports a psychiatric case history in which a 15-year-old boy has successfully been treated for such practice. This relative absence of firsthand information about social, psychological, and emotional aspects of the practice has not occurred because researchers failed to seek out practitioners. Indeed, Resnik (1972) reported that he had been unable to locate a single living practitioner of sexual asphyxia in the 10 years he had been interested in the subject. Medical researchers Litman and Swearingen (1972), however, were able to locate and interview nine such individuals. In an effort to reach such persons, they had an article ("Whips, Chains, and Leather") published in the *Los Angeles Free Press,* a weekly underground newspaper with a circulation of about 90,000. As a result of that article and an advertisement, they received about 30 responses. After screening the respondents, face-to-face interviews were conducted with nine men and three women. Only the men, however, were engaged in the type of behavior that they were trying to study.

All of the subjects of the Litman and Swearingen study were white middle-class males who were characterized as "intelligent, verbal, and cooperative." The researchers felt that their volunteer subjects had been motivated to respond to their advertisement by loneliness, a wish to share their interest with others, and a need to find legitimacy for their underground practices. In addition, the researchers felt that for some of them the response was a cry for help.

In general, the sexual orientation of these men was deemed to be homosexual. Even though many of the men had previous heterosexual experiences and several even indicated that they preferred women, Litman and Swearingen classified only two of the men as heterosexual. They also reported a trend toward increasing homosexuality with increasing age. Indeed, the researchers felt that the majority of the subjects hoped that they would find partners by responding to the ad. Finding such a partner is extremely important to them, because such a partner plays a protective role. This desire for partners, moreover, may help to explain the tendency toward homosexuality, in that homosexual partners are easier to find for unusual activity than are heterosexual ones. But despite this search for partners, the essence of the practice remains narcissistic. The focus of attention remains on themselves, even when others are participating; the practitioner of sexual asphyxia is preoccupied with his own fantasies and with his own view and sensations of the world.

Three of the subjects were married, and each had difficulty with his wife. The authors felt that this was the result of "basic personality flaws" rather than the result of their peculiar sexual behaviors. Although all of the subjects were aware of pornographic literature, only one-third indicated that they were strongly influenced by it. Moreover, transvestite elements were "surprisingly infrequent" in this group.

Litman and Swearingen saw the defining characteristics of this group of men to be extreme loneliness and isolation. The felt that all of these men were deeply depressed and death oriented: Six of the nine gave histories of serious depression—often accompanied by suicide attempts. They further posit that men used their perversion to fight off the death trend and to defend themselves against suicide. In spite of the researchers' belief that these men were experiencing both psychological and emotional pain, their study revealed no consistent patterns of family pathology or specific traumata in childhood.

THE ABSENCE OF A DEVIANT SUBCULTURE

One additional important point should be stressed here: No deviant subculture has grown up around the practice of sexual asphyxia. There is no underlying thread that connects one practitioner with another, and participants do not, however subtly, advertise what they do. The solitary nature of the act means that locating one practitioner, even a deceased one, is not likely to lead us to others among his friends and associates. This is unlike other forms of sexual deviance, such as prostitution and homosexuality, where subcultures are well defined and have been extensively examined in the literature. It is even true of other sexually deviant behaviors such as bondage and sadomasochism that have not been extensively studied by sociologists. People who participate in these subcultures routinely advertise in underground publications for like-minded partners; and browsing any "adult" bookstore leads to the discovery that there is a special B&D (bondage and discipline) section for those who enjoy such practices. Indeed, the coroner of San Francisco, where rates of sexual deviance are high, recently began holding clinics in safe sadomascochistic practices because of concern about the number of homicides related to the activity (*Washington Star,* March 13, 1981: 1).

The additional obstacle that this lack of a subculture presents to the researcher should not, however, preclude systematic and rigorous examination of sexual asphyxia. Even when the researcher is unable to implement "ideal" research design and uses volunteer subjects or must rely on "accidental deaths" that bring practitioners to the attention of local authorities, such studies can be useful so long as their sampling limitations are noted and kept in mind. By using information about those individuals unfortunate enough to die in the process, perhaps the living can learn enough from death investigations to illuminate the scope and dimensions of this potentially lethal sexual practice.

THE DADE COUNTY DATA

It was possible to study a number of cases of lethal sexual asphyxia taken from the files of the Medical Examiner Office of Dade County (Miami), Florida. That office is responsible for the investigation of all deaths within the county that are not obviously the result of natural causes. The purpose of the investigation is to determine the cause, manner, and mechanism of death (see Wright and Wetli, 1981). The scene and circumstances of death are initially evaluated by police agencies and, in cases of apparent lethal sexual asphyxia, by a forensic pathologist as well. Medical and social histories are subsequently obtained by both police and forensic investigators. Finally, a complete autopsy is performed. The data here reported were abstracted from the official reports prepared by these authorities.

Previous medical researchers (e.g., Resnik, 1972; Walsh et al., 1977) have outlined a number of criteria that are believed to indicate that a death may have been the result of the practice of sexual asphyxia. Although each of these indicators may not appear in every case, they collectively define the sexual asphyxia syndrome. These criteria include the following:

1. The act is solitary.
2. There is evidence of sexual activity.
3. There is no well-defined evidence of suicidal intent.
4. The deceased is completely or partially unclothed.
5. Transvestism may be present.

6. There may be evidence of previous episodes.

7. Often the extremities and sometimes the genitals are bound.

8. Erotic materials, especially pictures, are often present.

A detailed examination of the following case from Dade County illustrates the unusual nature of the sexual asphyxia syndrome and provides a concrete example of some of the criteria associated with it.

A hotel maid discovered a 39-year-old caucasian man hanging by his neck. He was naked, and his feet touched the floor. His hands were looped in a rope behind his back. The ligature consisted of a white nylon rope fashioned as a hangman's noose and threaded through two large eye-bolts which had been inserted into the wall. A full-length mirror was positioned in front of and to the side of the victim to permit complete self-viewing. A camera was positioned in front of the victim several feet away, and a remote shutter release device was nearby. A large sheet of dull black paper (approximately six feet wide by ten feet long) was on the floor to one side of the victim, and had apparently fallen from an overhang to which it had been attached with tape. On a nearby table were erotic photographs of women which had been cut out of the latest issue of *Playboy* magazine. On an adjacent chair was an ice bucket partially filled with urine. In the bathroom was a kit for developing 35mm slide film. In the bedroom were some hand tools (pliers, screw driver, etc.) as well as a plaster patching compound.

The film in the camera was removed and subsequently developed. It contained a series of self-taken photographs in which the victim was in various poses. The first revealed the victim clad in a shirt and bath towel imitating female attire. The subsequent pictures revealed a progressive removal of clothing until he was standing naked and holding the noose-end of the rope. The photographs were taken against a dull black background (the paper found on the floor). The penultimate photograph depicted him standing on a low stool and with the ligature mildly constricting the neck. In the final photograph the victim is pictured in nearly the exact position in which he was found at the scene. In none of the photographs was penile erection evident.

A subsequent investigation revealed that this man was a highly competent professional person with a high degree of intelligence. He was married and had two children. At one time he had been in psychotherapy where it was learned that he had marital problems related in part to his spouse's sexual withholding. His wife was aware of the episodes of photo-fantasies. It was learned that she disapproved of them and hence, for more than twelve years, he engaged in this activity outside the home (usually in hotel rooms). The victim's personality was described as narcissistic. He was obsessive-compulsive ("workaholic") and revealed a self-image of being both heroic and sacrificial, qualities which were evident in the self-taken photographs.

It is evident that the self-taken photographs were to subsequently induce sexual stimulation. It must be presumed that, in this sequence of
photographs, death intervened before the intended masturbatory activity
could take place.

DISCUSSION

The preceding description of some of the sociological variables in our data associated with lethal sexual asphyxia (in the argot of the bondage underground,
terminal sex) leads us to the following conclusions. The data support the observation that, unlike most forms of sexual deviance, sexual asphyxia is practiced by young, white, middle- or upper-middle-class males. It is almost
exclusively practiced in the home or some safe place near the home. We do
not know the number of practitioners, the frequency with which they perform the act, or the actual distribution of their ages.

Perhaps the young are victims of the practice so frequently because they
are, like youth everywhere, often oblivious of the dangers involved in what
they do. They are straightforward about it—they simply hang themselves.
Moreover, because the act is solitary, when something goes wrong and they
lose control, they lose their lives as well.

As practitioners become older, they probably become more cautious; they
use more elaborate devices and many search for partners. Perhaps then it is
merely adaptive behavior that leads to homosexual liaisons. Such unusual practices are indeed more likely to be accepted or at least tolerated by homosexual
partners than by heterosexual ones. Our data indicate that when the practice
is fatal among older practitioners some complicating factor such as the use of
alcohol or drugs is present that may have impaired their judgment enough to
have contributed to the accident costing them their lives.

When we examine the social origin of the act and the actor, it is clear that
many of the victims have severe difficulties obtaining sufficient sexual gratifications by other means. They lacked sufficient sexual outlets and generally
had difficulty interacting with females. Although some researchers (Resnik,
1972; Rosenblum and Faber, 1979) believe that the act generally becomes
more elaborate as the practitioner ages, our data indicate that the introduction
of fetishes and transvestism may occur early or late. One plausible explanation
for the elements of bondage and sadism that may enter into the practice is that
the only source of willing partners the practitioner can find is among members of the bondage community—whether they are heterosexual or homosexual. In order to receive the gratifications of sexual asphyxia with safety, they
submit to other forms of sexual deviance, bondage, and so forth. They do not
seek the painful pleasures of sadomasochism but rather endure them for the
sexual pleasures of sexual asphyxia that they seek. There is a great difference,
thus, between the practitioner of sadomasochism and the sexual asphyxiate.

In the end, however, many of these men return to solitary practice or never
find willing partners; and when an accident occurs, their search for a satisfac-

tory sexual outlet for their sexual needs becomes terminal. Our data confirm that these deaths were not suicides and that there was no suicidal intent. We caution, however, that the data presented in this paper are limited; in all of the cases we report, the individuals died while practicing sexual asphyxia. This group of practitioners therefore may not be representative of the general population of practitioners of sexual asphyxia. Thus, it is evident that additional research is needed to clarify many of the issues raised in this paper, especially the scope and dimensions of the problem—who practices sexual asphyxia, how often, and how frequently does it result in death.

NOTES

1. Werner Simon (1973) has argued that teenage suicides may not have increased as much as it appears. He posits that a number of these tragedies are actually accidental deaths resulting from the practice of sexual asphyxia.

2. For example, a well-known professional whose death clearly resulted from the practice of sexual asphyxia was reported to have died of a heart attack in his hometown newspaper.

3. It should be noted, however, that if a practitioner of sexual asphyxia decides to commit suicide, it is probable that he will choose this method. Litman and Swearingen (1972) report such a case.

REFERENCES

Danto, Bruce L. 1980. "A case of female auto-erotic death." *American Journal of Forensic Medicine and Pathology* 1: 117–121.

DeSade, Marquis. 1964. *Justine*. New York: Castle Books.

Enos, William F., Jr. 1973. "Commentary." *Medical Aspects of Human Sexuality* 7(11): 184–189.

Lewman, Larry V. 1978. "Case of the month." *Office of the Medical Investigator* 5: 11–13.

Litman, Robert E., and Charles Swearingen. 1972. "Bondage and suicide." *Archives of General Psychiatry* 27: 80–85.

———. 1973. "Bondage and suicide." *Medical Aspects of Human Sexuality* 7(11): 164–181.

Resnik, H. L. P. 1972. "Eroticized repetitive hangings: A form of self-destructive behavior." *American Journal of Psychotherapy* 26: 4–21.

Rosenblum, Stephen, and Myron M. Faber. 1979. "The adolescent sexual asphyxia syndrome." *Journal of the American Academy of Child Psychiatry* 19: 546–558.

Sass, F. A. 1974. "Sexual asphyxia in the female." *Journal of Forensic Sciences* 20: 181–185.

Simon Werner. 1973. "Commentary." *Medical Aspects of Human Sexuality* 7(11): 189–193.

Sullivan, William B., Jr., and Steve Wray. 1981. A Case of Sexual Asphyxia of a Female. Paper presented at the 33rd Annual Meetings of the American Academy of Forensic Sciences, Los Angeles, February.

Walsh, F. M., Charles J. Stahl, H. Thomas Unger, Oscar C. Lilienstern, and Robert G. Stephens. 1977. "Auto-erotic asphyxial deaths: A medicolegal analysis of forty-three cases." *Legal Medicine Annual* 1977: 157–182.

Wright, Ronald K., and Charles V. Wetli. 1981. "A guide to the forensic autopsy—conceptual aspects." *Pathology Annual* 16: 273–288.

32

Tearoom Trade
Homosexual Behavior in Public Restrooms

LAUD HUMPHREYS

O.K., here goes—no self-respecting homosexual in his right mind should condone sex in public places, but let's face it, it's fun. . . . The danger adds to the adventure. The hunt, the cruise, the rendezvous, a great little game. Then more likely than not, "instant sex." That's it.[1]

T he nature of sexual activity presents two severe problems for those who desire impersonal one-night-stands. In the first place, except for masturbation, sex necessitates collective action; and all collective action requires communication. Mutually understood signals must be conveyed, intentions expressed, and the action sustained by reciprocal encouragement. Under normal circumstances, such communication is ritualized in those patterns of word and movement we call courtship and love-making. Verbal agreements are reached and intentions conveyed. Even when deception is involved in such exchanges, as it often is, self-revelation and commitment are likely by-products of courtship rituals. In the search for impersonal, anonymous sex, however, these ordinary patterns of collective action must be avoided.

A second problem arises from the cultural conditioning of Western man. For him, sex is invested with personal meanings: interpersonal relationship, romantic love, and an endless catalogue of sentiments. Sex without "love" meets with such general condemnation that the essential ritual of courtship is almost obscured in rococo accretions that assure those involved that a respectable level of romantic intent has been reached. Normal preludes to sexual action thus encourage the very commitment and exposure that the tearoom participant wishes to avoid. Since ordinary ways reveal and involve, special ritual is needed for the impersonal sex of public restrooms.

Both the appeal and the danger of ephemeral sex are increased because the partners are usually strangers to one another. The propositioning of strangers for either heterosexual or homosexual acts is dangerous and exciting—so much so that it is made possible only by concerted action, which progresses in stages of increasing mutuality. The special ritual of tearooms, then, must be both noncoercive and noncommital.

APPROACHING

The steps, phases, or general moves I have observed in tearoom games all involve somatic motion. As silence is one of the rules of these encounters, the strategies of the players require some sort of physical movement: a gesture with the hands, motions of the eyes, manipulation and erection of the penis, a movement of the head, a change in stance, or a transfer from one place to another.

The approach to the place of encounter, although not a step within the game, resembles moves of the latter sort. Although occurring outside the interaction membrane, the approach may affect the action inside. An automobile may circle the area a time or two, finally stopping in front of the facility. In what I estimate to be about a third of the cases, the driver will park a moderate distance away from the facility—sometimes as far as 200 feet to the side or in back, to avoid having his car associated with the tearoom.

Unless hurried (or interested in some particular person entering, or already inside, the facility), the man will usually wait in his auto for five minutes or longer. While waiting, he looks the situation over: Are there police cars near? Does he recognize any of the other autos? Does another person waiting look like a desirable partner? He may read a newspaper and listen to the radio, or even get out and wipe his windshield, invariably looking up when another car approaches. The purpose here is to look as natural as possible in this setting, while taking the opportunity to "cruise" other prospective players as they drive slowly by.

Sometimes he will go into the restroom on the heels of a person he has been watching. Should he find the occupant of another auto interesting, he may decide to enter as a signal for the other man to follow. If no one else approaches or leaves, he may enter to see what is going on inside. Some will wait in their autos for as long as an hour, until they see a desirable prospect approaching or sense that the time is right for entry.

From the viewpoint of those already in the restroom, the action of the man outside may communicate a great deal about his availability for the game. Straights do not wait; they stop, enter, urinate, and leave. A man who remains in his car while a number of others come and go—then starts for the facility as soon as a relatively handsome, young fellow approaches—may be revealing both his preferences and his unwillingness to engage in action with anyone "substandard."

Whatever his behavior outside, any man who approaches an occupied tearoom should know that he is being carefully appraised as he strides up the path. While some are evaluating him from the windows, others may be engaged in "zipping the fly."

POSITIONING

Once inside the interaction membrane, the participant has his opportunity to cruise those already there. He will have only the brief time of his passage across the room for sizing-up the situation. Once he has positioned himself at the urinal or in a stall, he has already begun his first move of the game. Even the decision as to which urinal he will use is a tactical consideration. If either of the end fixtures is occupied, which is often the case, an entering party who takes his position at the center of the three urinals is "coming on too strong." This is apt to be the "forward" sort of player who wants both possible views. Should both ends be occupied, it is never considered fair for a new arrival to take the middle. He might interrupt someone else's play. For reasons other than courtesy, however, the skilled player will occupy one of the end urinals because it leaves him more room to maneuver in the forthcoming plays.

If the new participant stands close to the fixture, so that his front side may not easily be seen, and gazes downward, it is assumed by the players that he is straight. By not allowing his penis to be seen by others, he has precluded his involvement in action at the urinals. This strategy, followed by an early departure from the premises, is all that those who wish to "play it straight" need to know about the tearoom game. If he makes the positioning move in that manner, no man should ever be concerned about being propositioned, molested, or otherwise involved in the action. (For defecation, one should seek a facility with doors on the stalls.)

A man who knows the rules and wishes to play, however, will stand comfortably back from the urinal, allowing his gaze to shift from side to side or to the ceiling. At this point, he may notice a man in the nearest stall peer over the edge at him. The next step is for the man in the stall (or someone else in the room) to move to the urinal at the opposite end, being careful to leave a "safe" distance between himself and the other player.

My data indicate that those who occupy a stall upon entering (or who move into a stall after a brief stop at the urinal) are playing what might be called the Passive-Insertee System. By making such an opening bid, they indicate to other participants their intention to serve as fellator. In the systematic observation of fifty encounters ending in fifty-three acts of fellatio, twenty-seven of the insertees opened in this manner (twenty-five sitting on stools, two standing). Only two insertors opened by sitting on stools and four by standing in stalls.

Positioning is a far more "fateful" move for those who wish to be insertees than for others. In sixteen of the observed encounters, the fellator made no further move until the payoff stage of the game. The strategy of these men was to sit and wait, playing a distinctly passive role. Those who conclude as insertors, however, are twice as apt to begin at the urinals as are the insertees. A few of each just stood around or went to a window during the positioning phase of the game.

SIGNALING

The major thesis of Scott's work on horse racing is that "the proper study of social organization is the study of the organization of information."[2] To what extent this holds for all organizations is not within my realm of knowledge. For gaming encounters, however, this is undoubtedly true, with "skill" inhering in the player's ability to convey, interpret, assimilate, and act upon the basis of information given and received. Every move in the gaming encounter is not only a means of bettering one's physical position in relation to other participants but also a means of communication.

Whereas, for most insertees, positioning is vital for informing others of their intentions, about half of the eventual insertors convey such information in the signaling phase. The primary strategy employed by the latter is playing with one's penis in what may be called "casual masturbation."

> **Respondent:** The thing he [the potential insertee] is watching for is "handling," to see whether or not the guy is going to play with himself. He's going to pretend like he is masturbating, and this is the signal right there. . . .
>
> **Interviewer:** So the sign of willingness to play is playing with oneself or masturbation?
>
> **Respondent:** Pseudomasturbation.

The willing player (especially if he intends to be an insertor) steps back a few inches from the urinal, so that his penis may be viewed easily. He then begins to stroke it or play with the head of the organ. As soon as another man at the urinals observes this signal, he will also begin autoerotic manipulation. Usually, erection may be observed after less than a minute of such stimulation.

The eyes now come into play. The prospective partner will look intently at the other's organ, occasionally breaking his stare only to fix directly upon the eyes of the other. "This mutual glance between persons, in distinction from the simple sight or observation of the other, signifies a wholly new and unique union between them."[3] A few of the players have been observed to move directly from positioning to eye contact, but this seems to happen in only about 5 percent of the cases.

Through all of this, it is important to remember that showing an erection is, for the insertor, the one essential and invariable means of indicating a willingness to play. No one will be "groped" or otherwise involved in the directly sexual play of the tearooms unless he displays this sign. This touches on the rule of not forcing one's intentions on another, and I have observed no exceptions to its use. On the basis of extensive and systematic observation, I doubt the veracity of any person (detective or otherwise) who claims to have been "molested" in such a setting without first having "given his consent" by showing an erection. Conversely, anyone familiar with this strategy may become involved in the action merely by following it. He need not be otherwise skilled to play the game.

Most of those who intend to be insertors will engage in casual masturbation at a urinal. Others will do so openly while standing or sitting in a stall. Rarely, a man will begin masturbation while standing elsewhere in the room and then only because all other facilities are occupied.

In about 10 percent of the cases, a man will convey his willingness to serve as insertee by beckoning with his hand or motioning with his head for another in the room to enter the stall where he is seated. There are a few other signals used by men on the stools to attract attention to their interests. If there are doors on the stalls, foot-tapping or note-passing may be employed. If there is a "glory hole" (a small hole, approximately three inches in diameter, which has been carefully carved, at about average "penis height," in the partition of the stall), it may be used as a means of signaling from the stall. This has been observed occurring in three manners: by the appearance of an eye on the stool-side of the partition (a very strong indication that the seated man is watching you), by wiggling fingers through the hole, or by the projection of a tongue through the glory hole.

Occasionally, there is no need for the parties to exchange signals at this stage of the game. Others in the room may signal for a waiting person to enter the stall of an insertee. There may have been conversation outside the facility—or acquaintance with a player—which precludes the necessity of any such communication inside the interaction membrane. This was the case in about one-sixth of the acts I witnessed.

MANEUVERING

The third move of the game is optional. It conveys little information to other players and, for this reason, may be skipped. As the Systematic Observation Sheets show, twenty-eight of the eventual insertees and thirty-five of the insertors (out of fifty-three sexual acts) made no move during this phase of the interaction. This is a time of maneuvering, of changing one's position in relation to other persons and structures in the room. It is important at this point in the action, first, because it indicates the crucial nature of the next move (the contract) and, second, because it is an early means of discerning which men wish to serve as insertees.

Twenty of the thirty-three players observed in motion during this stage of the encounter later became insertees. Two-thirds of these used the strategy of moving closer to someone at the urinal:

> X entered shortly and went to third urinal. Y entered in about a minute and went to first urinal. . . . O stood and watched X and Y. Y was masturbating, as was X. X kept looking over shoulder at me. I smiled and moved over against far wall, lit cigarette. X moved to second urinal and took hold of Y's penis and began manipulating it. I moved to door to observe park policeman (in plain clothes with badge), who was seated on

park bench. Then I went back to position by wall. By this time, X was on knees in front of urinals, fellating Y. I went back to door, saw O approaching, and coughed loudly.

Others may use this stage to move closer to someone elsewhere in the room or to move from the urinal to an unoccupied stall. All of these strategies are implemental but nonessential to the basic action patterns. The restroom's floor plan, I have found, is the strongest determinant of what happens during this phase of the game. If there are only two urinals in the facility, the aggressor's maneuver might be no more than to take a half-step toward the prospective partner.

CONTRACTING

Positions having been taken and the signals called, the players now engage in a crucial exchange. Due to the noncoercive nature of tearoom encounters, the contract phase of the game cannot be evaded. Initially, participants have given little consent to sexual interaction. By means of bodily movements, in particular the exposure of an erect penis, they have signaled such consent. Now a contract must be agreed upon, setting both the terms of the forthcoming sexual exchange and the expression of mutual consent.

Eighty-eight percent of the contracts observed are initiated in one of two ways, depending upon the intended role of the initiator. One who wishes to be an insertee makes this move by taking hold of his partner's exposed and erect penis. One who wants to be an insertor under the terms of the contract steps into the stall where the prospective insertee is seated. If neither of these moves is rejected, the contract is sealed.

Manipulation of the other's organ is reciprocated in about half of the cases. Some respondents have indicated that they appreciate this gesture of mutuality, but it is not at all essential to the agreement reached. The lack of negative response from the recipient of the action is enough to seal the contract. It is interesting, in this connection, to note that such motions are seldom met with rejection (only one of my systematic observations records such a break in the action). By moving through gradual stages, the actors have achieved enough silent communication to guarantee mutuality.

In the positioning and signaling phases of the game, the players have already indicated their intentions. This stage, then, merely formalizes the agreement and sets the terms of the payoff. One should note, in this connection, that a party's relative aggressiveness or passivity in this phase of the game does not, in itself, indicate the role to be acted out at the climax of the interaction. In connection with the positioning of the first move, however, it does provide an indication of future role identification: the man who is seated in the stall *and* is the passive party to the contract will generally end up as an insertee; the man who stands at the urinal *and* is passive in the contract stage, however, will

Major Strategy Systems in Tearoom Encounters
(Source: Systematic observations of 48 encounters
ending in fellatio.)

	Insertee	Insertor
Active	position: urinal	position: urinal
	signal: casual masturbation	signal: casual masturbation
	contract: manipulates partner's penis	contract: steps into partner's stall
	(27%)* (41%)**	
Passive	position: sits in stall	position: urinal
	signal: masturbation (sometimes beckoning)	signal: masturbation (sometimes beckoning)
	contract: accepts partner's entry	contract: accepts partner's manipulation
	(50%)*	(27%)**
Totals	N = 37 (77%)*	N = 33 (68%)**

*Percentage of total insertees.

**Percentage of total insertors.

NOTE: Eleven insertees (23%) and fifteen insertors (32%) followed a variety of minor strategy systems, combining elements of the major systems with idiosyncratic moves.

usually be an insertor at the payoff. The more active insertees play from the urinal and initiate the contract by groping, but active insertors play from the urinal (or elsewhere in the room) and initiate the contract by entering the stall of a passive insertee (see table).

The systems of strategy illustrated in the table account for 77 percent of the patterns of play for the insertees observed and for 68 percent of the insertors' moves. (Again, the insertee role seems to be most stable in the encounters.) The Active-Insertee and Passive-Insertor Systems have already been illustrated and discussed in detail. An illustration from the systematic observation reports of the other two systems follows:

[This was the fourth encounter observed in this tearoom within an hour. The man here identified as X had been the fellator in the second of these actions and had remained seated on the stool throughout the third. I estimated his age at about fifty. He was thin, had grey hair, and wore glasses. Y was about thirty-five, wearing white jeans and a green sport shirt, and was described as "neat." He was well-tanned, had black hair, was balding, and drove up in a new, luxury-class automobile.]

Saw Y approaching tearoom from bridge, so I got into room just before him. I stood at first urinal for a minute, until he began to masturbate at other one—then I moved to the window and looked out on the road. I could see X peering through glory hole at Y when I was at the urinal. Y moved to opposite window and looked out. He then turned to look at X and me. He was stroking his penis through his pants, maintaining erection. I nodded to him to go ahead. He moved into first stall where X began to fellate him. This took less than five minutes. He wiped penis on tissue and left. X got up, zipped pants and left.

"X" played the Passive-Insertee System throughout these encounters. "Y" followed the Active-Insertor System, with some reassurance from the observer. The total time of this encounter—from entry of "Y" until his departure—was ten minutes. Note that the glory hole has three functions (the first two of which were employed in this encounter): as a peephole for observation, as a signaling device, and as a place of entry for the penis into the insertee's stall. The latter is very rare in the tearooms observed, and I have only twice seen these openings used in such a manner.

There may be forms of contracting other than the two I have described. I once observed a contract effected by the insertor's unzipping his pants directly in front of a prospective partner in the middle of the restroom. Another time, I noticed an active insertee grope a man whose erection was showing while his pants were still zipped. The move was not rejected, and the object of this strategy then played the insertor role. In a very few instances, my observations indicate that the insertee entered a stall where the eventual insertor was seated or that the insertor took hold of the partner's exposed penis at the urinals. These exceptions, while rare, make it necessary to withhold judgment as to what roles are being played until the payoff phase itself.

FOREPLAY

Although optional and quite variable, sexual foreplay may be seen as constituting a fifth phase of the tearoom encounters. Like maneuvering, it has very little communicative function and is not essential to production of the payoff. From positioning to payoff, nearly all players—and some of the waiters—engage in automanipulation. There is little need, therefore, to prepare the insertee for fellatio by any other means of stimulation.

Unlike coitus, oral-genital sex does not require rigidity of the organ for adequate penetration of the orifice. Whereas an erection is a necessary signal in the early phases of the game, interruptions and repositioning between the contract and payoff stages occasionally result in the loss of an erection by the prospective insertor. The observer has noted that it is not uncommon for a fellator to take the other man's penis in his mouth even in its flaccid state. The male sex organ is a versatile instrument. With the proper psychosocial circumstances (varying with the individual's prior conditioning), it can reach the orgasmic phase in less than a minute. The authors of *Human Sexual Response* briefly discuss the many factors that intersect in determining the length of the "sexual response cycle":

The first or excitement phase of the human cycle of sexual response develops from any source of somatogenic or psychogenic stimulation. The stimulative factor is of major import in establishing sufficient increment of sexual tensions to extend the cycle. If the stimulation remains adequate to individual demand, the intensity of response usually increases rapidly. In this manner the excitement phase is accelerated or shortened. If the stimulative approach is physically or psychologically objectionable, or is interrupted, the excitement phase may be prolonged greatly or even aborted. This first segment and the final segment (resolution phase) consume most of the time expended in the complete cycle of human sexual response.[4]

Foreplay may help in maintaining the level of stimulation required for advancing the response cycle. Such strategies as mutual masturbation and oral contact in the pubic area may not only add appreciably to the sensual pleasure of the players but may help to precipitate orgasm when the participants are operating under the pressure of time and threatened intrusion:

> It was now raining hard. O remained standing at window, saw X leave car and enter. Y drove up as X was walking toward tearoom, waited for about three minutes in car and run through rain to tearoom. X went to urinal nearest window. Y went to other urinal, urinated, and began to play with his penis, stroking head slowly. Couldn't see what X was doing. X then moved over by Y (they had both been looking at one another) and took hold of his penis. Y did not reciprocate or withdraw. Y then moved over to far stall, still masturbating. X went over and stood by him, taking hold of his penis again. X's pants were zipped and I could not see evidence of an erection. He unbuttoned Y's pants and slipped them down to his knees, as he did with his shorts. Playing with Y's testicles and stroking his legs, he began to fellate him. . . .

It should be noted that, due to the danger of interruption, participants in this gaming encounter seldom lower their pants to the floor or unbutton any other clothing. They generally remain ready to engage in covering action at a moment's notice. Perhaps the rain gave these men . . . a . . . sense of invulnerability.

THE PAYOFF

The action now moves into its culminating stage. As is illustrated by continuing the above narrative, intrusions may temporarily detach the payoff phase from the action that leads up to it, providing moments of incongruous suspense:

> Two kids, B and C, came running toward facility with fishing poles. I coughed. Y sat down on stool and X moved over to window. B and C entered, talking loudly and laughing at the rain. They rearranged some fishing gear in a box, then ran back outside, the rain having let up slightly.

D, an older boy around fourteen, came riding up on a bicycle from the bridge. I did not see him coming, but X and Y were still separated. D entered, went to urinal, urinated, looked out window by me and said, "Sure is raining out!" I remarked that it was letting up. "Guess I'll make a dash for it," he said. He left and rode off toward street. . . .

X and Y resumed activity with X working his head back and forth and rubbing his hands up under shirt of Y, who was again standing. I saw A coming up walk and coughed. X and Y broke—Y sat down—X moved back to window. A entered and I recognized him. I nodded to X and Y, who resumed fellating position. A peered around edge of stall to watch, then stood up and looked over partition to get a better view, masturbating as he stood. Y moaned at orgasm and pressed on back of X's head. X stood up and continued masturbating Y even after orgasm. Y withdrew in a minute and pulled up pants. X moved back to window. Y looked for paper in other stall but it had been used up. He tucked his penis in pants, zipped up and left. X came over to window by me and looked me over. I smiled and left. A remained on stool.

Among other things illustrated here, one may notice the importance of hand play in the sexual act itself. The observations indicate that body and hand movements carry the action through stages that, lacking conversation, might otherwise be awkward. Primarily by means of the hands, the structure of the encounter is well maintained, in spite of the absence of verbal encouragement. Caresses, friendly pats, relaxed salutes, support with the hands, and thrusting motions are all to be observed throughout the action. Normally, the man who takes the insertor role will sustain the action of the fellator by clasping the back of his head or neck or by placing his hands on the partner's shoulders. As a frequent insertee points out:

> When you are having sex, it's not just that the sexual organ is being activated. The whole body comes into it. And you want to use your whole body, and your hands are a very important part in sex. Next to the organs themselves, I think the hands are the most important part, even more important than the mouth for kissing. I really think the hands are more important—because you can do fantastic things, if you know how to do them. You can do fantastic things with your hands to another person's body.

Without the use of scientific instruments other than the human eye, it is impossible to say what proportion of the hand play during the sexual act is voluntary or involuntary. During the orgasmic phase, undoubtedly, there is a great deal of involuntary, spasmodic movement of the extremities, such as that described by Masters and Johnson:

> This involuntary spasm of the striated musculature of the hands and feet is an indication of high levels of sexual tension. Carpopedal spasm has been observed more frequently during male masturbatory episodes than during intercourse, regardless of body positioning.[5]

For physiological reasons, such spasmodic clutching of the hands is engaged in only by the insertor and confined mostly to the period surrounding orgasm.

The insertee, however, may have certain functional reasons for handling his partner. Some respondents have spoken of clutching the base of the penis with a hand, in order to ward off a thrust that may cause them to gag or choke:

> If the man has a very large piece of meat—I know from experience—I will not have somebody ram that thing down my throat. I'm sorry, but that hurts! It can cause a person to vomit. [Like if you put your fingers down your throat?] Exactly, and this can be very embarrassing. So, ordinarily, I will try to hold on. I know just about how much I can take. Then I am going to put my hand in a certain place, and I know it can't go any further than my hand. . . .

The same participant continues by describing another functional use of the hands during fellatio:

> Then I use my hands on the balls, too—on the scrotum. This can do wild things! [You said something about the hips. Or did I imagine that?] The hips or the backs of the legs. Now, there is one value in this which some people don't realize: these muscles contract first at the point of orgasm. This is one of the first signs of orgasm. When these muscles back in here begin to contract (the legs stiffen, these muscles contract or flex, or whatever—they get hard), you can tell at this point the orgasm is about to be reached. It is very helpful to know these things, especially if you are doing somebody. Because you can tell to go faster—or keep doing what you are doing. You can at least get ready and know not to pull away all of a sudden.

I suspect that, if one were to concentrate on observing peripheral matters in a study of heterosexual intercourse, he would find the same pattern of hand involvement: exploration of the partner's body, support of the head or pressure on the back, stimulation of the erotic zones, numerous caresses. At least in the payoff phase, silence in sexual encounters is not confined to the tearooms. When body communicates directly with body, spoken language is no longer essential. Thus far in history, the action of sex is the only universal language—perhaps because the tongue is but one among many members to convey the message, and the larynx is less important than the lips.

As has been indicated, it is a lack of such physical involvement—along with the silence—that tends to make tearoom sex less personal. When hand play does occur, therefore, it tends to raise the involvement level of the sexual action. Perhaps for that reason, some people attempt to avoid it:

> I saw X's hand as he motioned Y over into his stall. Y entered, stood facing X and unzipped. X ran his hands all over Y's buttocks, the back of his legs and up under his shirt while sucking. Y stood rather still with hands held out just far enough from his sides to give X freedom of movement without touching him.

The primary physical connection between the partners is that of the mouth, lips, and tongue of the fellator with the penis of the insertor. The friction and sucking action in the meeting of these organs is what produces the orgasm upon which the encounter focuses. A number of my respondents claim that the physical sensation of oral-genital copulation, while not unlike that of coitus, is actually more stimulating—or "exciting," as they generally word it. While some of the married men among the cooperating participants say that they actually prefer the sensations of fellatio to those of coitus, most agree that this is true only when certain other variables are held constant, when both acts take place in bed, for example. Many tend to look on tearoom sex as only a substitute for "the real thing."

> It's different—and I like both. I guess you could say I'd rather have sex with my wife. Getting a blow job isn't like having the real thing, but it has its points, too. I just don't know. I hadn't thought about it that way. I guess you really can't compare the two. Let's just say I like them both.

Some insertees retain the seminal fluid and swallow it: others clear their throats and spit it out. In one-fifth of the encounters observed, I noted that the insertee spit following the ejaculation of his partner. One respondent claimed that he only spits it out "when it tastes bad":

> The variety of tastes is unbelievable! You can almost tell what a person's diet is by what it tastes like. A person with a good, well-balanced ordinary diet, the fluid has a very mild, tangy, salty flavor. A person who has been drinking heavily—even if they aren't drunk or suffering from a hang-over, if they drink a lot—the stuff tastes like alcohol. And I mean pure, rot-gut alcohol, the vilest taste in the world!

I was unable to find any medical references to the taste of ejaculatory fluid, so I have not been able to verify this connoisseur's judgment. Other respondents will say only that they do think "some men taste different than others."

Acts of fellatio generally take place within the stall. This puts the insertee in a more comfortable position than crouching on his haunches elsewhere in the room. It also has an advantage in case of an intrusion, in that only one party to the action needs to move. There are certain tactical advantages as well. If the man who prefers the insertee position takes a stall and remains there for any length of time, he legitimizes himself, indicates the role he wishes to play, and needs only to wait for a partner to arrive.

Another twenty-nine percent of the observed acts took place in front of the urinals. From that position, both may turn to face the fixtures in case of an intrusion. The fellator is poorly braced for his action, however, and probably quite uncomfortable. Occasionally, the act will occur in front of a window. I am informed that this is especially true when no lookout is present, because it has the advantage of enabling the insertor to double as lookout. When the oral-genital contact takes place away from the stool, the insertee will generally squat or drop to his knees to make the necessary contact.

During oral copulation, other men may come from around the room into viewing position. Many will proceed to masturbate while watching, sometimes without opening their pants for the automanipulation. Seldom does the exchange that is the focal point of this attention last more than a few minutes. In looking over my data, I found indications that I had grossly overestimated the amount of time lapsed between insertion and orgasm. What seemed to me like "a long time" (sometimes recorded as five or ten minutes) was probably a reflection only of my nervousness during the payoff stage. Since I was attempting to pass as a voyeur-lookout (both aspects of the role requiring my closest attention during these moments), it was impossible for me to use my watch in timing. No true voyeur would glance at his watch in the middle of a sex act! Actually, I suspect that the oral penetration ranged from ten seconds to five minutes, not counting interruptions.

I have twice seen couples engaging in anal intercourse. This is a form of sexual activity rare in most tearooms, probably due to the great amount of time required and the drastic rearrangement of clothing involved, both of which tend to increase the danger of being apprehended in the act. Mutual masturbation is an occasional means of reaching orgasm, particularly by the urinals or elsewhere in a crowded room.

CLEARING THE FIELD

Once the sexual exchange is accomplished, most insertors step into a stall to use the toilet paper. After the penis is cleansed, clothes are rearranged and flies zipped. In those rare cases in the observed facilities where a workable wash basin is provided, the participants may wash their hands before leaving.

Nearly always in the observation records, when a man took the insertor role he left for his car immediately after cleansing. The insertee may leave, too, but he frequently waits in the tearoom for someone else to enter. Sometimes he becomes the insertor in a subsequent encounter, as in the following account:

> [X is about forty-five, wearing a green banlon shirt, light blue slacks, driving a red, late model sports coupe. Y is about thirty, driving a green Ford convertible. He wears a light blue shirt, dark blue slacks, and a conservative tie. He is described as being tan, masculine, well-dressed. B is about forty, balding, thin, tanned, wearing horn-rimmed glasses and a grey sport shirt. It is 2:25 on a beautiful Thursday afternoon, and there are few people in the park.]

> B was seated on stool when O entered. O stood at urinal a minute, noticed B watching him through glory hole. Crossed to far window, looked out and lit cigarette. Y entered and went to first urinal. X came in soon after Y and went to third urinal. They stood there for about five minutes. X kept peering over edge of stall at B and also at me. I crossed to opposite window and looked out missing pane. X was masturbating. Y went to second stall, lowered pants and sat down. I went back to window on right. Y

spread his legs and began masturbating. (He had removed his coat and hung it over the edge of stall.) He had slumped on the stool seat as if sitting in camp chair, legs stretched out almost straight in front of him but spread apart and was masturbating obviously. I went back to window overlooking street. X then moved to window by Y, stood there a minute, then leaned over and took hold of Y's penis and began stroking it. He then knelt on the floor to begin fellatio. B just sat in his stall and masturbated. Y moaned a bit at climax. He then wiped and X stood back by window and masturbated while watching Y. Y flushed toilet, put on his coat, zipped pants and left. X stepped in B's stall and was sucked by him. This didn't take more than a minute. X then went to urinal, cleared throat, spit and left. (I was able to see autos of both X and Y through window.)

This is what I have labeled a series encounter. Generally, in order to facilitate the eventual analysis, I have broken these up into "Encounter A," "Encounter B," etc. In the previous instance and in a few others among my systematic observations, I was not able to do so because of the rapid succession of events. During the hunting season, series encounters are the most common variety. Once the action begins to "swing," a series may last throughout the day, each group of participants trading upon the legitimation process of the previous game.

Another type of action that tends to swell the volume of sexual acts in a given facility is the simultaneous encounter, in which more than one sexual act is in process at the same time. The payoff phases are seldom reached at exactly the same time in these encounters, but they are staggered as in a round.

[It is a warm, humid, Friday afternoon. A few youngsters are playing ball in the park and some heterosexual couples are parked nearby. X is about thirty-five, tough looking, tattooed, dirty working clothes, drives an old Chevrolet. Y is about forty-five, lean and tanned, wearing tan work clothes. A is about thirty-two, neatly dressed with sport shirt and tie. I describe him as "masculine looking but wore pinky ring." He drove a new, foreign economy car. B is about fifty, heavy set, grey hair, sports clothes, rather unkempt. C is about forty, with a pot belly, wearing white sports shirt, dark blue pants.]

When O entered, X and Y were seated on stools with A standing by far window facing into room. While O urinated, he noticed that X was watching him through glory hole. O lingered at urinal for about four minutes, during which time A moved into stall with X (X is a noisy sucker, much "slurping," so I could tell what was happening but could not see). O crossed to far window, lit cigarette and peered out. A left first stall and stepped into space between Y and O. He stood there, masturbating both himself and Y, who had stood up. Meanwhile, C, who was sitting on bridge watching tearoom when O entered, came into the room. O saw him approaching through window and coughed. Y and A broke contact for the moment but, recognizing C, returned to action. C stood at urinal less than a minute, halted for another minute opposite stalls, then went into stall with X, who proceeded to fellate him. Y then stood on the toilet seat, watching X and C, while A sucked Y (A had to crouch but continued to masturbate). It was getting crowded on that side of the room, so O moved to opposite window. From this position, he could only see part of A's backside, Y's face and shoulders, and the backside of C. X kneaded C's buttocks and ran his hands up and down the backs of C's legs. When C finished, he left without wiping. A finished with Y about this time, went to urinal number three and spit. Y wiped and left. A stepped over to window by O, peered out through broken pane. His pants were still unzipped and he proceeded to masturbate and to look at O suggestively. O, feeling uncomfortable, went back to far window. B entered and A zipped up pants. B looked around as he went to middle urinal but stayed there for a brief time. He then moved over to stall with X. No one seemed to be made uncomfortable by B and seemed to recognize him. O then left, followed closely by A, who engaged him in conversation by water fountain. . . . All of this took place in twenty-five minutes.

The reader should be able to sense, at this point, that what I have described as a rather simple, six-step game (only four of which are essential to the action) may be acted out with infinite variety and confusing modifications. Every encounter reduces, ultimately, to the basic steps of positioning, signaling, contracting, and payoff; but no two of them are quite alike.

A pat on the shoulder, a wave of the hand, an occasional whispered "thanks" concludes the action. The departure ritual is simple and brief. Once the field is cleared, some individuals go to their homes or jobs, others return to their cars and await the arrival of fresh players, and a few may venture to a different tearoom to take their positions in another encounter.

The length of these games was observed to range between five and forty minutes, with an average duration of about eighteen minutes. Tearoom encounters, then, require relatively little time—a quarter of an hour if one knows where to go and how to play the game. Many suburban housewives may think their husbands delayed by the traffic when, in reality, the spouses have paused for a tearoom encounter. . . .

NOTES

1. Letter in "Open Forum: Sex in Public Places," edited by Larry Carlson in *Vector*, Vol. 3, No. 6 (May, 1967), p. 15. In my opinion, *Vector* is the best of the homophile journals. It is published monthly by the Society for Individual Rights, 83 Sixth Street, San Francisco, California.

2. Marvin B. Scott, *The Racing Game* (Chicago: Aldine, 1968), p. 3.

3. From Georg Simmel, *Soziologie*, as quoted in Goffman, *Behavior in Public Places* (New York: The Free Press, 1963), p. 93. For a thorough discussion of the use of eye contact in face-to-face engagements, see pp. 91–96 of Goffman's book.

4. William H. Masters and Virginia E. Johnson, *Human Sexual Response* (Boston: Little, Brown, 1966), pp. 5–6.

5. *Ibid.*, p. 173. See also pp. 296–297.

33

Turn-Ons for Money

Interactional Strategies of the Table Dancer

CAROL RAMBO RONAI
AND CAROLYN ELLIS

She swayed from side to side above him, her hands on his shoulders, her
knee brushing gently against the bulge in his pants. He looked up at the
bottom of her breasts, close enough to touch, but subtly forbidden. His
breath came in ever shorter gasps.

This is the world of the table dancer—a world where women exchange
titillating dances for money. Our study looks at the dynamic processes
of interaction that occur in the exchange. Previous studies (Carey et al.,
1974; Gonos, 1976; McCaghy and Skipper, 1969, 1972; Salutin, 1971; Skipper and McCaghy, 1970, 1971) have concentrated on "burlesque" or "go-go"
dancers, sometimes referring to them more generally as stripteasers. Dancers'
interactions with customers were restricted, for the most part, to the stage setting where they danced and received money from customers. Because investigators in these studies occupied positions as researchers or researchers as
customers, and relied to a large extent on survey and interview techniques,
this work led to a static description of this occupation.

Boles and Garbin (1974) have looked at customer-stripper interaction in a
setting where strippers sold drinks in addition to performing stage acts. Although they described interaction, they interpreted it in terms of norms, club
motif, and customer goals. They found that the conflict between customers'
goals and strippers' goals resulted in "counterfeit intimacy" (Foote, 1954), a situation in which an aura of intimacy masked mutually exploitative interactions.

Although counterfeit intimacy is a structural reality in such contexts, this
description created another model of behavior that ignored the interactive,
dynamic nature of the exchanges and set up in its place stiff caricatures behav-
ing in an unbending, cardboard manner. As actors get caught up in dialogue,
they exchange symbols, extract meanings, and modify expectations of what
goals they can reasonably expect to reach. Interaction has a tentative quality
(Blumer, 1969; Turner, 1962); goals are in a constant state of flux.

From "Turn-Ons for Money: Interactional Strategies of the Table Dancer," Carol Rambo
Ronai and Carolyn Ellis, *Journal of Contemporary Ethnography*, Vol. 18, No. 3, 1989.
Reprinted by permission of Sage Publications, Inc.

The nature of selling and performing table dances that we describe yields more opportunity for interaction between customer and dancer than in previous studies. A table dancer must be a charming and sexy companion, keep the customer interested and turned on, make him feel special, and be a good reader of character and a successful salesperson; at the same time, she must deal with her own negative feelings about the customer or herself, negotiate limits, and then keep him under control to avoid getting fired by management.

Much of the early research literature has described stripping as a deviant occupation. Later, Prus and Irini (1980) looked at stripping as conforming to the norms of a bar subculture. Demystifying this "deviant" activity even further, we show that bargaining strategies in the bar actually mirror "respectable" negotiation in mainstream culture.

We begin by discussing the methods we used to elicit in-depth understanding of strategies used by table dancers. After describing the dance club setting, we turn to a description and analysis of particular tactics used on the stage, at the tables between stage acts, and then during the table dances in the pits. Our conclusion analyzes how this exchange reflects buying and selling in service occupations as well as the negotiation of gender relationships in mainstream society.

METHODS

Our study approaches stripping from the point of view of dancers and the dancer as researcher, the people with the most access to the thoughts, feelings, and strategies of exotic dancers. Dancers concentrate on manipulating men as they pursue money in exchange for a turn-on. In order for their strategies to work, they must understand and coordinate them with the games of men.

Our information comes primarily from the experiences of the first author, who danced during 1984 and 1985 to pay her way through school. As a "complete-member-researcher" (Adler and Adler, 1987), she conducted opportunistic research (Riemer, 1977), that is, she studied a setting in which she was already a member. She interviewed dancers to find out how and why they began this occupation and kept a journal of events that happened while dancing. Later, she reconstructed, in chronological field notes, a retrospective account of her own dancing history, paying special attention to strategies, emotion work, and identity issues. She used "systematic sociological introspection" (Ellis, forthcoming) to put herself mentally and emotionally back into her experiences and record what she remembered (see Bulmer's, 1982, concept of "retrospective participant observation").

In May 1987, the first author danced in one strip bar for the explicit purpose of gathering data for a master's thesis, chaired by the second author. With approval of bar management, but without the knowledge of other dancers, she acted in the dual capacity of researcher and dancer. This time her primary identity was that of researcher, although as a complete-member-researcher she attempted to "become the phenomenon" (Adler and Adler, 1987; Jorgensen, 1989; Mehan and Wood, 1975). When she danced, she took on the identity of a dancer, suffered identity conflicts similar to those she had experienced during earlier dancing, and shared a common set of experiences and feelings with other dancers. She kept field notes of events, which were buttressed by "interactive introspection" (Ellis, 1988), whereas the second author talked her through her experiences, probing at and recording her feelings and thoughts. She conducted informal interviews in the dressing room with dancers and on the floor with customers. Sometimes she revealed her dual role to customers as a strategy to keep them interested in spending more money and to get them to introspect about their own motives for being in the bar.

Because this article is concerned with describing dancers' subtle manipulation strategies that occurred semiprivately, we pulled much of our material from episodes engaged in by the first author, in which process was most easily observed. Because we believe that sociologists should acknowledge the role of their own introspection in their research (Ellis, forthcoming), the first author reveals which of the experiences in the article are hers. Throughout this article, we refer to the first author by her dancer name, Sabrina.

We realize the bias inherent in using introspection primarily from one source. For example, Sabrina, more than most dancers, tended to attract customers interested in mental stimulation as well as physical turn-on. Yet we could not have gained an in-depth understanding of intimate exchange, for example during table dances, in any other way. To understand this bias, we compared Sabrina's strategies and experiences with those of other dancers we observed and other bar participants with whom we talked. Later in 1987, we conducted interviews with four strippers, eight customers, four managers, three bar owners, and a law officer. This article then uses a triangulated method (Denzin, 1978; Webb et al., 1965) to present typical responses from field work and in-depth ones from current and retrospective introspection.

SETTING

An exotic dance club located in the Tampa Bay area of Florida provided the setting for this study. Since liquor was served, full nudity was prohibited by state law. Appearing individually in full costume on stage, each stripper gradually removed her clothing during a dance routine. By the end of the act, the dancer wore pasties that concealed her nipples and panties that covered genitals, pubic hair, and the cheeks of her derriere. Men handed out tips to dancers during performances.

Between acts, dancers strolled around the floor, making themselves available to spend time with customers. They made money if customers bought them drinks. However, the main attraction and source of income in this bar was the table dance. A dancer "sold" dances in a complicated negotiation process through which she convinced the client that he was turned on to her and/or that she was turned on to him. At the same time, she controlled the situation so that she was not caught disobeying "house" rules, many of which corresponded to what county authorities considered illegal. For example, since "charging" for a table dance was considered soliciting, the dancer, using word games similar to those used by the masseuse studied by Rasmussen and Kuhn (1976), suggested that there was "generally a contribution of $5."

After a dancer successfully sold a dance, she led her customer to one of the two elevated corners of the bar, known generically as the "The Pit," and affectionately nicknamed by customers as "Horny Holler" and "The Passion Pit." Railings and dim lights offered an artificial boundary between this area and the rest of the bar. Clothed in a bralike top and full panties or other revealing costume, the dancer leaned over a seated patron, her legs inside his, and swayed suggestively in rhythm to the music playing in the bar. Theoretically, customers were allowed to touch only the hips, waist, back, and outside of a dancer's legs. Many men tried and some succeeded in doing more. Disobeying rules prohibiting direct sexual stimulation or touching meant more money for dancers, but it also meant risking that management might reprimand them or that a "customer" would turn out to be an undercover officer or a representative looking for infractions on behalf of club management.

ELEMENTS OF STRATEGY

On the Stage

A dancer used symbols that appealed to her audience. At the same time, these symbols distanced her from customers and denoted that the stage was a performance frame (Goffman, 1974; Mullen, 1985). Her appearance, eye contact, manner, and choice of music made up her main expressive equipment.

Having a "centerfold" figure was an obvious asset for dancers. But the best looking woman did not always make the most money. A dancer's presentation of self was also a crucial factor in a customer's decision to tip her. Similar to strippers described by Gonos (1976) and Robboy (1985), women often portrayed exaggerated stereotypes through their clothing style and movement. For instance, a "vamp style" dancer wore suggestive street clothing such as a leather micro-mini skirt, spike-heeled boots, and a halter-style top while strutting around the stage displaying overt sexual mannerisms such as "flushing" (opening her shirt to reveal her pasty-clad breasts). Others had a "gimmick." For example, one woman was an acrobat; another stood on her head while twirling her large breasts. In contrast, a more sensual dancer dressed in sexy bedroom clothing such as a corset and garters or a teddy, and displayed subtle sensual behavior such as slow undulation of the hips.

A dancer chose symbols that drew a certain type of customer to her. Dressing the part of the vamp, for example, reflected an extroverted attitude that attracted customers out to have a good time. Overtly sexual dancers were more likely to perform sexual favors in the bar or meet a man for sex outside the bar. The sensual presentation of self attracted customers who were interested in a "serious," private interaction. Customers interpreted each dancer's symbols as cues to what it might be like to interact with her or, specifically, to have sex with her. For example, Jim, a regular customer, discussed Samantha, a sensual dancer: "She is nothing to look at. God, she's only twenty-six, and we both know she looks like forty. But the way she moves, man! She promises the moon and stars in bed."

Most dancers used eye contact to "feel out" a patron. Managing frequent eye contact while dancing on stage usually meant a tip for the dancer and made a customer feel as if a dancer was specifically interested in him.

A dancer's first close contact with a customer often occurred while accepting a tip. During the exchange, the dancer formed impressions about how the customer was reacting to her, and the customer decided whether he was attracted to the woman. The customer stood at the side of the stage holding currency, which signaled the dancer that he wanted to tip her. The dancer greeted him while accepting the tip in her garter and said "thanks," perhaps giving him a "special" look.

At this point, a dancer might choose from several courses of action, such as "coming on" to a customer, doting on a customer, and using humor. When dancers "came on" to customers, they grinned, wiggled their breasts, spread their legs, struck their buttocks, suggestively sucked their fingers, talked dirty, or French kissed.

Others, such as the sensual dancer, doted on a customer for a few seconds. She caressed his arm, wrapped her arms around his neck, and smiled while he tipped her. If she felt confident of his interest, typical comments she might make were: "I would love a chance to get to know you," or "I look forward to sitting with you," which meant accompanying him to his table after her stage performance.

Humor was an effective and safe tool for generating a good impression while accepting a tip on stage. Customers generally construed a funny statement made by a dancer as friendly and spontaneous. Often it made a nervous client more at ease. Sabrina noted lines she used: "What's a nice guy like you doing in a dump like this?" or "I bet you'd look better up here than I do."

Familiar with the usual "acts" of dancers, such as coming on and showing phony interest, customers were pleased when they thought a woman had "dropped the routine." Often this meant only that she had staged a less frequently displayed one. A dancer had to be careful not to use the same line more than once on the same person, or let a customer overhear it being used on another man. No matter a customer's taste, he wanted a sincere performance.

Dick, a customer who was feeling jilted one evening, commented to Sabrina: "The thing with that chick, Dana, is that she makes a big deal out of you while she is onstage, but if you watch her real close, you notice she looks at everyone who tips her 'that way.'" Another customer reported he did not like a dancer in the bar named Tammy because she was insincere: "She frenched me and told me to insert my dollar deeply (in her garter). Now I ain't stupid. I know a come on like that is a fake."

A dancer's music affected how a customer viewed her. This was reflected in Tim's comment about Jessica: "That girl has a great body, but every time I hear her music [heavy metal] I get the creeps thinking about what she must be like." While most women danced to top-40 music, some used other music to attract a tip from a particular kind of client. Mae, an older dancer in her late thirties, played country music and presented herself as a country woman. Bikers and blue-collar workers were loyal to Mae, expressing sentiments like: "She's the only *real* woman in the bar."

On the Floor

Offstage, interaction was even more complex. Between stage performances, a dancer circulated among customers and offered her company. Body language, expressions, and general appearance helped define each customer's interest in her and the difficulty of being with him. Once a dancer located an interested customer and introduced herself, or followed up on a contact made while performing on stage, she then had to convince him that he wanted to spend time with her. Ordinarily, her eventual goal was to sell a table dance.

CHOOSING A CUSTOMER

The ideal customer had a pleasant disposition, was good looking, had time and money to spend, and was sitting at one of the tables on the floor. Most customers did not meet all these criteria. Dancers weighed these features for each customer and also compared them against the circumstances of the evening. Sabrina often asked herself: "What do I want more right now? Money or someone nonthreatening to sit with?" Her answer was different depending upon time of night, how much money she had made already, and how she felt at the moment. Other dancers made the same calculations. For example, three hours before the bar closed one night, Naomi said, "I know this guy I'm sitting with doesn't have a lot of money, but I've made my hundred for the night so I can afford to take it easy." Another time, Vicky said, "God! I know I should be out there hustling instead of drinking with Jim, but I just can't get into it. I guess I'll just get fucked-up and blow it off today." Darcy displayed a more typical attitude, "It's twelve thirty already and I haven't made shit! This guy I'm sitting with better cough it up or I'm taking off." Negotiations with oneself and with the customer were always in process. Throughout the interaction, each participant tried to ascertain what she or he was willing to give and how much could be acquired from the other.

Attractive customers appeared, at first, more appealing. They were pleasant to look at and the dancer could pretend to be on a date while sitting with them. But these men seemed to know they were more desirable than others in the bar and were more likely to bargain with those resources. The end result was that the dancer spent most of the interaction trying to convince the customer to spend money while he tried to persuade her to go out on a date.

When Sabrina was new to the profession, she decided one evening to sit with a good looking, blonde-haired man. She reported the following:

I started talking to him and eventually led the conversation to the point where I asked, "Would you care for a table dance?"

"Later," he replied.

I continued to make small talk. "Do you come in here often?"

"I stop in once every few months," he responded.

For the next 15 minutes we covered various topics of conversation such as his job and my schooling. Then I asked him again, "Do you want a table dance?"

"Are you going to go to 'le Bistro' with me tomorrow night?"

"I'll think about it," I responded, in hopes of getting a table dance out of him before I turned him down. "Do you want that table dance?"

"Will you go out with me?" he insisted.

"I'm still considering it," I lied.

We volleyed back and forth for 30 minutes. Finally, he told me, "I don't want a dance. I just want to know if you will go out with me."

This customer was aware that Sabrina would not stay with him unless she thought he might want a dance. Both used strategies and gambled time hoping one would give in to the other's goals. Each lost a bet.

Sometimes customers who were old, heavy, unattractive, or otherwise weak in social resources came into the bar. Many women avoided these men, while others, like Sabrina, realized unattractive men were eager for company and tended to treat a dancer better and spend more money than their more attractive competitors would. With the right strategies, dancers could control these men. For example, a dancer might corner a customer into treating her as he would his granddaughter by acting polite and addressing him as "sir." This insinuated that, of course, he would never act inappropriately. Some accepted the role to such an extent that they acted like grandfathers. One man told Scarlet that she was cute, tweaked her cheek, and compared her to his granddaughter.

When scanning the bar and deciding whom to approach first, a dancer tried to find the man who appeared to have the most money. Logically, the better a customer was dressed, the more likely he was to have money. However, he also had a higher probability of already being in the company of another dancer.

Making sure a customer was not spoken for by another dancer was important. It was considered dangerous (one could get into an argument) and rude to sit with another dancer's customer. Some regular customers, for instance, visited the bar to see particular dancers. These customers often turned down another dancer's offer of company by saying they were "waiting for someone." When a dancer entered the bar, she immediately scanned the room, paying particular attention to which women were seated with which customers. If she noticed later that a woman had left a table for a long period of time, she then asked her if it was okay to sit with that customer. This served the dual purpose of following tacit rules (i.e., being polite) and gave the dancer an opportunity to gather information about the customer in question.

Sabrina was warned about a customer in this manner. Upon asking Debbie if she was finished with "the old man in the corner wearing a hat," Debbie replied, "Sure, you can have him. That's 'Merv the perv.' He has lots of money, but he'll want to stick his finger up your asshole for twenty bucks a feel."

A dancer might ignore all other customers to sit with one of her "regulars." When two or more of her regulars were in the bar, she had to juggle them, first sitting with one and then the other. It was difficult to table dance for both of them and still portray "special attachment." Eventually, she had to offer an account (Scott and Lyman, 1968) to one of them. One excuse was to appeal to the principle of fairness: "I really want to be with you, but he came in first and now I have to be with him." Or she might appeal to higher loyalties (Sykes and Matza, 1957), insinuating that the decision was out of her control: "I have to go sit with another customer now. My bosses know I avoid him and they're watching me."

Time in the bar correlated with decreased spending. If a customer had been spending for a while, it was fair to assume that he would run out of money or would soon decide to leave, that is, unless he was intoxicated and freely using a credit card. Dancers in this situation risked having to deal with and control a problematic person who did not remember or pay for the correct number of dances purchased. On the other hand, a dancer might convince a drunk credit card customer to pay for more dances than he actually bought.

A customer's location in the bar indicated his attitude toward female company. In this club, sitting at the bar meant little interest in interacting with dancers. Patrons near the stage wanted to see the show. Being seated at one of the tables in the floor area was conducive to interaction with dancers and to inquiries about table dances.

AT THE TABLES

Once a customer accepted an offer of company, a dancer sat with him and introduced herself. Her overall goal remained fairly consistent—money with no hassle. Many women also enjoyed the attention they received and got an exhibitionist thrill out of being desired and told how beautiful they were. Others believed the compliments were just part of the game. Some liked the feeling of conquering and being in control. Others felt degraded and out of control.

The customer's manifest goal was impersonal, sexual turn-ons for money; a close examination showed other objectives that shadowboxed with and sometimes transcended this more obvious goal. Although most customers initially focused on the pursuit of sex in or outside the bar, they also came looking for a party, to feel good about themselves, to find a friend or companion, or to develop a relationship. A dancer's strategies varied depending on her personality and her perception of the customer.

Some women said nothing. A customer who wanted passive indifference from an attractive female willing to turn him on liked this approach. Sex, not conversation, was his goal. The dancer did not have to initiate activity nor get to know the customer. Her role was to respond as a sexual nonperson by allowing him to kiss and fondle her body. Verbal interaction potentially endangered the continuance of the exchange.

Most customers wanted a dancer to interact with them. Seduction rhetoric (Rasmussen and Kuhn, 1976) became part of the dancer's sexual foreplay before the table dance as well as a vehicle for the customer to persuade the dancer to see him outside the bar. By talking "dirty" and acting "like a whore"—for example, telling stories about kinky sex in her life outside the bar—a dancer could keep a customer "going," eager to buy the next dance, ready to believe the dancer might have sex with him later.

If a customer wanted a prostitute, he dropped hints such as, "Do you do work on the side?" or "Where does a guy go for a good time around here?" or "Do you date?" Sometimes he propositioned outright: "Will you go to bed with me for a hundred dollars?" The more blatant proposals told the dancer that the customer was not a police officer; all of the requests informed her he had money to spend and opened up the possibility of using strategies to extract it.

One strategy dancers used in this situation was to mislead a customer into thinking she might meet him later if he bought table dances from her now. From the first author:

> Ted bought dances from me two at a time. After several of these, he asked, "Are you going to see me at the Holiday Inn tonight?"
>
> "Why should I?" I responded.
>
> "Because I am new in town and have lots of money."
>
> "I don't go out with strange men," I said.
>
> "Well, why don't you get to know me then," he said. He bought two more dances, then asked, "Do you know me now?" I smiled at him.
>
> He continued, "Why don't you meet me after you're done working. What time do you get off?"
>
> In an effort to shift the focus of the conversation, I said suggestively, "When do you get off?"
>
> "I get off on you baby!" He exclaimed. "I'm in room 207. Will you be there?"
>
> To keep him going while not committing myself, I said, "I don't know."
>
> We talked a while, and then he asked again. I replied, "I've never turned a trick in my life. I'm not sure I ever will."
>
> "So we won't do it for money," he said. "Come see me tonight." He buys two more dances and we sit down again. I start the conversation first this time to keep him interested yet deter him from bringing up my meeting him. "Tell me, Ted, what is the kinkiest thing you have ever done in bed." This conversation kept us busy for a while, until, sixty dollars later, he asks, "Do I go to the bank machine or not?"

"What do you mean?" I ask.

"If you are going to see me tonight, I need to go to the teller. I'm out of money."

I had a big grin on my face and asked, "Will you be back here after the teller?"

"Probably not," he replied.

"Too bad," I said.

"Would you see me if I bought more dances?" he asked. I was tempted to say maybe, but I thought at this point I was being too obvious.

"Probably not," I said.

He stood to leave. "You show up tonight at room 207 if you want. It was fun."

Similar to the strippers discussed by Prus and Irini (1980), a few women used the bar setting as a place to make contacts for their prostitution careers, while many more had sex occasionally outside the bar to augment their incomes. Before accepting an offer, a woman usually asked other dancers about the customer or spent time getting to know him. Interacting with him then gave her an opportunity to make money table dancing. Most women claimed they had sex "only for the money." A few, such as Sasha, seemed to enjoy sexual contact in and out of the bar. Sasha's enthusiasm—"I'm so horny, I want a cock tonight"—was deemed deviant by the other dancers, who ostracized her—usually avoided her and talked behind her back—for her overt enjoyment.

The customer who wanted a date outside the bar could be handled in a similar manner to the customer looking for a prostitute. Often a dancer conveyed the impression, "if only I knew you were safe" by saying: "You could be Jack-the-Ripper," "You could be a cop," "It's not safe to date everyone you meet in here." Then she suggested interest by saying, "I need a chance to get to know you better." The logical way for a dancer to get to know the customer was for him to spend time and money buying drinks and table dances from her. Lured by the offer of expensive dinners or vacations, and sometimes attracted by a man she liked, most dancers occasionally accepted dates.

If customers were in the bar "to party" (to be entertained) in groups, such as bachelor parties, a dancer wasted no time on interaction. She asked immediately if they wanted a dance. These men interacted mostly with each other, requiring dancers to be lively and entertaining hostesses while treating them like sex objects. Often they commented on her body—her big tits, nice ass, or ugly face—as though she were not there. Party groups purchased dances with the same attitude and frequency as they bought rounds of drinks.

Most men who came to the bar seemed to want to find a friend or companion, or in some other way be treated as a special person. One of Sabrina's customers left the bar twice during an evening to change shirts, just to see if she recognized him when he returned. The best ploy in this situation was for the dancer to put on an honest front, altercasting (Weinstein and Deutschberger, 1963) her customer into the role of being special and "different" from other men.

Most successful dancers were able to hold conversations with these men. Asking his name, where he lived, occupation, and what he did with his spare time provided initial interaction. Finding common ground helped conversation run smoothly. Asking questions at a leisurely pace, making comments, and showing interest both verbally and nonverbally afforded a semblance of credibility to the conversational process. This dialogue helped the dancer to "check out" (Rasmussen and Kuhn, 1976) the customer to make sure he was not a police officer, determine how much money he had to spend and which of her interactional strategies might make him willing to part with it. Giving the customer an opportunity to talk about himself and to demonstrate whatever expertise he had made him feel good about himself. A customer pleased with his presentation of self was more apt to spend money. Sabrina told this story:

> In the field, I had a regular customer, Ray, who was a systems' analyst. I was shopping for a computer at the time, so I enlisted Ray's assistance. Ray had an opportunity to show off his expertise, and feel like he was helping. He turned-on to the contrast of seeing me as intellectual and a sex object.

The best way for a dancer to convince a customer that she found him appealing and unique was to find a likable characteristic about the customer and continually tell him how impressed she was with him and with that trait. For example, some men liked to be praised for their appearance, success, intelligence, sexual desirability, trustworthiness, or sensitivity. The dancer had to convey to him directly that she preferred his company to others in the bar, or indirectly through such statements as "You're not as vulgar as the rest of these guys in here"; "You're more intelligent than most men I meet in here"; "You're not just another one of these assholes," or "I appreciate your spending time with me. When I'm sitting with you I'm safe from those animals out there." The message was that because of his specialness, she could be "straight" with him, be who she really was, instead of putting on one of her usual acts.

This tactic worked best with customers the dancer liked and enjoyed talking to; otherwise, it was difficult to muster up and maintain the sincerity necessary for a believable performance. When this strategy worked, the dancer had close to total control of the interaction. Then the customer tried hard to meet the dancer's expectations, spending money and treating her like a date or friend to avoid disappointing her. If he stopped spending money, the dancer might say, and sometimes mean, "I'll see you later. Don't get angry with me. I know you understand that I have to make money, although I would rather spend time with you. If I don't find anything, I'll come back and visit." Sometimes the customer responded by spending more money to keep the dancer around. If not, he was forced to "understand" her leaving because he and the dancer had an honest relationship and she had been "straight" with him about the nature of her job. This strategy was an effective way to cultivate regular customers.

Sometimes a dancer did not have anything in common with a customer. Over time, most dancers worked up routine questions to keep conversation flowing. Sabrina frequently used lines such as: "What do you look for in a woman?" "Why do you visit strip bars?" "What is your opinion of that dancer over there?" "I try," she said, "to get the customer to share something personal with me. I like for him to feel like there is something more solid than a salesperson-customer relationship."

Some regular customers acted as if they were involved in a long-term, serious relationship with a dancer. They bought her expensive gifts such as diamonds, minks, cars, and flowers. These customers seemed to forget the businesslike nature of the bar setting.

Dancers in these interactions appeared involved with the customers. However, most did not take the relationship outside the bar, since this would have cut off a source of income. But they tried to convince the men of their desire to leave the bar scene and be saved by them, even though it was impossible now. Sabrina, for example, had many offers from men who wanted to rescue her from the bar. She developed a routine to solicit this desire from men—it usually meant more money for her in the bar—but that allowed her to reject their proposals without causing anger. She explained:

> I presented myself as attractive and intelligent, but helpless, trapped by circumstances. When they asked me to leave the bar, I told them I had to work to pay for school. When they suggested setting me up in a place of my own, I told them I was independent and wanted to do it on my own. This put them off, but kept them interested and earned their respect.

Mae, a dancer mentioned earlier, seemed to have a knack for cultivating these types of relations. Sabrina describes a discussion with Mae while sharing a ride home.

> Mae had been given a mink coat that night by a customer and she had given it back to him.
>
> Always intimidated by this woman, I took a moment to get up some nerve and finally asked, "Why did you give back the mink?" "I couldn't hock it for very much, and I won't use it here in Florida. I'd rather get money," she stated.
>
> "How are you going to get money?" I asked.
>
> "I'll get more money from him by being the type of person who gives this stuff back than if I keep it. I have lots of customers who give me nicer stuff than that mink."
>
> She spoke to the driver, "Hey, do you remember that necklace Tom gave me?"
>
> The driver replied, "It's true, Mae can really get them going. That necklace was a grand, easy."
>
> "Did you keep the necklace?" I asked.
>
> "Hell yes!" she responded.

Mae had a routine that could "really get them going." But she and other dancers, usually the older ones, who used this technique often, took some aspect of the relationship seriously. They saw these men as "options" or possibilities for a life change. On the other hand, they felt this was too good to be true, or were unsure about making the change because of other factors in their lives, such as a husband or children. Keeping the interaction going, yet not allowing it to take place outside the bar, meant they were able to have romance, feel appreciated, and, to some extent, have a relationship while they continued making money in their occupations. However, the occasional relationship that did work out in the bar kept everyone hoping. Sabrina, for example, met her husband there.

CLOSING THE SALE

A dancer rapidly closed a sale on a table dance to a man who wanted sexual favors in the bar. But since these men often violated rules regarding touching and sexual stimulation, some dancers did not feel that they were worth the trouble. For example, one night Annette came into the dressing room and announced, "I just left this old geezer who wanted me to rub him off with my knee. I'm not into it. If someone else wants to, go for it."

The same problems existed after a quick sale to men in the bar for a party. In this situation, a dancer had to concentrate on not acting offended long enough to perform table dances and collect her money. For some dancers, the money was not worth the degradation. As a result, they avoided the bachelor parties.

The customer who wanted to be treated as special took more time. Questioning allowed time for the dancer to convince him that he wanted a table dance from her. It was important that she not appear pushy, yet she needed to determine quickly whether she could make money from this person. Would he buy table dances? Did he want to spend time getting to know a dancer or go directly to a dance? Answers to such questions guided the dancer in constructing her behavior toward the customer.

If a customer purchased a drink for a dancer, she then knew that he was interested enough to spend some time with her. Some customers, however, bought drinks for dancers but refused to purchase table dances, claiming table dances got them "worked up for nothing." If a customer acknowledged that right away, a dancer then had to make a decision about staying or leaving based on the availability of other moneymaking opportunities in the bar. If the action in the club was slow, she might stay with him since she made $1 on every drink he bought for her. Regular customers were always good for a drink: "I'll go sit with Jim today," said Sharon. "At least I know he'll buy me a drink if nothing else." Often a dancer gave the waitress a secret signal indicating that no liquor should be put in her glass. The waitress brought the drink in a special glass, placed a dollar under the dancer's napkin and the drink on top of it.

Most women closed on a dance after the first drink had arrived and it was apparent that the customer liked her. If the customer said no, most dancers left fairly quickly. But in rare cases a customer paid $50–$100 for a dancer to sit with him for a while. This guaranteed the dancer money without trouble and bought the customer companionship. Customers who saw themselves in an involved relationship with a dancer generally rejected table dances in favor of company. When these customers bought dances they treated the dancer gently, barely touching her for fear of offending her.

Even when a dancer was not paid for her company, it was not always a good idea for her to leave immediately when a man refused a table dance. As a rare and novel routine, staying made the dancer appear sincere in her interest and less concerned about making money. Sabrina occasionally used this approach:

> "Why are you still sitting here?" the customer asked immediately after he had turned me down for a table dance.
> "I'm finishing my drink," I replied.
> "Then you are leaving?" he asked.
> "Oh, sir, I had no idea you wanted me to go. You must be waiting for someone. Forgive me for being so rude," I said tongue in cheek. I stood to leave.
> "Hold it, hold it. Sit back down. I don't necessarily want you to leave. The girls always leave after you say no to a dance. You must be new here. You really should leave when customers say no. You won't make any money this way." During this exchange he was clutching my arm. He loosened his grip. "Wouldn't that be rude to just up and walk off?" I asked incredulously. He stares at me a minute, and then smiles. "Lady," he says. "You are a card. I want a table dance." He bought four.

In the Pits

Once a customer agreed to a table dance, another set of complex exchanges took place. Although interaction varied with the particular dancer and customer, common routines offered promise of what was to come. Leading the customer to the pit, one acrobatic dancer followed a routine of bending from the waist and peering at her customer from between straight legs. Ascending the stairs to the pit, she performed various kicks and other gestures to demonstrate her flexibility. Another dancer sashayed gracefully in an elegant and poised, yet seductive, manner. Sabrina's style was to talk in a sexy way as she walked: "See that corner. That's my corner. I love to take my men there."

Once in the pit, a woman sat close to the man. Often she put her hand on his leg, draped an arm on his shoulder, or swung a leg over his lap. Some girls necked with their customers, French kissing with a frenzied passion. Other dancers allowed kisses only on the cheek.

If a customer tried to French kiss when a dancer did not want it, she had several "routines" to control him. Leveling a questioning look at the customer and then backing away from him was enough to stop most men. When a client voiced dissatisfaction over the limitation—"What did you do that for?" or "What's your problem? Why are you so cold?"—it usually indicated an aggressive and potentially problematic customer. Sabrina's response to this was, "Imagine if I kissed every guy in the bar like that before I kissed you. Would that be a turn-on for you?" Most customers backed off then with comments such as, "You're absolutely right. I never thought of that before." By their continuous attempts, however, it was apparent that some were being insincere, assuming, like the dancer, that if they moved more slowly, they would get more of what they wanted. But sometimes the restriction reflected positively on the customer's impression of the dancer. One customer stated to Sabrina after she used this routine: "You have a lot of respect for yourself. I like that."

While some women danced immediately, many waited one or two songs before actually starting a table dance. Sabrina noted that she rarely danced on the first available song because it gave off the impression that she was just interested in making money quickly. She preferred to sit with a customer for a while, talk, drink, and get to know him better. This created a sexual or intimate atmosphere and convinced him that she liked spending time with him. Often this cultivated customers who were likely to buy a greater number of dances, and return to visit her later.

At the beginning of a new song, a dancer might say: "Would you like that table dance now?" or "Let's go for it, baby," depending on the type of interaction in which they were involved. Sexually oriented behavior on the part of the customer called for aggressive behavior from the dancer; less sexually overt actions required more subtle requests.

TABLE DANCES

Strategy became important during a table dance; close quarters meant a dancer's presentation could be difficult to maintain and a customer hard to control. Normally, a dancer attempted to maintain eye contact with a patron, operating on the premise that it demonstrated interest and that if he had his eyes on her, he wouldn't have his hands on her as much. Sabrina hypothesized that a customer confronting a dancer's eyes was forced to acknowledge her "personhood," and that he then was less likely to violate it. Another impression given off (Goffman, 1959) by the dancer's body language was that the intimate exchange demonstrated by this eye contact might be impinged upon by the customer's groping at her body. Sometimes eye contact was difficult if a customer caused the dancer to laugh or feel disgusted (for example, if he was ugly or panting). In this situation, a dancer could turn away from him and make an impersonal shaking of her derriere part of her dance.

Sexual activity was illegal during table dances, but it sometimes occurred. Customers and dancers acknowledged that "hand jobs," oral sex, and intercourse happened, although infrequently. Once a customer requested that Sabrina wear a long skirt during a table dance so that intercourse could take place unobserved.

More common were body-to-penis friction and masturbation. The most frequent form consisted of the customer sliding down to the end of his seat, spreading his legs, and pulling the dancer in close to him where she could then use her knees discreetly to rub his genitals while she danced. Customers sometimes wore shorts without underwear to allow their genitals to hang out the side, or they unzipped their pants to bare their genitals, or masturbated themselves by hand while watching the dancer.

If a customer insisted on violating rules—putting his fingers inside the dancer's briefs or touching her breasts—a dancer might dance much faster than normal, or sway quickly side to side, to escape the wandering hands. If he was insistent, a dancer might grab his wrists teasingly, but firmly, and say, "No, no," addressing him as if he were a misbehaving child.

These attempts to control the customer could not be too aggressive at the outset, or the customer would be turned off. A subtle game was being played: The customer attempted to get the dancer to go as far as she would, and bend the rules, without antagonizing her so much that she stopped dancing; the dancer attempted to keep him in line, but in such a way that he still wanted to buy dances from her. A particularly good strategy at this point was for the dancer to make it look as if she was interested in what he wanted to do, but, because of management, was unable to oblige him: "Look, this would be fine, but I'm going to get in trouble with management. They're going to catch us if you keep acting like this." This disclaimer (Hewitt and Stokes, 1975) shifted the focus of the patron's annoyance to management and away from her and reasserted the idea that this was a respectable occupation with rules (see Hong et al., 1975).

If a man continued to act inappropriately, the dancer most likely lost her money and the negotiation process broke down. If the customer did not pay after the dance, the dancer had no recourse. Her only power was her seductiveness or ability to persuade the customer subtly that he "owed" it to her. Fights between customers and dancers started occasionally because a man did not want to pay a woman who "didn't give him a good dance." Management quickly squelched these and fired or fined dancers who were involved.

Most dances, however, were successful. After one of these, a dancer might give the customer a reward for "being good." Sabrina reported that she kissed the customer, closed mouthed, on the cheek or on the corner of his mouth. By gently resting her fingers on his chin, tilting up his head, and delivering a kiss, she left the impression, "I'm involved with you. I like you."

After a table dance had been completed, the next goal was to keep the interaction going so that the customer would buy more dances. If a customer continued to hold onto a dancer after the song ended, it usually signaled that he wanted her to dance through the next song. If he let her go, a dancer might look inquisitively at the customer and ask, "Is that all for now? Do you want to continue?" or "Will you want a dance later?" The questions asked depended on the dancer's impression of how involved the customer was with the dance. At the least, she encouraged him to look her up the next time he returned to the bar.

EXCHANGE FROM THE BOTTOM UP

Interaction in strip bars reflects negotiation in "respectable" society. What is being exchanged—economic resources for sexual titillation, ego gratification, and submission—is viewed in our society as honorable (Lasch, 1977; Lipman-Blumen, 1984; Safilios-Rothschild, 1977). The strategies dancers use to sell their product are similar to those used by sellers in reputable service occupations (Bigus, 1972; Browne, 1973; Davis, 1959; Henslin, 1968; Prus, 1987; Katovich and Diamond, 1986). Unlike many deviant sales (Luckenbill, 1984), dancers and customers normally are protected by a structured, bureaucratic setting with formal rules.

Interaction in a strip club represents negotiation in a buyer's market: sexual turn-on is available for the asking. Although men show some interest in being customers simply by walking through the door, they must be persuaded to "buy" from a certain dancer. To establish control, women use facilitating (Prus, 1987) or cultivating techniques (Bigus, 1972), much like those used by service workers trying to sell a product directly to a client. To acquire customers, a dancer must develop mutual trust. The most important weapon in the arsenal of interaction is to present oneself as sincere: be warm and imply realness, appear spontaneous, give out insider information to show loyalty, accentuate honesty, demonstrate that one is different from others in similar positions, or tell hard-luck stories. At the same time, a dancer must attempt to determine the trustworthiness of her customer: Will he pay for the dance, and will he hassle her later?

Once trust is established, the dancer must promote repeat patronage and customer loyalty (Prus, 1987). This is done by calling on the norm of reciprocity (Gouldner, 1960). The expectation is that the customer will repay friendship, special attention, and favors with money. Thus a hard sell often is not as productive as other more indirect techniques, such as taking personal interest in customers (Prus, 1987), nurturing pseudo friendships, or effecting obligation (Bigus, 1972). Much like any business relationship, the seller must gauge time spent in an encounter to pay-off potential.

Interaction in the bar also reflects power dynamics in mainstream society. As a subordinate group, women in general have responded to men's macromanipulation of societal institutions by using micromanipulation—interpersonal behaviors and practices—to influence the power balance (Lipman-Blumen, 1984). Women in the bar play a game that they know well; in some form, they have been forced to play it for years. They are accustomed to anticipating male behavior, pleasing and charming men, appearing to be what they want, and following their rules. At the same time, dancers are skilled at manipulating to get their own needs met. The bar is a haven for them; they are old hands.

Women who dance for a living have fewer resources or opportunities to manipulate the macrostructure than do most women. Many come from broken homes where fathers often were absent. They frequently had distant relations with parents and left home at an early age. They had sexual experience earlier than other females had. Financial crisis often served as the impetus for starting this occupation. Few have sufficient training or education to make as much money in other occupations (Carey et al., 1974; Skipper and McCaghy, 1971).

Although dancers often have few resources, they are used to taking care of themselves. The occupation of stripping demands that they be tough. It provides them with a context of control. Being the purveyors and gatekeepers of sexuality has always provided powerful control for women (Safilios-Rothschild, 1977); it serves this function even more for those women who make sexual turn-on into an occupation.

In male-female relationships, sex is "shrouded in romantic mystique" (Salutin, 1971). It has been okay for women to exchange sex for financial security (Salutin, 1971), as long as they confined the exchange to the context of love and marriage (Safilios-Rothschild, 1977). On this level, the activity in the bar is deviant. There this shroud is removed, revealing the rawness of the exchange, the unequal distribution of macropower, and the often cold, calculating nature of the microstrategies. There, sexuality is carried out in public between people who are often strangers. The dancers use sex as a direct currency of exchange: turn-ons for money. They are not likely to have illusions of love. For them, this is a job. When they are tempted to redefine the situation, their histories with men or the realities of their lives remind them otherwise.

For some dancers then, there is a feeling that they have won the ultimate game in American society, which continues to judge the value of women by their attractiveness and seductiveness (Chernin, 1982). Dancers get validation, attention, and money for displaying these characteristics and argue that they are doing nothing more than most women do, not as much as some.

Yet, this world is not a haven for women. If they could make the same money and have the same freedom in another occupation, most dancers would pursue an alternative to table dancing, but they cannot (Prus and Irini, 1980; Robboy, 1985). Most also have internalized "honorable" exchange, and, without the shroud of romance, outright trading of their bodies sometimes breaks through as degrading (Prus and Irini, 1980; Salutin, 1971; Skipper and McCaghy, 1971). They suffer identity problems as they take on the negative attitudes of mainstream society toward their occupation (Rambo [Ronai], 1987; Skipper and McCaghy, 1970, 1971). Many are disillusioned with males to the point that they characterize their audience as degenerates (McCaghy and Skipper, 1969), yet these same degenerates decide their take-home pay.

The negotiation process we have described then is a case study of exchange between those differentially empowered. As in other occupations in which a person's job requires emotion management, stripping has high emotional costs (Hochschild, 1983). Stripping, as a service occupation, pays well, but costs dearly.

REFERENCES

Adler, P. A., and P. Adler. (1987). *Membership Roles in Field Research*. Newbury Park, CA: Sage.

Bigus, O. (1972). "The milkman and his customer: A cultivated relationship." *Urban Life and Culture* 1: 131–165.

Blumer, H. (1969). *Symbolic Interactionism: Perspective and Method*. Englewood Cliffs, NJ: Prentice-Hall.

Boles, J., and A. P. Garbin. (1974). "The strip club and customer-stripper patterns of interaction." *Sociology and Social Research* 58: 136–144.

Browne, J. (1973). *The Used-Car Game: A Sociology of the Bargain*. Lexington, MA: Lexington Books.

Bulmer, M. (1982). "When is disguise justified? Alternatives to covert participant observations." *Qualitative Sociology* 5: 251–264.

Carey, S. H., R. A. Peterson, and L. K. Sharpe. (1974). "A study of recruitment and socialization in two deviant female occupations." *Soc. Symposium* 11: 11–24.

Chernin, K. (1982). *The Obsession: Reflections on the Tyranny of Slenderness*. New York: Harper Collophon.

Davis, F. (1959). "The cab driver and his fare: Facets of a fleeting relationship." *Amer. J. of Sociology* 65: 158–165.

Denzin, N. K. (1978). *The Research Act*. New York: McGraw-Hill.

Ellis, C. (1988). "Keeping emotions in the sociology of emotions." University of South Florida. (unpublished).

———. (forthcoming). "Sociological introspection and emotional experience." *Symbolic Interaction* 13.1.

Foote, N. N. (1954). "Sex as play." *Social Problems* 1: 159–163.

Goffman, E. (1959). *The Presentation of Self in Everyday Life*. Garden City, NY: Doubleday.

———. (1974). *Frame Analysis: An Essay on the Organization of Experience*. Cambridge, MA: Harvard Univ. Press.

Gonos, G. (1976). "Go-Go dancing: A comparative frame analysis." *Urban Life* 9: 189–219.

Gouldner, A. (1960). "The norm of reciprocity." *Amer. Soc. Rev.* 25: 161–178.

Henslin, J. (1968). "Trust and the cab driver." In M. Truzzi. (ed.), *Sociology and Everyday Life* (pp. 138–155). Englewood Cliffs, NJ: Prentice-Hall.

Hewitt, J., and R. Stokes. (1975). "Disclaimers." *Amer. Soc. Rev.* 40: 1–11.

Hochschild, A. (1983). *The Managed Heart: Commercialization of Human Feeling*. Berkeley: Univ. of California Press.

Hong, L. K., W. Darrough, and R. Duff. (1975). "The sensuous rip-off: Consumer fraud turns blue." *Urban Life and Culture* 3: 464–470.

Jorgensen, D. L. (1989). *Participant Observation*. Newbury Park, CA: Sage.

Katovich, M. A., and R. L. Diamond. (1986). "Selling time: Situated transactions in a noninstitutional environment." *Soc. Q.* 27: 253–271.

Lasch, C. (1977). *Haven in a Heartless World*. New York: Basic Books.

Lipman–Blumen, J. (1984). *Gender Roles and Power*. Englewood Cliffs, NJ: Prentice-Hall.

Luckenbill, D. F. (1984). "Dynamics of the deviant sale." *Deviant Behavior* 5: 337–353.

McCaghy, C. H., and J. K. Skipper. (1969). "Lesbian behavior as an adaptation to the occupation of stripping." *Social Problems* 17: 262–270.

———. (1972). "Stripping: Anatomy of a deviant life style." In S. D. Feldman and G. W. Thielbar (eds.), *Life Styles: Diversity in American Society* (pp. 362-373). Boston: Little, Brown.

Mehan, H., and H. Wood. (1975). *The Reality of Ethnomethodology*. New York: John Wiley.

Mullen, K. (1985). "The impure performance frame of the public house entertainer." *Urban Life* 14: 181–203.

Prus, R. (1987). "Developing loyalty: Fostering purchasing relationships in the marketplace." *Urban Life* 15: 331–366.

Prus, R., and S. Irini. (1980). *Hookers, Rounders, and Desk Clerks: The Social Organization of the Hotel Community*. Salem, WI: Sheffield.

Rambo. (Ronai)., C. (1987). "Negotiation strategies and emotion work of the stripper." University of South Florida. (unpublished).

Rasmussen, P., and L. Kuhn. (1976). "The new masseuse: Play for pay." *Urban Life* 5: 271–292.

Riemer, J. W. (1977). "Varieties of opportunistic research." *Urban Life* 5: 467–477.

Robboy, H. (1985). "Emotional labor and sexual exploitation in an occupational role." Presented at the annual meetings of the Mid South Sociological Society, Little Rock, AK.

Safilios-Rothschild, C. (1977). *Love, Sex, and Sex Roles*. Englewood Cliffs, NJ: Prentice-Hall.

Salutin, M. (1971). "Stripper morality." *Transaction* 8: 12–22.

Scott, M. B., and S. M. Lyman. (1968). "Accounts." *Amer. Soc. Rev.* 33: 46–62.

Skipper, J. K., and C. H. McCaghy. (1970). "Stripteasers: The anatomy and career contingencies of a deviant occupation." *Social Problems* 17: 391–405.

———. (1971). "Stripteasing: A sex oriented occupation." In J. Henslin (ed.), *The Sociology of Sex* (pp. 275–296). New York: Appleton Century Crofts.

Sykes, G., and D. Matza. (1957). "Techniques of neutralization: A theory of delinquency." *Amer. Soc. Rev.* 22: 664–670.

Turner, R. (1962). "Role-taking: Process versus conformity." In A. M. Rose (ed.), *Human Behavior and Social Process* (pp. 20–40). Boston: Houghton Mifflin.

Webb, E. J., D. T. Campbell, R. D. Schwartz, and L. Sechrest. (1965). *Unobtrusive Measures*. Chicago: Rand McNally.

Weinstein, Eugene A., and Paul Deutschberger. (1963). "Some dimensions of altercasting." *Sociometry* 26: 454–466.

34

Fraternities and Rape
on Campus

PATRICIA YANCEY MARTIN
AND ROBERT A. HUMMER

apes are perpetrated on dates, at parties, in chance encounters, and in specially planned circumstances. That group structure and processes, rather than individual values or characteristics, are the impetus for many rape episodes was documented by Blanchard (1959) 30 years ago (also see Geis 1971), yet sociologists have failed to pursue this theme (for an exception, see Chancer 1987). A recent review of research (Muehlenhard and Linton 1987) on sexual violence, or rape, devotes only a few pages to the situational contexts of rape events, and these are conceptualized as potential risk factors for individuals rather than qualities of rape-prone social contexts.

Many rapes, far more than come to the public's attention, occur in fraternity houses on college and university campuses, yet little research has analyzed fraternities at American colleges and universities as rape-prone contexts (cf. Ehrhart and Sandler 1985). Most of the research on fraternities reports on samples of individual fraternity men. One group of studies compares the values, attitudes, perceptions, family socioeconomic status, psychological traits (aggressiveness, dependence), and so on, of fraternity and nonfraternity men (Bohrnstedt 1969; Fox, Hodge, and Ward 1987; Kanin 1967; Lemire 1979; Miller 1973). A second group attempts to identify the effects of fraternity membership over time on the values, attitudes, beliefs, or moral precepts of members (Hughes and Winston 1987; Marlowe and Auvenshine 1982; Miller 1973; Wilder, Hoyt, Doren, Hauck, and Zettle 1978; Wilder, Hoyt, Surbeck, Wilder, and Carney 1986). With minor exceptions, little research addresses the group and organizational context of fraternities or the social construction of fraternity life (for exceptions, see Letchworth 1969; Longino and Kart 1973; Smith 1964).

Gary Tash, writing as an alumnus and trial attorney in his fraternity's magazine, claims that over 90 percent of all gang rapes on college campuses involve fraternity men (1988, p. 2). Tash provides no evidence to substantiate this claim, but students of violence against women have been concerned with fraternity men's frequently reported involvement in rape episodes (Adams and

From "Fraternities and Rape on Campus," Patricia Y. Martin and Robert A. Hummer, *Gender & Society*, Vol. 3, No. 4, 1989. Reprinted by permission of Sage Publications, Inc.

Abarbanel 1988). Ehrhart and Sandler (1985) identify over 50 cases of gang rapes on campus perpetrated by fraternity men, and their analysis points to many of the conditions that we discuss here. Their analysis is unique in focusing on conditions in fraternities that make gang rapes of women by fraternity men both feasible and probable. They identify excessive alcohol use, isolation from external monitoring, treatment of women as prey, use of pornography, approval of violence, and excessive concern with competition as precipitating conditions to gang rape (also see Merton 1985; Roark 1987).

The study reported here confirmed and complemented these findings by focusing on both conditions and processes. We examined dynamics associated with the social construction of fraternity life, with a focus on processes that foster the use of coercion, including rape, in fraternity men's relations with women. Our examination of men's social fraternities on college and university campuses as groups and organizations led us to conclude that fraternities are a physical and sociocultural context that encourages the sexual coercion of women. We make no claims that all fraternities are "bad" or that all fraternity men are rapists. Our observations indicated, however, that rape is especially probable in fraternities because of the kinds of organizations they are, the kinds of members they have, the practices their members engage in, and a virtual absence of university or community oversight. Analyses that lay blame for rapes by fraternity men of "peer pressure" are, we feel, overly simplistic (cf. Burkhart 1989; Walsh 1989). We suggest, rather, that fraternities create a sociocultural context in which the use of coercion in sexual relations with women is normative and in which the mechanisms to keep this pattern of behavior in check are minimal at best and absent at worst. We conclude that unless fraternities change in fundamental ways, little improvement can be expected.

METHODOLOGY

Our goal was to analyze the group and organizational practices and conditions that create in fraternities an abusive social context for women. We developed a conceptual framework from an initial case study of an alleged gang rape at Florida State University that involved four fraternity men and an 18-year-old coed. The group rape took place on the third floor of a fraternity house and ended with the "dumping" of the woman in the hallway of a neighboring fraternity house. According to newspaper accounts, the victim's blood-alcohol concentration, when she was discovered, was .349 percent, more than three times the legal limit for automobile driving and an almost lethal amount. One law enforcement officer reported that sexual intercourse occurred during the time the victim was unconscious: "She was in a life-threatening situation" (*Tallahassee Democrat*, 1988b). When the victim was found, she was comatose and had suffered multiple scratches and abrasions. Crude words and a fraternity symbol had been written on her thighs (*Tampa Tribune*, 1988). When law enforcement officials tried to investigate the case, fraternity members refused to cooperate. This led, eventually, to a five-year ban of the fraternity from campus by the university and by the fraternity's national organization.

In trying to understand how such an event could have occurred, and how a group of over 150 members (exact figures are unknown because the fraternity refused to provide a membership roster) could hold rank, deny knowledge of the event, and allegedly lie to a grand jury, we analyzed newspaper articles about the case and conducted open-ended interviews with a variety of respondents about the case and about fraternities, rapes, alcohol use, gender relations, and sexual activities on campus. Our data included over 100 newspaper articles on the initial gang rape case; open-ended interviews with Greek (social fraternity and sorority) and non-Greek (independent) students (N = 20); university administrators (N = 8, five men, three women); and alumni advisers to Greek organizations (N = 6). Open-ended interviews were held also with judges, public and private defense attorneys, victim advocates, and state prosecutors regarding the processing of sexual assault cases. Data were analyzed using the grounded theory method (Glaser 1978; Martin and Turner 1986). In the following analysis, concepts generated from the data analysis are integrated with the literature on men's social fraternities, sexual coercion, and related issues.

FRATERNITIES AND THE SOCIAL CONSTRUCTION OF MEN AND MASCULINITY

Our research indicated that fraternities are vitally concerned—more than with anything else—with masculinity (cf. Kanin 1967). They work hard to create a macho image and context and try to avoid any suggestion of "wimpishness," effeminacy, and homosexuality. Valued members display, or are willing to go along with, a narrow conception of masculinity that stresses competition, athleticism, dominance, winning, conflict, wealth, material possessions, willingness to drink alcohol, and sexual prowess vis-à-vis women.

Valued Qualities of Members

When fraternity members talked about the kind of pledges they prefer, a litany of stereotypical and narrowly masculine attributes and behaviors was recited and feminine or woman-associated qualities and behaviors were expressly denounced (cf. Merton 1985). Fraternities seek men who are "athletic," "big guys," good in intramural competition, "who can talk college sports." Males "who are willing to drink alcohol," "who drink socially," or "who can hold their liquor" are sought. Alcohol and activities associated with the recreational use of alcohol are cornerstones of fraternity social life. Nondrinkers are viewed with skepticism and rarely selected for membership.[1]

Fraternities try to avoid "geeks," nerds, and men said to give the fraternity a "wimpy" or "gay" reputation. Art, music, and humanities majors, majors in traditional women's fields (nursing, home economics, social work, education), men with long hair, and those whose appearance or dress violate current norms are rejected. Clean-cut, handsome men who dress well (are clean, neat, conforming, fashionable) are preferred. One sorority woman commented that "the top ranking fraternities have the best looking guys."

One fraternity man, a senior, said his fraternity recruited "some big guys, very athletic" over a two-year period to help overcome its image of wimpiness. His fraternity had won the interfraternity competition for highest grade-point average several years running but was looked down on as "wimpy, dancy, even gay." With their bigger, more athletic recruits, "our reputation improved; we're a much more recognized fraternity now." Thus a fraternity's reputation and status depends on members' possession of stereotypically masculine qualities. Good grades, campus leadership, and community service are "nice" but masculinity dominance—for example, in athletic events, physical size of members, athleticism of members—counts most.

Certain social skills are valued. Men are sought who "have good personalities," are friendly, and "have the ability to relate to girls" (cf. Longino and Kart 1973). One fraternity man, a junior, said: "We watch a guy [a potential pledge] talk to women . . . we want guys who can relate to girls." Assessing a pledge's ability to talk to women is, in part, a preoccupation with homosexuality and a conscious avoidance of men who seem to have effeminate manners or qualities. If a member is suspected of being gay, he is ostracized and informally drummed out of the fraternity. A fraternity with a reputation as wimpy or tolerant of gays is ridiculed and shunned by other fraternities. Militant heterosexuality is frequently used by men as a strategy to keep each other in line (Kimmel 1987).

Financial affluence or wealth, a male-associated value in American culture, is highly valued by fraternities. In accounting for why the fraternity involved in the gang rape that precipitated our research project had been recognized recently as "the best fraternity chapter in the United States," a university official said: "They were good-looking, a big fraternity, had lots of BMWs [expensive, German-made automobiles]." After the rape, newspaper stories described the fraternity members' affluence, noting the high number of members who owned expensive cars (*St. Petersburg Times*, 1988).

The Status and Norms of Pledgeship

A pledge (sometimes called an associate member) is a new recruit who occupies a trial membership status for a specific period of time. The pledge period (typically ranging from 10 to 15 weeks) gives fraternity brothers an opportunity to assess and socialize new recruits. Pledges evaluate the fraternity also and decide if they want to become brothers. The socialization experience is structured partly through assignment of a Big Brother to each pledge. Big Brothers are expected to teach pledges how to become a brother and to support them as they progress through the trial membership period. Some pledges are repelled by the pledging experience, which can entail physical abuse; harsh discipline; and demands to be subordinate, follow orders, and engage in demeaning routines and activities, similar to those used by the military to "make men out of boys" during boot camp.

Characteristics of the pledge experience are rationalized by fraternity members as necessary to help pledges unite into a group, rely on each other, and join together against outsiders. The process is highly masculinist in execution as well as conception. A willingness to submit to authority, follow orders, and do as one is told is viewed as a sign of loyalty, togetherness, and unity. Fraternity pledges who find the pledge process offensive often drop out. Some do this by openly quitting, which can subject them to ridicule by brothers and other pledges, or they may deliberately fail to make the grades necessary for initiation or transfer schools and decline to reaffiliate with the fraternity on the new campus. One fraternity pledge who quit the fraternity he had pledged described an experience during pledgeship as follows:

> This one guy was always picking on me. No matter what I did, I was wrong. One night after dinner, he and two other guys called me and two other pledges into the chapter room. He said, "Here, X, hold this 25 pound bag of ice at arms' length 'til I tell you to stop." I did it even though my arms and hands were killing me. When I asked if I could stop, he grabbed me around the throat and lifted me off the floor. I thought he would choke me to death. He cussed me and called me all kinds of names. He took one of my fingers and twisted it until it nearly broke. . . . I stayed in the fraternity for a few more days, but then I decided to quit. I hated it. Those guys are sick. They like seeing you suffer.

Fraternities' emphasis on toughness, withstanding pain and humiliation, obedience to superiors, and using physical force to obtain compliance contributes to an interpersonal style that de-emphasizes caring and sensitivity but fosters intragroup trust and loyalty. If the least macho or most critical pledges drop out, those who remain may be more receptive to, and influenced by, masculinist values and practices that encourage the use of force in sexual relations with women and the covering up of such behavior (cf. Kanin 1967).

Norms and Dynamics of Brotherhood

Brother is the status occupied by fraternity men to indicate their relations to each other and their membership in a particular fraternity organization or group. Brother is a male-specific status; only males can become brothers, although women can become "Little Sisters," a form of pseudomembership. "Becoming a brother" is a rite of passage that follows the consistent and often lengthy display by pledges of appropriately masculine qualities and behaviors. Brothers have a quasi-familial relationship with each other, are normatively said to share bonds of closeness and support, and are sharply set off from non-members. Brotherhood is a loosely defined term used to represent the bonds that develop among fraternity members and the obligations and expectations incumbent upon them (cf. Marlowe and Auvenshine [1982] on fraternities' failure to encourage "moral development" in freshman pledges).

Some of our respondents talked about brotherhood in almost reverential terms, viewing it as the most valuable benefit of fraternity membership. One senior, a business-school major who had been affiliated with a fairly high-status fraternity throughout four years on campus, said:

> Brotherhood spurs friendship for life, which I consider its best aspect, although I didn't see it that way when I joined. Brotherhood bonds and unites. It instills values of caring about one another, caring about community, caring about ourselves. The values and bonds [of brotherhood] continually develop over the four years [in college] while normal friendships come and go.

Despite this idealization, most aspects of fraternity practice and conception are more mundane. Brotherhood often plays itself out as an overriding concern with masculinity and, by extension, femininity. As a consequence, fraternities comprise collectivities of highly masculinized men with attitudinal qualities and behavior norms that predispose them to sexual coercion of women (cf. Kanin 1967; Merton 1985; Rapaport and Burkhart 1984). The norms of masculinity are complemented by conceptions of women and femininity that are equally distorted and stereotyped and that may enhance the probability of women's exploitation (cf. Ehrhart and Sandler 1985; Sanday 1981, 1986).

Practices of Brotherhood

Practices associated with fraternity brotherhood that contribute to the sexual coercion of women include a preoccupation with loyalty, group protection and secrecy, use of alcohol as a weapon, involvement in violence and physical force, and an emphasis on competition and superiority.

Loyalty, Group Protection, and Secrecy   Loyalty is a fraternity preoccupation. Members are reminded constantly to be loyal to the fraternity and to their brothers. Among other ways, loyalty is played out in the practices of group protection and secrecy. The fraternity must be shielded from criticism. Members are admonished to avoid getting the fraternity in trouble and to bring all problems "to the chapter" (local branch of a national social fraternity) rather than to outsiders. Fraternities try to protect themselves from close scrutiny and criticism by the Interfraternity Council (a quasi-governing body composed of representatives from all social fraternities on campus), their fraternity's national office, university officials, law enforcement, the media, and the public. Protection of the fraternity often takes precedence over what is procedurally, ethically, or legally correct. Numerous examples were related to us of fraternity brothers' lying to outsiders to "protect the fraternity."

Group protection was observed in the alleged gang rape case with which we began our study. Except for one brother, a rapist who turned state's evidence, the entire remaining fraternity membership was accused by university and criminal justice officials of lying to protect the fraternity. Members consistently failed to cooperate even though the alleged crimes were felonies, involved only four men (two of whom were not even members of the local chapter), and the victim of the crime nearly died. According to a grand jury's findings, fraternity officers repeatedly broke appointments with law enforcement officials, refused to provide police with a list of members, and refused to cooperate with police and prosecutors investigating the case (*Florida Flambeau*, 1988).

Secrecy is a priority value and practice in fraternities, partly because full-fledged membership is premised on it (for confirmation, see Ehrhart and Sandler 1985; Longino and Kart 1973; Roark 1987). Secrecy is also a boundary-maintaining mechanism, demarcating in-group from out-group, us from them. Secret rituals, handshakes, and mottoes are revealed to pledge brothers as they are initiated into full brotherhood. Since only brothers are supposed to know a fraternity's secrets, such knowledge affirms membership in the fraternity and separates a brother from others. Extending secrecy tactics from protection of private knowledge to protection of the fraternity from criticism is a predictable development. Our interviews indicated that individual members knew the difference between right and wrong, but fraternity norms that emphasize loyalty, group protection, and secrecy often overrode standards of ethical correctness.

Alcohol as Weapon Alcohol use by fraternity men is normative. They use it on weekdays to relax after class and on weekends to "get drunk," "get crazy," and "get laid." The use of alcohol to obtain sex from women is pervasive—in other words, it is used as a weapon against sexual reluctance. According to several fraternity men whom we interviewed, alcohol is the major tool used to gain sexual mastery over women (cf. Adams and Abarbanel 1988; Ehrhart and Sandler 1985). One fraternity man, a 21-year-old senior, described alcohol use to gain sex as follows: "There are girls that you know will fuck, then some you have to put some effort into it. . . . You have to buy them drinks or find out if she's drunk enough."

A similar strategy is used collectively. A fraternity man said that at parties with Little Sisters: "We provide them with 'hunch punch' and things get wild. We get them drunk and most of the guys end up with one." "'Hunch punch,'" he said, "is a girls' drink made up of overproof alcohol and powdered Kool-Aid, no water or anything, just ice. It's very strong. Two cups will do a number on a female." He had plans in the next academic term to surreptitiously give hunch punch to women in a "prim and proper" sorority because "having sex with prim and proper sorority girls is definitely a goal." These women are a challenge because they "won't openly consume alcohol and won't get openly drunk as hell." Their sororities have "standards committees" that forbid heavy drinking and easy sex.

In the gang rape case, our sources said that many fraternity men on campus believed the victim had a drinking problem and was thus an "easy make." According to newspaper accounts, she had been drinking alcohol on the evening she was raped; the lead assailant is alleged to have given her a bottle of wine after she arrived at his fraternity house. Portions of the rape occurred in a shower, and the victim was reportedly so drunk that her assailants had difficulty holding her in a standing position (*Tallahassee Democrat*, 1988a). While raping her, her assailants repeatedly told her they were members of another fraternity under the apparent belief that she was too drunk to know the difference. Of course, if she was too drunk to know who they were, she was too drunk to consent to sex (cf. Allgeier 1986; Tash 1988).

One respondent told us that gang rapes are wrong and can get one expelled, but he seemed to see nothing wrong in sexual coercion one-on-one. He seemed unaware that the use of alcohol to obtain sex from a woman is grounds for a claim that a rape occurred (cf. Tash 1988). Few women on campus (who also may not know these grounds) report date rapes, however; so the odds of detection and punishment are slim for fraternity men who use alcohol for "seduction" purposes (cf. Byington and Keeter 1988; Merton 1985).

Violence and Physical Force   Fraternity men have a history of violence (Ehrhart and Sandler 1985; Roark 1987). Their record of hazing, fighting, property destruction, and rape has caused them problems with insurance companies (Bradford 1986; Pressley 1987). Two university officials told us that fraternities "are the third riskiest property to insure behind toxic waste dumps and amusement parks." Fraternities are increasingly defendants in legal actions brought by pledges subjected to hazing (Meyer 1986; Pressley 1987) and by women who were raped by one or more members. In a recent alleged gang rape incident at another Florida university, prosecutors failed to file charges but the victim filed a civil suit against the fraternity nevertheless (Tallahassee Democrat, 1989).

Competition and Superiority   Interfraternity rivalry fosters in-group identification and out-group hostility. Fraternities stress pride of membership and superiority over other fraternities as major goals. Interfraternity rivalries take many forms, including competition for desirable pledges, size of pledge class, size of membership, size and appearance of fraternity house, superiority in intramural sports, highest grade-point averages, giving the best parties, gaining the best or most campus leadership roles, and, of great importance, attracting and displaying "good looking women." Rivalry is particularly intense over members, intramural sports, and women (cf. Messner 1989).

FRATERNITIES'
COMMODIFICATION OF WOMEN

In claiming that women are treated by fraternities as commodities, we mean that fraternities knowingly, and intentionally, *use* women for their benefit. Fraternities use women as bait for new members, as servers of brothers' needs, and as sexual prey.

Women as Bait Fashionably attractive women help a fraternity attract new members. As one fraternity man, a junior, said, "They are good bait." Beautiful, sociable women are believed to impress the right kind of pledges and give the impression that the fraternity can deliver this type of woman to its members. Photographs of shapely, attractive coeds are printed in fraternity brochures and videotapes that are distributed and shown to potential pledges. The women pictured are often dressed in bikinis, at the beach, and are pictured hugging the brothers of the fraternity. One university official says such recruitment materials give the message: "Hey, they're here for you, you can have whatever you want," and, "We have the best looking women. Join us and you can have them too." Another commented: "Something's wrong when males join an all-male organization as the best place to meet women. It's so illogical."

Fraternities compete in promising access to beautiful women. One fraternity man, a senior, commented that "the attraction of girls [i.e., a fraternity's success in attracting women] is a big status symbol for fraternities." One university official commented that the use of women as a recruiting tool is so well entrenched that fraternities that might be willing to forgo it say they cannot afford to unless other fraternities do so as well. One fraternity man said, "Look, if we don't have Little Sisters, the fraternities that do will get all the good pledges." Another said, "We won't have as good a rush [the period during which new members are assessed and selected] if we don't have these women around."

In displaying good-looking, attractive, skimpily dressed, nubile women to potential members, fraternities implicitly, and sometimes explicitly, promise sexual access to women. One fraternity man commented that "part of what being in a fraternity is all about is the sex" and explained how his fraternity uses Little Sisters to recruit new members:

> We'll tell the sweetheart [the fraternity's term for Little Sister], "You're gorgeous; you can get him." We'll tell her to fake a scam and she'll go hang all over him during a rush party, kiss him, and he thinks he's done wonderful and wants to join. The girls think it's great too. It's flattering for them.

Women as Servers The use of women as servers is exemplified in the Little Sister program. Little Sisters are undergraduate women who are rushed and selected in a manner parallel to the recruitment of fraternity men. They are affiliated with the fraternity in a formal but unofficial way and are able, indeed required, to wear the fraternity's Greek letters. Little Sisters are not full-fledged fraternity members, however; and fraternity national offices and most universities do not register or regulate them. Each fraternity has an officer called Little Sister Chairman who oversees their organization and activities. The Little Sisters elect officers among themselves, pay monthly dues to the fraternity, and have well-defined roles. Their dues are used to pay for the fraternity's social events, and Little Sisters are expected to attend and hostess fraternity parties and hang around the house to make it a "nice place to be." One fraternity man, a senior, described Little Sisters this way: "They are very social girls, willing to join in, be affiliated with the group, devoted to the fraternity." Another member, a sophomore, said: "Their sole purpose is social—attend parties, attract new members, and 'take care' of the guys."

Our observations and interviews suggested that women selected by fraternities as Little Sisters are physically attractive, possess good social skills, and are willing to devote time and energy to the fraternity and its members. One undergraduate woman gave the following job description for Little Sisters to a campus newspaper:

> It's not just making appearances at all the parties but entails many more responsibilities. You're going to be expected to go to all the intramural games to cheer the brothers on, support and encourage the pledges, and just be around to bring some extra life to the house. [As a Little Sister] you have to agree to take on a new responsibility other than studying to maintain your grades and managing to keep your checkbook from bouncing. You have to make time to be a part of the fraternity and support the brothers in all they do. (The Tomahawk, 1988)

The title of Little Sister reflects women's subordinate status; fraternity men in a parallel role are called Big Brothers. Big Brothers assist a sorority primarily with the physical work of sorority rushes, which, compared to fraternity rushes, are more formal, structured, and intensive. Sorority rushes take place in the daytime and fraternity rushes at night so fraternity men are free to help. According to one fraternity member, Little Sister status is a benefit to women because it gives them a social outlet and "the protection of the brothers." The gender-stereotypic conceptions and obligations of these Little Sister and Big Brother statuses indicate that fraternities and sororities promote a gender hierarchy on campus that fosters subordination and dependence in women, thus encouraging sexual exploitation and the belief that it is acceptable.

Women as Sexual Prey   Little Sisters are a sexual utility. Many Little Sisters do not belong to sororities and lack peer support for refraining from unwanted sexual relations. One fraternity man (whose fraternity has 65 members and 85 Little Sisters) told us they had recruited "wholesale" in the prior year to "get lots of new women." The structural access to women that the Little Sister program provides and the absence of normative supports for refusing fraternity members' sexual advances may make women in this program particularly susceptible to coerced sexual encounters with fraternity men.

Access to women for sexual gratification is a presumed benefit of fraternity membership, promised in recruitment materials and strategies and through brothers' conversations with new recruits. One fraternity man said: "We always tell the guys that you get sex all the time, there's always new girls. . . . After I became a Greek, I found out I could be with females at will." A university official told us that, based on his observations, "no one [i.e., fraternity men] on this campus wants to have 'relationships.' They just want to have fun [i.e., sex]." Fraternity men plan and execute strategies aimed at obtaining sexual gratification, and this occurs at both individual and collective levels.

Individual strategies include getting a woman drunk and spending a great deal of money on her. As for collective strategies, most of our undergraduate interviewees agreed that fraternity parties often culminate in sex and that this outcome is planned. One fraternity man said fraternity parties often involve sex and nudity and can "turn into orgies." Orgies may be planned in advance, such as the Bowery Ball party held by one fraternity. A former fraternity member said of this party:

> The entire idea behind this is sex. Both men and women come to the party wearing little or nothing. There are pornographic pinups on the walls and usually porno movies playing on the TV. The music carries sexual overtones. . . . They just get schnockered [drunk] and, in most cases, they also get laid.

When asked about the women who come to such a party, he said: "Some Little Sisters just won't go. . . . The girls who do are looking for a good time, girls who don't know what it is, things like that."

Other respondents denied that fraternity parties are orgies but said that sex is always talked about among the brothers and they all know "who each other is doing it with." One member said that most of the time, guys have sex with their girlfriends "but with socials, girlfriends aren't allowed to come and it's their [members'] big chance [to have sex with other women]." The use of alcohol to help them get women into bed is a routine strategy at fraternity parties.

CONCLUSIONS

In general, our research indicated that the organization and membership of fraternities contribute heavily to coercive and often violent sex. Fraternity houses are occupied by same-sex (all men) and same-age (late teens, early twenties) peers whose maturity and judgment is often less than ideal. Yet fraternity houses are private dwellings that are mostly off-limits to, and away from scrutiny of, university and community representatives, with the result that fraternity house events seldom come to the attention of outsiders. Practices associated with the social construction of fraternity brotherhood emphasize a macho conception of men and masculinity, a narrow, stereotyped conception of women and femininity, and the treatment of women as commodities. Other practices contributing to coercive sexual relations and the cover-up of rapes include excessive alcohol use, competitiveness, and normative support for deviance and secrecy (cf. Bogal-Allbritten and Allbritten 1985; Kanin 1967).

Some fraternity practices exacerbate others. Brotherhood norms require "sticking together" regardless of right or wrong; thus rape episodes are unlikely to be stopped or reported to outsiders, even when witnesses disapprove. The ability to use alcohol without scrutiny by authorities and alcohol's frequent association with violence, including sexual coercion, facilitates rape in fraternity houses. Fraternity norms that emphasize the value of maleness and masculinity over femaleness and femininity and that elevate the status of men and lower the status of women in members' eyes undermine perceptions and treatment of women as persons who deserve consideration and care (cf. Ehrhart and Sandler 1985; Merton 1985).

Androgynous men and men with a broad range of interests and attributes are lost to fraternities through their recruitment practices. Masculinity of a narrow and stereotypical type helps create attitudes, norms, and practices that predispose fraternity men to coerce women sexually, both individually and collectively (Allgeier 1986; Hood 1989; Sanday 1981, 1986). Male athletes on campus may be similarly disposed for the same reasons (Kirshenbaum 1989; Telander and Sullivan 1989).

Research into the social contexts in which rape crimes occur and the social constructions associated with these contexts illumine rape dynamics on campus. Blanchard (1959) found that group rapes almost always have a leader who pushes others into the crime. He also found that the leader's latent homosexuality, desire to show off to his peers, or fear of failing to prove himself a man are frequently an impetus. Fraternity norms and practices contribute to the approval and use of sexual coercion as an accepted tactic in relations with women. Alcohol-induced compliance is normative, whereas, presumably, use of a knife, gun, or threat of bodily harm would not be because the woman who "drinks too much" is viewed as "causing her own rape" (cf. Ehrhart and Sandler 1985).

Our research led us to conclude that fraternity norms and practices influence members to view the sexual coercion of women, which is a felony crime, as sport, a contest, or a game (cf. Sato 1988). This sport is played not between men and women but between men and men. Women are the pawns or prey in the interfraternity rivalry game; they prove that a fraternity is successful or prestigious. The use of women in this way encourages fraternity men to see women as objects and sexual coercion as sport. Today's societal norms support young women's right to engage in sex at their discretion, and coercion is unnecessary in a mutually desired encounter. However, nubile young women say they prefer to be "in a relationship" to have sex while young men say they prefer to "get laid" without a commitment (Muehlenhard and Linton 1987). These differences may reflect, in part, American puritanism and men's fears of sexual intimacy or perhaps intimacy of any kind. In a fraternity context, getting sex without giving emotionally demonstrates "cool" masculinity. More important, it poses no threat to the bonding and loyalty of the fraternity brotherhood (cf. Farr 1988). Drinking large quantities of alcohol before having sex suggests that "scoring" rather than intrinsic sexual pleasure is a primary concern of fraternity men.

Unless fraternities' composition, goals, structures, and practices change in fundamental ways, women on campus will continue to be sexual prey for fraternity men. As all-male enclaves dedicated to opposing faculty and administration and to cementing in-group ties, fraternity members eschew any hint of homosexuality. Their version of masculinity transforms women, and men with womanly characteristics, into the out-group. "Womanly men" are ostracized; feminine women are used to demonstrate members' masculinity.

Encouraging renewed emphasis on their founding values (Longino and Kart 1973), service orientation and activities (Lemire 1979), or members' moral development (Marlowe and Auvenshine 1982) will have little effect on fraternities' treatment of women. A case for or against fraternities cannot be made by studying individual members. The fraternity qua group and organization is at issue. Located on campus along with many vulnerable women, embedded in a sexist society, and caught up in masculinist goals, practices, and values, fraternities' violation of women—including forcible rape—should come as no surprise.

NOTE

1. Recent bans by some universities on open-keg parties at fraternity houses have resulted in heavy drinking before coming to a party and an increase in drunkenness among those who attend. This may aggravate, rather than improve, the treatment of women by fraternity men at parties.

REFERENCES

Adams, Aileen, and Gail Abarbanel. 1988. *Sexual Assault on Campus: What Colleges Can Do*. Santa Monica, CA: Rape Treatment Center.

Allgeier, Elizabeth. 1986. "Coercive Versus Consensual Sexual Interactions." G. Stanley Hall Lecture to American Psychological Association Annual Meeting, Washington, DC, August.

Blanchard, W. H. 1959. "The Group Process in Gang Rape." *Journal of Social Psychology* 49: 259–66.

Bogal-Allbritten, Rosemarie B., and William L. Allbritten. 1985. "The Hidden Victims: Courtship Violence Among College Students." *Journal of College Student Personnel* 43: 201–4.

Bohrnstedt, George W. 1969. "Conservatism, Authoritarianism and Religiosity of Fraternity Pledges." *Journal of College Student Personnel* 27: 36–43.

Bradford, Michael. 1986. "Tight Market Dries Up Nightlife at University." *Business Insurance* (March 2): 2, 6.

Burkhart, Barry. 1989. Comments in Seminar on Acquaintance/Date Rape Prevention: A National Video Teleconference, February 2.

Burkhart, Barry R., and Annette L. Stanton. 1985. "Sexual Aggression in Acquaintance Relationships." In *Violence in Intimate Relationships*, edited by G. Russell (pp. 43–65). Englewood Cliffs, NJ: Spectrum.

Byington, Diane B., and Karen W. Keeter. 1988. "Assessing Needs of Sexual Assault Victims on a University Campus." In *Student Services: Responding to Issues and Challenges* (pp. 23–31). Chapel Hill: University of North Carolina Press.

Chancer, Lynn S. 1987. "New Bedford, Massachusetts, March 6, 1983–March 22, 1984: The 'Before and After' of a Group Rape." *Gender & Society* 1: 239–60.

Ehrhart, Julie K., and Bernice R. Sandler. 1985. *Campus Gang Rape: Party Games?* Washington, DC: Association of American Colleges.

Farr, K. A. 1988. "Dominance Bonding through the Good Old Boys Sociability Network." *Sex Roles* 18: 259–77.

Florida Flambeau. 1988. "Pike Members Indicted in Rape." (May 19): 1, 5.

Fox, Elaine, Charles Hodge, and Walter Ward. 1987. "A Comparison of Attitudes Held by Black and White Fraternity Members." *Journal of Negro Education* 56: 521–34.

Geis, Gilbert. 1971. "Group Sexual Assaults." *Medical Aspects of Human Sexuality* 5: 101–13.

Glaser, Barney G. 1978. *Theoretical Sensitivity: Advances in the Methodology of Grounded Theory.* Mill Valley, CA: Sociology Press.

Hood, Jane. 1989. "Why Our Society Is Rape-Prone." *New York Times,* May 16.

Hughes, Michael J., and Roger B. Winston, Jr. 1987. "Effects of Fraternity Membership on Interpersonal Values." *Journal of College Student Personnel* 45: 405–11.

Kanin, Eugene J. 1967. "Reference Groups and Sex Conduct Norm Violations." *The Sociological Quarterly* 8: 495–504.

Kimmel, Michael, ed. 1987. *Changing Men: New Directions in Research on Men and Masculinity.* Newbury Park, CA: Sage.

Kirshenbaum, Jerry. 1989. "Special Report, an American Disgrace: A Violent and Unprecedented Lawlessness Has Arisen Among College Athletes in all Parts of the County." *Sports Illustrated* (February 27): 16–19.

Lemire, David. 1979. "One Investigation of the Stereotypes Associated with Fraternities and Sororities." *Journal of College Student Personnel* 37: 54–57.

Letchworth, G. E. 1969. "Fraternities Now and in the Future." *Journal of College Student Personnel* 10: 118–22.

Longino, Charles F., Jr., and Cary S. Kart. 1973. "The College Fraternity: An Assessment of Theory and Research." *Journal of College Student Personnel* 31: 118–25.

Marlowe, Anne F., and Dwight C. Auvenshine. 1982. "Greek Membership: Its Impact on the Moral Development of College Freshmen." *Journal of College Student Personnel* 40: 53–57.

Martin, Patricia Yancey, and Barry A. Turner. 1986. "Grounded Theory and Organizational Research." *Journal of Applied Behavioral Science* 22: 141–57.

Merton, Andrew. 1985. "On Competition and Class: Return to Brotherhood." *Ms.* (September): 60–65, 121–22.

Messner, Michael. 1989. "Masculinities and Athletic Careers." *Gender & Society* 3: 71–88.

Meyer, T. J. 1986. "Fight Against Hazing Rituals Rages on Campuses." *Chronicle of Higher Education* (March 12): 34–36.

Miller, Leonard D. 1973. "Distinctive Characteristics of Fraternity Members." *Journal of College Student Personnel* 31: 126–28.

Muehlenhard, Charlene L., and Melaney A. Linton. 1987. "Date Rape and Sexual Aggression in Dating Situations: Incidence and Risk Factors." *Journal of Counseling Psychology* 34: 186–96.

Pressley, Sue Anne. 1987. "Fraternity Hell Night Still Endures." *Washington Post* (August 11): B1.

Rapport, Karen, and Barry R. Burkhart. 1983. "Personality and Attitudinal Characteristics of Sexually Coercive College Males." *Journal of Abnormal Psychology* 93: 216–21.

Roark, Mary L. 1987. "Preventing Violence on College Campuses." *Journal of Counseling and Development* 65: 367–70.

St. Petersburg Times. 1988. "A Greek Tragedy." (May 29): 1F, 6F.

Sanday, Peggy Reeves. 1981. "The Socio-Cultural Context of Rape: A Cross-Cultural Study." *Journal of Social Issues* 37: 5–27.

———. 1986. "Rape and the Silencing of the Feminine." In *Rape,* edited by S. Tomaselli and R. Porter (pp. 84–101). Oxford: Basil Blackwell.

Sato, Ikuya. 1988. "Play Theory of Delinquency: Toward a General Theory of 'Action.'" *Symbolic Interaction* 11: 191–212.

Smith, T. 1964. "Emergence and Maintenance of Fraternal Solidarity." *Pacific Sociological Review* 7: 29–37.

Tallahassee Democrat. 1988a. "FSU Fraternity Brothers Charged." (April 27): 1A, 12A.

———. 1988b. "FSU Interviewing Students About Alleged Rape." (April 24): 1D.

———. 1989. "Woman Sues Stetson in Alleged Rape." (March 19): 3B.

Tampa Tribune. 1988. "Fraternity Brothers Charged in Sexual Assault of FSU Coed." (April 27): 6B.

Tash, Gary B. 1988. "Date Rape." *The Emerald of Sigma Pi Fraternity* 75(4): 1–2.

Telander, Rick, and Robert Sullivan. 1989. "Special Report, You Reap What You Sow." *Sports Illustrated* (February 27): 20–34.

The Tomahawk. 1988. "A Look Back at Rush, A Mixture of Hard Work and Fun" (April/May): 3D.

Walsh, Claire. 1989. Comments in Seminar on Acquaintance/Date Rape Prevention: A National Video Teleconference, February 2.

Wilder, David H., Arlyne E. Hoyt, Dennis M. Doren, William E. Hauck, and Robert D. Zettle. 1978. "The Impact of Fraternity and Sorority Membership on Values and Attitudes." *Journal of College Student Personnel* 36: 445–49.

Wilder, David H., Arlyne E. Hoyt, Beth Shuster Surbeck, Janet C. Wilder, and Patricia Imperatrice Carney. 1986. "Greek Affiliation and Attitude Change in College Students." *Journal of College Student Personnel* 44: 510–19.

35

Pool and Golf Hustling

ROBERT PRUS
AND C. R. D. SHARPER

HUSTLING POOL

As pool hustling[1] involves a much greater element of playing skill than does card or dice hustling, pool hustlers are much more attuned to the relative playing skill of the opposition. Thus, although their reference terms may vary, a major concern is that of accurately rating players on a skill dimension.

> You can start by putting pool hustlers into three categories. First, you have the class A or really top-notch players. They are the elites, the players who really take on the big money or get involved in big games and they will usually win. They usually manage themselves very well and either they back themselves up or they have backers. Now the B-type player is a very good player, but he's liable to get into trouble because there are a lot of B or B+ players and you often run up against each other or maybe an A player. You see everyone gets to a peak and very few get to that elite peak. I don't care how much you play. If everyone kept improving, there would be no place where you could level off. Then there's the C players. In pool hustling, they're usually the suckers. They think that they can play, but they're really not that talented.

In addition to playing well and accurately gauging one's competition relative to oneself, another important feature is that of negotiating the game parameters to insure oneself an edge:

> Regardless of who you play, you have to watch yourself, because you can only spot a guy so many points before you're getting hustled yourself. You give them enough to tempt them, but not enough for them to beat you and sometimes that's a fine line. If you go over it, you're finished. So managing yourself is a big thing. . . . Like this one guy I know, he's a C player, but he's such a good manager that he will often beat B-level players. He will convince them that he needs a lot of points, and I have had it happen myself. I've spotted him too many points and blown money. I started out playing him and I spotted him ten points, then twelve, then fifteen, and that was too many—he ended up beating me for the money I

From Robert Prus and C. R. D. Sharper, *Road Hustler,* 1991. Reprinted by permission of Robert Prus.

had won off him and more besides. Then he refused to play me with less than the fifteen points. Now if I had managed that better, I wouldn't have spotted him any more than ten and he would have kept coming back to me for action. So you try to give the man no more points that you can help, because it can escalate and then if you give him too much and you miss a shot or he gets a little lucky, you're in trouble. It puts too much pressure on you. . . .

Now the other part of managing the game is that of shooting up the bets. Say you eventually want to play for fifty a game, well you try not to start off playing for five. You might say "I'll play you for twenty." He may say, "Let's play for ten." Now he may be the type of person that if you beat him over a few games, he'll want to increase the bet to get even. If you know this, you may try to get three games ahead and then start saying, "Gee, I have to go." You know, leaving him stuck thirty bucks. That usually gets these guys hot. So okay, so you agree to one more game for twenty but you know damn well that you are going to be there for more than one game. So you beat him. Then you just rack the balls up again and casually say, "You want to play another for twenty." Then you have him playing for twenty. You adjust to what you figure he can handle, so you start at ten a game or five, whatever you think he might go for.

In addition to managing the game, another critical aspect of career hustling hinges on one's ability to arrange viable matches:

In pool, it's always a set up. You just can't walk in, like Paul Newman in the movie *The Hustler*. It's not that easy. Sure, if you're well known like he supposedly was, you could do it, but to just go in and bang up against people, very few pool hustlers could do that. If you're one of these elites, you might do that but otherwise it's pretty rough. If you don't take the time to find out who you can beat and who you can't, you will lose a lot of money. . . . Now you might meet someone and set them up yourself, you know, entice them into playing for bigger stakes and then take them off. Or maybe some bird dog will meet one of these action people and steer the guy into you, but to just run in and say, "I want to challenge the best player in this place!" gets kind of risky. If you run in cold like that, they are more apt to get leery and think that you're a hustler right away, so you might have to really open up just to make a few bucks.

While we will be discussing other aspects of pool hustling later in this section, the reader may find the following discussion on "bar pool" hustling particularly valuable in illustrating some principles of pool hustling:

Now there's another kind of hustle involving pool tables at bars, and what a hustle that is! Here's the idea. You'll work with maybe one or two other guys, something like a crap crew, but it's not as organized and expenses are hard to hold down because a lot of these guys are lushes. These people are always looking for a new face, a half-decent player, to put in the bar.

Now, usually they work in a factory or hillbilly neighborhood and they'll put you in this bar during the week. So you may go in Wednesday and Thursday night, then Friday night you may take the joint off. Now, the guys you're working with will have already looked the bar over, so they know what speed the bar is at and they'll know the top players. They'll explain everything to you, who's good, who's not, and how the different guys play.

So you go in and pretty soon you are playing the best players and beating them, but not for any kind of money. You go in there to try and establish credibility as being a real good player. You also show them that you don't really gamble that much, that you like to gamble, but nothing big. So you might only make your drinking expenses the first few nights, but you warm up to these people, so they figure that you're the best, right! Now comes Friday or Saturday and you'll be there and pretty soon your partners will come in. Of course, you don't know them, but eventually one thing leads to another and now they're challenging you. They've played the other guys and beat them, now they challenge you. So here you play the shit ass, you don't want to play and you're just sitting there drinking. They have beaten all the guys that you beat, so now the locals say, "Hey, why don't you want to play this guy, we'll bet on you!" That will happen ninety percent of the time. You say, "Now look, you know I don't like to gamble too much." "Never mind, don't worry about those things, we'll bet on you." So you make the game look good and usually you win a good buck, like two or three hundred on that night is nothing. . . . So you lose the game, but you have to make it look good, because these people are rough. Like you have them coax you into it. You don't run up and say, "I'll play him!" or you'll blow the whole thing. . . .

Now your partners have to be able to really play well for the thing to work or you'll be in big trouble when you lose. And it helps cool out the locals if your partner beats everyone else. Then you say, "Well, I quit, I'm not betting any more. I don't want to lose your money." Like you wouldn't do that until you figure that it's pretty well the end of the night or if it looks like it's going to get pretty rough. Then you just get up and leave. You say, "That's it, you guys are crazy, I didn't want to bet that much money." You're constantly cooling them out and you even invest a little of your money, so you can take off a little pressure. There's many ways of cooling them out, but a lot of times you just can't do it. Now, it's not so much bolting out the door, it's more yelling and screaming. It gets a little rough sometimes because the guys are drinking, but a lot of hustlers swear by it. And they will have their own little spots like road hustlers will in cards or craps and they will say this area's good, this one's bad, this spot is always good on Saturdays, or this spot is only good on every second Thursday, like when the guys get paid. They have their own little files, maybe not down on paper but in their heads.

While pool hustling may seem a relatively uncomplicated way of making money, this is not the case. Not only are hustlers continually trying to manipulate their "clients" but also one another:

Pool hustling steadily is frustrating. You can't show your top speed. So you make a shot, then you have to avoid making a shot, but make it look like you really tried. Like you might be partners with a guy that you're double-crossing and if you make a certain shot, you win. But you are trying to take him off with these other two guys, so you have to make it look real close, say to where the ball jiggles in the horns and jumps out, something like that. You can't miscue, like in the movies. You're not supposed to miscue! You're a good player, right, so if you are shooting game ball and miscue, the guy isn't going to play for too long. He's going to get hostile and you're liable to get a cue over your head. So you have to miss the shot, but just barely or sometimes you might even make it, because it has to look good. You're cooling out that sucker constantly. Like you might make game ball and scratch. It might be a fairly easy shot and if you don't make it the sucker is going to pull up, so you make the ball and scratch. So you say to the guy, "Gee, you make a good shot and look what happens." And maybe he'll say, "You're hard luck, you know, you're hard luck." So the three of us are conspiring to take this guy, right. Whatever he loses we are cutting up. Say we are betting fifty a man, so the three of us are cutting up fifty dollars every game, which isn't bad. A game takes about ten minutes, and say you know a guy's got a thousand or five hundred, it doesn't take long to beat him for his money. So if you miss a shot in the game, it had better look good or he will pull up and it's going to cost you money. . . . That's why I don't really like pool hustlers, they are constantly doing business. You really can't tell when you are in a good game or a bad game. That's why when I play, usually I play one guy head up and that's it. If I lose I lose, if I win I win. Because when you play with four guys, two and two as partners, you really don't know what's going on.

In addition to its other drawbacks, pool hustling also involves a mobile life-style and unpredictable income:

Now if you are going to hustle steadily, you pretty well have to be mobile, because you get known in your area and when the locals know what you can do, it's very difficult to take anyone off, except in a set up. If you're an A player, okay you don't have to run around that much. When you get that reputation, you can locate at one pool room and people will come from all over the country to try beating you, because they know that you're action. They know that if they beat you they can win a big buck because all the locals are going to bet on you. But a B player pretty well has to move around if he wants to hustle and he's apt to get into trouble because he's going to run into some topnotchers, you just can't know them all.

Like myself, I was a B player and I didn't like big cities because you run into too many B+ and A players. And with them, unless they're poor managers, it's like knocking your head against the wall. I found it more lucrative to go to these smaller communities. Like you might go in a pool room and not play too well at first, just well enough to beat a few guys. Then everyone's saying, "Hey, this guy beat so and so" and you go up the ladder. You may play one guy that's not too good and you beat him, then they get another guy that's a little better, you beat him, and by the end of the night you are pretty well playing the best in the town. And, while you're there, you'll maybe try to stir up a little interest so that they'll go get the best. Now playing the best, you have to show a little class, a little speed. But you're not sure how the guy plays, so you have to weigh him up and if you find that you can beat him, you beat him.

Now some guys are good losers as well as good winners, but some guys do not lose gracefully! And part of the reason that it sometimes gets physical is that you aren't just betting with one guy, you're betting with maybe eight or nine guys. You might be betting this guy ten, this guy twenty, this guy five, the player twenty, and maybe the owner twenty. Usually these guys will lose their money gracefully, but I have had it happen that some of the guys are ready to fight you for the money they owe you, so I've lost a little money that way, because I'd rather not fight with the guy. You fight the guy and he may break your nose or your arm or something, and how can you go and hustle pool like that! . . .

At one point, I ran around with this muscle man. Then I could go real strong. Like at the end of the night, if I knew I could beat the guy, we would bet as much as we possibly could, whereas if I was alone, I would take it easier. But this man was big and he let everyone know that he was betting. When I felt that things were right, that I was stroking good and pretty sure of beating the guy, I would tell him to bet all that he could and he would be in there like a dirty old shirt. But in those years, I was just getting into cards and we weren't smart enough to know how to find enough stags or we'd go and have a good time more than actually beating them. Like we'd try and beat them and maybe we couldn't. So it was mostly pool hustling and skuffing around for whatever we could.

HUSTLING GOLF

Although less extensive and more prestigious than hustling pool, the principal themes in hustling golf are very similar. Thus, while game skill is basic to hustling golf, so are the other elements that characterize successful pool hustling: being able to accurately assess the relative talents of one's opposition and oneself; setting up targets; keeping the game interesting; and avoiding showing one's top speed:

In many ways, it's like pool. You try to figure out what the guy can do relative to you before he does. And this is where hustling in your own home town is rough because they get to know exactly what you can do. Like say they've seen you in a few tournaments, then it's like pulling teeth. They don't want to bet with you, so if you want to hustle golf, say with some regularity, you pretty well have to go on the road. Here again, it's better to have that setup. A lot of times another hustler, or maybe a sponsor, will get connected with a sucker and give you a call, so that's good. It's more difficult if you have to introduce yourself to the sucker, and it wastes time, because you have to find the suckers and stir up some action and establish credibility all on your own. . . . Of course, a lot depends on the way you manage the situation. Say your sponsor warmed up the guy, and set up a little match. So you play. Again you have to take your time and communicate with the guy. You don't get up and play your best game. If you smack the ball down the fairway, get on the green, and then putt it home, the guy would quit on you. So you have to take your time, and if you do that, you may be able to get three plays out of him. . . .

A big thing in hustling golf is being able to take the pressure of gambling for money. It's funny, but people think that they are better than they are. But once you start betting money, the pressure is on. What I would do is bet each individual hole. I wouldn't bet on the outcome of the whole game or a nine-hole total unless I had to. It's better to put pressure on the man at every hole rather than say betting the eighteen-holes total or playing Nassau, which means you bet each nine and the eighteen. You might bet him fifty or a hundred dollars every hole, double on eagles, whatever you can get out of the guy. You have be careful how you handle the man though, because you want to beat him, but you also want to keep him coming so that he still feels he has a chance to take you off. And it's tough, because you have to avoid showing your speed. You win a hole, lose a hole, win a hole, lose a hole. Then, when the money is bet, you have to be able to get your head together and play the way you are supposed to play. It's not that golf is so lucky, but a man can get a break where he's approaching the green, and maybe get closer than you, or maybe make an exceptionally long putt. You can't overcome something like that on the hole, but in the long run, if you're a consistent golfer, you will overcome it. . . . And, you don't beat the guy too badly. And, it's funny, you see, if the guy's a half-good player, your scores will be about the same, so he can't understand it, "Well gee, he didn't beat me by too much you know." But, if he just looks at the holes, he'll see, "Well, he beat me on the holes, where he was supposed to, and that's where it counts!" But it doesn't always work out; he may walk away and you've blown a little money. Like you can beat the guy outright, but you want to get him stuck to where you can increase the bet. And you might have a made a special trip to play this guy and now you lose a little or maybe end up with a hundred dollars.

The Ringer

Occasionally, like the pool hustler running into an A-level player, the golf hustler may run into a "ringer," a top-level golfer. This encounter may be a (unfortunate) match of the hustler's own choosing or may represent an attempt by a previously beaten player to even the score:

This guy may be a top amateur in the state or the country. Sure you know the big names, but you don't know everyone. So we hit a few holes and I play well enough that the other guy has to play well too, he has to show a little bit of speed. If you see the guy hit the ball a ton off the tee, he's good on his approach, and he's a good putter, then you forget about him. You say, "Well, that's it. Here's your hundred or two-hundred and I quit." Because, if you bet each individual hole, you can quit at any time, you don't have to play the eighteen or the nine. If the guy looks too strong, you just quit. Now a sucker that you beat previously might laugh, "Ah I got someone that could beat you!" you know. You say "Yeah, well I'll play you and give you a shot at nine," or something. You try to get back to him. Now you might spot the guy a few strokes to where he thinks he's getting enough and he's not and you take him off again. . . .

But you see, it's kind of hard feelings when you beat these people, more so than if you are in a crap game because a crap game isn't as personal. If you're golfing, you are with him and him alone and if you beat him, then he sort of figures maybe somebody set him up. . . . And with a sucker, you have to be a gentleman, really. You can try a few things if you are playing a topnotcher, because he's a ringer and he knows it, but it's tough to con people in golf. Now, you talk to the guy a little, but what's more important is that the guy may be a little anxious betting money.

This is true for suckers generally, and it's also true for a lot of these good amateurs or ringers. Say some member you beat knows a local who is a terrific amateur, and he sets you up okay, but what they usually don't understand is that once you get this amateur betting money, he's under a lot of pressure.

Now, sometimes, if the guy's a high-liner, he'll say to the amateur, "Don't worry about it, I'll bet my money, you don't have to worry about a thing. If we win, I'll take care of you!" So now, there isn't that much pressure on the amateur. Then what you do is you try to get this amateur to bet with you on the side. You try to get this amateur knowing that he's gambling, because then it's a different situation altogether. It's not like going to an amateur tournament. What have you got to lose? Nothing, maybe some prestige, but there's no money involved. Then there's different things you might do in the game. The guy might be addressing the ball and you may cough a little bit or clear your throat, just when he's about to hit. He's going to get real hot you know, but you may unnerve him, where he won't play up to par. If the guy is really good, you pay him off and quit.

Managing the Game

As in pool, managing the game is a critical aspect of hustling golf. Unless one is able to give the other player the impression that he has a chance to win, it is unlikely that he will be able to maximize his other talents. In this regard, the more successful hustlers seem better able to subtly disarm their opponents:

This one guy I know is no better at golfing than I am, but he is a terrific manager. Like I could beat him in a tournament, but I can't beat him on a handicap or managing myself. He manages himself in such a way that he looks like a real lark, he looks like a real duffer, and usually he'll get to play another duffer. But, you see, he's the type of guy who'll knock the ball down the fairway and knock it down to the green and knock it into the hole in fewer strokes than the other guy does, and just enough to beat him. And that's the whole gaff. If you can manage to play people you know you can beat and have them feel that you are less talented than them, that's a hell of an edge. He gives people the impression that he's not as good as they are and he takes them off. And he does the same thing with pool and bowling. . . .

Hustling golf is a lot like hustling pool; it's a real rat race. You have to keep looking for action and when you find it, you have to finesse around. You can't show your top speed; you have to keep trying to entice the other player. Now you meet a better class of people on the golf course, like professional and business people, but you still have to be able to size up the sucker and watch how you manage the game. So, it's much the same, you try to get the guy involved and maybe get him stuck a few bucks to where he wants to play for more to get even, but you're always on that fine edge.

THE DOUBLE STEER

Of the hustles discussed thus far, the "double steer" most clearly approximates the classic notion of the confidence game (Maurer, 1940). Its principles are fairly general and one, thus, finds it used in the context of card, dice, and pool hustling. In the double steer, a hustler and a target conspire to beat a third person (usually another hustler) who, unbeknown to the target, has conspired with the first hustler to beat the target for his money. In attempting to cheat the other person, the target, himself, is beaten.

As the double steer is contingent on locating a single suitable target, it represents an occasional rather than a regular involvement for most career hustlers. Typically, a hustler will "happen upon" a viable target. If the target does not already know that the hustler is a hustler of sorts, the hustler will inform the target that he can "do a few things," and together they devise a plan for beating a third person. Although this scheme has unlimited applications, the following account, portraying the double steer in the context of card and dice hustling, may be illustrative:

There are a lot of ways you can work the double steer. I'll give you an example. Like this one day I bumped into this fellow who had just won eight or nine hundred at the race track. He knew that I was a half-assed card mechanic and we worked out a plan to beat a third man, who was really my partner, with a cooler. So the three of us got together and my partner flashes a lot of money looking like he's good action. Now before this, I gave this sucker a cooler to put in his jacket and told him at the proper time he should hand it to me under the table when we're playing. So, fine, he does that and I deal. But instead of him winning, my partner wins. Now he's blown four hundred or so and he's looking at you with real bad eyes.

About then your partner goes to the washroom or something and you have a chance to be alone with the sucker. He says, "What the hell happened?" You say, "You dumb son of a bitch, when you handed me the cards I think one of the cards got misplaced or maybe there's a card in your jacket or something." And before your partner returns you tell him, "We'll get your money back, but only bet when I put in the cooler." Like your partner wouldn't stay away too long, just long enough for you to convince the sucker that something went wrong and it was his fault, that you made up the cooler and it was right! Now here's where the kiss-off comes. The guy is stuck a few bucks and he's susceptible, so you double duke him. You give him three of a kind and your partner something better. So usually the sucker will get eager and bet big here, so that when the hand is over he has blown another couple or three hundred. But he can't really say much to you because you told him, "Only bet when the cooler comes in." So you sort of get the guy in the middle like this and there are lots of things you might do. Then comes the cool out. You quit and signal him to leave with you while he still has a few bucks left. Then you tell him, "Why did you bet on that hand? I told you only to bet when I put in a cooler. I had one already to go, but I had to wait until the right time. I told you not to bet until I made my move!" And he'll say, "Well, it was a good hand." But you stick with him and maybe take him for a beer or a hamburger or something, and after a bit he'll calm down a little. Like you shift the blame to him and then you try to help him feel better. . . .

Now usually you would work this double steer with a businessman or supposedly an honest person. Someone who likes to take an edge if he can find it, but not a card hustler because they wouldn't go for it. You might do it to a hep man, say a booster or whatever, but you're best to go to a legitimate man. And you don't start ripping him off right away. You let him win a little to entice him. . . . You do the same thing in craps. Like you might let the man know that you have some small percentage dice. Then in the game you might switch in the opposite percentage. Pretty soon, he gets stuck a few bucks and wants to get his money back, so he makes larger bets. Now, you might switch in tops, so that he loses. Again you cool him out. You let him know that you have lost some money too and you say, "I told you that with small percentage dice, they're not sure things. You have to bet the same amount consistently or you can get in trouble. But you just wouldn't listen. You had to try making it all back in one shot. You can't do it that way, you have to take your time." It's a thing where you have to be able to relate to people and try to explain things to them in a way that they can understand. You see in the game situation, they get caught up in the betting and after you have to help them explain it to themselves where you're not getting the blame. . . .

But the double steer is the last thing I would work on because these are more likely to be people you know and they may start having bad feelings towards you, so it's not good in that way. Now if you have a set up with another hustler, that's okay, but where it's people that you might be bumping into, it's not so good. But if the money is there and you know the guy is just going to blow it anyways, well you might try to get it and if it's done right, the guy will still be, let's say, a friend when he sees you.

IN PERSPECTIVE

Pool [and] golf . . . by no means exhaust the field of hustles in which professional card and dice hustlers may, at some point in their lives, find themselves. Thus, it would not be unusual to find these people in a variety of social hustles, such as boosting, "laying the note" (shortchanging cashiers), or fencing. These activities reflect the opportunistic orientation of the hustlers and the nature of their contacts:

> After you're connected, you start to learn about all these gaffs. Like, this laying the note. For some hustlers, it's almost standard procedure to make ten or twenty bucks laying the note when they first leave the hotel. That way, they figure that they're sure to have something for the day. This one guy in our crew did it now and then, but it always seemed too awkward to me, so I never got into it. But these guys have so many gaffs and you're in that environment where you are continually bumping into people so you have the chance to get into this or that. You might meet these people and when they see that you're solid, they start to open up to you, so you have a lot of contacts in different hustles. . . .
>
> Some of these hustles are rough, but some of them require a lot of skill and training to where it's not worth it. You would have to start almost as a novice again, but, several times, I've been tempted to set myself up as a fence. Maybe some booster will call you Friday or Saturday night, "We've got forty suits, fifty pairs of slacks, give me $500 for the lot." Which is a bargain, eh! But you see, they've been drinking or are on drugs, or maybe they're gambling and they want the money now, kind of thing. Many times I've had the urge, you know. You have the money and it's a hell of a deal, but then you think, "What am I going to do with all this stuff?" Some of these fences have legitimate businesses and they can absorb all these bargains, but how many items could I handle? Now you may pick up a few things and unload these to people you know, but to go into fencing at this level is too demanding of your time.

Although these side hustles represent part-time, off-season, or previous activities for professional card and dice hustlers, they reflect a level of involvement in the thief subculture; and whether one is discussing card and dice hustling, pool hustling, booking, boosting, or other hustles, in order to become a *proficient* practitioner of any professional hustle, some general themes are operative. First one has to get connected with an "insider," someone already in that hustle. Second, the specific techniques of each hustle have to be learned from insiders; while one may acquire a general aptitude for conning people around from almost any hustle, each hustle has specific features that must be learned. Third, from association with an insider, one finds that he encounters justifications for hustling and establishes personal contacts that promote his hustling involvements; without this social support, it would be difficult for persons to maintain themselves in this activity. Fourth, the mastery of any hustle requires practice and dedication. Fifth, it is easier to hustle when one has a supporting cast; partnerships are not only conducive to developing more sophisticated routines, but two or more persons are better able to establish target confidence than solitary operators. Finally, once a person becomes involved in one hustle, he is likely to encounter people in related hustles, so that new hustling opportunities open up to him.

NOTE

1.   Given the abbreviated nature of the discussion, we feel fortunate in being able to refer the reader to Ned Polsky's (1967) excellent discussion of pool hustling for additional background material.

PART VIII

Deviant Careers

One of the fascinating things about people's involvement in deviance is that it evolves, yielding a shifting and changing experience. Doing something for the first time is very different from doing it for the hundredth time. It is fruitful, then, to consider involvement in deviance from a career perspective, to see what the nature of deviance is and how it develops over the course of people's involvement with it. Sociologists have documented various stages of people's participation in such things as drug use, drug dealing, fencing, carrying out a professional hit, engaging in prostitution, and shoplifting. Although these activities are very different in character, they have structural similarities in the way that people experience them according to the stage of their involvement. In fact, the career analogy has been applied fruitfully to the study of deviance because people go through many of the same cycles of entry, upward mobility, achieving career peaks, aging in the career, burning out, and getting out of deviance as they do in legitimate work. The career phases most commonly analyzed are those at the beginning and the end, when participants are involved in the transition between deviance and the conventional world, but we will also look at some features of intermediary career deviance. In examining deviant careers we note the limitation of the comparison to enterprises in the legitimate realm; whereas legitimate work has several varieties of structure, the patterns for deviant careers are more flexible (Luckenbill and Best 1981). Entry can take many shapes and lengths of time. Once a person is in deviance, behavior shifts can be lateral and downward as well as upward, precipitous as well as gradual and controlled, repetitive as well as dissimilar, and involve continuity or complete shift into other venues. Exits are problematic, varying in degree of coercion or voluntariness, being temporary or lasting, and involving anything from going out on top to slinking away in debt and disgrace. Let us consider these aspects of the deviant career one by one.

Sociologists have been most fascinated by the process through which people **enter deviance.** They have looked at "push factors," "pull factors," and "subcultural factors." Although some people venture into deviance on their own, the vast majority do it with the encouragement and assistance of others, often joining cooperative deviant enterprises. The turning points that mark significant phases in their transitions have been explored, as well as their changing self-identities. Most commonly, people who become involved in deviance do so through a process of shifting their circle of friends. They drift into new peer groups as they are drifting into deviance, or their whole peer group drifts into deviance together as the members enter a new phase of the life cycle.

The selections in this section highlight two different ways of entering deviance: alone or with a group. In "Marks of Mischief: Becoming and Being Tattooed," Clinton Sanders considers why people get permanent body designs. Although this is largely an individual decision, potential tattooees, influenced by others they know who are already similarly decorated, see a way of affiliating themselves with a larger, somewhat nebulous, deviant community. Having a tattoo has traditionally put individuals out of the mainstream of American middle-class society, and people wanting to show their disaffiliation with this normative position and lifestyle are often attracted to the rebellious symbolism of tattoos. Sanders shows with irony that although tattoos last permanently, they usually result from an impulse decision that people have been mulling over only loosely.

Next, Martín Sánchez Jankowski, in "Joining a Gang," details and contrasts the many complex factors that induce individuals to enter a highly cohesive and illicit group. These range from the physical to the economic and social. But joining a gang is not a one-way decision; not every individual is acceptable as a member. Gangs have organizational needs that must be met through the continuous recruitment of new people, people who will mesh with existing members and be reliable in uncertain or dangerous situations. They may choose to entice new members with their attractiveness, to pressure them into joining through a sense of duty, or to coerce them through threats. Sánchez Jankowski also addresses the question of why some individuals do not join gangs. This selection serves as an interesting counterpoint to the other discussions of gangs in the readings on motorcycle gangs and international organized crime.

Being deviant holds different challenges. Participants must manage their deviance, their relationships within deviant communities, and their safety from agents of social control, and they must evolve a personal style for their deviance, balancing their deviance with the nondeviant aspects of their lives, such as their relationships with family members, people in the community, and those on whom they rely to meet their legitimate needs.

Jody Miller outlines some of the ways streetwalkers evade, confront, and avenge themselves against the violence that serves as a continuous challenge to their working lives in "Victimization and Resistance Among Street Prostitutes." Miller traces the kinds of violence to which prostitutes are exposed and describes the different strategies women use to recognize and deal with it, alone and with the help of others.

This section also includes an article by James Myers on "Nonmainstream Body Modification: Genital Piercing, Branding, Burning, and Cutting." These decorations, forms of deviant body adornment growing in popularity on the outskirts of conventional living, are sported by people circulating in a variety of subcultures. Myers takes us on a journey through a series of workshops offered by expert body modifiers who discuss and illustrate the intricacies of these practices. He then discusses some of the motivations underlying people's participation in this form of body embellishment, grounding his explanation within a cross-cultural and historical perspective.

Although people tend to think that getting into deviance represents the more difficult end of the career span, sociologists have found that there can be greater problems associated with **exiting deviance.** People change during their involvement with deviance and the easily obtained money they often find there; returning to a more restricted base of funds is not always easy. They also become accustomed to the free-wheeling lifestyle and open value system associated with a deviant community, in which conventional norms are disdained. Reentering the straight world with its morality may chafe. They may have difficulty earning a living if they have been involved in occupational deviance, where they were making money through illicit means. They may have difficulty putting together a résumé that accounts for their gap in legitimate employment and finding someone who will hire them. They may have difficulty adhering to the structure of the 9-to-5 straight world. They may have difficulty finding legitimate skills through which they can support themselves.

Yet most people do not want to spend their whole lives engaged in deviance. We discuss the process by which people burn out of deviance in "Shifts and Oscillations in Deviant Careers: Upper-Level Drug Dealers and Smugglers." After spending several years in the upper echelons of the drug trade, many marijuana smugglers find that the drawbacks of the lifestyle exceed the rewards. The initial excitement and thrills turn to paranoia, people whom they know get busted all around them, and their risk of arrest grows. Years of consuming drugs to excess takes its toll on them physically, and they come to regard in a new light the straight life that they formerly rejected as boring. Yet they cannot easily quit dealing; they have developed a high-spending lifestyle that they are loathe to abandon. When they try to retire from trafficking, they quickly use up all their money and are drawn back into the business. Thus, their patterns of exiting often resemble a series of quittings and restartings, as they move out of deviance with great difficulty.

Ira Sommers, Deborah Baskin, and Jeffrey Fagan consider a broader swath of exiting deviance in their research on how women leave the deviant arena in "Getting Out of the Life: Crime Desistance by Female Street Offenders." Working with a sample of women who had been involved in violent street crime, Sommers, Baskin, and Fagan consider the special circumstances faced by women in deciding to go straight and following through on that determination. Their model for the desistance process illustrates the varied factors women experience in street deviance that ultimately lead them to wanting to get out, the means by which they make the decision to strive for closure on their deviant careers, how they do it, and how, once out, they stay away from deviance (a critical problem, as we learned in our last reading). This work highlights the intricate and complex process associated with desistance, and its reverse mirror parallels the means by which most people enter into lives of deviance.

36

Marks of Mischief

Becoming and Being Tattooed

CLINTON R. SANDERS

A person's physical appearance is a central element affecting his or her self-definition, identity, and interaction with others (Cooley, 1964: 97–104, 175–178, 183; Stone, 1970; Zurcher, 1977: 44–45, 175–178). People use appearance to place each other into categories which aid in the anticipation and interpretation of behavior and to make decisions about how best to coordinate social activities (Goffman, 1959: 24–25; McCall and Simmons, 1982: 214–216; Ruesch and Kees, 1972: 40–41, 57–65).

How closely one meets the cultural criteria for beauty is an appearance factor of key social and personal import. Being defined as attractive has considerable impact on our social relationships. We think about attractive people more often, define them as being more healthy, express greater appreciation for their work, and find them to be more appealing interactants (Jones et al., 1984: 53–56). Attractive people are more adept at establishing relationships (Brislin and Lewis, 1968) and enjoy more extensive and pleasant sexual interactions than do those who are not as physically appealing (Hatfield and Sprecher, 1986). Their chances of economic success are greater (Feldman, 1975), and they are consistently defined by others as being of high moral character (Needleman and Weiner, 1977).

Enjoying more frequent positive interactions, attractive people have correspondingly more positive self-definitions. In general, they express more feelings of general happiness (Berscheid et al., 1973), have higher levels of self esteem and are less likely than the relatively unattractive to expect that they will suffer from mental illness in the future (Napoleon et al., 1980).

Clothing, cosmetics, and hair styling are mechanisms for altering appearance that have in common the relative ease with which one can change one's social "vocabulary" if the message communicated becomes outdated, undesirably stigmatizing, or otherwise worthy of reconsideration. In general, the cross-cultural literature on adornment and body alteration indicates that non-permanent decorative forms (principally costume and body paint) are most commonly associated with transitional statuses or specific and limited social

From "Marks of Mischief: Becoming and Being Tattooed," Clinton R. Sanders, *Journal of Contemporary Ethnography*, Vol. 16, No. 4, 1988. Reprinted by permission of Sage Publications, Inc.

situations. The major forms of permanent alteration—body sculpture, infibulation (piercing), cicatrization (scarification), and tattooing—are, on the other hand, connected to permanent statuses (e.g., gender, maturity), life-long social connections (e.g., clan or tribal membership), or conceptions of beauty that show considerable continuity from generation to generation (see Polhemus, 1978: 149–173).

Those who choose to permanently modify their bodies in ways that violate prevailing appearance norms—or who reject culturally prescribed alterations—risk being defined as socially or morally inferior. Public display of voluntarily acquired, symbolic physical deviance effectively communicates a wealth of information that shapes the social situation in which interaction takes place (Goffman, 1963a; Lofland, 1973: 79–87).

This article focuses on tattooing as a form of permanent body alteration in contemporary society. Choosing to mark one's body in this way changes the tattooee's experience of his or her physical self and has significant potential for altering social interaction. Because of the historical course of tattooing in the West the tattoo is conventionally defined as an indication of the bearer's alienation from mainstream norms and social networks. It is *voluntary stigma* that symbolically isolates the bearer from "normals." Since tattooees are deemed to be responsible for their "deviant" physical condition, the mark is especially discrediting (Jones et al., 1984: 56–65).

Like most stigmatizing conditions, however, tattooing also has an affiliative effect; it identifies the bearer as a member of a select group. When publically displayed the tattoo may act as a source of mutual accessibility (Goffman, 1963b: 131–139). Fellow tattooees commonly recognize and acknowledge their shared experience, decorative tastes, and relationship to conventional society. Tattooing also has affiliative impact in that it is routinely employed to demonstrate one's indelible connection to primary associates (e.g., name tattoos) or groups whose members share specialized interests and activities (e.g., motorcycling, use of illegal drugs, or involvement with a specific youth gang). . . .

METHOD

I first became interested in tattooing in San Francisco in 1979. Having a bit of time on my hands, I decided to explore the more obscure museums listed in the telephone directory. Climbing the dingy stairway leading to Lyle Tuttle's Tattoo Art Museum, I found myself in a new and fascinating world of cultural production. After looking at the sizeable collection of tattoo memorabilia, I entered the tattoo studio adjacent to the museum and, like many first-time visitors to tattoo establishments, impulsively decided to join the ranks of the tattooed. After choosing a small scarab design from the wall "flash," I submitted to the unexpectedly painful tattoo experience. Although the resident tattooist was not very forthcoming in response to the questions I forced out between clinched teeth, I did realize that this was a phenomenon which combined my interests in both social deviance and art worlds and offered a research experience which would provide a much-needed escape from the polite confines of academia.

Returning to the east coast I visited a small "street shop" in a "transitional neighborhood" located a few minutes from my office. The owner was flattered that a "professor" would want to hang out and listen to him talk about himself, and I soon became a regular participant in the shop, observing the work, talking to the participants, and—despite my original vow to never again undergo the pain of indelible body alteration—eventually receiving considerable tattoo "work" from a variety of renowned tattoo artists with whom I came into contact during the subsequent seven years.

The following discussion is based primarily on data collected during participant observation in four tattoo "studios" located in or near major urban centers in the east. Three were traditional shops specializing in the formulaic images favored by military personnel, bikers, laborers, and occasional groups of college students and secretaries. One establishment was a "custom" studio, in which a tattooist with extensive professional experience in a variety of artistic media created original and unique works of art for a more select, monied, and aesthetically sophisticated clientele.

For the most part, my role in the settings was that of one of a number of regular hangers-on who either lived in the neighborhood or were friends of the local artist. My participation in the establishment to which I originally gained access was considerably more extensive. In addition to (apparently) just standing around and chatting, I helped with the nontattooing business of the shop. I made change for the amusement games, provided information about cost and availability of designs, stretched the skin of customers who were receiving tattoos on body areas other than arms or legs, calmed the anxiety of first-time recipients, and made myself generally useful.

In addition to the field data, this discussion is based on a series of lengthy, semi-structured tape-recorded interviews conducted with tattoo recipients encountered during the course of the research. I collected interviews with 16 people (10 men and 6 women) who were representative of the sex, age, and social status categories I encountered in the field settings. Their average age was 24 (from 17 to 39); as a group they carried 35 tattoos (9 had one, 3 had two, and 4 had three or more).

A somewhat more structured body of data was drawn from 163 four-page questionnaires completed by tattooees contacted in three separate settings. Fifty-six were filled out by tattoo "enthusiasts" attending the 1984 convention of the National Tattoo Association in Philadelphia, 44 were returned by clients in the "artistic" studio, and 63 questionnaire respondents completed the instrument following their tattoo experience in the street shop in which I began to collect field data. Sixty-eight percent of the questionnaire respondents were men and 32 percent were women. They ranged in age from 17 to 71, with an average of 30 years. Sixty-two percent of the respondents had received some education past high school and 5% had graduate degrees. Skilled craftwork, machine operation, and general laboring were the most common occupations pursued by the men; service and clerical work was most heavily represented among the women. Twelve percent of the men and 6% of the women were involved in professional or technical occupations.

THE PROCESS OF BECOMING A TATTOOED PERSON

Initial Motives

Becoming tattooed is a highly social act. The decision to acquire a tattoo (and, as we will see in a later section, the image that is chosen), like most major consumer products is motivated by how the recipient defines him or her self. The tattoo becomes an item in the tattooee's personal "identity kit" (Facetti and Fletcher, 1971; Goffman, 1961: 20–21), and in turn it is used by those with whom the individual interacts to place him or her into a particular, interaction-shaping social category (see Csikszentmihalyi and Rochberg-Halton, 1981; Solomon, 1983).

When asked to describe how they decided to get a tattoo, the vast majority of respondents made reference to another person or group. Family members, friends, business associates, and other people with whom they regularly interacted were described as being tattooed. Statements such as "Everyone I knew was really into tattoos. It was a peer decision. Everyone had one, so I wanted one" and "My father got one when he was in the war and I always wanted one, too" were typical. Entrance into the actual tattooing "event," however, has all of the characteristics of an impulse purchase. It typically is based on very little information or previous experience (58% of the questionnaire respondents reported *never* having been in a studio prior to the time they received their first tattoo). While tattooees commonly reported having "thought about getting (a tattoo) for a long time," they usually drifted into the actual experience when they "didn't have anything better to do," had sufficient money to devote to a nonessential purchase, and were, most importantly, in the general vicinity of a tattoo establishment. The following accounts were fairly typical.

> We were up in Maine and a bunch of us were just talking about getting tattoos—me and my friends and my cousins. One time my cousin came back from the service with one and I liked it. . . . The only place I knew about was ———— ————'s down in Providence. We were going right by there on our way back home, so we stopped and all got them.

> My friends were goin' down there to get some work, you know. That was the only place I knew about, anyway. My friends said there was a tattoo parlor down by the beach. Let's go! I checked it out and seen something I liked. I had some money on me so I said, "I'll get this little thing and check it out and see how it sticks." I thought if I got a tatty it might fade, you know. You never know what's goin' to happen. I don't want anything on my body that is goin' to look fucked up.

The act of getting the tattoo itself is usually, as seen in these quotes, a social event experienced with close associates. Sixty-nine percent of the interviewees (11 of 16) and 64% of the questionnaire sample reported having received their first tattoo in the company of family members or friends. These close associates act as "purchase pals" (Bell, 1967). They provide social support for the decision, help to pass anxiety-filled waiting time, offer opinions regarding the design and body location, and commiserate with or humorously ridicule the recipient during the tattoo experience (see Becker and Clark, 1979; Sanders, 1985a).

The tattoo event frequently involves a ritual commemoration of a significant transition in the life of the recipient (compare Brain, 1979: 174–184; Ebin, 1979: 39–56; Van Gennep, 1960). The tattooee conceives of the mark as symbolizing change—especially achieving maturity and symbolically separating the self from individuals or groups (parents, husbands, wives, employers, etc.) who have been exercising control over the individual's personal choices. A tattoo artist related his understanding of his clients' motivations in this way:

> I do see that many people get tattooed to find out again . . . to say, "Who was I before I got into this lost position?" It's almost like a tattoo pulls you back to a certain kind of reality about who you are as an individual. Either that or it transfers you to the next step in your life—the next plateau. A woman will come in and say, "Well, I just went through a really ugly divorce. My husband had control of my fucking body and now I have it again. I want a tattoo. I want a tattoo that says that I have the courage to get this, that I have the courage to take on the rest of my life. I'm going to do what I want to do and do what I have to do to survive as a person." That's a motivation that comes through the door a lot.

One interviewee expressed her initial reason for acquiring her first tattoo in almost exactly the same terms:

> (My friend and I) both talked semiseriously about getting (a tattoo). I mentioned it to my husband and he was adamantly opposed—only certain seedy types get tattoos. He didn't want someone else touching my body intimately, which is what a tattoo would involve . . . even if it was just my arm. He was against it, which made me even more for it. . . . I finally really decided some time last year when my marriage was coming apart. It started to be a symbol of taking my body back. I was thinking that about the time I got divorced would be a good time to do it.

Locating a Tattooist

Like the initial decision to get a tattoo, the tattooist one decides to patronize commonly is chosen through information provided by members of the individual's personal networks. The shop in which they received their first tattoo was located by 58% of the questionnaire respondents through a recommendation provided by a friend or family member. Since in most areas establishments that dispense tattoos are not especially numerous, many first-time tattooees choose a studio on a very practical basis—it is the only one they know about or it is the studio which is closest to where they live (20% of the questionnaire sample chose the shop on the basis of location, 28% because it was the only one they knew about).

The central importance of personal recommendation as the source of tattoo clients is well-known to tattooists. All tattooists have business cards that they hand out quite freely (one maintained that he had dispensed over 50,000 cards in the past two years). Listing one's services in the telephone directory is the other major means employed to draw customers, since it provides locating information for those who, for a variety of reasons, do not have interpersonal sources.

Most first-time tattooees enter the tattoo setting with little information about the process or even the relative skill of the artist. Rarely do recipients spend as much time and effort acquiring information about a process that is going to indelibly mark their bodies as they would were they preparing to purchase a TV set or other far less significant consumer item.

Consequently, tattooees usually enter the tattoo setting ill-informed and experiencing a considerable degree of anxiety. Their fears center around the anticipated pain of the process and the permanence of the tattoo. Here, for example, is an interaction that took place while a young man received his first tattoo.

> **Recipient:** Is this going to fade out much? There's this guy at work that has these tattoos all up and down his arms and he goes back to the guy that did them every couple of months and gets them recolored because they fade out. (general laughter)
>
> **Sanders:** Does this guy work in a shop or out of his house?
>
> **R:** He just does it on the side.
>
> **Tattooist:** He doesn't know what the fuck he's doing.
>
> **R:** This friend of mind told me that getting a tattoo really hurts. He said there would be guys in here hollering and bleeding all over the place.
>
> **S:** Does he have any tattoos?
>
> **R:** No, but he says he wants to get some. . . . Hey, this really doesn't hurt that much. It doesn't go in very deep, does it? It's like picking a splinter out of your skin. I was going to get either a unicorn or a Pegasus. I had my sister draw one up because I thought they just drew the picture on you or something. I didn't know they did it this way (with an acetate stencil). I guess this makes a lot more sense.
>
> **S:** You ever been in a tattoo shop before this?
>
> **R:** No, this is my first time. Another guy was going to come in with me, but he chickened out.

For the most part, tattooists are quite patient about answering the questions clients ask with numbing regularity (pain, price, and permanence). This helps to put the recipient more at ease, smooths the service delivery interaction, and increases the chances that a satisfied customer—who will recommend the shop to his/her friends and perhaps return again for additional work—will leave the establishment (for extended discussions of in-shop interaction see Becker and Clark, 1979; Govenar, 1977; St. Clair and Govenar, 1981; and Sanders, 1983 and 1985b).

CHOOSING A DESIGN
AND BODY LOCATION

Tattooees commonly stated their basic motivations for becoming tattooed in very general terms. Wearing a tattoo connected the person to significant others who were similarly marked, made one unique by separating him or her from those who were too convention-bound to so alter their bodies, symbolized freedom or self-control, and satisfied an aesthetic desire to decorate the physical self.[1]

The image one chooses, on the other hand, is usually selected for a specific reason. Typically, design choice is related to the person's connection to other people, his or her definition of self or, especially in the case of women, the desire to enhance and beautify the body.

One of the most common responses which tattoo clients gave to my routine question, "How did you go about deciding on this particular tattoo?" was to make reference to a personal association with whom they had a close emotional relationship. Some chose a particular tattoo because it was like that worn by a close friend or a member of their family. Others chose a design which incorporated the name of their boy/girlfriend, spouse or child or a design associated with that person (e.g., zodiac signs):

> I had this homemade cross and skull here and I needed a coverup. [The tattooist] couldn't just do anything, so I thought to myself, "My daughter was born in May, and that's the Bull." I'm leaving the rest of this arm clean because it is just for my daughter. If I ever get married, I'll put something here [on the other arm]. I'll get a rose or something for my wife.
>
> This tattoo is a symbol of friendship. Me and my best friend—I've known him since I could walk—came in together and we both got bluebirds to have a symbol that when we do part we will remember each other by it.

The ongoing popularity of "vow tattoos," such as the traditional heart with "MOM" or flowers with a ribbon on which the loved one's name is written, attests to the importance of tattooing as a way of symbolically expressing love and commitment (see Anonymous, 1982).

Similarly, tattoos are used to demonstrate connection and commitment to a group. For example, military personnel pick tattoos which relate to their particular service, motorcycle gang members choose club insignia, and members of sports teams enter a shop en masse and all receive the same design.

Tattoos are also employed as symbolic representations of how one conceives of the self or interests and activities which are key features of self definition. Tattooees commonly choose their birth sign or have their name or nickname inscribed on their bodies. Others choose more abstract symbols of the self.

> I put a lot of thought into this tattoo. I'm an English lit major, and I thought that the medieval castle had a lot of significance. I'm an idealist, and I thought that was well expressed by a castle with clouds. Plus, I'm blond and I wanted something blue.

> [Quote from field notes] Two guys in their twenties come in and look at the flash. After looking around for a while one of the guys comes over to me and asks if we have any bees. I tell him to look through the book [of small designs] because I have seen some bees in there. I ask, "Why do you want a bee? I don't think I have ever seen anyone come in here for one." He replies, "I'm allergic to bees. If I get stung by one again I'm going to die. So I thought I'd come in here and have a big, mean looking bee put on. I want one that has this long stinger and these long teeth and is coming in to land. With that, any bee would think twice about messing with me."

Tattooees commonly represent the self by choosing designs which symbolize important personal involvements, hobbies, occupational activities, and so forth. In most street shops, the winged insignia of Harley-Davidson motorcycles and variants on that theme are the most frequently requested images. During one particularly busy week in the major shop in which I was observing, a rabbit breeder acquired a rabbit tattoo, a young man requested a cartoon frog because the Little League team he coached was named the "Frogs," a fireman received a fire fighter's cross insignia surrounded by flame, and an optician chose a flaming eye.

No matter what the associational or self-definitional meaning of the chosen tattoo, the recipient is commonly aware of the decorative-aesthetic function of the design. When I asked tattooees to explain how they went about choosing a particular design, they routinely made reference to aesthetic criteria—they "liked the colors" or they "thought it was pretty."

> [I didn't get this tattoo] because of being bad or cool or anything like that. It's like a picture. You see a picture you like and you put it in your room or your house or something like that. It's just a piece of work that you like. I like the art work they do here. I like the color [on my tattoo]. It really brings it out—the orange and the green. I like that—the colors.

On their part, tattooists tend to recognize the aesthetic importance of their work as seen by their clients. One tattooist, for example, observed:

> If you ask most people why they got (a particular tattoo) they aren't going to have any deep Freudian answers for you. The most obvious reason that someone gets a tattoo is because they like it for some reason and just want it. I mean, why do people wear rings on their fingers or any sort of non-

functional decorative stuff—put on makeup or dye their hair? People have the motivation to decorate themselves and be different and unique. . . .

Tattooing is really the most intimate art form. You carry it on your body. The people that come in here are really mostly just "working bumpkins." They just want to have some art they can understand. This stuff in museums is bullshit. Nobody ever really sees it. It doesn't get to "the people" like tattoo art.[2]

A number of factors determine a tattooee's decision about where on the body the tattoo will be located. The vast majority of male tattooees choose to have their work placed on the arm. In his study of the tattoos carried by 2,000 members of the Royal Navy, Scutt found that 98% had received their tattoo(s) on the arm (Scutt and Gotch, 1974: 96). In my own research, 55% of the questionnaire respondents received their first tattoo on the arm or hand (71% of the males and 19% of the females). The 16 interviewees had, altogether, 35 tattoos, 27 of which were carried by the 10 males. Eighty-one percent (22) of the men's tattoos were on their arms (of the remainder 2 were on hips, one was on the back, one on the face, and one on the recipient's chest). The 6 women interviewees possessed 8 tattoos—3 on the back or the shoulder area, 3 on the breast, 1 on an arm, and 1 on the lower back. Thirty-five percent of female questionnaire respondents received their first tattoo on the breast, 13% on the back or shoulder, and 10% on the hip.

Clearly, there is a definite convention affecting the decision to place the tattoo on a particular part of the body—men, for the most part, choose the arm while women choose the breast, hip, lower abdomen or back/shoulder. To some degree the tendency for male tattooees to have the tattoo placed on the arm is determined by technical features of the tattoo process. Tattooing is a two-handed operation. The tattooist must stretch the skin with one hand while inscribing the design with the other. This operation is most easily accomplished when the tattoo is being applied to an extremity. Tattooing the torso is more difficult and, commonly, tattooists have an assistant who stretches the client's skin when work is being done on that area of the body. Technical difficulty, in turn, affects price. Most tattooists charge 10 to 25% more for tattoos placed on body parts other than the arm or leg. The additional cost factor probably has some effect on the client's choice of body location.

Pain is another factor shaping the tattooee's decision. The tattoo machine contains needle groups which superficially pierce the skin at high speed, leaving small amounts of pigment in the tiny punctures. Obviously, this process will cause more or less pain depending on the sensitivity of the area being tattooed. In general, tattooing arms or legs is less painful than marking body areas with a higher concentration of nerve endings or parts of the body where the bones are not cushioned with muscle tissue.[3]

The different symbolic functions of the tattoo for males versus females appears to be a major issue affecting the sex-based conventions regarding choice of body site. Women tend to regard the tattoo (commonly a small, delicate design) as a permanent body decoration primarily intended for personal pleasure and the enjoyment of those with whom they are most intimate. The chosen tattoos are, therefore, placed on parts of the body most commonly seen by those with whom women have primary relationships. Since tattoos on women are especially stigmatizing, placement on private parts of the body allows women to retain unsullied identities when in contact with casual associates or strangers (see Goffman, 1963a: 53–55, 73–91). Here, for example, is a portion of a brief conversation with a young woman who carried an unconventional design (a snake coiled around a large rose) on what is, for women, an unconventional body location (her right bicep).

Sanders: How did you decide on that particular design?

Woman: I wanted something really different and I'd never seen a tattoo like this on a woman before. I really like it, but sometimes I look at it and wish I didn't have it.

S: That's interesting. When do you wish you didn't have it?

W: When I'm getting real dressed up in a sleeveless dress and I want to look . . . uh, prissy and feminine. People look at a tattoo and think you're real bad . . . a loose person. But I'm not.

Another interviewee described the decision-making process she had gone through in choosing to acquire a small rose design on her shoulder, emphasizing aesthetic issues and stigma control.

> The only other place I knew of that women got tattoos was on the breast. I didn't want it on the front of my chest because I figured if I was at work and had an open blouse or a scoop neck, then half would show and half wouldn't. I wanted to be able to control when I wanted it to show and when I didn't. If I go for a job interview I don't want a tattoo on my breast. I didn't want it, like, on my thigh or on the lower part of my stomach. I didn't like how they look there. I just thought it would look pretty on my shoulder. . . . The main reason is that I can cover it up if I want to.

Men, on the other hand, typically are less inclined than women to define the tattoo primarily as a decorative and intimate addition to the body. Instead, the male tattoo is an identity symbol—a more public display of interests, associations, separation from the normative constraints of conventional society and, most generally, masculinity. The designs chosen by men are usually larger than those favored by women and, rather than employing the gentle imagery of nature and mythology (flowers, birds, butterflies, unicorns and so forth), they frequently symbolize more violent impulses. Snakes, bloody daggers, skulls, dragons, grim reapers, black panthers, and birds of prey are dominant images in the conventional repertoire of tattoo designs chosen by men. Placement of the image on the arm allows both casual public display and, should the male tattooee anticipate a critical judgment from someone whose negative reaction could have untoward consequences (mostly commonly, an employer), the tattooed arm can be easily hidden with clothing. One male interviewee spoke about the public meaning of tattoos and expressed his understanding of the difference between male and female tattoos as follows:

> You fit into a style. People recognize you by your hair style or by your tattoo. People look at you in public and say, "Hey, they got a tattoo. They must be a particular kind of person," or, "He's got his hair cropped short (so) he must be a different kind of person." The person with a tattoo is telling people that he is free enough to do what he wants to do. He says, "I don't care who you think I am. I'm doing what I want to do." (The tattoo) symbolizes freedom. It says something about your personality. If a girl has a skull on her arm—it's not feminine at all—that would symbolize vengeance. If a woman gets a woman's tattoo, that's normal. If she gets a man's tattoo symbolizing vengeance or whatever, I feel that is too far over the boards. A woman should act like a woman and keep her tattoos feminine. Those vengeance designs say, "Look out." People see danger in them.

THE INTRAPERSONAL AND INTERPERSONAL EXPERIENCE OF WEARING A TATTOO

Impact on Self-Definition

As indicated in the foregoing presentation of the initial motives which prompt the decision to acquire a tattoo, tattooees consistently conceive of the tattoo as having impact on their definition of self and demonstrating to others information about their unique interests and social connections. Interviewees commonly expressed liking their tattoo(s) because it (they) made him or her "different" or "special" (see Goffman, 1963a: 56–62):

> Having a tattoo changes how you see yourself. It is a way of choosing to change your body. I enjoy that. I enjoy having a tattoo because it makes me different from other people. There is no one in the whole world who has a right arm that looks anything like mine. I've always valued being different from other people. Tattooing is a way of expressing that difference. It is a way of saying, "I am unique."

In describing his understanding of his client's motives, one tattoo artist employed the analogy of the customized car.

> Tattooing is really just a form of personal adornment. Why does someone get a new car and get all of the paint stripped off of it and paint it candy apple red? Why spend $10,000 on a car and then spend another $20,000 to make it look different from the car you bought? I associate it with ownership. Your body is one of the things you indisputably own. There is a tendency to adorn things that you own to make them especially yours.

Interviewees also spoke of the pleasure they got from the tattoo as related to having gone through the mysterious and moderately painful process of being tattooed. The tattoo demonstrated courage to the self ("For some people it means that they lived through it and weren't afraid"). One woman, when asked whether she intended to acquire other tattoos in the future, spoke of the excitement of the experience as the potential motivator of additional work.

> [Do you think you will have more work done after you add something to the one you have now?] Oh God! I don't know why, but my initial reaction is, "I hope I don't, but I think I'm going to." I think getting a tattoo is so exciting and I've always been kind of addicted to excitement. It's fun. While it hurt and stuff it was a new experience and it wasn't that horrible for me. It was new and different.

In a poignant statement, another woman spoke similarly of the tattoo as memorializing significant aspects of her past experience.

> [In the future] when I'm sitting around and bored with my life and I wonder if I was ever young once and did exciting things, I can look at the tattoo and remember.

Interactional Consequences

In general, tattooees' observations concerning the effect of having a tattoo and the process of being tattooed on their self-definitions were rather basic and off-hand. In contrast, all interviewees spoke at some length about their social experiences with others and how the tattoo affected their identities and interactions. Some stressed the affiliational consequences of being tattooed—the mark identified them as belonging to a special group.

> I got tattooed because I had an interest in it. My husband is a chef and our friends tend to be bikers, so it gets me accepted more into that community. They all think of me as "the college girl" and I'm really not. So this (tattoo) kind of brings the door open more. . . . The typical biker would tell you that you almost have to have tattoos to be part of the group.

Most took pleasure in the way the tattoo enhanced their identities by demonstrating their affiliation with a somewhat more diverse group—tattooed people.

> Having a tattoo is like belonging to a club. I love seeing tattoos on other people. I go up and talk with other people with tattoos. It gives me an excuse because I'm not just going up to talk with them. I can say, "I have one, too." I think maybe subconsciously I got (the tattoo) to be part of that special club.
>
> Having tattoos in some ways does affect me positively because people will stop me on the street and say, "Those are really nice tattoos," and show me theirs. We kind of . . . it is a way of having positive contact with strangers. We have something very much in common. We can talk about where we got them and the process of getting them and that sort of thing.

Given the symbolic meaning carried by tattoos in conventional social circles, all tattooees have the experience of being the focus of attention because of the mark they carry. The positive responses of others are, of course, the source of the most direct pleasure.

> People seem to notice you more when you walk around with technicolor arms. I don't think that everyone who gets tattooed is basically an exhibitionist, someone that walks down the street and says, "Hey, look at me!" you know. But it does draw attention to yourself. [How do people respond when they see your tattoos?] Well, yesterday we were sitting in a bar and the lady brings a beer over and she says, "That's gorgeous," and she's looking at the wizard and she's touching them and picking up my shirt. Everyone in the bar was looking and it didn't bother me a bit.

Not all casual encounters are as positive as this one. Revelation of the tattoo is also the source of negative attention when defined by others as a stigmatizing mark.

> Sometimes at these parties the conversation will turn to tattoos and I'll mention that I have some. A lot of people don't believe it, but if I'm feeling loose enough I'll roll up my sleeve and show my work. What really aggravates me is that there will almost always be someone who reacts with a show of disgust. "How could you do that to yourself?" No wonder I usually feel more relaxed and at home with bikers and other tattooed people.

> I think tattoos look sharp. I walk down the beach and people look at my tattoos. Usually they don't say anything. [When they do] I wish they would say it to my face . . . like, "Tattoos are ugly." But, when they say something behind my back . . . "Isn't that gross." Hey, keep your comments to yourself! If you don't like it, you don't like it. I went to the beach with my father and I said, "Hey, let's walk down the beach," and he said, "No, I don't feel like it." What are you, embarrassed to walk with me?

Given the negative responses that tattooees encounter with some frequency when casual associations or strangers become aware of their body decorations, most are selective about to whom they reveal their tattoos. This is particularly the case when the "other" is in a position to exercise control over the tattooee.

> Usually I'm fairly careful about who I show my tattoos to. I don't show them to people at work unless they are really close friends of mine and I know I won't get any kind of hassle because of them. I routinely hide my tattoos. . . . I generally hide them from people who wouldn't understand or people who could potentially cause me trouble. I hide them from my boss and from a lot of the people I work with because there is no reason for them to know.

Tattooees commonly use the reactions of casual associates or relative strangers as a means of categorizing them. A positive reaction to the tattoo indicates social and cultural compatibility, while a negatively judgmental response is seen as signifying a narrow and convention-bound perspective.

I get more positive reactions than I do negative reactions. The negative reactions come from people who aren't like me—who have never done anything astray. It is the straight-laced, conservative person who really doesn't believe that this is acceptable in their set of norms.

It seems as though I can actually tell how I'm going to get along with people and vice-versa by the way they react to my tattoo. It's more or less expressive of the unconventional side of my character right up front. Most of the people who seem to like me really dig the tattoo too [quoted in Hill, 1972: 249].

While it is fairly easy to selectively reveal the tattoo in public settings when interacting with strangers or casual associates, hiding the fact that one is tattooed, thereby avoiding negative social response, is difficult when the "other" is a person with whom the tattooee is intimately associated. The majority of those interviewed recounted incidents in which parents, friends, and, especially for the women, lovers and spouses reacted badly when they initially became aware of the tattoo.

[What did your husband say when he saw your tattoo?] He said he almost threw up. It grossed him out. I had asked him years ago, "What would you think if . . . " and he didn't like the idea. So, I decided not to tell him. It seemed a smart thing to do. He just looked rather grossed out by the whole thing; didn't like it. Now it is accepted, but I don't think he would go for another one.

Another woman interviewee recounted a similar post-tattoo experience with her boyfriend.

I got a strange reaction from my boyfriend. We had a family outing to go to and there was going to be swimming and tennis and all this stuff and I was real excited about going. He said, "Are you going to go swimming?" I said, "Yeah." I was psyched, because I love to swim. He looked at me and said, "You know, your tattoo is going to show if you go swimming." Probably. He didn't want me to go swimming because he didn't want his parents to know that I had a tattoo. Lucky for him it was cloudy that day and nobody swam. I told him, "I'm sorry, but I know your parents can handle this kind of news." To boot, he's got a shamrock on his butt! So he has a tattoo—a real double standard there. He didn't say anything for a while after I first got it. It was subtle. He let me know he didn't like it but that because it was on me he could excuse it. He's got adjusted to it, though. He just let me know that he's never dated a girl who's got a tattoo before. He would prefer that I didn't have it, but there isn't much he can do about that now.

Given the negative social reaction often precipitated by tattoos, it would be reasonable to expect that tattooees who regretted their decision would have emphasized the unpleasant interactional consequences of the tattoo. Interviewees and questionnaire respondents rarely expressed any doubts about their decision to acquire a tattoo. Those that did indicate regret, however, usually did not focus on the stigmatizing effect of the tattoo. Instead, regretful tat-tooees most commonly were dissatisfied with the *technical quality* of the tattoo they purchased.[4] (See also Sanders, 1985a.) . . .

NOTES

1. Questionnaire respondents were given an open-ended question which asked them to speculate as to why people get tattooed. Of the 163 respondents, 135 provided some sort of reply to this item. Forty-four percent of those responding emphasized that becoming tattooed was motivated by a desire for self-expression (e.g., "vanity," "it's a personal preference," "a statement of who you are"), 21% emphasized tattooing as a mechanism for asserting uniqueness and individuality (e.g., "people like to be different," "personal originality," "it makes you special"), and 28% made some form of aesthetic statement (e.g., "because it is beautiful," "a form of art that lasts forever," "body jewelry").

2. Of the 35 tattoos worn by the 16 interviewees, 14% (5) represented birds, 6% (2) represented mammals, 14% (5) represented mythical animals, 9% (3) represented insects, 3% (1) represented a human female, 17% (6) represented human males, 14% (5) were noncommercial symbols (hearts, crosses, military insignia, etc.), 14% (5) were floral, 3% (1) were names or vow tattoos, and 6% (2) were some other image. Questionnaire respondents were asked to indicate the design of their first tattoo. They were: 14% (23) bird, 11% (18) mammal, 12% (19) mythical animal, 10% (17) insect, 1% (2) human female, 6% (10) human male, 4% (6) commercial symbols, 8% (13) noncommercial symbols, 21% (34) floral/arborial, 4% (7) name/vow, and 9% (14) other.

3. The painfulness of the tattoo process is the most unpleasant element of the tattoo event. Only 33% (54) of the questionnaire respondents maintained that there was something about the tattoo experience that they disliked. One third of these (18) said that the pain was what they found most unpleasant. Fourteen of the 16 interviewees mentioned pain as a troublesome factor. Numerous observations of groups of young men discussing pain or stoically expressing little regard for the pain as they were receiving tattoos made it difficult not to see the tattoo event as having ritualized initiatory aspects. In some cases the tattoo process provides a situation in which the male tattooee can demonstrate his "manliness" to his peers. Here, for example, is a description of an incident in which five members of a local college football team acquired identical tattoos on their hips:

I asked the guy nearest to me if they are all getting work done. "Yeah. He [indicates friend] was so hot for it he would have done it himself if we couldn't get it done today." [You all getting hip shots?] "Yeah, that's where all jocks get them. The coach would shit if he found out." The conversation among the jocks turns to the issue of pain. They laugh as the guy being worked on grimaces as W [artist] finishes the outline and wipes the piece down with alcohol. The client observes that this experience isn't bad compared to the time "I fucked up my hand in a game and had to have steel pins put in the knuckles. One of them got bent and the doctor had to cut it out. That was bad. I got the cold sweats." Some of the others join in by telling their "worst pain I ever experienced" stories. The guy being worked on is something of a bleeder and the others kid him about this. As W begins shading one of them shouts, "Come on, really grind it in there."

The cross-cultural literature on body alteration indicates that the pain of the process is an important factor. Ebin (1979: 88–89), for example, in discussing tattooing in the Marquesas Islands, states:

The tattoo was not only an artistic achievement: it also demonstrated that its recipient could bear pain. On one island, the word to describe a person who was completely covered with tattoos is ne'one'o, based on a word meaning either "to cry for a long time" or "horrific." One observer in the Marquesas noted that whenever people discussed the tattoo design, they emphasized the pain with which it was acquired.

See also Becker and Clark, 1979: 10, 19; Brain, 1979: 183–184; Ross and McKay, 1979: 44–49, 67–69; St. Clair and Govenar, 1981: 100–135.

4.   Other than simply accepting the regretted mark, there are a few avenues of resolution open to dissatisfied tattooees. At the most extreme, the tattooee may try to obliterate the offending mark with acid or attempt to cut it off. A somewhat more reasoned (and considerably less painful) approach entails seeking the aid of a der-

matologist or plastic surgeon who will medically remove the tattoo. However, the most common alternative chosen by regretful tattooees is to have the technically inferior piece redone or covered with another tattoo created by a more skilled practitioner. Tattooists estimate that 40 to 50% of their work entails reworking or applying cover-ups to poor-quality tattoos. See Goldstein et al., 1979 and Hardy, 1983.

REFERENCES

Anonymous. (1982). "The name game." *Tattootime* 1: 50–54.

Becker, N., and R. Clark. (1979). "Born to raise hell: An ethnography of tattoo parlors." Presented at the meetings of the Southwestern Sociological Association, March.

Bell, G. (1967). "Self-confidence, persuasability and cognitive dissonance among automobile buyers." In D. Cox (ed.), *Risk-Taking and Information Handling in Consumer Behavior* (pp. 442–468). Boston: Harvard University Graduate School of Business Administration.

Berscheid, E., et al. (1973). "Body image, physical appearance and self-esteem." Presented at the annual meetings of the American Sociological Association.

Brain, D. (1979). *The Decorated Body.* New York: Harper & Row.

Brislin, R., and S. Lewis. (1968). "Dating and physical attractiveness: A replication." *Psych. Reports* 22: 976–984.

Cooley, C. H. (1964 [1902]). *Human Nature and the Social Order.* New York: Schocken.

Csikszentmihalyi, M., and E. Rochberg-Halton. (1981). *The Meaning of Things.* Cambridge: Cambridge Univ. Press.

Ebin, V. (1979). *The Body Decorated.* London: Thames & Hudson.

Facetti, G., and A. Fletcher. (1971). *Identity Kits: A Pictorial Survey of Visual Signals.* New York: Van Nostrand Reinhold.

Feldman, S. (1975). "The presentation of shortness in everyday life." In S. Feldman and G. Thielbar (eds.), *Life Styles* (pp. 437–442). Boston: Little, Brown.

Goffman, E. (1959). *Presentation of Self in Everyday Life.* Garden City, NY: Doubleday.

———. (1961). *Asylums.* Garden City, NY: Doubleday.

———. (1963a). *Stigma.* Englewood Cliffs, NJ: Prentice-Hall.

———. (1963b). *Behavior in Public Places.* New York: Free Press.

Goldstein, N., et al. (1979). "Techniques of removal of tattoos." *J. of Dermatological Surgery and Oncology* 5: 901–910.

Govenar, A. (1977). "The acquisition of tattooing competence: An introduction." *Folklore Annual of the University Folklore Association* 7 and 8: 43–63.

Hardy, D. (1983). "Inventive cover work." *Tattootime* 2: 12–17.

Hatfield, E., and S. Sprecher. (1986). *Mirror, Mirror . . .* Albany: State Univ. of New York Press.

Hill, A. (1972). "Tattoo renaissance." In G. Lewis (ed.), *Side-Saddle on the Golden Calf* (pp. 245–249). Pacific Palisades CA: Goodyear.

Jones, E., et al. (1984). *Social Stigma: The Psychology of Marked Relationships.* New York: Freeman.

Lofland, L. (1973). *A World of Strangers.* New York: Basic Books.

McCall, G., and J. Simmons. (1982). *Social Psychology*. New York: Free Press.

Napoleon, T., et al. (1980). "A replication and extension of 'physical attractiveness and mental illness.'" *J. of Abnormal Psychology* 89: 250–253.

Needleman, B., and N. Weiner. (1977). "Appearance and moral status in the arts." Presented at the annual meetings of the Popular Culture Association.

Polhemus, T. (ed.). (1978). *The Body Reader*. New York: Pantheon.

Ross, R., and H. McKay. (1979). *Self-Mutilation*. Lexington, MA: Lexington Books.

Ruesch, J., and W. Kees. (1972). *Nonverbal Communication*. Berkeley: Univ. of California Press.

St. Clair, L., and A. Govenar. (1981). *Stoney Knows How: Life as a Tattoo Artist*. Lexington: Univ. Press of Kentucky.

Sanders, C. (1983). "Drill and fill: Client choice, client typologies and interactional control in commercial tattoo settings." Presented at the Art of the Body Symposium, UCLA.

———. (1985a). "Tattoo consumption: Risk and regret in the purchase of a socially marginal service." In E. Hirshman and M. Holbrook (eds.), *Advances in Consumer Research,* Vol. XII (pp. 17–22). New York: Association for Consumer Research.

———. (1985b). "Selling deviant pictures: The tattooist's career and occupational experience." Presented at the Conference on Social Theory, Politics and the Arts, Adelphi University, October.

Scutt, R., and R. Gotch. (1974). *Art, Sex and Symbol*. New York: Barnes.

Solomon, M. (1983). "The role of products as social stimuli: A symbolic interactionist perspective." *J. of Consumer Research* 10 (December): 319–329.

Stone, G. (1970). "Appearance and the self." In G. Stone and H. Faberman (eds.), *Social Psychology Through Symbolic Interaction* (pp. 394–414). Waltham, MA: Xerox.

Van Gennep, A. (1960). *The Rites of Passage*. Chicago: Univ. of Chicago Press.

Zurcher, L. (1977). *The Mutable Self*. Newbury Park, CA: Sage.

37

Joining a Gang

MARTÍN SÁNCHEZ JANKOWSKI

*Now it is thought to be the mark of a man of practical wisdom to be able to
deliberate well about what is good and expedient for himself, not in some
particular respect, e.g. about what sorts of thing conduce to health or to strength,
but about what sorts of thing conduce to the good life in general.*

ARISTOTLE

THE NICOMACHEAN ETHICS

We are looking for a few good men.

U.S. MARINE CORPS RECRUITING POSTER

[have] argued that one of the most important features of gang members was
their defiant individualist character. I explained the development of defiant
individualism by locating its origins in the material conditions—the competition and conflict over resource scarcity—of the low-income neighborhoods of most large American cities. These conditions exist for everyone who
lives in such neighborhoods, yet not every young person joins a gang. Although I have found that nearly all those who belong to gangs do exhibit defiant individualist traits to some degree, not all those who possess such traits
join gangs. This chapter explores who joins a gang and why in more detail.

Many studies offer an answer to why a person joins a gang, or why a group
of individuals start a gang. These studies can be divided into four groupings.
First, there are those that hold the "natural association" point of view. These
studies argue that people join gangs as a result of the natural act of associating
with each other.[1] Their contention is that a group of boys, interrelating with
each other, decide to formalize their relationship in an attempt to reduce the
fear and anxiety associated with their socially disorganized neighborhoods.
The individual's impetus to join is the result of his desire to defend against
conflict and create order out of the condition of social disorganization.

The second group of studies explains gang formation in terms of "the subculture of blocked opportunities": gangs begin because young males experience persistent problems in gaining employment and/or status. As a result,
members of poor communities who experience the strain of these blocked

From Martín Sánchez Jankowski, *Islands in the Street: Gangs and American Urban Society*
(Berkeley: University of California Press, 1991). Copyright © 1991 The Regents of the
University of California. Reprinted by permission of the publisher.

opportunities attempt to compensate for socioeconomic deprivation by joining a gang and establishing a subculture that can be kept separate from the culture of the wider society.[2]

The third group of studies focuses on "problems in identity construction." Within this broad group, some suggest that individuals join gangs as part of the developmental process of building a personal identity or as the result of a breakdown in the process.[3] Others argue that some individuals from low-income families have been blocked from achieving social status through conventional means and join gangs to gain status and self-worth, to rebuild a wounded identity.[4]

A recent work by Jack Katz has both creatively extended the status model and advanced the premise that sensuality is the central element leading to the commission of illegal acts. In Katz's "expressive" model, joining a gang, and being what he labels a "badass," involves a process whereby an individual manages (through transcendence) the gulf that exists between a sense of self located within the local world (the here) and a reality associated with the world outside (the there). Katz argues that the central elements in various forms of deviance, including becoming involved in a gang and gang violence, are the moral emotions of humiliation, righteousness, arrogance, ridicule, cynicism, defilement, and vengeance. "In each," he says, "the attraction that proves to be most fundamentally compelling is that of overcoming a personal challenge to moral—not material—existence."[5]

Most of these theories suffer from three flaws. First, they link joining a gang to delinquency, thereby combining two separate issues. Second, they use single-variable explanations. Third, and most important, they fail to treat joining a gang as the product of a rational decision to maximize self-interest, one in which both the individual and the organized gang play a role. This is especially true of Katz's approach, for two reasons. First, on the personal level, it underestimates the impact of material and status conditions in establishing the situations in which sensual needs/drives (emotions) present themselves, and overestimates/exaggerates the "seductive" impact of crime in satisfying these needs. Second, it does not consider the impact of organizational dynamics on the thought and action of gang members.

In contrast, the data presented here will indicate that gangs are composed of individuals who join for a variety of reasons. In addition, while the individual uses his own calculus to decide whether or not to join a gang, this is not the only deciding factor. The other deciding factor is whether the gang wants him in the organization. Like the individual's decision to join, the gang's decision to permit membership is based on a variety of factors. It is thus important to understand that who becomes a gang member depends on two decision-making processes: that of the individual and that of the gang.

THE INDIVIDUAL AND THE
DECISION TO BECOME A MEMBER

Before proceeding, it is important to dismiss a number of the propositions that have often been advanced. The first is that young boys join gangs because they are from broken homes where the father is not present and they seek gang membership in order to identify with other males—that is, they have had no male authority figures with whom to identify. In the ten years of this study, I found that there were as many gang members from homes where the nuclear family was intact as there were from families where the father was absent.[6]

The second proposition given for why individuals join gangs is related to the first: it suggests that broken homes and/or bad home environments force them to look to the gang as a substitute family. Those who offer this explanation often quote gang members' statements such as "We are like a family" or "We are just like brothers" as indications of this motive. However, I found as many members who claimed close relationships with their families as those who denied them.

The third reason offered is that individuals who drop out of school have fewer skills for getting jobs, leaving them with nothing to do but join a gang. While I did find a larger number of members who had dropped out of school, the number was only slightly higher than those who had finished school.

The fourth reason suggested, disconfirmed by my data, is a modern version of the "Pied Piper" effect: the claim that young kids join gangs because they are socialized by older kids to aspire to gang membership and, being young and impressionable, are easily persuaded. I found on the contrary that individuals were as likely to join when they were older (mid to late teens) as when they were younger (nine to fifteen). I also found significantly more who joined when they were young who did so for reasons other than being socialized to think it was "cool" to belong to a gang. In brief, I found no evidence for this proposition.

What I did find was that individuals who live in low-income neighborhoods join gangs for a variety of reasons, basing their decisions on a rational calculation of what is best for them at that particular time. Furthermore, I found that they use the same calculus (not necessarily the same reasons) in deciding whether to stay in the gang, or, if they happen to leave it, whether to rejoin.

Reasons for Deciding to Join a Gang

Most people in the low-income inner cities of America face a situation in which a gang already exists in their area. Therefore the most salient question facing them is not whether to start a gang or not, but rather whether to join an existing one. Many of the reasons for starting a new gang are related to issues having to do with organizational development and decline—that is, with the existing gang's ability to provide the expected services, which include those that individuals considered in deciding to join. . . . This section deals primarily, although not exclusively, with the question of what influences individuals to join an existing gang. However, many of these are the same influences that encourage individuals to start a new gang.

Material Incentives

Those who had joined a gang most often gave as their reason the belief that it would provide them with an environment that would increase their chances of securing money. Defiant individualists constantly calculate the costs and benefits associated with their efforts to improve their financial well-being (which is usually not good). Therefore, on the one hand, they believe that if they engage in economic ventures on their own, they will, if successful, earn more per venture than if they acted as part of a gang. However, there is also the belief that if one participates in economic ventures with a gang, it is likely that the amount earned will be more regular, although perhaps less per venture. The comments of Slump, a sixteen-year-old member of a gang in the Los Angeles area, represent this belief:

> Well, I really didn't want to join the gang when I was a little younger because I had this idea that I could make more money if I would do some gigs [various illegal economic ventures] on my own. Now I don't know, I mean, I wasn't wrong. I could make more money on my own, but there are more things happening with the gang, so it's a little more even in terms of when the money comes in. . . . Let's just say there is more possibilities for a more steady amount of income if you need it.

It was also believed that less individual effort would be required in the various economic ventures in a gang because more people would be involved. In addition, some thought that being in a gang would reduce the *risk* (of personal injury) associated with their business ventures. They were aware that if larger numbers of people had knowledge of a crime, this would increase the risk that if someone were caught, others, including themselves, would be implicated. However, they countered this consideration with the belief that they faced less risk of being physically harmed when they were part of a group action. The comments of Corner, a seventeen-year-old resident of a poor Manhattan neighborhood, represent this consideration. During the interview, he was twice approached about joining the local gang. He said:

> I think I am going to join the club [gang] this time. I don't know, man, I got some things to decide, but I think I will. . . . Before I didn't want to join because when I did a job, I didn't want to share it with the whole group—hell, I was never able to make that much to share. . . . I would never have got enough money, and with all those dudes [other members of the gang] knowing who did the job, you can bet the police would find out. . . . Well, now my thinking is changed a bit 'cause there's more people involved and that'll keep me safer. [He joined the gang two weeks later.]

Others decided to join the gang for financial security. They viewed the gang as an organization that could provide them or their families with money in times of emergency. It represented the combination of a bank and a social security system, the equivalent of what the political machine had been to many new immigrant groups in American cities.[7] To these individuals, it provided both psychological and financial security in an economic environment of scarcity and intense competition. This was particularly true of those who were fifteen and younger. Many in this age group often find themselves in a precarious position. They are in need of money, and although social services are available to help during times of economic hardship, they often lack legal means of access to these resources. For these individuals, the gang can provide an alternative source of aid. The comments of Street Dog and Tomahawk represent these views. Street Dog was a fifteen-year-old Puerto Rican who had been in a New York gang for two years:

> Hey, the club [the gang] has been there when I needed help. There were times when there just wasn't enough food for me to get filled up with. My family was hard up and they couldn't manage all of their bills and such, so there was some lean meals! Well, I just needed some money to help for awhile, till I got some money or my family was better off. They [the gang] was there to help. I could see that [they would help] before I joined, that's why I joined. They are there when you need them and they'll continue to be.

Tomahawk was a fifteen-year-old Irishman who had been in a gang for one year:

> Before I joined the gang, I could see that you could count on your boys to help in times of need and that meant a lot to me. And when I needed money, sure enough they gave it to me. Nobody else would have given it to me; my parents didn't have it, and there was no other place to go. The gang was just like they said they would be, and they'll continue to be there when I need them.

Finally, many view the gang as providing an opportunity for future gratification. They expect that through belonging to a gang, they will be able to make contact with individuals who may eventually help them financially. Some look to meet people who have contacts in organized crime in the hope of entering that field in the future. Some hope to meet businessmen involved in the illegal market who will provide them with money to start their own illegal businesses. Still others think that gang membership will enable them to meet individuals who will later do them favors (with financial implications) of the kind fraternity brothers or Masons sometimes do for each other. Irish gang members in New York and Boston especially tend to believe this.

Recreation

The gang provides individuals with entertainment, much as a fraternity does for college students or the Moose and Elk clubs do for their members. Many individuals said they joined the gang because it was the primary social institution of their neighborhood—that is, it was where most (not necessarily the biggest) social events occurred. Gangs usually, though not always, have some type of clubhouse. The exact nature of the clubhouse varies according to how much money the gang has to support it, but every clubhouse offers some form of entertainment. In the case of some gangs with a good deal of money, the clubhouse includes a bar, which sells its members drinks at cost. In addition, some clubhouses have pinball machines, soccer-game machines, pool tables, Ping-Pong tables, card tables, and in some cases a few slot machines. The clubhouse acts as an incentive, much like the lodge houses of other social clubs.[8]

The gang can also be a promoter of social events in the community, such as a big party or dance. Often the gang, like a fraternity, is thought of as the organization to join to maximize opportunities to have fun. Many who joined said they did so because the gang provided them with a good opportunity to meet women. Young women frequently form an auxiliary unit to the gang, which usually adopts a version of the male gang's name (e.g., "Lady Jets"). The women who join this auxiliary do so for similar reasons—that is, opportunities to meet men and participate in social events.[9]

The gang is also a source of drugs and alcohol. Here, most gangs walk a fine line. They provide some drugs for purposes of recreation, but because they also ban addicts from the organization, they also attempt to monitor members' use of some drugs.[10]

The comments of Fox and Happy highlight these views of the gang as a source of recreation.[11] Fox was a twenty-three-year-old from New York and had been in a gang for seven years:

> Like I been telling you, I joined originally because all the action was happening with the Bats [gang's name]. I mean, all the foxy ladies were going to their parties and hanging with them. Plus their parties were great.
>
> They had good music and the herb [marijuana] was so smooth. . . . Man, it was a great source of dope and women. Hell, they were the kings of the community so I wanted to get in on some of the action.

Happy was a twenty-eight-year-old from Los Angeles, who had been a gang member for eight years:

> I joined because at the time, Jones Park [gang's name] had the best clubhouse. They had pool tables and pinball machines that you could use for free. Now they added a video game which you only have to pay like five cents for to play. You could do a lot in the club, so I thought it was a good thing to try it for awhile [join the gang], and it was a good thing.

A Place of Refuge and Camouflage

Some individuals join a gang because it provides them with a protective group identity. They see the gang as offering them anonymity, which may relieve the stresses associated with having to be personally accountable for all their actions in an intensely competitive environment. The statements of Junior J. and Black Top are representative of this belief. Junior J. was a seventeen-year-old who had been approached about becoming a gang member in one of New York's neighborhoods:

I been thinking about joining the gang because the gang gives you a cover, you know what I mean? Like when me or anybody does a business deal and we're members of the gang, it's difficult to track us down 'cause people will say, oh, it was just one of those guys in the gang. You get my point? The gang is going to provide me with some cover.

Black Top was a seventeen-year-old member of a Jamaican gang in New York:

Man, I been dealing me something awful. I been doing well, but I also attracted me some adversaries. And these adversaries have been getting close to me. So joining the brothers [the gang] lets me blend into the group. It lets me hide for awhile, it gives me refuge until the heat goes away.

Physical Protection

Individuals also join gangs because they believe the gang can provide them with personal protection from the predatory elements active in low-income neighborhoods. Nearly all the young men who join for this reason know what dangers exist for them in their low-income neighborhoods. These individuals are not the weakest of those who join the gang, for all have developed the savvy and skills to handle most threats. However, all are either tired of being on the alert or want to reduce the probability of danger to a level that allows them to devote more time to their effort to secure more money. Here are two representative comments of individuals who joined for this reason. Chico was a seventeen-year-old member of an Irish gang in New York:

When I first started up with the Steel Flowers, I really didn't know much about them. But, to be honest, in the beginning I just joined because there were some people who were taking my school [lunch] money, and after I joined the gang, these guys laid off.

Cory was a sixteen-year-old member of a Los Angeles gang:

Man I joined the Fultons because there are a lot of people out there who are trying to get you and if you don't got protection you in trouble sometimes. My homeboys gave me protection, so hey, they were the thing to do. . . . Now that I got some business things going I can concentrate on them and not worry so much. I don't always have to be looking over my shoulder.

A Time to Resist

Many older individuals (in their late teens or older) join gangs in an effort to resist living lives like their parents'. As Joan Moore, Ruth Horowitz, and others have pointed out, most gang members come from families whose parents are underemployed and/or employed in the secondary labor market in jobs that have little to recommend them.[12] These jobs are low-paying, have long hours, poor working conditions, and few opportunities for advancement; in brief, they are dead ends.[13] Most prospective gang members have lived through the pains of economic deprivation and the stresses that such an existence puts on a family. They desperately want to avoid following in their parents' path, which they believe is exactly what awaits them. For these individuals, the gang is a way to resist the jobs their parents held and, by extension, the life their parents led. Deciding to become a gang member is both a statement to society ("I'll not take these jobs passively") and an attempt to do whatever can be done to avoid such an outcome. At the very least, some of these individuals view being in a gang as a temporary reprieve from having to take such jobs, a postponement of the inevitable. The comments of Joey and D.D. are representative of this group. Joey was a nineteen-year-old member of an Irish gang in Boston:

> Hell, I joined because I really didn't see anything in the near future I wanted to do. I sure the hell didn't want to take that job my father got me. It was a shit job just like his. I said to myself, "Fuck this!" I'm only nineteen, I'm too young to start this shit. . . . I figured that the Black Rose [the gang] was into a lot of things and that maybe I could hit it big at something we're doing and get the hell out of this place.

D.D. was a twenty-year-old member of a Chicano gang in Los Angeles:

> I just joined the T-Men to kick back [relax, be carefree] for a while. My parents work real hard and they got little for it. I don't really want that kind of job, but that's what it looked like I would have to take. So I said, hey, I'll just kick back for a while and let that job wait for me. Hey, I just might make some money from our dealings and really be able to forget these jobs. . . . If I don't [make it, at least] I told the fuckers in Beverly Hills what I think of the jobs they left for us.

People who join as an act of resistance are often wrongly understood to have joined because they were having difficulty with their identity and the gang provided them with a new one. However, these individuals actually want a new identity less than they want better living conditions.

Commitment to Community

Some individuals join the gang because they see participation as a form of commitment to their community. These usually come from neighborhoods where gangs have existed for generations. Although the character of such gangs may have changed over the years, the fact remains that they have continued to exist. Many of these individuals have known people who have been in gangs, including family members—often a brother, but even, in considerable number of cases, a father and grandfather. The fact that their relatives have a history of gang involvement usually influences these individuals to see the gang as a part of the tradition of the community. They feel that their families and their community expect them to join, because community members see the gang as an aid to them and the individual who joins as meeting his neighborhood obligation. These attitudes are similar to attitudes in the larger society about one's obligation to serve in the armed forces. In a sense, this type of involvement represents a unique form of local patriotism. While this rationale for joining was present in a number of the gangs studied, it was most prevalent among Chicano and Irish gangs. The comments of Dolan and Pepe are representative of this line of thinking. Dolan was a sixteen-year-old member of an Irish gang in New York:

> I joined because the gang has been here for a long time and even though the name is different a lot of the fellas from the community have been involved in it over the years, including my dad. The gang has helped the community by protecting it against outsiders so people here have kind of depended on it. . . . I feel it's my obligation to the community to put in some time helping them out. This will help me to get help in the community if I need it some time.

Pepe was a seventeen-year-old member of a Chicano gang in the Los Angeles area:

> The Royal Dons [gang's name] have been here for a real long time. A lot of people from the community have been in it. I had lots of family in it so I guess I'll just have to carry on the tradition. A lot of people from outside this community wouldn't understand, but we have helped the community whenever they've asked us. We've been around to help. I felt it's kind of my duty to join 'cause everybody expects it. . . . No, the community doesn't mind that we do things to make some money and raise a little hell because they don't expect you to put in your time for nothing. Just like nobody expects guys in the military to put in their time for nothing.

In closing this section on why individuals join gangs, it is important to reemphasize that people choose to join for a variety of reasons, that these reasons are not exclusive of one another (some members have more than one), that gangs are composed of individuals whose reasons for joining include all those mentioned, that the decision to join is thought out, and that the individual believes this was best for his or her interests at the moment.

ORGANIZATIONAL RECRUITMENT

Deciding whether or not to join a gang is never an individual decision alone. Because gangs are well established in most of these neighborhoods, they are ultimately both the initiators of membership and the gatekeepers, deciding who will join and who will not.

Every gang that was studied had some type of recruitment strategy. A gang will frequently employ a number of strategies, depending on the circumstances in which recruitment is occurring. However, most gangs use one particular style of recruitment for what they consider a "normal" period and adopt other styles as specific situations present themselves. The three most prevalent styles of recruitment encountered were what I call the fraternity type, the obligation type, and the coercive type.

The Fraternity Type of Recruitment

In the fraternity type of recruitment, the gang adopts the posture of an organization that is "cool," "hip," the social thing to be in. Here the gang makes an effort to recruit by advertising through word of mouth that it is looking for members. Then many of the gangs either give a party or circulate information throughout the neighborhood, indicating when their next meeting will be held and that those interested in becoming members are invited. At this initial meeting, prospective members hear a short speech about the gang and its rules. They are also told about the gang's exploits and/or its most positive perks, such as the dances and parties it gives, the availability of dope, the women who are available, the clubhouse, and the various recreational machinery (pool table, video games, bar, etc.). In addition, the gang sometimes discusses, in the most general terms, its plans for creating revenues that will be shared among the general membership. Once this pitch is made, the decision rests with the individual. When one decides to join the gang, there is a trial period before one is considered a solid member of the group. This trial period is similar, but not identical, to the pledge period for fraternities. There are a number of precautions taken during this period to check the individual's worthiness to be in the group. If the individual is not known by members of the gang, he will need to be evaluated to see if he is an informant for one of the various law enforcement agencies (police, firearms and alcohol, drug enforcement). In addition, the individual will need to be assessed in terms of his ability to fight, his courage, and his commitment to help others in the gang.

Having the *will* to fight and defend other gang members or the "interest" of the gang is considered important, but what is looked upon as being an even more important asset for a prospective gang member is the *ability* to fight and to carry out group decisions. Many researchers have often misinterpreted this preference by gangs for those who can fight as an indication that gang members, and thus gangs as collectives, are primarily interested in establishing reputations as fighters.[14] They interpret this preoccupation as being based on adolescent drives for identity and the release of a great deal of aggression. However, what is most often missed are the functional aspects of fighting and its significance to a gang. The prospective member's ability to fight well is not looked upon by the organization simply as an additional symbol of status. Members of gangs want to know if a potential member can fight because if any of them are caught in a situation where they are required to fight, they want to feel confident that everyone can carry his or her own responsibility. In addition, gang members want to know if the potential gang member is disciplined enough to avoid getting scared and running, leaving them vulnerable. Often everyone's safety in a fight depends on the ability of every individual to fight efficiently. For example, on many occasions I observed a small group of one gang being attacked by an opposing gang. Gang fights are not like fights in the movies: there is no limit to the force anybody is prepared to use—it is, as one often hears, "for all the marbles." When gang members were attacked, they were often outnumbered and surrounded. The only way to protect themselves was to place themselves back to back and ward off the attackers until some type of help came (ironically, most often from the police). If someone cannot fight well and is overcome quickly, everyone's back will be exposed and everyone becomes vulnerable. Likewise, if someone decides to make a run for it, everyone's position is compromised. So assessing the potential member's ability to fight is not done simply to strengthen the gang's reputation as "the meanest fighters," but rather to strengthen the confidence of other gang members that the new member adds to the organization's general ability to protect and defend the collective's interests. The comments of Vase, an eighteen-year-old leader of a gang in New York, highlight this point:

> When I first started with the Silk Irons [gang's name], they checked me out to see if I could fight. After I passed their test, they told me that they didn't need anybody who would leave their butts uncovered. Now that I'm a leader I do the same thing. You see the guy over there? He wants to be in the Irons, but we don't know nothing about whether he can fight or if he got no heart [courage]. So we going to check out how good he is and whether he going to stand and fight. 'Cause if he ain't got good heart or skills [ability to fight], he could leave some of the brothers [gang members] real vulnerable and in a big mess. And if [he] do that, they going to get their asses messed up!

As mentioned earlier, in those cases where the gang has seen a prospective member fight enough to know he will be a valuable member, they simply admit him. However, if information is needed in order to decide whether the prospective gang member can fight, the gang leadership sets up a number of situations to test the individual. One favorite is to have one of the gang members pick a fight with the prospective member and observe the response. It is always assumed that the prospective member will fight; the question is, how well will he fight? The person selected to start the fight is usually one of the better fighters. This provides the group with comparative information by which to decide just how good the individual is in fighting.[15] Such fights are often so intense that there are numerous lacerations on the faces of both fighters. This test usually doubles as an initiation rite, although there are gangs who follow up this test phase with a separate initiation ritual where the individual is given a beating by all those gang members present. This beating is more often than not symbolic, in that the blows delivered to the new members are not done using full force. However, they still leave bruises.

Assessing whether a prospective gang member is trustworthy or not is likewise done by setting up a number of small tests. The gang members are concerned with whether the prospective member is an undercover agent for law enforcement. To help them establish this, they set up a number of criminal activities (usually of medium-level illegality) involving the individual(s); then they observe whether law enforcement proceeds to make arrests of the specific members involved. One gang set up a scam whereby it was scheduled to commit an armed robbery. When a number of the gang members were ready to make the robbery, the police came and arrested them—the consequence of a new member being a police informer. The person responsible was identified and punished. Testing the trustworthiness of new recruits proved to be an effective policy because later the gang was able to pursue a much more lucrative illegal venture without the fear of having a police informer in the organization.

Recruiting a certain number of new members who have already established reputations as good fighters does help the gang. The gang's ability to build and maintain a reputation for fighting reduces the number of times it will have to fight. If a gang has a reputation as a particularly tough group, it will not have as much trouble with rival gangs trying to assume control over its areas of interest. Thus, a reputation acts as an initial deterrent to rival groups. However, for the most part, the gang's concern with recruiting good fighters for the purpose of enhancing its reputation is secondary to its concern that members be able to fight well so that they can help each other.

Gangs who are selective about who they allow in also scrutinize whether the individual has any special talents that could be useful to the collective. Sometimes these special talents involve military skills, such as the ability to build incendiary bombs. Some New York gangs attempted to recruit people with carpentry and masonry skills so that they could help them renovate abandoned buildings.

Gangs that adopt a "fraternity recruiting style" are usually quite secure within their communities. They have a relatively large membership and have integrated themselves into the community well enough to have both legitimacy and status. In other words, the gang is an organization that is viewed by members of the community as legitimate. The comments of Mary, a 53-year-old garment worker who was a single parent in New York, indicate how some community residents feel about certain gangs:

> There are a lot of young people who want in the Bullets, but they don't let whoever wants to get in. Those guys are really selective about who they want. Those who do get in are very helpful to the whole community. There are many times that they have helped the community . . . and the community appreciates that they have been here for us.

Gangs that use fraternity style recruitment have often become relatively prosperous. Having built up the economic resources of the group to a level that has benefited the general membership, they are reluctant to admit too many new members, fearing that increased numbers will not be accompanied by increases in revenues, resulting in less for the general membership. Hackman, a twenty-eight-year-old leader of a New York gang, represented this line of thought:

> Man, we don't let all the dudes who want to be let in. We can't do that, or I can't 'cause right now we're sitting good. We gots a good bank account and the whole gang is getting dividends. But if we let in a whole lot of other dudes, everybody will have to take a cut unless we come up with some more money, but that don't happen real fast. So you know the brothers ain't going to dig a cut, and if it happens, then they going to be on me and the rest of the leadership's ass and that ain't good for us.

The Obligation Type of Recruitment

The second recruiting technique used by gangs is what I call the "obligation type." In this form, the gang contacts as many young men from its community as it can and attempts to persuade them that it is their duty to join. These community pressures are real, and individuals need to calculate how to respond to them, because there are risks if one ignores them. In essence, the gang recruiter's pitch is that everyone who lives in this particular community has to give something back to it in order to indicate both appreciation of and solidarity with the community. In places where one particular gang has been in existence for a considerable amount of time (as long as a couple of generations), "upholding the tradition of the neighborhood" (not that of the gang) is the pitch used as the hook. The comments of Paul and Lorenzo are good examples. Paul was a nineteen-year-old member of an Irish gang in New York:

> Yeah, I joined this group of guys [the gang] because they have helped the community and a lot of us have taken some serious lumps [injuries] in doing that. . . . I think if a man has any sense of himself, he will help his community no matter what. Right now I'm talking to some guys about joining our gang and I tell them that they can make some money being in the gang, but the most important thing is they can help the community too. If any of them say that they don't want to get hurt or something like that, I tell'm that nobody wants to get hurt, but sometimes it happens. Then I tell them the bottom line, if you don't join and help the community, then outsiders will come and attack the people here and this community won't exist in a couple of years.

Lorenzo was a 22-year-old Chicano gang member from Los Angeles. Here he is talking to two prospective members:

> I don't need to talk to you dudes too much about this [joining a gang]. You know what the whole deal is, but I want you to know that your barrio [community] needs you just like they needed us and we delivered. We all get some battle scars [he shows them a scar from a bullet wound], but that's the price you pay to keep some honor for you and your barrio. We all have to give something back to our community.[16]

This recruiting pitch is primarily based on accountability to the community. It is most effective in communities where the residents have depended on the gang to help protect them from social predators. This is because gang recruiters can draw on the moral support that the gang receives from older residents.

Although the power of this recruiting pitch is accountability to the community, the recruiter can suggest other incentives as well. Three positive incentives generally are used. The first is that gang members are respected in the community. This means that the community will tolerate their illegal business dealings and help them whenever they are having difficulty with the police. As Cardboard, a sixteen-year-old member of a Dominican gang, commented:

Hey, the dudes come by and they be putting all this shit about that I should do my part to protect the community, but I told them I'm not ready to join up. I tell you the truth, I did sometimes feel a little guilty, but I still didn't think it was for me. But now I tell you I been changing my mind a little. I thinking more about joining. . . . You see the dudes been telling me the community be helping you do your business, you understand? Hey, I been thinking, I got me a little business and if they right, this may be the final straw to get me, 'cause a little help from the community could be real helpful to me. [He joined the gang three weeks later.]

The second incentive is that some members of the community will help them find employment at a later time. (This happens more in Irish neighborhoods.) The comments of Andy, a seventeen-year-old Irish-American in Boston, illustrate this view:

The community has been getting squeezed by some developers and there's been a lot of people who aren't from the community moving in, so that's why some of the Tigers [gang's name] have come by while we've been talking. They want to talk to me about joining. Just like they been saying, the community needs their help now and they need me. I really was torn because I thought there might be some kind violence used and I don't really want to get involved with that. But the other day when you weren't here, they talked to me and told me that I should remember that the community remembers when people help and they take care of their own. Well, they're right, the community does take care of its own. They help people get jobs all the time 'cause they got contacts at city hall and at the docks, so I been thinking I might join. [He joined three weeks later.]

The third incentive is access to women. Here the recruiter simply says that because the gang is a part of the community and is respected, women look up to gang members and want to be associated with them. So, the pitch continues, if an individual wants access to a lot of women, it will be available through the gang. The comments of Topper, a fifteen-year-old Chicano, illustrate the effectiveness of this pitch:

Yeah, I was thinking of joining the Bangers [a gang]. These two homeboys [gang members] been coming to see me about joining for two months now. They've been telling me that my barrio really needs me and I should help my people. I really do want to help my barrio, but I never really made up my mind. But the other day they were telling me that the *mujeres* [women] really dig homeboys because they do help the community. So I was checking that out and you know what? They really do! So, I say, hey, I need to seriously check the Bangers out. [One week later he joined the gang.]

In addition to the three positive incentives used, there is a negative one. The gang recruiter can take the tack that if a prospective member decides not to join, he will not be respected as much in the community, or possibly even within his own family. This line of persuasion can be successful if other members of the prospective recruit's family have been in a gang and/or if there has been a high level of involvement in gangs throughout the community. The suggestion that people (including family) will be disappointed in him, or look down on his family, is an effective manipulative tool in the recruiting process in such cases. The comments of Texto, a fifteen-year-old Chicano, provide a good example:

> I didn't want to join the Pearls [gang's name] right now cause I didn't think it was best for me right now. Then a few of the Pearls came by to try to get me to join. They said all the stuff about helping your barrio, but I don't want to join now. I mean I do care about my barrio, but I just don't want to join now. But you heard them today ask me if my father wanted me to join. You know I got to think about this, I mean my dad was in this gang and I don't know. He says to me to do what you want, but I think he would be embarrassed with his friends if they heard I didn't want to join. I really don't want to embarrass my dad, I don't know what I'm going to do. [He joined the gang one month later.]

The "obligation method of recruitment" is similar to that employed by governments to secure recruits for their armed services, and it meets with only moderate results. Gangs using this method realized that while they would not be able to recruit all the individuals they made contact with, the obligation method (sometimes in combination with the coercive method) would enable them to recruit enough for the gang to continue operating.

This type of recruitment was found mostly, although not exclusively, in Irish and Chicano communities where the gang and community had been highly integrated. It is only effective in communities where a particular gang or a small number of gangs have been active for a considerable length of time.

The Coercive Type of Recruitment

A third type of recruitment involves various forms of coercion. Coercion is used as a recruitment method when gangs are confronted with the need to increase their membership quickly. There are a number of situations in which this occurs. One is when a gang has made a policy decision to expand its operations into another geographic area and needs troops to secure the area and keep it under control. The desire to build up membership is based on the gang's anticipation that there will be a struggle with a rival gang and that, if it is to be successful, it will be necessary to be numerically superior to the expected adversary.

Another situation involving gang expansion also encourages an intense recruitment effort that includes coercion. When a gang decides to expand into a geographic area that has not hitherto been controlled by another gang, and is not at the moment being fought for, it goes into the targeted area and vigorously recruits members in an effort to establish control. If individuals from this area are not receptive to the gang's efforts, then coercion is used to persuade some of them to join. The comment of Bolo, a seventeen-year-old leader of a New York gang, illustrates this position:

> Let me explain what just happened. Now you might be thinking, what are these dudes doing beating up on somebody they want to be in their gang? The answer is that we need people now, we can't be waiting till they make up their mind. They don't have to stay for a long time, but we need them now. . . . We don't like to recruit this way 'cause it ain't good for the long run, but this is necessary now because in order for us to expand our business in this area we got to get control, and in order to do that we got to have members who live in the neighborhood. We can't be building no structure to defend ourselves against the Wings [the rival gang in the area], or set up some communications in the area, or set up a connection with the community. We can't do shit unless we got a base and we ain't going to get any base without people. It's that simple.

A third situation where a gang feels a need to use a coercive recruiting strategy involves gangs who are defending themselves against a hostile attempt to take over a portion of their territory. Under such conditions, the gang defending its interests will need to bolster its ranks in order to fend off the threat. This will require that the embattled gang recruit rapidly. Often, a gang that normally uses the fraternity type of recruitment will be forced to abandon it for the more coercive type. The actions of these gangs can be compared to those of nation-states when they invoke universal conscription (certainly a form of coercion) during times when they are threatened and then abrogate it when they believe they have recruited a sufficient number to neutralize the threat, or, more usually, when a threat no longer exists. The comments of M.R. and Rider represent those who are recruited using coercion. M.R. was a nineteen-year-old ex-gang member from Los Angeles:[17]

> I really didn't want to be in any gang, but one day there was this big blowout [fight] a few blocks from here. A couple of O Streeters who were from another barrio came and shot up a number of the Dukes [local gang's name]. Then it was said that the O Streeters wanted to take over the area as theirs, so a group of the Dukes went around asking people to join for a while till everything got secure. They asked me, but I still didn't want to get involved because I really didn't want to get killed over something that I had no interest in. But they said they wanted me and if I didn't join and help they were going to mess me up. Then the next day a couple of them pushed me around pretty bad, and they did it much harder the following day. So I thought about it and then decided I'd join. Then after some gun fights things got secure again and they told me thanks and I left.

Rider was a sixteen-year-old member of an Irish gang from New York:

> Here one day I read in the paper there was fighting going on between a couple of gangs. I knew that one of the gangs was from a black section of

the city. Then some of the Greenies [local Irish gang] came up to me and told me how some of the niggers from this gang were trying to start some drugs in the neighborhood. I didn't want the niggers coming in, but I had other business to tend to first. You know what I mean? So I said I thought they could handle it themselves, but then about three or four Greenies said that if I didn't go with them that I was going to be ground meat and so would members of my family. Well, I know they meant business because my sister said they followed her home from school and my brother said they threw stones at him on his way home. So, they asked me again and I said OK. . . . Then after we beat the niggers' asses, I quit. . . . Well, the truth is that I wanted to stay, but after the nigger business was over, they didn't want me. They just said that I was too crazy and wouldn't work out in the group.

This last interview highlights the gang's movement back to their prior form of recruitment after the threat was over. Rider wanted to stay in the gang but was asked to leave. Many of the members of the gang felt Rider was too crazy, too prone to vicious and outlandish acts, simply too unpredictable to trust. The gang admired his fighting ability, but he was the kind of person who caused too much trouble for the gang. As T.R., an eighteen-year-old leader of the gang:

There's lots of things we liked about Rider. He sure could help us in any fight we'd get in, but he's just too crazy. You just couldn't tell what he'd do. If we kept him, he'd have the police on us all the time. He just had to go.

There is also a fourth situation in which coercion is used in recruiting. Sometimes a gang that has dominated a particular area has declined to such an extent that it can no longer control all its original area. In such situations, certain members of this gang often decide to start a new one. When this occurs, the newly constituted gang often uses coercive techniques to recruit members and establish authority over its defined territory. Take the comments of Rob and Loan Man, both of whom were leaders of two newly constituted gangs. Rob was a sixteen-year-old leader of the gang:

There was the Rippers [old gang's name], but so many of their members went to jail that there really wasn't enough leadership people around. So a number of people decided to start a new gang. So then we went around the area to check who wanted to be in the gang. We only checked out those we really wanted. It was like pro football scouts, we were interested in all those that could help us now. Our biggest worry was getting members, so when some of the dudes said they didn't want to join, we had to put some heavy physical pressure on them; because if you don't get members, you don't have anything that you can build into a gang. . . . Later after we got established we didn't need to pressure people to get them to join.

Loan Man was a twenty-five-year-old member of a gang in New York:

I got this idea to start a new gang because I thought the leaders we had were all fucked up. You know, they had shit for brains. They were ruining everything we built up and I wasn't going to go down with them and lose everything. So I talked to some others who didn't like what was going on and we decided to start a new club [gang] in the neighborhood we lived in. So we quit. . . . Well, we got new members from the community, one way or the other. . . . You know we had to use a little persuasive muscle to build our membership and let the community know we were able to take control and hold it, but after we did get control, then we only took brothers who wanted us [they used the fraternity type of recruiting].

In sum, the coercive method of recruitment is used most by gangs that find their existence threatened by competitor gangs. During such periods, the gang considers that its own needs must override the choice of the individual and coercion is used to induce individuals to join their group temporarily. . . .

WHO DOES NOT JOIN A GANG?

Who does not become involved with a gang, and why? There are two answers. First, individuals who see no personal advantages in participating in a gang do not become involved. These individuals can be separated into two distinct groups. The first is those who possess all the characteristics associated with defiant individualism, but have decided that participation in a gang is not to their advantage at the present. Most of these individuals are involved in a variety of economic ventures (usually illegal) that they hope will make them rich and do not perceive any advantage to becoming involved with the gang's activities, but the vast majority of them will become involved with a gang at some time in the future. The comment of Cover, a seventeen-year-old who lives in New York, illustrates this point:

Right now, man, there ain't no reason for me to join the Black Widows. I got some good business going and I'm making some decent money. If the police don't mess me up, I can get some good cash flow going. So right now there ain't any incentive to join, you know what I mean? Now I ain't saying that I won't join sometime, 'cause they gots some good business going themselves. But right now I want to keep with what I'm doing and see where it goes. [He joined the gang one year later.]

There are also those who not only see no advantage to becoming involved in a gang, but also see significant disadvantages: the risks of being killed or imprisoned associated with gang life are too great in their view. They want to get out of the area they live in, but they have developed a strategy for doing so that does not involve the gang. Some of these people will seek socioeconomic mobility through sports, placing their hopes for a better life in their ability to become professional athletes. Ironically, however, there is probably less likelihood of their achieving such mobility through sports than through becoming involved in some illegal type of business venture.

Others from low-income neighborhoods who seek socioeconomic mobility but do not want to join a gang are those prepared to take the risk that investing their time and money in some type of formal training (formal education or the trades) will produce mobility for them. The comments of Phil, the eighteen-year-old son of a window washer in Los Angeles, represent this group:

> No, I don't want to join any gang. I know you can make money by being in, but frankly I don't want to take the risks of being killed or something. I mean some of the dudes in the gang make a whole lot of money, but they take some big risks too. I just don't want to do that. I want to get out of this neighborhood, so I'll just take my chances trying to get out by studying and trying to go to college. I know there are risks with that too. I mean even if you go to college don't mean you going to make a fortune. My cousin went to college and he started a business and it failed, so I know there is risks that I won't make doing it my way, but at least they don't include getting shot or going to prison.

The individuals in both these groups (those seeking mobility through sports and those who seek it through training) possess some of the characteristics associated with defiant individualism, but not the full defiant individualist character structure. Why some people from low-income areas have only a few of the characteristics of defiant individualism and others have them all has to do with a number of contingent factors related to exactly how each individual has experienced his or her social environment.

The second answer to the question of who does not become involved in gangs and why has to do with the fact that gangs do not want everyone who seeks membership. They will reject people if they are not good fighters, cannot be trusted, or are unpredictable (cannot be controlled), as well as if the gang itself already has too many members (in which case additional members would create difficulties in terms of social control and/or a burden in providing the services expected by the membership). Take the comment of Michael, the eighteen-year-old son of a street sweeper in New York:

> Sure, I wanted to join the Spears and I hung out with them, but they never invited me to formally join. You have to get a formal invitation to join, and they didn't give me one. They just told me that they had too many members right now, maybe sometime later.

WHAT HAPPENS TO GANG MEMBERS?

What is the trajectory of the individual who is involved in gangs? Thrasher and most of the subsequent studies on gangs believed that individuals who were in gangs simply matured out of them as they got older.[18] However, evidence from the present study suggests that the real story is more complicated. I found there are seven possible outcomes, some of which are not necessarily exclusive of each other. First, some people stay in the gang. As of this writing some individuals in their late thirties were still members of gangs. What will happen to them is open to question.

Second, some members will drop out of their gangs and pursue various illegal economic activities on their own.

Third, a number of gang members will move on to another type of organization or association. Many of the Irish will join Irish social clubs. Others will move to various branches of organized crime.

Fourth, there will be individuals who move from gangs and become involved in smaller groups like "crews" where they can receive a larger take of the money they have stolen than if they were in a gang.

Fifth, some will be imprisoned for a considerable part of their lives. While this will negate their involvement in the gangs of the streets, they will remain involved in the prison gangs.

Sixth, there will be those whose fate will be death as a consequence of a drug overdose, a violent confrontation, or the risks of lower-class life.[19]

Seventh, a large number will take the jobs and live the lives they were trying to avoid. While this may appear to be what Thrasher and others have previously reported, it is hardly accurate to think of it as "maturing out of the gang."

What do these future paths mean for the gang? Gangs are composed of individuals with defiant individualist characters who make decisions on the basis of what is good for them. On an individual level, this means that gang members will often come and go throughout long periods of their lives. Take the comment of Clip, a thirty-six-year-old member of a Los Angeles gang:

> I've been in gangs since I was fifteen. I joined and then quit and joined again. I did different things, I been married twice, but I come back to the gang 'cause there is always a chance if you get some business going, you can make some big money and live in leisure. That's been my goal and always will be.

On the aggregate level this means that coming and going is merely an integral part of the organizational environment.

CONCLUSION

Who joins a gang and why have been central concerns of many studies having to do with gangs. One set of studies concentrates on delinquency, asking why individuals are inclined to engage in illegal acts. Another incorporates gangs into a larger analysis of community. Neither approach directly addresses the gang itself. If one begins not from delinquency or community but from the defiant individualist character of gang recruits, one sees that defiant individuals make rational decisions as to what is best for them. Although previous studies have argued that all individuals have the same reason for becoming involved with gangs, prospective members in fact have a variety of reasons for doing so. However, this [paper] also shows that the varying motives of defiant individuals do not suffice to explain who joins a gang and why. Gang involvement is also determined by the needs and desires of the organization. The answer to the question of who joins a gang and why depends on the complex interplay between the individual's decision concerning what is best for him and the organization's decision as to what is best for it.

NOTES

1.   See Frederic Thrasher, *The Gang* (Chicago: University of Chicago Press, 1928); Gerald D. Suttles, *The Social Order of the Slum* (Chicago: University of Chicago Press, 1968); John Hagedorn, *People and Folks* (Chicago: Lakeview Press, 1988).

2.   Of course, some of the studies cited here overlap these categories, and I have therefore placed them according to the major emphasis of the study. See Richard A. Cloward and Lloyd B. Ohlin, *Delinquency and Opportunity* (New York: Free Press, 1960); Hagedorn, *People and Folks*; Joan Moore, *Homeboys* (Philadelphia: Temple University Press, 1978); James F. Short, Jr., and Fred L. Strodtbeck, *Group Process and Gang Delinquency* (Chicago: University Press, 1965).

3.   Here again it is important to restate that many of these studies overlap the categories I have created, but I have attempted to identify them by what seems to be their emphasis. See Herbert A. Block and Arthur Niederhoffer, *The Gang* (New York: Philosophical Library, 1958); Lewis Yablonsky, *The Violent Gang* (New York: Macmillan, 1966).

4.   See the qualifying statement in nn. 2 and 3 above. See Ruth Horowitz, *Honor and the American Dream* (New Brunswick, N.J.: Rutgers University Press, 1983); Albert Cohen, *Delinquent Boys* (Glencoe, Ill.: Free Press, 1955); Walter B. Miller, "Lower Class Culture as a Generating Milieu of Gang Delinquency," *Journal of Social Issues* 14 (1958): 5–19; James Diago Vigil, *Barrio Gangs* (Austin: University of Texas Press, 1988).

5.   See Jack Katz, *The Seduction of Crime: Moral and Sensual Attractions in Doing Evil* (New York: Basic Books, 1988), p. 9.

6.   Although the present study is not a quantitative study, the finding reported here and the ones to follow are based on observation of, and conversations and formal interviews with, hundreds of gang members.

7.   For a discussion of the political machine's role in providing psychological and financial support for poor immigrant groups, see Robert K. Merton, *Social Theory and Social Structure* (New York: Free Press, 1968), pp. 126–36. Also see William L. Riordan, *Plunkitt of Tammany Hall* (New York: Dutton, 1963).

8.   There are numerous examples throughout the society of social clubs using the lodge or clubhouse as one of the incentives for gaining members. There are athletic clubs for the wealthy (like the University Club and the Downtown Athletic Club in New York), social clubs in ethnic neighborhoods, the Elks and Moose clubs, the clubs of various veterans' associations, and tennis, yacht, and racket ball clubs.

9.   See Anne Campbell, *Girls in the Gang* (New York: Basil Blackwell, 1987).

10. For the use of drugs as recreational, see Vigil, *Barrio Gangs*; and Jeff Fagan, "Social Organization of Drug Use and Drug Dealing Among Urban Gangs," *Criminology* 27 (1986): 633–70, who reports varying degrees of drug use among various types of gangs. For studies that report the monitoring and/or prohibition of certain drugs by gangs, see Vigil, *Barrio Gangs*, on the prohibition of heroin use in Chicano gangs; and Thomas Mieczkowski, "Getting Up and Throwing Down: Heroin Street Life in Detroit," *Criminology* 24 (November 1986): 645–66.

11. See Thrasher, *The Gang*, pp. 84–96. He also discusses the gang as a source of recreation.

12. See Moore, *Homeboys*, ch. 2; Horowitz, *Honor and the American Dream*, ch. 8; Vigil, *Barrio Gangs*; Hagedorn, *People and Folks*.

13. For a discussion of these types of jobs, see Michael J. Piore, *Notes for a Theory of Labor Market Stratification*, Working Paper no. 95 (Cambridge, Mass.: Massachusetts Institute of Technology, 1972).

14. See Horowitz, *Honor and American Dream*; and Ruth Horowitz and Gary Schwartz, "Honor, Normative Ambiguity and Gang Violence," *American Sociological Review* 39 (April 1974): 238–51. There are many other studies that could have been cited here. These two are given merely as examples.

15. The testing of potential gang members as to their fighting ability was also observed by Vigil. See his *Barrio Gangs*, pp. 54–55.

16. This quotation was recorded longhand, not tape-recorded.

17. I first met M.R. when he was in one of the gangs that I was hanging around with. He subsequently left the gang, and I stayed in touch with him by talking to him when our paths crossed on the street. This quotation is from a long conversation that I had with him during one of our occasional encounters.

18. See Thrasher, *The Gang*, pp. 66–67. A great number of studies take a similar position. I shall mention but a few. See Suttles, *Social Order of the Slum*, and William F. Whyte, *Street Corner Society* (Chicago: University of Chicago Press, 1943). The work of Ruth Horowitz represents a modified exception to the other findings. While she does imply that gang members grow old and abandon their street lives, she also reports that involvement in gangs may last well into the thirties for individuals. See her *Honor and the American Dream*, pp. 177–97.

19. Some of these people have died from illnesses, food poisoning, and various accidents.

38

Victimization and Resistance
Among Street Prostitutes

JODY MILLER

Feminist research has tended to dichotomize women's experiences, presenting women either as victims or as active agents (Gordon 1986; Harding 1987; Lather 1991; Walkerdine 1990). According to Harding, focusing only on women's victimization is problematic because it "tend[s] to create the false impression that women have *only* been victims, that they have never successfully fought back, that women cannot be effective social agents on behalf of themselves or others" (1987, p. 5). On the other hand, focusing upon women's agency is often misconstrued as blaming the victim (Gordon 1986, p. 23). These tensions within feminist research are magnified when studying women marginalized by additional forms of oppression.

The goal of the current research is to bring these theoretical issues into a discussion of concrete research, drawing on interviews with street prostitutes about their experiences with violence. Because women engaged in street prostitution have consistently been the brunt of myth and misunderstanding, the aim of this paper is to learn about violence against street prostitutes from the perspectives of the women who encounter it, and to examine their resistance to violence on the streets. My research on prostitutes' experiences with violence provides a complex example of the ways in which victimization and agency exist simultaneously and in multidimensional ways in women's lives.

Previous research on violence against prostitutes has indicated that women working at the lower levels of prostitution (i.e. streetwalking) not only face the most violence, but also the most social control by the police and police abuse (Carmen and Moody 1985; Diana 1985; James 1978; James et al. 1977; Perkins and Bennett 1985). In the first major attempt to investigate systematically various forms of violence against street prostitutes, Silbert and Pines (1982) found that 70 percent of the 200 females interviewed had been raped by clients, 65 percent reported being beaten by clients, and 66 percent were physically abused by pimps. In addition, the theme of violence frequently surfaces in research about the other aspects of street prostitution, as well as in writings by women in the industry (Delecoste and Alexander 1988; Diana 1985; Jaget 1980; Miller 1986; Pheterson 1989; Reynolds 1986; Weisberg

From "'Your Life Is on the Line Every Time You're on the Streets,'" Jody Miller,
Humanity & Society, V. 17, No. 4, 1993, pp. 422–442. Reprinted by permission of the
Association for Humanist Sociology.

1985). My study, while based on a small number of interviews, will add to the literature by examining not just the extent of victimization, but also prostitutes' strategies for coping with and resisting victimization.

While there were many structural constraints on the lives of the women I interviewed and most had been victimized frequently, they were anything but passive victims, and employed a number of strategies to resist and fight back at violence against them. I will begin by discussing the types and frequency of violence encountered by street prostitutes, as well as their perceptions about the prevalence of violence on the streets.[1] Then I will discuss the variety of strategies developed by women to avoid violent attacks. These included categorization and selectivity in choosing the men they would engage in sex-for-money exchanges with, as well as methods for staying in control during the encounter. In addition, the women developed a number of means for coping with violent encounters when they occurred, including ways of fighting back or escaping, using street justice to get revenge for violent assaults, and, occasionally, calling on the police for protection.

METHOD

My research is based on semi-structured in-depth interviews with sixteen prostitutes incarcerated at the county jail in a midwestern city during December 1990 and January 1991. Because most of the women working the streets in this city were addicted to crack, interviews in jail provided an environment to talk to women who were not high, and also provided longer periods of uninterrupted time than the streets would have permitted (Agar 1977). In addition, when prostitutes are on the streets, time is money. I was unable to compensate them financially, so I felt approaching women on the streets would be unethical. The jail setting allowed me to speak with them at no monetary cost, and provided the women who chose to talk with me a break from the boredom of incarceration.

While it is potentially problematic to generalize from persons within jail settings, there are reasons to be optimistic about my sampling. Researchers have found extremely high rates of arrest among street prostitutes, and this is especially the case for drug-addicted prostitutes (Diana 1985; James 1978; Miller 1986). In fact every woman in Diana's (1985) study had been arrested for loitering. This was especially likely to be the case in the city I conducted my research in, because the vice division of the police department and the judicial system were actively and openly committed to incarcerating women involved in street prostitution.

According to jail officials, of the approximately two hundred women incarcerated at the jail at any one time, around ten percent were serving time for prostitution-related charges. Women were drawn for the sample from a computer generated list of solicitation and loitering cases, and from the social workers' knowledge of "known prostitutes" incarcerated on other charges. They were told about the study and invited to participate by a jail worker.

Women who agreed to participate were brought to a lawyer's booth in the jail where I conducted the interviews. Most of the women approached were willing to meet with me; only two refused to be interviewed, and another asked that the interview be terminated part way through because the subject matter was too emotionally charged.

The women ranged in age from 20 to 38, with the majority in their early and late 20s. Nine were African American, seven were white. Five had been working since their early to mid-teens. I began by collecting demographic information. I then asked general questions about the women's involvement in prostitution, their perceptions of prostitution and their commitment to the work. Once we talked more generally about their involvement in prostitution, I asked questions about their perceptions of violence and danger on the streets, and finally, about their personal experiences with violence. Most of the women constructed our relationship as one between an experienced streetwise individual and a somewhat naive "straight" person.

THE EXTENT OF VIOLENCE
ON THE STREETS

Street prostitutes face widespread victimization. In fact, it is part of the very fabric of the street environment. Of the sixteen women interviewed, fifteen (93.8 percent) had experienced some form of sexual assault, attempted or completed. Nearly half of the women interviewed (43.8 percent) reported being forced or coerced into having sex with men who identified themselves as police officers. Three quarters of the women had been raped by one or more tricks,[2] and 62.6 percent (ten women) had been raped in other contexts on the streets. In addition, nine women (56.3 percent) had their money stolen from them by tricks after their sexual transactions were completed. In addition to sexual assault, the women in this study reported a variety of physical attacks, which often (though not always) took place in the context of sexual attacks. Fourteen women experienced some form of physical assault, ranging from being punched or kicked (five women), beaten up (ten women), stabbed or slashed with a razor or knife (five women). Four had been struck with objects, including a baseball bat, stick, crutch, and brick. Six had been kidnapped and held captive. Two were choked; three suffered serious injuries such as broken bones; one had her head rammed through a glass door; and one was tortured with electric shock.

In an important sense, a major finding from my interviews was that the violence against these women could not be quantified. Many of them indicated that their experiences with violence were simply too numerous to count. Cissy, a 29-year-old white woman, said she was raped or robbed in the context of turning tricks "at least once a week, once a week, twice a week." Lacy, a 21-year-old white woman, responded with a variety of answers about the widespread existence of violence: "All the time, all the time. It happens to us girls all the time." "I could tell you, we could sit here all day and talk about it." When I asked her if she had ever been stabbed, she held out her arms and said "yeah, take your pick" referring to the numerous scars on her arms.

> We have alot of really sick people out there, you know, and when they get ahold of you, they do a whole bunch of crazy stuff to you. I've been in the hospital before and I been shot, I been stabbed eight times, I been kidnapped, I been tied up, all kinds of crazy things. There's just some really sick people out there.

None of the women I interviewed considered the streets a safe place for prostitutes to work. Ginger, a 27-year-old African American woman, had worked previously in houses and felt the streets were much more dangerous. "[W]hen you work on the inside, you have no, you have no question. A man comes there, he's coming there for one thing. When you're on the streets, it's skeptical." According to Candy, a 20-year-old white woman, "you know, its a 50-50 chance every time you get in a car." While violence was an ever-present part of street prostitution, constantly expected or experienced and frequently unavoidable, the women I spoke with did not passively accept this aspect of their work, but instead employed a variety of tactics to stay safe, protect one another, and fight back against violence. In the following sections, I will examine these strategies, beginning with prostitutes' perceptions and selections of tricks.

STREET PROSTITUTES' STRATEGIES
TO COMBAT VIOLENCE

The women I spoke with tended to have a specific picture of the type of man who constituted danger on the streets. Usually this involved perceptions of individual pathology. Blondie, a 23-year-old white woman, explained: "I mean, you got alot of weirdos out there. You have to jump out of cars and you know, give up the money that you already made. They just wanna do all kinds of stuff, you know?" Likewise, Jessie, a 29-year-old African American woman, told me "It's like, oh, you know you don't have to be doin' nothing, just standing on the corner, a psycho's gonna grab you and throw you in a car if you're not alert." Kay Kay, a 21-year-old white woman, was more adamant: "Men are so sick. If you just really knew. There is some sick men out there. Very sick . . . wanna beat you up, for pleasure, so they can come. That's just, that's sick. Men are sick."

Their perceptions of danger—as a threat caused by sick men—shaped the strategies used to keep themselves safe. There was a consistent pattern among the women I interviewed of categorizing and labeling potential tricks according to a variety of criteria, based on interpretations of what constituted the likelihood of a man being a "psycho." These interpretations will be the basis of the following section. In addition, the women discussed a variety of tactics used during their day to day encounters with men to "stay in control." I will conclude by discussing how women handled the inevitable violent encounters that arose. They utilized a number of strategies, both during and after the encounters, including fighting back or running away, calling upon street justice, and less frequently, going to the police.

My focus on street prostitutes' strategies for combating violence is in no way meant to present them as invincible women, able to maintain control on the streets. The structures of street prostitution do not allow for this, but instead perpetuate an unsafe environment in which working prostitutes are always vulnerable to attack. Rather, my goal is to further illuminate the complex nature of women's experiences of danger. While the women I spoke with were exploited and victimized on the streets, sometimes brutally, sometimes with incredible frequency, they did not passively accept this aspect of their job, but actively resisted their victimization. Some showed incredible courage and strength; others survived. This is an aspect of violence against women that needs to be examined in more detail, in order to keep from framing women solely as disempowered victims.

SELECTION OF TRICKS

Knowing What to Look For

For street prostitutes, the men who approach them offering to exchange money for sex pose both the potential for earning quick money and the potential for danger. A large amount of the violence street prostitutes encounter is at the hands of tricks. As a result, the women I spoke with tried—to the extent possible given their need for money or their need for drugs at the time, both of which varied from woman to woman and encounter to encounter—to be selective about the men they dated. They employed a number of strategies for determining what made a safe date. For some women, long-time experience on the streets provided them with what they felt was a special knowledge for reading tricks. Other less experienced women sometimes went by instincts, refusing dates with men who gave them bad vibes.

Jessie, a 29-year-old African American woman with several years' experience on the streets, believed the streets were dangerous, but felt that her knowledge and experience provided her with an edge that other women didn't have. She explained, "It depends on you. For me, it's just somewhat dangerous. You know what I mean, 'cause I know what to look for. For some people, it's probably very very dangerous." Dee, a 28-year-old white woman who had been working the streets for twelve years, also felt she was experienced enough to tell which clients were dangerous. She explained, "I mean, you gotta be able to know, gotta be able to read people, you know. Look at somebody and tell. Just, you, I can talk to a person and know in five minutes if he's got a problem or not. I learned that." Sugar, a 31-year-old African American woman who had been working on the streets since her late teens, told me, "You can almost, I guess you can't read the person, but you can tell by the way they talk and the way they be actin'. That's how I do 'em. Read 'em."

Sometimes, subtle clues within the interaction tipped the women off that something was about to happen. According to Princess, a 38-year-old African American woman with a number of years' experience, she was infrequently victimized because:

> Well, for me, I have sensed the danger, and it helped. 'Cause I've been in danger several times. But I had sensed it. Guy pulled a machete on me, and I already had my knife out. And uh, I opened the car door and got out. He had his machete and I had my knife. I said "we'll be cuttin' each other up."

She discussed the clues within the encounter that tipped her off and allowed her to escape unharmed. By watching his moves and noticing when he seemed to stall or contradict himself, and by having a weapon on hand, she was prepared to confront him and get away:

> Oh, he was nice, he said, "I wanna date you." I said "well I don't date black guys." 'Cause he lookeded real creepy. He said "well I wanna date you. I'll pay whatever you want." And I said "I want fifty dollars." He said "ok, I'll give it to you. I, I'm gonna pay you." And I said uh, "well alright," I parked him . . . in an alley. When we get down, he say "you wanna smoke a joint?" And I said "No. I'm here to take care of business. I don't, you know, time is money." So then he says "ok, how much did you say you wanted?" And that's when I felt, you know, the hanky panky was about to happen. He did like, he put his hand down in between the car and the door, the seat and the door. And he looked over at me and he said "how much do you want?" Lookin' all gloomy like that. And I said "how much did you tell me you was gonna give me?" And at that time I seen his hand come up. And I had my knife in my hand and I opened the door and backed out.

Younger women and women with less experience as street prostitutes did not discuss their ability to use street knowledge to read clients. Instead, some described following their instinct, refusing to date a man or terminating the date when they got a bad feeling about it. For example, Jane, a 28-year-old African American woman who had only been on the streets for six months, explained her decision to terminate a date: "There was just something about him, and I always follow my first mind. And I say, 'never mind.'" Veronica, a 28-year-old African American woman who had begun working less than a year before we spoke, explained, "if I get a leery gut feeling I don't go with them. Uh, I usually go with my instinct." At the same time, however, Veronica did not feel instinct or street knowledge were enough to protect prostitutes from violence. She continued, "it's not what you know. You're just taking a chance."

Age, Race, and Class Characteristics of Tricks

Many times, tricks' demographic characteristics were drawn upon in deciding which ones to date. The women I spoke with consistently categorized men based on their perceptions of the men's race and age, and some based their choices on perceptions of class as well. Older men and white men were consistently preferred to younger and/or African American men. Kate, a 30-year-old African American woman, explained, "Like um, young black guys you don't go out with. Older black guys, ok. Young white guys, some of 'em are ok. Middle aged guys are always cool." Similarly, Lisa, a 33-year-old African American woman, told me "I don't date black guys. I don't date young guys. It was usually older white guys."

Some women could afford to be more choosy than others, depending on their economic situation and whether they were addicted to drugs. Ginger, a 27-year-old African American woman, saw only six to eight clients a night, and felt she was able to be more selective than other women because she did not use drugs and she earned additional money working as a bartender. She explained, "It pays to be selective. Even though you're not always sure, but uh, you have more than that approaching you with it but uh, I have to say no sometimes because I don't wanna get in the car with 'em or they're young."

Choosing to date older men had specific advantages. They were perceived as less likely to pull something, and they also posed less physical threat. Dee, a 28-year-old white woman, said "most, I try to date real old men. . . . Some of the younger ones, you can't trust 'em." And Blondie, a 23-year-old white woman, explained, "I get in cars with older guys that I know I can handle if they try to do anything to me. I'll beat 'em up. (laughs)" Likewise, Pepper, a 22-year-old African American woman, explained, "if they're not old men period I don't mess with 'em."

When race was called upon to characterize clients, it often involved the use of widespread cultural stereotypes about African American men, sexuality and violence. Veronica, a 28-year-old African American woman, told me "I don't mess with black clients. Because most black clients, they wanna give you hell, they wanna screw you to death or they want you to give 'em head for two or three hours and only wanna give you twenty dollars. I don't go for that." Dee, a 28-year-old white woman, said "you can't trust a black man," and Cissy, a 29-year-old white woman, explained "a lot of times I wouldn't get in the car with a black guy. I don't date blacks."

These attitudes about African American men were extremely pervasive among both the African American and white women I talked to. There was a tendency to define all African American men as violent based on bad experiences with a few. It is difficult to judge the extent to which their perceptions came from cultural mythology (Davis 1981, 1985), and how much came from actual experiences. It seemed that no matter how much violence was encountered at the hands of white men, the women were never willing to make the statement "I don't date whites" although they frequently made the statement, as Jane, a 28-year-old African American woman did, "I'm not prejudice or nothing like that but I don't get in the car with no black dates." Kay Kay, a

21-year-old white woman, even recognized that her tendency to view all black men as dangerous stemmed from only one bad experience two years prior to our interview. She had been lured into an abandoned house by a man who said he had some cocaine, and when she got there, he stabbed her six times and raped her repeatedly. She told me:

> I won't go out with black men. 'Cause I got this by a black man [refers to scars on her body]. I will not go out with them. Unless I know 'em. Ok. I mean, some of 'em I do, not all. It took me a long time to um, like blacks. 'Cause when I got stabbed I hated every black, every black person was a nigger. They was all a nigger to me. I was prejudiced. Women, men, babies. I didn't care. Because one black man hurt me.

Yet even though she recognized that her prejudice stemmed from one bad experience, she still maintained stereotyped notions of African American men as dangerous. "Oh, I don't fuck with black people. They hurt you more than whites. They do. I don't care what everybody says. I mean, there's some white retards, but there's some black killers." By contrast, Sugar, a 31-year-old African American woman, was the only interview subject who preferred African American men to white men. She explained, "I'd say, me, I prefer to date a black man, 'cause you don't uh have too many problems. Some of 'em you do, but most of 'em you don't. It be the white guys that be tryin' to pull the knifes and things like that. That's mostly white."

Often the women's experiences with violence contradicted the assumptions they made about who was safe and who was dangerous, especially when they called upon categories such as "white" or clues that indicated middle class status. For example, Candy, a 20-year-old white woman, was stabbed by a white trick in his thirties that she felt safe with because "he looked like he was alright. He didn't look crazy. He was dressed nice." Similarly, Dee, a 28-year-old white woman, was forced to give a free blow job to an African American man who "looked like he had money. You know, he didn't look like a street person, looked like a businessman." This is why Ginger, a 27-year-old African American woman, believed the idea of "knowing what to look for" was a trap women fell into that made them more vulnerable.

> [Y]ou never go by looks, looks are very deceiving . . . never go by looks. You can never say, "well he looks like this, he looks. . . ." Looks are very deceiving. And women still say that, which I don't understand that either, I mean, they'll still say "well he looks like he'll be. . . ." You can never, you never know.

At the same time, however, she explained that there was a small population of young African American men on the streets that preyed on street prostitutes:

> "I'm not trying to be hypocritical, I'm not contradicting myself when I say that about young black guys, it's just that young black guys riding around out here on the streets, they really are not into dating too much . . . be-cause they think all the girls out there are into drugs, geekers is what they

call them . . . and they're really out just to take something. They might pay you and then try to rob you so basically the bottom line is really stay away from 'em. You know, don't even subject yourself to 'em. And that is a fact about young black guys out there. Riding [the strip]. Because that's what they're looking for is someone to take advantage of. . . . Not to say that an old black guy can't do it or a white guy couldn't do it.

Sharing Information

In addition to judging tricks based on their characteristics or based on instinct or street knowledge, the women I spoke with also discussed a code of ethics among street prostitutes that involved sharing information about tricks with one another. While many of the women were quick to point out that they didn't consider most of the other women they worked around friends, they nonetheless felt a sense of obligation to pass on knowledge about dangerous men. Cissy, a 29-year-old white woman, explained:

> if we share dates or something, like say [somebody I know] is out there with me . . . and they'll go by and maybe he's eyeing me and uh, you know, I'll try to get him to come back, and like one of 'em will go, "he's a forty [dollar date], he's cool, he's a forty" or "don't go with him, he'll fuck you up" or uh, you know what I mean, we're like, they come around so much everybody knows 'em, we just, we tell each other if one's cool or not.

Cissy had been able to avoid some violent tricks because other women had warned her about them. "You know, there's a lot of 'em I haven't been with, because other girls, other girls told me and I just don't go with them." Blondie, a 23-year-old white woman, was driven to a cornfield and raped by a trick. One of the reasons she agreed to go with the man was because she had "seen him out there a lot of times . . . [a]nd none of the girls ever said he wasn't a good guy." Because the women usually "pass it on if somebody does something," she thought she was safe.

Dee described an incident in which a man pulled a knife on her and she had to fight her way out of his car. She explained, "I told my girlfriends watch out for a little blue Honda, told 'em what he looked like. Said he had a grey jogging suit on, I described him to my girlfriends. Cause most of the girls'll do that, if they run into a fool out there, they'll tell the other girls what to look for." According to Jessie, she tended to keep her experiences with violence to herself rather than talking to anyone about them. However, she explained:

> um, I would have pointed either one of them out though, to a girlfriend. If he would've rode by I would've said "that's the one that did this." Then I would've brought it up. You know, I would've looked out for you like that. If I see you gettin' ready to get in a car with somebody that I know has done something really rank and crazy to me, I'll tell you. That's the least I could do, 'cause that's the least I would expect from you. Whether you like me or not, you know, is to tell me if I'm in a potential dangerous situation.

STAYING IN CONTROL

Once women chose to get in the car with a date, they continued to employ strategies to avoid violence. Even when the man passed their criteria of what constituted a safe date, most women did not let down their guard when the transaction began. Instead, they called upon a variety of tactics to maintain their safety. These tactics included a number of specific rituals and actions. Jessie said "Um, get paid up front. If they won't give me money before I have any kind of sexual contact with them, I won't date 'em." Similarly, Cissy explained the importance of maintaining control over situations in order to stay safe:

> Don't act stupid. I mean, you know. The guy can tell if you're stupid or not. Get in a car, always be in control. Have condoms available all of the time. Don't never go out your area. I don't ever go to a dude's house. Unless I really know him. I'll pull 'em down in the alley in a minute, nearest alley. Um, don't take 'em nowhere. Don't go nowhere with 'em. Stay in control at all times, you know.

For many women, staying alert to the man's actions and being aware of the surroundings were the most important means of trying to maintain control over the situation and remain safe. According to Pepper, "I just know that, if I'm gonna get hurt it's gonna be because my carelessness, not because I'm just gonna walk in and trust somebody to, like, not hurt me. 'Cause you don't know, maybe your friend next to you's gonna hurt you. I don't trust anybody." Sugar explained "I had a person try to turn an alley on me once before. One tried to catch the freeway with me. You got to look out for all that kind of stuff." A common ritual was to make sure the car door was not rigged to lock them in. Lisa, a 33-year-old African American woman, explained:

> I never got in a car and had no way out of it. I kept my hand on the door handle at all times. If I got in a car, I always acted like, I would never close the door all the way, to have to, to give me reason to have to open this door back up. You know, I'd close the door and it's like, wait I gotta open this door and close it back up. If that door didn't open than I was out the window or something.

Some women were able to be more careful than others. As with Ginger's description of the number of dates she turned down to stay safe (see above), Lisa described the contrast between her method of car dating and those of the other women she encountered on the streets. Like Ginger, Lisa was not a drug user, and was less desperate for money than many women.

> I'd see girls get into cars, ten minutes they were back. I wasn't, I don't know what they were doing, ok, other than they weren't too concerned with who they were with, other than the money. . . . [I]f you got in the car at the same time I got in the car, by the time I come back, you turned three dates and I've just been, and I'm still getting back from that one, and you would swear, you must have made a thousand dollars. No. My precaution is you're gonna have to talk to me first buddy. You know, you're

not just gonna take me and drive me around some corner and do what you want and give me money and it's gonna be done.

Other women felt it was better to get transactions over with quickly, both because it allowed them to make more money, and because it was perceived as putting them at less risk of attack. In addition to having to watch out for the men they were dating, street prostitutes also had to stay alert to people in the area who would attack both the women and the men they were engaged in sex with. Princess tried to engage in sex quickly, and to convince the men to pay attention to the surroundings when she couldn't, advising them to "pay attention to the rearview mirror." She described her frustration when the tricks didn't recognize this danger:

> Ordinarily it'd take me about five minutes to make thirty, forty dollars off a blow job. When it goes to fifteen minutes, I'm pretty pissed. You know. 'Cause you're a white man, you're down here in the ghetto, you know, I don't want one of these thugs to walk around the corner and see you and wanna knock your window out and take your money. And mines too. So just do what we gotta do. And get out of here. I'm tryin' to save your life and mine. They have no understanding of that. "Ain't nobody comin'.". . . I know a couple of girls that it has happened to. They'll come down the street with a crow bar, bust the windows out. You know, he's sittin' there with the car runnin', lookin' down at you.

HANDLING VIOLENT ENCOUNTERS

Fighting Back in the Situation

No matter how careful the women were when they dated, they all sometimes ran into violent men on the streets. The women I spoke with recognized that violent encounters were sometimes inevitable. Cissy explained:

> There's guys like that. Out there now. Well everyone's gonna find him, everyone's gonna find him the first time. Everyone's gonna find all of 'em the first time. You know what I mean? If you just remember who they are the next time they're out. . . . But that's just part of the business. That's just part of working, that's just what happens. You have to accept it.

However, when it was possible, many of the women found ways to fight back against some acts of violence against them, in a variety of ways. Some carried weapons to protect themselves, but there were perceived risks involved in doing so. Because the vice officers in the city heavily targeted prostitutes, the likelihood of being arrested when armed with a weapon was considerable, and most of the women knew of other prostitutes who were sent to state prison under such circumstances. Lacy, a 21-year-old white woman, carried a weapon until she had a violent encounter that made her stop. She explained:

> Well, I used to carry a knife with me, like a six inch switchblade and I used to, you know, carry it up my coat sleeve. But uh, this date one night,

he picked me up and he had a knife and he stabbed me right here and right here [points to scars], and I stabbed him back, and I guess, you know I was so scared and I was so mad, you know, because you know, you, a lot of stuff builds up in you and stuff when you're out there and when something like that happens, you fight for your life, you know, you try to kill him, and uh, I tried to stab him in his heart, and I missed his heart by four inches and I got his lung, and they tried to get me for um, aggravated assault and everything. He spent eight days in the hospital, and they tried to get me in trouble for it. So I basically quit carrying anything.

A more common strategy for prostitutes was to try to get away from violent men. Women described fighting their way out of cars, distracting men so they could run away, and jumping from moving vehicles. For example, Pepper, a 22-year-old African American woman, told of an encounter in which the trick demanded his money back from her after they'd had sex:

> I snatched his keys out his car and threw 'em out the window. . . . He jumped out of the car to go get his keys and I took off runnin'. Then he come runnin' behind me. . . . I start hollerin' my friend's name, I said "girl" and this dude that came down the steps from these other apartments, and they start chasin' him. So he took off runnin' back towards his car and pulled out.

Princess, a 38-year-old African American woman, described a situation in which she was almost raped, and escaped by jumping out of a moving vehicle:

> He said "I'm gonna take you. And I'm gonna go in every hole you got until I get tired.". . . He used to come and watch me dance at the bar. So I thought he was alright. Shit. We got way out to [the north side] and I decided I'm gonna jump out. So I jumped out. . . . It took all the skin off my face, off my arms and everything.

Kay Kay, a 21-year-old white woman, told of when a trick held a gun on her:

> I looked out the car "shoot me motherfucker, make it loud." I, I don't care. You gotta kill me, because uh, I ain't riskin' my life. I, I risk it every day, I mean, either shoot me, you ain't rapin' me, you ain't stickin' no gun up in me. . . . I jumped out his car. Yes I did, with my pants down to my knees. I don't care. Wanna see my bootie? I mean, I'd rather show my bootie than get hurt.

STREET JUSTICE

In addition to trying to fight against or escape from violence when it occurred, the women I spoke with also discussed their ability to sometimes call upon street justice, whereby at least some of the time, some drug dealers provided protection or revenge for women for acts of violence against them. Lacy, a 21-year-old white woman, explained, "we got a thing called street justice. . . .

You know, like when somebody does something to us, we go get somebody to do something to them." For example, she described an incident in which she was beat up "real real bad" and robbed by an ex-boyfriend:

> I was goin' with this, this drug dealer at the time. . . . I was crying, I was all bloody and beat up. My boyfriend comes in his car. And he stopped in the middle of the street and he jumped out and everything, and carried me to the back and he said "what happened?" I said "he beat me up and took my money." And uh, my boyfriend picked him up, was throwin' him on the concrete on his head, and gave him thirty stitches in his head.

Usually, the women I spoke with discussed their ability to call upon street justice in the context of encounters with other street people, rather than tricks. For example, Jessie and a friend robbed one of her tricks and her friend in turn robbed her and put her head through a glass door. To retaliate, she "had some boys go and beat him up and rob him again that night." According to Jessie, "there's a lot of people out there, 'cause of how I am, they see I'm hurt, will wanna know why. You know, 'cause it's like, we hang around Jessie, we get high with her, she's not like, she don't deserve that. You know, and um, they go take care of him for me."

Street justice could sometimes be used to provide women increased safety when dating. For example, Kay Kay often parked her dates near dope houses to afford herself some measure of protection. "I park right in back of the dope-man's house. I don't care. . . . 'Cause you hurt me, that's your ass. I got a lot of friends out there. A lot of people that care about me." Once a date picked her up and then refused to pay, saying that another woman he knew never made him pay.

> "Well you gotta pay with me," I said. And, he says, we get to where we was parking, "oh, I'm not gonna um, pay you." "Motherfucker let me out of this car." I got out the car, "Rape! Rape!" He let me go. He had to. I was in the dope neighborhood, I know every dope boy there is.

GOING TO THE POLICE

The women I spoke with did not make a habit of going to the police when they were victimized. Usually, this was because they felt the police would not take their complaints seriously or would harass them for being on the streets. Lacy explained, "I've seen situations and stuff where they haven't cared. Or that they've seen, they've seen situations, you know, with their own eyes and chose to look the other way." She also believed that the police:

> just consider us a statistic. . . . I don't know, it's just, I mean they figure like, well maybe if we do die at least they won't have to fuck with us no more. You know, they won't have to worry about nothin' with us and takin' us to jail. Or havin' us out there on the streets, you know, so that their boss will be mad at them.

Cissy said of incidents in which she's been raped by men, "I don't never report none of that stuff to the police. I don't know why. They'll look at me like I'm stupid 'cause I'm a prostitute. 'Well bitch you shouldn't be out there' you know, or something like that." Similarly, Blondie said that street prostitutes "get raped all the time and it doesn't matter to the police because you're a prostitute, you know. You really don't have any case against 'em."

However, the women I spoke with did occasionally turn to the police, especially when it was to report a man they considered particularly menacing, or someone who was targeting street prostitutes regularly. Usually when they did so, it was not to press charges, which they perceived as futile, but to call upon the police to stop the man from further violence. While the police were not perceived as an avenue women could go through individually in order to combat violent men, they did believe that the police could be useful in scaring potentially dangerous men away. Both Cissy and Blondie had gone to the police under such circumstances. Cissy described one incident:

> I've had a guy, one tried to rip my pocket off, he tried to get his money back. I climbed out the window and shit. I puked on him when he tried to cram it down my throat right? I puked all over. He said "Ah, goddamn you, give me my money back blah, blah, blah." I said "Oh no bud" you know, and I went out the window, and uh he was acting crazy on me and shit. And I went and I flagged down a police and I told him about it. I meant I can't be havin' him around here beatin' us up, takin' our money and shit. And I gave him the license plates and stuff.

When Blondie was raped by the man who drove her to a cornfield instead of taking her where she had agreed to go with him, she decided to tell the police about the man, but she "didn't tell them what he'd done" to her. Instead, she "told 'em there was a guy in a red truck that was goin' around, you know, harrassin' the girls. . . ." She explained:

> finally they pulled him over and caught him but that was the only one. You know, we just don't even see no reason in tryin' to report 'em anymore because they don't do anything. That one time they did something, I guess he must of hurt one of the girls or something. The police down there are pretty nice though, you know, I mean they're friendly. If you say "that guy in that blue car went and he just tried to, you know, grab me and put me in his car" they'd go after him. Whether they'd do anything or not is a different story. They'd probably scare him bad enough not to try it again.

DISCUSSION

Women engaged in street prostitution face widespread violence. Of the sixteen women I interviewed, twelve had been raped by tricks, ten had been raped by others on the streets; in all, fifteen of the sixteen women interviewed experienced some form of sexual attack. In addition, fourteen women had been physically attacked, some in the most brutal fashion imaginable. For most, violence against them was seen as immeasurable. These are important facts to know and research. We need to understand what in our culture makes this widespread victimization possible, why street prostitutes face more abuse than women in other contexts in the United States (Bracey 1979; Hatty 1989; Silbert and Pines 1982).

At the same time, however, it is crucial that we not paint prostitutes as mere victims of abuse. . . . The women I spoke with resisted violence and fought back against it in a variety of ways when they worked on the streets. It is important to recognize these strategies without losing sight of the various ways in which prostitutes are oppressed. The women called upon an expansive repertoire in choosing which men to date, using street knowledge and instinct as well as demographic characteristics such as race, age, and class. In addition, the women I spoke with described street networks which they could call upon both to gain information about and protection from potential dates and to seek revenge against violent offenders. In addition, there were a variety of strategies called upon during the transaction to stay in control, including getting money upfront, making sure car doors were not rigged, and staying alert to men's behaviors and aware of the environment.

As a result of these various strategies, many women were able to avoid some of the violent encounters they otherwise would have experienced. In addition, when the inevitable violent trick was encountered, women were sometimes able to fight back and get away either unharmed or with fewer injuries than might have been the case otherwise. They also occasionally called upon either street or police justice in order to maintain or restore some amount of safety on the streets. While these findings in no way diminish our understanding of the amount of violence prostitutes are the target of, they do provide important knowledge for feminist researchers, and warn us against the simplistic portrayal of women as victims. My goal in this paper has been to further illuminate the complex nature of prostitutes' experiences with violent victimization. This is an aspect of violence against women that needs to be examined in more detail, in order to keep from framing women solely as disempowered victims.

NOTES

1. The term *violence* has evoked many definitions. My focus is on robbery, physical violence, and sexual violence which has specifically taken place in the context of working the streets. Under the category robbery, I include situations in which money is exchanged for sex then taken back by tricks after the transaction is completed, in addition to robbery on the streets or robbery by men who pretend to be tricks, then rob women of their money without demanding sex. To measure physical violence, I asked the following questions: In the last year, has anyone pushed, grabbed, shoved, or slapped you? kicked, bit, or hit you with their fist? hit you with something, beat you up or attacked you? choked you? threatened you with a knife or gun? used a knife or gun against you? I asked the following ques-

tions to examine their experiences with sexual violence, all of which I classify as some form of rape: Has anyone raped you? made you participate in sexual acts by using physical force or physically threatening you? made you participate in sexual acts by using a weapon or threatening you with a weapon? made you participate in sexual acts by threatening you in some other way, such as threatening to turn you in to the police or threaten to arrest you, for example? made you participate in sexual acts when you were unable to consent because you were drugged or passed out, for example?

2.   This was the language employed by most of the women to talk about sex-for-money exchanges. They referred to the men they exchanged with as either dates or tricks. I will use these terms, as well as the term *client* interchangeably.

REFERENCES

Agar, Michael H. 1977. "Ethnography in the Streets and in the Joint: A Comparison." In *Street Ethnography: Selected Studies of Crime and Drug Use in Natural Settings,* edited by R. S. Weppner (pp. 23–38). Beverly Hills: Sage Publications.

Bell, Laurie, ed. 1987. *Good Girls, Bad Girls: Feminists and Sex Trade Workers Face to Face.* Seattle: The Seal Press.

Bracey, Dorothy H. 1979. *Baby Pros.* New York: The John Jay Press.

Carmen, Arlene, and Howard Moody. 1985. *Working Women: The Subterranean World of Street Prostitution.* New York: Harper & Row.

Davis, Angela Y. 1981. *Women, Race and Class.* New York: Vintage Books.

Davis, Angela Y. 1985. *Violence Against Women and the Ongoing Challenge to Racism.* Latham, NY: Kitchen Table Women of Color Press.

Delacoste, Frédérique, and Priscilla Alexander, eds. 1988. *Sex Work.* London: Virago Press.

Diana, Lewis. 1985. *The Prostitute and Her Clients.* Springfield, Illinois: Charles C. Thomas.

Gordon, Linda. 1986. "What's New in Women's History." In *Feminist Studies, Critical Studies,* edited by T. de Lauretis (pp. 20–30). Bloomington: Indiana University Press.

Hatty, Suzanne. 1989. "Violence Against Prostitute Women: Social and Legal Dilemmas." *Australian Journal of Social Issues.* 2(24): 235–248.

Jaget, Claude, ed. 1980. *Prostitutes: Our Lives.* Bristol, England: Falling Wall Press.

James, Jennifer. 1978. "The Prostitute as Victim." In *The Victimization of Women,* edited by J. R. Chapman and M. Gates (pp. 175–202). Beverly Hills: Sage Publications.

James, Jennifer, Jean Withers, Marilyn Haft, Sara Theiss, and Mary Owen. 1977. *The Politics of Prostitution.* Social Research Associates.

Miller, Eleanor M. 1986. *Street Woman.* Philadelphia: Temple University Press.

Perkins, Roberta, and Garry Bennett. 1985. *Being a Prostitute.* Sydney, Australia: George Allen & Unwin.

Pheterson, Gail, ed. 1989. *A Vindication of the Rights of Whores.* Seattle: Seal Press.

Reynolds, Helen. 1986. *The Economics of Prostitution.* Springfield, Illinois: Charles C. Thomas.

Silbert, Mimi H., and Ayala M. Pines. 1982. "Victimization of Street Prostitutes." *Victimology: An International Journal.* 7: 122–133.

Weisberg, D. Kelly. 1985. *Children of the Night.* Lexington, Massachusetts: Lexington Books.

39

Nonmainstream Body Modification

Genital Piercing, Branding, Burning, and Cutting

JAMES MYERS

The term *body modification* properly includes cosmetics, coiffure, ornamentation, adornment, tattooing, scarification, piercing, cutting, branding, and other procedures done mostly for aesthetic reasons. It is a phenomenon possibly as old as genus *Homo,* or at least as ancient as when an intelligent being looked down at some clay on the ground, daubed a patch of it on each cheek, and caught the pleasing reflection on the surface of a pond. Appropriate to my overall topic is Thevos's (1984) observation that a "self retouching impulse" distinguishes humans from other animals (p. 3).

At the outset, it is important to distinguish between the two main types of body modification: permanent (or irreversible) and temporary. Permanent modifications, such as tattooing, branding, scarification, and piercing result in indelible markings on the surface of the body. With the exception of branding, these marks involve the application of sharp instruments to the skin. Dental alterations, skull modeling, and modern plastic surgery are also forms of permanent body modification. Temporary modifications include body painting, cosmetics, hair styling, costume, ornamentation, and any other alteration that can be washed off, dusted away, or simply lifted off the body. This article focuses on permanent body modifications in contemporary United States, especially genital piercing, branding, and cutting.[1]

The literature of anthropology abounds with descriptive and analytical accounts of body modification among humans, but almost all of it emanates from people living or who had lived in the non–Western traditional societies of the world. From Mayan tongue piercing to Mandan flesh skewering, Ubangi lip stretching to Tiv scarification, there is a vast and incredibly varied body of literature that seeks to explain it all—anthropologically, psychologically, sociologically, and biologically. Curiously, very little research has been

From "Nonmainstream Body Modification: Genital Piercing, Branding, Burning, and Cutting," James Myers, *Journal of Contemporary Ethnography,* Vol. 21, No. 3, 1992. Reprinted by permission of Sage Publications, Inc.

done on contemporary, nonmainstream American body modification. When one considers the huge amount of literature devoted to the subject among traditional non-Western peoples, this paucity of data becomes glaringly evident. For example, one of the best recent sources on body modification is Rubin's (1988) *Marks of Civilization,* but even in this excellent publication, most of the articles deal with tattoos and cicatrization and none are devoted to such contemporary Euro-American practices as multiple piercing, scarification, cutting, and branding.

That a tattoo renaissance has been occurring in the United States since the late 1960s is now quite evident in the popular media and to a growing extent in scholarly publications and papers presented at professional conferences (see especially Govenar 1977; Rubin 1988; Sanders 1986, 1988a, 1988b, 1989; St. Clair and Govenar 1981).[2] This void in the literature is probably due more to a simple lack of awareness of the practice than it is a lack of interest, as the population of people involved in multiple piercing, scarification, branding, and cutting is minuscule compared to tattooing. In addition, because the modification and/or jewelry involved typically creates even greater revulsion in the general public's eye than tattoos, much of the work is kept secret among recipients and their intimates.

My observations and conclusions regarding nonmainstream body modification run counter to the general public's assessment that people so involved are psychological misfits bent on disfigurement and self-mutilation. None of the people I interviewed, however deep and varied their involvement in body alteration, fit the standard medical models for "self-mutilation." . . .

METHOD AND POPULATION

My original plan was to concentrate my research efforts on tattooing, but 4 months into the 24-month study period, I shifted my focus almost entirely to piercing, cutting, burning, and branding. The change was brought about by my increasing awareness of the growing popularity of nonmainstream modification other than tattoos and the realization that research on the subject was scant. I was also intrigued by the deep feelings of revulsion and resentment held by mainstream American society against these forms of body modification.[3]

Using participant observation and interviews as primary data-gathering techniques, I involved myself in six workshops organized especially for the San Francisco SM (sadomasochist) community by Powerhouse (fictitious name), a San Francisco Bay Area SM organization.[4] Tattoo and piercing studios were also a rich source of data, as was the 5th Annual Living in Leather Convention held in Portland, Oregon in October 1990. I gathered additional data from a small but dedicated group of nonmainstream body modifiers at my university and the city in which it is located. Interviews with several medical specialists and an examination of pertinent medical literature provided an important perspective, as did solicited and unsolicited commentary from hundreds of mainstream society individuals who viewed my body modification slides and/or heard me lecture on the topic. Finally, chance encounters with devotees served to broaden my awareness and understanding of the various forms of nonmainstream body modification.

Entree to the workshops was of paramount importance to the study; thus early in the fieldwork, I contacted the primary organizer of Powerhouse and introduced myself as a straight, male anthropologist interested in attending the workshops in order to gather data on nonmainstream body modification for use in my university classroom and publication in a scholarly journal. Her response was immediate:

> Good God, yes! You're welcome to come. We need people to see that just because we're kinky doesn't mean we're crazy, too. You'll see people here with all kinds of sexual interests. We learn from each other and have a heckuva lot of fun while we're at it.

As is true for most ethnographic participant observation situations, the largest amount of my data from the workshops were gathered from observation. I participated in the true sense of the word on two occasions, once during a play piercing demonstration and again during a playing with fire demonstration.[5] The rest of my participation involved such typical "interested involvement" as mingling, asking questions as an audience member, introducing myself around, helping arrange chairs, setting up demonstration paraphernalia, taking photographs, conducting interviews, and generally lending a hand whenever possible. At the Living in Leather Convention in Portland, I was able to expand my involvement by showing my body modification slides to several people, attending parties, and helping out at the host organization's hospitality suite.

The population of body modifiers in my study included males and females, heterosexuals and homosexuals (lesbians and gays), bisexuals, and SMers. It is important to note that the single largest group was composed of SM homosexuals and bisexuals. Although this skewing likely resulted from my extended contact with the Powerhouse workshops and several SM body modifiers whom I interviewed at the Living in Leather Convention in Portland, it is supported by a 1985 piercing profile of subscribers to *Piercing Fans International Quarterly* (Nichols 1985). The survey determined that 37% of the group was gay, 15% bisexual, and 57% involved in dominant-submissive play, a keystone of SM activity. Also of interest from the *PFIQ* profile, 83% had attended college, 24% had college degrees, and 33% had undertaken postgraduate study. Caucasians represented 93% of the survey.

Like any fieldwork, this research had its pleasant and difficult aspects. On the positive side was the subject matter itself. Body modification is inherently fascinating to human beings. In addition, there was the relative ease with which I was able to gather empirical data on the topic. The people I interviewed and observed were for the most part barely subdued exhibitionists who took joy in displaying and discussing their body and its alterations. This was especially true when a group was together and a sense of trust pervaded the room. On such occasions, an exuberant "show and tell" was the order of the day. To field-workers accustomed to tight-lipped, monosyllabic responses and other forms of "informant lockjaw" from people they are studying, and who have been advised on occasion what they could do with their camera, it should be understandable why it was a pleasure to work with this uninhibited, communicative population.[6]

Such rapport presupposes that an element of trust has been achieved between the field-worker and the individuals or group being studied. Many people whom I interviewed were keenly aware that because mainstream society regarded them as deviants, there was a high probability that harm to themselves or their life-style was never far away. Thus interaction between the field-worker and the individuals being studied must occur early to establish the trust necessary to conduct a worthwhile study. . . .

BODY MODIFICATION WORKSHOPS

Most of the ethnographic data in my study were derived from the body modification SM workshops I observed and the contacts I made while in attendance. These workshops were part of a series of continuing programs sponsored by Powerhouse and were designed to "enhance the SM experience."[7] Taught by individuals who were regarded as professional practitioners of various nonmainstream body modifications, the workshops were limited to a top enrollment of 50 people. The six workshops I attended were on male piercing, female piercing, branding and burning, cutting, play piercing, and playing with fire.[8] The audience at each workshop was markedly homogeneous. With the exception of myself and perhaps a half-dozen others, each session was typically attended by SM-oriented lesbians, gays, and bisexuals. Participants ranged in age from their late teens to their late 50s, with most attendees in their mid-20s and 30s. Leather was predominant—jackets, trousers, skirts, chaps, trucker's caps, gloves, boots, arm bands, wrist bands, and gauntlets. Heavily laden key rings, hunting knives in leather scabbards, slave collars, and T-shirts with sexual preference messages were also omnipresent. Tattoos, lip and nasal septum piercings, and multiple pierced ears were quickly visible, whereas more intimate piercings, such as nipple, navel, and genital would become evident as the workshops proceeded. It is fair to say that the groups would have attracted some attention were they to have gathered in a suburban shopping mall.

The four workshops described here were held on Saturday afternoons in an upstairs room of a liberal church in San Francisco.[9] Two other workshops I attended but do not describe in this article were conducted in a small room above a popular San Francisco gay bar.

Male Piercing

The first workshop in the series was on male piercing. Jim Ward, the teacher, was the president of Gauntlet, Inc., one of the few firms in the world that manufactures nonmainstream piercewear. Recognized as a "master piercer," Ward has been piercing since the mid-1970s and has estimated that he has done 15,000 piercings in the 14 years between 1975 and 1989. He is also the editor and publisher of *Piercing Fans International Quarterly (PFIQ),* a successful glossy publication devoted exclusively to the subject of piercing. . . .

Ward arrived at the workshop early to set up his piercing equipment and a massage table that would serve as a piercing couch. He was wearing Levi's, a studded belt, black boots, and a black T-shirt that had the logo "Modern Primitives" (see Vale and Juno 1989) printed above 12 white-bordered rectangles, each of which contained a graphic drawing of one of the most popular genital piercings. Ward's lover and assistant set out several jewelry display cases and arranged chairs for the audience. He had multiple ear piercings, a bonelike tusk in his nasal septum, and a Gauntlet button on his T-shirt that proclaimed "We've got what it takes to fill your hole."

Ward's popularity and fame were evident as several arrivees paid their respect by shaking his hand or hugging and kissing him. Even though the workshop was on male piercing, one third of the audience was women, a crossing-over evident at each of the workshops regardless of the gender-specific body modification being highlighted. Ward welcomed the group, confirmed that his prearranged volunteers were present, and began his discussion of male piercing. It was evident that he had been through the routine many times, which he had, both before live audiences and in his continuing series in *PFIQ,* "Piercing with a Pro." His presentation was divided into halves, the first of which was a general discussion of the topic, or as he said, "the ins and outs of piercings," and the second consisting of actual demonstrations. As was true of each of the Powerhouse workshops, there was much emphasis on safety, cleanliness, sterilization, and proper hygiene after the procedure. Assuming that most of his audience was already involved in or at least aware of piercing, Ward dispensed without definition such esoteric piercing terminology as ampallang, dydoe, frenum, Prince Albert, guiche, and so on. Questions were asked about autoclave temperatures, rubber gloves, anesthetics, antiseptics, play piercing versus permanent piercing, the dangers of AIDS and hepatitis, body rejection, jewelry selection, and the like.

During the break, I asked Ward about his own piercings:

> Well, you can see them in my ear lobe and tragus, but I also have a Prince Albert in my cock and a nipple ring on each tit. Oh, I've got a guiche with a piece of cord in it, too. I'm wearing all that stuff right now. I've been piercing myself for 20 years, but I don't wear jewelry in most of the holes. I travel all over the country, and I can tell you it's a real mind-fuck to get on an airplane and sit next to some hunk knowing you've got all this sexy stuff on.

I also talked with an audience member who was not interested in getting pierced but wanted to see what the attraction was for his pierced friends:

> I don't feel the need to get pierced. Actually, I'm deathly afraid of needles. I don't think I have to look like a pin cushion in order to look sexy. When I'm out cruising I might stuff my balls through a coupla cock rings. Gives me a great feeling and enough basket to turn a few heads. Best of all, no artificial holes in my body to get infected.

Ward's first volunteer after the break was a leather-clad male who wanted his left nipple repierced. He sat shirtless on the table as his companion offered him a reassuring hug. Ward examined the nipple and told the group the scar tissue from the previous piercing would make this one more difficult. He also took advantage of the audience's concern to note the difference between pain and sensation in piercing and that he preferred the latter term to best describe the feeling. The volunteer's facial expression gave the impression that he had some doubts about Ward's evaluation. Before starting the piercing, Ward summoned his second volunteer and explained to the group that before he did the nipple job he needed to prep Number 2 for his forthcoming Prince Albert, a procedure that requires the application of a local anesthetic because the needle pierces the urethra, a particularly sensitive area. Number 2 dropped his leather trousers to his ankles, and Ward casually tamped a xylacane-coated cotton swab into the urethra about 1 inch. A male in the audience teased, "I bet he wishes there wasn't any anesthetic on that Q-tip®." Laughter. Ward directed the volunteer to step over to one side of the room and wait for the anesthetic to numb the area. The volunteer, leather trousers still at his ankles and undershorts dropped below his knees, hopped over to the wall where he waited patiently, with the Q-tip® jauntily protruding from the tip of his penis.

Ward returned to the first volunteer and spent several moments discussing different types of male nipples and the particular piercing technique warranted by each. Then he scrubbed the volunteer's nipple with Hibiclens and Betadine, marked each side of the nipple with a dot to guide the needle path, clamped a Pennington forceps on the nipple to keep it from retracting and to afford better manageability, and expertly pushed a needle through the guide dots. An audible sharp gasp and a rigid tensing of the volunteer's body confirmed Ward's earlier comment about the likelihood of tougher tissue in repiercings. There was an immediate sigh of relief from the group accompanied by applause and congratulatory whoops. One end of the jewelry was used to push the needle the rest of the way through the nipple, thus resulting in the needle being expulsed and the jewelry attached in one continuous movement. The entire procedure had taken less than 3 minutes.

The second volunteer was invited back to the table, Q-tip® still in place as he waddled across the room. Ward had him sit on the table, then decided that it would be better if he stood on the table. There was some concern in the group about this stance, as the volunteer was visibly trembling, a circumstance that was all the more worrisome because the table itself began to shake. It was not clear whether the bare-legged volunteer was simply cold or whether he was suffering from pre-op jitters. Nevertheless, Number 2 balanced precariously atop the uncertain table while Ward, who had now gained an eye-level view of his work site, examined the about-to-be-pierced penis with his eyes and his fingers. As he worked, Ward maintained a running commentary on the history of the Prince Albert, noting that "it was originally designed to tether the penis to either the right or left pant leg for a neater looking appearance, but now it's strictly erotic." After completing the usual prepping around the piercing area, he deftly pushed the needle into the underside of the penis just behind the head, into the urethra, up toward the tip of the penis and the still lodged Q-tip®. As he pushed, the Q-tip® suddenly popped out of the urethra—"a sure sign I'm on course"— followed by the tip of the gleaming needle. The volunteer gazed warily down at the sight while being steadied by a friend. Applause and cheers. The jewelry was attached and Number 2 was eased down from the table. Still shaking, he pulled up his shorts and trousers. Ward peeled off his rubber gloves and disposed of them while discussing his thoughts on abstinence during the healing process.

The third volunteer was to receive a dydoe, a piercing that would pass through both sides of the upper edge of the glans. This volunteer, in his late 40s, removed his trousers and undershorts and stretched out calmly on the table. With more than 2 hours of discussion and demonstrations behind him, Ward was now much quieter. The usual preliminaries were undertaken while the volunteer chattered about his piercing history. The group was only mildly interested in the disclosures, but full attention resumed when Ward began the actual piercing. As with the first two volunteers, this one emitted a controlled but audible gasp, then relaxed. The jewelry was attached and the volunteer hopped off the table and dressed while the group applauded. Later, this person expressed his feelings about piercing to me:

> I like the jewelry very much, but the real turn-on comes from having my body penetrated. Everytime I see that sharp, shiny needle heading towards my flesh I know I'm going to get either a dick orgasm or a head orgasm or maybe both.[10]

Several of the audience congratulated the new piercees and expressed their appreciation to Jim Ward. The first workshop was over.

Branding and Burning

The second workshop of the series was devoted to branding and burning and was taught by Fakir Musafer. Musafer, who pierced his own penis at age 13, was 58 years old at the time of the workshop. Recognized by many as the doyen of "modern primitivism" in the United States, there is little in the practice of body modification and "body play" that he has not experienced on his own body. He has been tattooed, burned, cut, pierced, skewered, and electrically shocked. He has fasted, deprived himself of sleep, rolled on beds of thorns, and constricted and compressed various parts of his body with belts and corsets. He has also gilded, flagellated, punctured, and manacled himself. In addition to frequently suspending himself with fleshhooks à la the Sun Dance, he has reclined on a bed of nails or blades, "negated" his scrotum and penis by sealing them in plaster, and conversely "enhanced" the same by such practices as scrotum enlargement and penis elongation (a procedure accomplished by regular stretch workouts with 3-pound weights, or, what he refers to as "rock on a cock" exercises). His fame continues to spread in the United States and abroad through his continuing involvement with Jim Ward's *PFIQ* and the release of "Dances Sacred and Profane," a widely distributed videotape that highlights his

Sun Dance ritual. Musafer undoubtedly represents the distant end on any scale of contemporary nonmainstream body modifiers, the great percentage of whom are content with their tattoos and/or multiple piercings.

Musafer arrived 30 minutes before the start of the workshop. He was wearing a black T-shirt and baggy khaki cotton trousers with large flapped mid-leg pockets. Puffing on a cigarette, his hair dyed sable brown, and bereft of any visible piercing jewelry, he looked like any other middle-aged ad executive enjoying his weekend. I volunteered to help him unload his van. The contents of the box I carried up the stairs vaguely hinted of his workshop's topic— acetylene torch, metal snips, needle-nose pliers, wire, matches, incense sticks, candles, several strips of copper and tin, mirror, two potatoes, and various other oddities.

Like Jim Ward at the previous workshop, Musafer was hugged and kissed by many arrivees. And, as in all the workshops, a spirit of *bonhomie* prevailed as arriving couples and singles of both sexes hugged and kissed acquaintances, chatted amiably with others, and shared their latest body art. A few moments before the starting time, Musafer shed his T-shirt, revealing a small ring in each nipple. He removed the rings and deftly replaced them with hollow metal tubes, each, according to his admission, 7/8 of an inch in diameter by 1 inch in length. Then he quickly inserted white teflon tubes through the holes on each side of his chest that had been created several years earlier for his Sun Dance hooks. Each tube was about the size of a king-sized cigarette and ran vertically through the flesh behind each nipple. Finally he reached into a pocket and pulled out a large nasal septum ring, which, with the aid of an audience member, was installed in his nose. He pulled his T-shirt back on and commented, "A-h-h, that's more like it!" Now, with open arms he welcomed everyone.

"The Fakir," as he frequently refers to himself, is an old pro at this sort of presentation and glibly but knowledgeably discussed his first topic of the day: branding. With the aid of an easel and predrawn charts, we learned about technique ("Don't go too deep, you're not doing a 'Mighty Dog' brand"), patterns ("The simpler the better"), important reminders ("Remember, each mark in the final scar will be two to four times thicker than the original imprint"), desirable locations ("The flatter the surface the better. Try to stick with the chest, back, tummy, thighs, butt, leg, upper arm"), and tools and materials and where to get them.

A prearranged volunteer indicated a preference for a 2-inch skull to be branded on the calf of her left leg. She reclined on the table with her skirt pulled up over her knees and her left leg extended toward Musafer's work site. A previously drawn pattern was transferred onto her left calf. As we watched, he fashioned the skull shape out of a strip of metal cut from a coffee can and showed us how to heat and apply the brand. Holding the brand in his needle-nosed pliers, he heated it in the acetylene torch until it was red hot, then quickly applied it to a piece of cardboard to check the design's appearance. Musafer also cut a potato in half and applied the reheated brand to the cut surface, noting that this was a good way for beginners to practice depth control before doing the real thing on human skin. The volunteer, a professional piercer and cutter in her early 30s, laughed with the audience as the potato hissed and smoked from Musafer's strike. "The Fakir" was now ready and alerted his client. The brand was heated, precisely positioned over the desired part of the pattern, and struck. There was an immediate hiss and a crackling sound, followed by a wisp of smoke and the odor of scorched flesh. The volunteer scarcely twitched. Musafer examined his first strike and proceeded to do six more to finish the skull, complete with stylized eyes, mouth, and teeth. The only time the volunteer reacted to the hot brand, and intensively so, was when Musafer inadvertently brushed the edge of her left foot with a "cooled" brand as he returned it to the torch for renewed heating. Musafer rubbed some Vaseline on the brand and the foot burn, and the volunteer sat up and put her low-cut boot back on. The audience applauded.

Musafer's next demonstration was on burning.[11] Displaying a row of seven or eight self-imposed circular burns on the front of his right thigh, each about the size of a penny and resembling inoculation scars, he discussed different types of burnings ("Cigarette burns are nasty but nice"), and techniques to cause them ("I prefer incense sticks because they work real well and smell so delightful during the burn"). There was no prearranged volunteer for burning, but a woman in her early 20s volunteered from the audience. Her friends cheered her as she removed her Levi's and sat on the table. Musafer touched her gently on the legs and softly said, "You are giving your flesh to the gods." He also instructed her on the importance of deep breathing and visualization as he glued a 3-inch length of incense stick to her left thigh and lit it. Within 4 or 5 minutes, the stick had burned down to skin level and extinguished itself. Although the stick was slightly less than a pencil in diameter, the circular burn mark quickly expanded to the penny sized marks I had viewed earlier on Musafer's thigh. Throughout the burning, the volunteer kept her eyes tightly closed and followed Musafer's instructions on deep breathing. She now opened her eyes and the audience applauded. Several of the audience members came up to where she was sitting and examined the burn, while one of her friends asked Musafer for some extra incense sticks to take home. Musafer doled out some sticks and jokingly admonished, "Watch it, this stuff is catching!" Many chuckles. Musafer responded to several last-minute questions while packing up his paraphernalia. This workshop had ended.

Female Piercing

The female piercing workshop was taught by Raelynn Gallina, a woman recognized throughout the Bay Area as a professional "total body" piercer. In addition to her popularity as a piercer and cutter (some of her clients glowingly refer to has as "Queen of the Blood Sports"), Raelynn is a successful designer and manufacturer of jewelry. Although she pierces males ("above the navel only"), the greatest proportion of her clientele is female, most of whom are, like herself, lesbian. She had been piercing for 7 years at the time of the workshop. Two thirds of the workshop's audience of 45 people were women, all but a few of whom were lesbians involved in sadomasochism. Most of the men present were gay SMers. Raelynn announced that she had four volunteers for the afternoon and would pierce a nipple, a clitoris hood, an inner labia, and a nasal septum.

The first half of the workshop was devoted to a "do it yourself" clinic on technique, tools, antiseptics, and various do's and don'ts and ended with the admonition that it was safer and better to be pierced by a pro:

> I get a lot of people who want me to fix up their bungled piercings. Usually turns out they were heavy into a torrid scene when someone says, "Heh! Wouldn't it be hot if we pierced each other!" All I'm saying is if you do get carried away, make sure you know what you're doing.

After the break, Raelynn's spiritual-psychological bent revealed itself as she emphasized the importance of centering, grounding, visualization, client-practitioner compatibility, and the relationship between individual personality and type of piercing jewelry to be worn. She later commented to me:

> Piercing is really a rite of passage. Maybe a woman is an incest victim and wants to reclaim her body. Maybe she just wants to validate some important time in her life. That's why I like to have a ceremony to go along with my piercing and why I do it in a temple—my home. Most of my clientele are bright, sensitive women. It's not as if a bunch of "diesel dykes" are busting into my place to prove how tough they are by getting their boobs punched through with needles.

Raelynn's first volunteer was an achondroplastic dwarf in her 30s. Dressed in leather trousers and field boots, it was obvious that she was extremely popular with many in the group. At least three different women lifted her up and danced merrily around with her in their arms during the break, while numerous others bent down to kiss her on the cheeks or lips. As Raelynn described the nipple piercing she was about to perform, the volunteer peeled off her blouse and climbed up on a long-legged director's chair. Clearly visible across her left breast in what appeared to be a recent cutting were the words "The bottom from Hell." There was much laughter, expressions of encouragement, and joking about anticipated pain, needles, second doubts, and the like. Raelynn scrubbed the volunteer's left breast and nipple with Hibiclens, applied a good coating of Betadine, and clamped a Pennington forcep on the nipple. The exact penetration and exit points for the needle were marked with a pen, and Raelynn quickly and expertly forced a needle through the nipple and into a small cork at the exit point. The forceps were removed, and a gold ring was inserted in the place of the needle. The only sign of pain from the volunteer was a short gasp as the needle pierced the nipple. As in the previous piercing procedures, there was a collective sigh of relief from the group, followed by applause and various congratulatory remarks. The volunteer climbed down from her perch and hugged Raelynn around the hips. Raelynn scooped her up, returned the hug, and kissed her.

By the time the first volunteer had put her blouse back on, the second volunteer had already removed her jeans and underpants and was sitting on the chair. In her mid 20s, this volunteer would receive a clitoris hood piercing and jewelry. During most of the procedure, she held a hand mirror over her pubic area to better monitor the procedure. Raelynn advised the client to close her eyes and visualize the process. The piercing and jewelry attachment was completed within 3 minutes, again with the client showing minimal reaction to the actual piercing. As the volunteer pulled her tight jeans on, Raelynn reminded the group that it was important to wear loose-fitting clothes when getting a genital piercing.

The third volunteer, in her mid-20s, had spent the first half of the workshop curled up in the laps of three different women. She removed her cotton skirt and hopped onto the chair. Pantyless and clean-shaven, two labia rings and a clitoris hood ring were easily visible. Raelynn announced that this client would be getting a third labia ring today and with a theatrical leer, added, "And she has asked me to do it real-l-l slow and with a twist." The audience responded with mock moaning and various teasing expressions. While Raelynn talked about genital piercing in general, the client sat spread-legged and observed with the hand mirror her present labia piercings and jewelry. Raelynn noted that a clamp was not usually needed for labia piercings and started the procedure. An audible intake of air and a slight tensing of the body were the only signs that the needle had pierced the flesh. The jewelry was quickly attached and the usual applause delivered.

I was unable to remain for the last piercing of the session, a nasal septum procedure through the nose of the first volunteer.

Cutting

Raelynn was also the teacher for the workshop on cutting. She had arrived in the room an hour before the scheduled starting time to set up her equipment and a videocamera. As people entered the room, she greeted them, occasionally examining a piercing or cutting that she had apparently performed at a previous occasion. By the starting time, 43 people had arrived and there was the usual happy buzz of chatter. Three fourths of the group were women, most of whom were wearing leather. As in the other workshops, the majority of the group members were gay and lesbian. Couples held hands, snuggled, kissed, and engaged in animated conversations. Raelynn sat on a table with knees crossed and officially greeted everyone. Although she welcomed the group by saying "Hello fellow blood sluts" and there was a button attached to her equipment case that read "I'm hungry for your blood," she quickly stated she does her cutting for aesthetic reasons and not just for the joy of blood. Some of the audience responded in unison, "O-h-h, su-u-re!" She told the group she had been cutting for approximately 8 years and that she got her start while "caught up in some heavy SM scenes."

During the 2 1/2 hour workshop, she discussed where cuttings should be done on the body ("Fleshy areas like the butt, thighs, back—not on the neck, joints, places where there are veins"), cleanliness ("Cutting is a clean procedure, not a sterile one"), use of rubber gloves, and concern about AIDS and hepatitis, depth of cut, design, tools, and various other bits of information regarding her subject. She also distinguished between her style of cutting and the types of scarification and cicatrization done in several preliterate populations of the world.

After a short break, Rosie, a prearranged volunteer in her 40s, stripped to the waist to receive her cutting. She and Raelynn had decided earlier on a design that consisted of a pattern of stylized animal scales in a triangular shape. The design was large and would be cut into the upper left area of the back near the shoulder blade. Raelynn scrubbed Rosie's back with the usual antiseptics, dried it off, and covered the area with stick deodorant to facilitate the transfer of the design. Using a No. 15 disposable scalpel ("Toss it after it's been used"), she started her first incision. As she cut, she explained that cutting was not really a painful procedure because of the sharp scalpel and the shallow cuts. Someone in the group wondered aloud, "Is it bleeding yet?" to which someone else reported, "I certainly hope so!" Raelynn also urged would-be cutters to remember to start cutting at the bottom of the design so that the dripping blood would not wash away the uncut design. Both the cutter and the client were obviously moved by the procedure. Daubing away some blood, Raelynn told the group, "Once you start cutting someone, you get a very high, heady experience." Rosie, her eyes closed and mouth sensuously open, emitted several soft sighs during the 10-minute procedure. At one point, Rosie squeezed her companion's hand and whispered, "This is intense, wonderfully intense."

When the cutting was completed, Raelynn blotted the design several times to soak up the still bleeding incisions. Rosie was alerted to brace herself for the alcohol rinsings, which were done several times over the cutting. A towel around Rosie's waist kept the alcohol from dribbling further down her body. Raelynn then ignited a fresh rinsing of alcohol with her cigarette lighter. A loud poof was heard, and a bluish flame danced across the entire left side of Rosie's back. The flame was quickly doused as Raelynn announced, "Rosie asked me to do that because she's into fire." Referred to as "slash and burn" by Raelynn, the fire event was repeated two more times as the audience oohed and aahed. Finally, Raelynn rubbed black ink over the entire design and the wound was covered with a protective surgical wrap. In a few days, the excess ink would be scrubbed away, leaving the lines of the cutting colored black.

Kay, the second prearranged volunteer, wanted the fish cutting already on her back touched up. The original cutting had been done by Raelynn 9 months earlier, and although the design was still easily discernible, Kay liked the idea of having it redone. Raelynn prepared the area, unpackaged a new scalpel, and recut the design in less than 10 minutes. No ink was rubbed into the wound as Kay preferred the natural look.

Raelynn ended the workshop with a discussion of different body reactions to cutting, explaining that some people scarred nicely, well enough that there was no evidence of any cutting, whereas some keloided into large amorphous bumps.

While people were socializing after the session, I talked with a woman I recognized from earlier workshops. She told me Raelynn had pierced her labia, navel, and both nipples. She also compared Fakir Musafer and Jim Ward unfavorably with Raelynn:

> Those guys are out on the edge! They're an embarrassment. I mean,
> bones through the noses, the branding, the fleshhooks, the pain. You

heard them—"If a client is in pain, you just keep on pushing and jabbing. It'll be over before you know it." Raelynn is great because she is gentle and looks for any special aspects of your personality that will help her do a better piercing or cutting. . . .

CONCLUSIONS

Taken as a whole, the responses from my informants portray a group of individuals who for a variety of reasons enthusiastically involve themselves in nonmainstream body modification. They readily admit that their body modification interests are statistically outside the average range, but none transfer this conclusion to a statement regarding a deficiency in mental health. The medical literature on the topic presents a picture of deeply disturbed individuals engaging in self-mutilation for various psychopathological reasons (for an understanding of the medical interpretation of self-mutilation, see American Psychiatric Association 1987; Eckert 1977; Greilsheimer and Grover 1979; Pao 1969; Phillips and Muzaffer 1961; Tsunenari et al. 1981). This view is supported by the general nonparticipating public. My empirical observations lead me to disagree with the latter assessment. The overwhelming number of people in my study appear to be remarkedly conventional sane individuals. Informed, educated, and employed in good jobs, they are functional and successful by social standards. . . .

The number of contemporary Americans who have become involved with nonmainstream body modification is presently small. However, it is important to remember that the practices discussed in this article are a relatively new phenomenon in this culture. Each year, American society is bombarded with new body alterations, many of which are quickly assessed as unacceptable for one reason or another and fail to enter mainstream society. However, recent history also shows that some initially rejected alterations may take hold in a subculture and eventually catapult their way into the larger society. For example, ear piercing in America moved from nonmainstream to mainstream society in less than a decade and multiple ear piercing among both males and females is now relatively common. Lip and nose piercing is increasingly tolerated, but whether nipple piercing will follow suit remains to be seen.

A growing number of people in American culture believe that the penis and the clitoris are just as deserving of gilding as are earlobes. These individuals, like the style setters in earlier times who defied American society's strictures on body alteration by experimenting with such daring embellishments as lipstick, rouge, painted nails, eye makeup, and radical hairstyles, join human beings around the world in using their bodies to express a symbolic language that reveals their sentiments, dispositions, and desired alliances. Through adornment, the naked skin moves one from the biological world to the cultural world. As Claude Lévi-Strauss observed in Vale and Juno's (1989) book,

> The unmarked body is a raw, inarticulate, mute body. It is only when the body acquires the "Marks of Civilization" that it begins to communicate and becomes an active part of the social body. (p. 158)

NOTES

1. Of course, most so-called "irreversible" body modifications are not truly so. Tattoos fade, scars may flatten, cutting may heal without scars, and piercings, if not tended, will close.

2. A *Newsweek* magazine article (January 7, 1991) recognized the current popularity of tattooing in the United States and noted, "It's the most painful trend since whalebone corsets: tattooing, the art of the primitive and the outlaw, has been moving steadily into the fashion mainstream." This same article also observed that in the past 20 years, the number of professional tattoo studios had jumped from 300 to 4,000.

3. Interestingly, today many tattooed people regard piercing, branding, and scarification as repugnant. For example, the following warning was displayed prominently on the wall of a Northern California tattoo studio (it was apparently part of a registration form for a 1982 national tattoo convention): "This convention is for Tattoo Artists and Fans who care about the Tattoo Profession. Anyone breaking the following rules will be asked to leave with no refunds. Facial tattoos other than cosmetic (eyebrows, lines, etc.) not permitted. Piercing of the private parts of the anatomy not permitted to be shown at any time. Any facial piercing with bones, chains, etc. must be removed during entire convention."

4. I was accompanied at each workshop by Craig Moro, a Berkeley resident who had served several months as a volunteer for the San Francisco Sexual Information Switchboard (SIS), a call-in telephone service for people seeking sexual information.

5. The first occasion was a workshop on play piercing, a procedure that involves brief piercing with hypodermic or sewing needles, fishhooks, staples, and the like for fun and enjoyment and, unlike "permanent" piercing, is not done with the intention of installing some type of jewelry in the hole. After a lengthy introduction to the techniques, hygiene, and materials involved, the workshop leader divided audience volunteers into piercers and piercees. As a piercer, I selected an experienced partner to pierce, donned my rubber gloves, popped a hypodermic needle from its protective capsule using the recommended technique, and was within inches of making the jab when it suddenly occurred to me that I had some serious doubts about what I was doing. Even though the teacher was exquisitely clear on the need for care and safety during the "scene," my congenital clumsiness and concern about the blood being splattered here and there caused me to back out as gracefully as possible at that late instant. Another volunteer happily replaced me, and I don't believe my rapport was damaged by the event.

The second participation occurred during the workshop on playing with fire. Here it was simply a matter of overcoming my innate fear of being burned and joining the group of eager volunteers brushing each other with a lighted small torch soaked in a 70% solution of isopropyl alcohol. To my surprise, the activity was enjoyable and caused some of my fellow volunteers to wonder if I was considering changing my sexual proclivities.

6. The intellect of many SMers whom I interviewed and observed during my fieldwork was confirmed to me at one San Francisco workshop entitled "Playing with Fire." During the workshop, the instructor's knowledgeable discourse on the ignition points of various isopropyl alcohols was interrupted by several audience members who had extemporaneously launched into an animated conversation on such matters as flammability versus combustionability, chemical structures of alcohol and gasoline, and the medical definitions and implications of the various types of skin burns. After listening to the display a few moments, the Instructor broke in by expressing her astonishment at the esoteric outpouring, only to be sharply reminded by one audience member, "There are no dumb SMers!"

7. Because SM practices are so varied, it is difficult to provide a satisfactory single definition of SM behavior. Charles Moser, a psychotherapist who specializes in SM clients and wrote his Ph.D. dissertation on

sadomasochism, uses five criteria to identify people involved in SM (in Truscott 1989): (1) appearance of dominance and submission, (2) role-playing, (3) consensuality, (4) sexual content, and (5) mutual definition (i.e., people involved recognize that what they are doing is different from the "norm"). Townsend (1983) suggested "a short list of characteristics" that he believed are present in most scenes that he would classify as SM: a dominant-submissive relationship, a giving and receiving of pain that is pleasurable to both parties, fantasy and role-playing, humiliation, fetish involvement, and the acting out of one or more ritualized interactions (bondage, flagellation, etc.).

SM does not have to involve pain. There are many people who prefer a gentler approach to what otherwise would be considered SM; thus one sees the letters "D and S" for "dominance and submission," or "B and D" for "bondage and discipline." For an extended discussion of the pain issue, see Baumeister (1988), Gebhard (1969), Reik (1957), and Weinberg (1987).

For a scholarly presentation of SM behavior, see Weinberg's (1987) review of recent sociological literature on sadomasochism in the United States. For a nonacademic but informative introduction to SM behavior, I recommend two popular books by insiders: Urban Aboriginals (Mains 1984) and The Leatherman's Handbook 2 (Townsend 1983). Because these sources are male oriented (very little academic work has appeared regarding female SM), the reader interested in female SM will find helpful the Sandmutopia Guardian and Dungeon Journal or Dungeonmaster magazine. Pat Califia has also written knowledgeably of the lesbian SM community (see Califia 1987).

Not surprising, the SM people I interviewed were infuriated with the standard psychological characterization of SM as aberrant behavior. For example, The DSM–III–R, the official diagnostic manual for the American Psychiatric Association, lists both sadism and masochism as psychosexual disorders.

8.   Other Powerhouse workshops scheduled during my fieldwork period were Creative Bondage, Electrical Toys, Clothespins and Staples, Male Tit and Genitorture, Tit Play, Cock and Ball Torture, Whipping and Caning, Mummification, and Equestrian Restraints.

9.   The church's liberal reputation was confirmed to me one sunny Saturday afternoon as our group huddled over a spread-legged woman undergoing a clitoris hood piercing to the accompaniment of an a cappella choir rehearsing Handel's "Hallelujah Chorus" in a downstairs room—a surrealistic scene that would not have escaped Van Gennep (1960).

10.   It is not unusual to hear SM people remark on whether a particular body modification or play technique would produce a genital orgasm or an equal thrill in the mind ("head orgasm") or both.

11.   A distinction may be made between two different types of burning that I witnessed during my fieldwork with the SM body modifiers. One type, as described in the Musafer burning demonstration, was intended to leave a mark. Another type, "play burning," capitalizes on the classic SM goals of trust and the threat of pain and/or injury, but does so without leaving intentional burn marks. The ritual use of fire in SM scenes is not uncommon. The cross-cultural use of fire as a means of "cooking" a person symbolically was discussed by Lévi Strauss (1970). See also Tonkinson (1978) on the use of fire on Mardudjara initiates and Warner (1964) on fire jumping among the Ngoni.

REFERENCES

American Psychiatric Association. 1987. *Diagnostic and statistical manual of mental disorders*. Washington, DC: American Psychiatric Association.

Baumeister, R. 1988. Masochism as escape from self. *Journal of Sex Research* 25: 29.

Califia, P. 1987. A personal view of the history of the lesbian community and movement in San Francisco. In *Samois: Coming to power* (pp. 243–87). Boston, MA: Alyson.

Eckert, G. 1977. The pathology of self-mutilation and destructive acts: A forensic study and review. *Journal of Forensic Sciences* 22: 54.

Gebhard, P. 1969. Fetishism and sadomasochism. In *Dynamics of deviant sexuality,* edited by J. H. Masserman (pp. 71–80). New York: Grune & Stratton.

Govenar, A. 1977. The acquisition of tattooing competence: An introduction. *Folklore Annual of the University Folklore Association* 7 and 8.

Greilsheimer, H., and J. Grover. 1979. Male genital self-mutilation. *Archives of General Psychiatry* 36: 441.

Lévi-Strauss, C. 1970. *The raw and the cooked*. New York: Harper.

Mains, G. 1984. *Urban aboriginals*. San Francisco: Gay Sunshine Press.

Nichols, M. 1985. The piercing profile evaluated. *Piercing Fans International Quarterly* 24: 14–15.

Pao, P. 1969. The syndrome of delicate self-cutting. *British Journal of Medical Psychiatry* 42: 195.

Phillips, R., and A. Muzaffer. 1961. Aspects of self-mutilation in the population of a large psychiatric hospital. *Psychiatric Quarterly* 35: 421.

Reik, T. 1957. *Masochism in modern man*. New York: Grove.

Rubin, A. 1988. *Marks of civilization*. Los Angeles: Museum of Cultural History.

Sanders, C. 1986. Tattooing as fine art and client work: The art work of Carl (Shotsie) Gorman. *Appearances* 12: 12–13.

———. 1998a. Drill and fill: Client choice, client typologies and interactional control in commercial tattoo settings. In *Marks of civilization,* edited by A. Rubin (pp. 219–31). Los Angeles: Museum of Cultural History.

———. 1988b. Marks of mischief: Becoming and being tattooed. *Journal of Contemporary Ethnography* 16: 395–432.

———. 1989. *Customizing the body: The art and culture of tattooing*. Philadelphia: Temple University Press.

St. Clair, L., and A. Govenar. 1981. *Stoney knows how: Life as a tattoo artist*. Lexington: University Press of Kentucky.

Thevos, M. 1984. *The painted body*. New York: Rizzoli.

Tonkinson, R. 1978. *The Mardudjara aborigines: Living the dream in Australia's desert*. New York: Holt, Rinehart & Winston.

Townsend, L. 1983. *The leatherman's handbook 2*. New York: Modernismo.

Truscott, C. 1989. Interview with a sexologist: Dr. Charles Moser. *Sandmutopia Guardian and Dungeon Journal* 4: 23–24.

Tsunenari, S., et al. 1981. Self-mutilation: Plastic spherules in penile skin in *yakuza,* Japan's racketeers. *American Journal of Forensic Medical Pathology* 2: 203.

Vale, V., and A. Juno. 1989. *Modern primitives*. San Francisco: Re/Search.

Van Gennep, A. 1960. *The rites of passage*. Chicago: University of Chicago Press.

Warner, W. L. 1964 *A black civilization: A study of an Australian tribe*. New York: Harper.

Weinberg, T. 1987. Sadomasochism in the United States: A review of recent sociological literature. *Journal of Sex Research* 23: 50–69.

40

Shifts and Oscillations in Deviant Careers

The Case of Upper-Level Drug Dealers and Smugglers

PATRICIA A. ADLER

AND PETER ADLER

The upper echelons of the marijuana and cocaine trade constitute a world which has never before been researched and analyzed by sociologists. Importing and distributing tons of marijuana and kilos of cocaine at a time, successful operators can earn upwards of a half million dollars per year. Their traffic in these so-called "soft"[1] drugs constitutes a potentially lucrative occupation, yet few participants manage to accumulate any substantial sums of money, and most people envision their involvement in drug trafficking as only temporary. In this study we focus on the career paths followed by members of one upper-level drug dealing and smuggling community. We discuss the various modes of entry into trafficking at these upper levels, contrasting these with entry into middle- and low-level trafficking. We then describe the pattern of shifts and oscillations these dealers and smugglers experience. Once they reach the top rungs of their occupation, they begin periodically quitting and re-entering the field, often changing their degree and type of involvement upon their return. Their careers, therefore, offer insights into the problems involved in leaving deviance.

Previous research on soft drug trafficking has only addressed the low and middle levels of this occupation, portraying people who purchase no more than 100 kilos of marijuana or single ounces of cocaine at a time (Anonymous, 1969; Atkyns and Hanneman, 1974; Blum et al., 1972; Carey, 1968; Goode, 1970; Langer, 1977; Lieb and Olson, 1976; Mouledoux, 1972; Waldorf et al., 1977). Of these, only Lieb and Olson (1976) have examined dealing and/or smuggling as an occupation, investigating participants' career

From "Shifts and Oscillations in Deviant Careers: The Case of Upper-Level Drug Dealers and Smugglers," Patricia A. Adler and Peter Adler, *Social Problems,* Vol. 31, No. 2, 1983. © 1983 Society for the Study of Social Problems. Reprinted by permission of University of California Press Journals.

developments. But their work, like several of the others, focuses on a population of student dealers who may have been too young to strive for and attain the upper levels of drug trafficking. Our study fills this gap at the top by describing and analyzing an elite community of upper-level dealers and smugglers and their careers.

We begin by describing where our research took place, the people and activities we studied, and the methods we used. Second, we outline the process of becoming a drug trafficker, from initial recruitment through learning the trade. Third, we look at the different types of upward mobility displayed by dealers and smugglers. Fourth, we examine the career shifts and oscillations which veteran dealers and smugglers display, outlining the multiple, conflicting forces which lure them both into and out of drug trafficking. We conclude by suggesting a variety of paths which dealers and smugglers pursue out of drug trafficking and discuss the problems inherent in leaving this deviant world.

SETTING AND METHOD

We based our study in "Southwest County," one section of a large metropolitan area in southwestern California near the Mexican border. Southwest County consisted of a handful of beach towns dotting the Pacific Ocean, a location offering a strategic advantage for wholesale drug trafficking.

Southwest County smugglers obtained their marijuana in Mexico by the ton and their cocaine in Colombia, Bolivia, and Peru, purchasing between 10 and 40 kilos at a time. These drugs were imported into the United States along a variety of land, sea, and air routes by organized smuggling crews. Southwest County dealers then purchased these products and either "middled" them directly to another buyer for a small but immediate profit of approximately $2 to $5 per kilo of marijuana and $5,000 per kilo of cocaine, or engaged in "straight dealing." As opposed to middling, straight dealing usually entailed adulterating the cocaine with such "cuts" as manitol, procaine, or inositol, and then dividing the marijuana and cocaine into smaller quantities to sell them to the next-lower level of dealers. Although dealers frequently varied the amounts they bought and sold, a hierarchy of transacting levels could be roughly discerned. "Wholesale" marijuana dealers bought directly from the smugglers, purchasing anywhere from 300 to 1,000 "bricks" (averaging a kilo in weight) at a time and selling in lots of 100 to 300 bricks. "Multi-kilo" dealers, while not the smugglers' first connections, also engaged in upper-level trafficking, buying between 100 to 300 bricks and selling them in 25 to 100 brick quantities. These were then purchased by middle-level dealers who filtered the marijuana through low-level and "ounce" dealers before it reached the ultimate consumer. Each time the marijuana changed hands its price increase was dependent on a number of factors: purchase cost; the distance it was transported (including such transportation costs as packaging, transportation equipment, and payments to employees); the amount of risk assumed; the quality of the marijuana; and the prevailing prices in each local drug market. Prices in the cocaine trade were much more predictable. After purchasing kilos of cocaine in South America for $10,000 each, smugglers sold them to Southwest County "pound" dealers in quantities of one to 10 kilos for $60,000 per kilo. These pound dealers usually cut the cocaine and sold pounds ($30,000) and half-pounds ($15,000) to "ounce" dealers, who in turn cut it again and sold ounces for $2,000 each to middle-level cocaine dealers known as "cut-ounce" dealers. In this fashion the drug was middled, dealt, divided and cut—sometimes as many as five or six times—until it was finally purchased by consumers as grams or half-grams.

Unlike low-level operators, the upper-level dealers and smugglers we studied pursued drug trafficking as a full-time occupation. If they were involved in other businesses, these were usually maintained to provide them with a legitimate front for security purposes. The profits to be made at the upper levels depended on an individual's style of operation, reliability, security, and the amount of product he or she consumed. About half of the 65 smugglers and dealers we observed were successful, some earning up to three-quarters of a million dollars per year.[2] The other half continually struggled in the business, either breaking even or losing money.

Although dealers' and smugglers' business activities varied, they clustered together for business and social relations, forming a moderately well-integrated community whose members pursued a "fast" lifestyle, which emphasized intensive partying, casual sex, extensive travel, abundant drug consumption, and lavish spending on consumer goods. The exact size of Southwest County's upper-level dealing and smuggling community was impossible to estimate due to the secrecy of its members. At these levels, the drug world was quite homogeneous. Participants were predominantly white, came from middle-class backgrounds, and had little previous criminal involvement. While the dealers' and smugglers' social world contained both men and women, most of the serious business was conducted by the men, ranging in age from 25 to 40 years old.

We gained entry to Southwest County's upper-level drug community largely by accident. We had become friendly with a group of our neighbors who turned out be heavily involved in smuggling marijuana. Opportunistically (Riemer, 1977), we seized the chance to gather data on this unexplored activity. Using key informants who helped us gain the trust of other members of the community, we drew upon snowball sampling techniques (Biernacki and Waldorf, 1981) and a combination of overt and covert roles to widen our network of contacts. We supplemented intensive participant–observation, between 1974 and 1980,[3] with unstructured, taped interviews. Throughout, we employed extensive measures to cross-check the reliability of our data, whenever possible (Douglas, 1976). In all, we were able to closely observe 65 dealers and smugglers as well as numerous other drug world members, including dealers' "old ladies" (girlfriends or wives), friends, and family members. . . .

SHIFTS AND OSCILLATIONS

. . . Despite the gratifications which dealers and smugglers originally derived from the easy money, material comfort, freedom, prestige, and power associated with their careers, 90 percent of those we observed decided, at some point, to quit the business. This stemmed, in part, from their initial perceptions of the career as temporary ("Hell, nobody wants to be a drug dealer all their life"). Adding to these early intentions was a process of rapid aging in the career: dealers and smugglers became increasingly aware of the restrictions and sacrifices their occupations required and tired of living the fugitive life. They thought about, talked about, and in many cases took steps toward getting out of the drug business. But as with entering, disengaging from drug trafficking was rarely an abrupt act (Lieb and Olson, 1976: 364). Instead, it more often resembled a series of transitions, or oscillations,[4] out of and back into the business. For once out of the drug world, dealers and smugglers were rarely successful in making it in the legitimate world because they failed to cut down on their extravagant lifestyle and drug consumption. Many abandoned their efforts to reform and returned to deviance, sometimes picking up where they left off and other times shifting to a new mode of operating. For example, some shifted from dealing cocaine to dealing marijuana, some dropped to a lower level of dealing, and others shifted their role within the same group of traffickers. This series of phase-outs and re-entries, combined with career shifts, endured for years, dominating the pattern of their remaining involvement with the business. But it also represented the method by which many eventually broke away from drug trafficking, for each phase-out had the potential to be an individual's final departure.

Aging in the Career

Once recruited and established in the drug world, dealers and smugglers entered into a middle phase of aging in the career. This phase was characterized by a progressive loss of enchantment with their occupation. While novice dealers and smugglers found that participation in the drug world brought them thrills and status, the novelty gradually faded. Initial feelings of exhilaration and awe began to dull as individuals became increasingly jaded. This was the result of both an extended exposure to the mundane, everyday business aspects of drug trafficking and to an exorbitant consumption of drugs (especially cocaine). One smuggler described how he eventually came to feel:

> It was fun, those three or four years. I never worried about money or anything. But after awhile it got real boring. There was no feeling or emotion or anything about it. I wasn't even hardly relating to my old lady anymore. Everything was just one big rush.

This frenzy of overstimulation and resulting exhaustion hastened the process of "burnout" which nearly all individuals experienced. As dealers and smugglers aged in the career they became more sensitized to the extreme risks they faced. Cases of friends and associates who were arrested, imprisoned, or killed began to mount. Many individuals became convinced that continued drug trafficking would inevitably lead to arrest ("It's only a matter of time before you get caught"). While dealers and smugglers generally repressed their awareness of danger, treating it as a taken-for-granted part of their daily existence, periodic crises shattered their casual attitudes, evoking strong feelings of fear. They temporarily intensified security precautions and retreated into near-isolation until they felt the "heat" was off.

As a result of these accumulating "scares," dealers and smugglers increasingly integrated feelings of "paranoia"[5] into their everyday lives. One dealer talked about his feelings of paranoia:

> You're always on the line. You don't lead a normal life. You're always looking over your shoulder, wondering who's at the door, having to hide everything. You learn to look behind you so well you could probably bend over and look up your ass. That's paranoia. It's a really scary, hard feeling. That's what makes you get out.

Drug world members also grew progressively weary of their exclusion from the legitimate world and the deceptions they had to manage to sustain that separation. Initially, this separation was surrounded by an alluring mystique. But as they aged in the career, this mystique became replaced by the reality of everyday boundary maintenance and the feeling of being an "expatriated citizen within one's own country." One smuggler who was contemplating quitting described the effects of this separation:

> I'm so sick of looking over my shoulder, having to sit in my house and worry about one of my non–drug world friends stopping in when I'm doing business. Do you know how awful that is? It's like leading a double life. It's ridiculous. That's what makes it not worth it. It'll be a lot less money [to quit], but a lot less pressure.

Thus, while the drug world was somewhat restricted, it was not an encapsulated community, and dealers' and smugglers' continuous involvement with the straight world made the temptation to adhere to normative standards and "go straight" omnipresent. With the occupation's novelty worn off and the "fast life" taken-for-granted, most dealers and smugglers felt that the occupation no longer resembled their early impressions of it. Once they reached the upper levels of the occupation, their experience began to change. Eventually, the rewards of trafficking no longer seemed to justify the strain and risk involved. It was at this point that the straight world's formerly dull ambiance became transformed (at least in theory) into a potential haven.

Phasing Out

Three factors inhibited dealers and smugglers from leaving the drug world. Primary among these factors were the hedonistic and materialistic satisfactions the drug world provided. Once accustomed to earning vast quantities of money quickly and easily, individuals found it exceedingly difficult to return to the income scale of the straight world. They also were reluctant to abandon the pleasure of the "fast life" and its accompanying drugs, casual sex, and power. Second, dealers and smugglers identified with, and developed a commitment to, the occupation of drug trafficking (Adler and Adler, 1982). Their self-images were tied to that role and could not be easily disengaged. The years invested in their careers (learning the trade, forming connections, building reputations) strengthened their involvement with both the occupation and the drug community. And since their relationships were social as well as business, friendship ties bound individuals to dealing. As one dealer in the midst of struggling to phase-out explained:

> The biggest threat to me is to get caught up sitting around the house with friends that are into dealing. I'm trying to stay away from them, change my habits.

Third, dealers and smugglers hesitated to voluntarily quit the field because of the difficulty involved in finding another way to earn a living. Their years spent in illicit activity made it unlikely for any legitimate organizations to hire them. This narrowed their occupational choices considerably, leaving self-employment as one of the few remaining avenues open.

Dealers and smugglers who tried to leave the drug world generally fell into one of four patterns.[6] The first and most frequent pattern was to postpone quitting until after they could execute one last "big deal." While the intention was sincere, individuals who chose this route rarely succeeded; the "big deal" too often remained elusive. One marijuana smuggler offered a variation of this theme:

> My plan is to make a quarter of a million dollars in four months during the prime smuggling season and get the hell out of the business.

A second pattern we observed was individuals who planned to change immediately, but never did. They announced they were quitting, yet their outward actions never varied. One dealer described his involvement with this syndrome:

> When I wake up I'll say, "Hey, I'm going to quit this cycle and just run my other business." But when you're dealing you constantly have people dropping by ounces and asking, "Can you move this?" What's your first response? Always, "Sure, for a toot."

In the third pattern of phasing-out, individuals actually suspended their dealing and smuggling activities, but did not replace them with an alternative source of income. Such withdrawals were usually spontaneous and prompted by exhaustion, the influence of a person from outside the drug world, or problems with the police or other associates. These kinds of phase-outs usually lasted only until the individual's money ran out, as one dealer explained:

> I got into legal trouble with the FBI a while back and I was forced to quit dealing. Everybody just cut me off completely, and I saw the danger in continuing, myself. But my high-class tastes never dwindled. Before I

knew it I was in hock over $30,000. Even though I was hot, I was forced to get back into dealing to relieve some of my debts.

In the fourth pattern of phasing out, dealers and smugglers tried to move into another line of work. Alternative occupations included: (1) those they had previously pursued; (2) front businesses maintained on the side while dealing or smuggling; and (3) new occupations altogether. While some people accomplished this transition successfully, there were problems inherent in all three alternatives.

(1) Most people who tried resuming their former occupations found that these had changed too much while they were away. In addition, they themselves had changed: they enjoyed the self-directed freedom and spontaneity associated with dealing and smuggling, and were unwilling to relinquish it.

(2) Those who turned to their legitimate front business often found that these businesses were unable to support them. Designed to launder rather than earn money, most of these ventures were retail outlets with a heavy cash flow (restaurants, movie theaters, automobile dealerships, small stores) that had become accustomed to operating under a continuous subsidy from illegal funds. Once their drug funding was cut off they could not survive for long.

(3) Many dealers and smugglers utilized the skills and connections they had developed in the drug business to create a new occupation. They exchanged their illegal commodity for a legal one and went into import/export, manufacturing, wholesaling, or retailing other merchandise. For some, the decision to prepare a legitimate career for their future retirement from the drug world followed an unsuccessful attempt to phase-out into a "front" business. One husband-and-wife dealing team explained how these legitimate side businesses differed from front businesses:

> We always had a little legitimate "scam" [scheme] going, like mail-order shirts, wallets, jewelry, and the kids were always involved in that. We made a little bit of money on them. Their main purpose was for a cover. But [this business] was different; right from the start this was going to be a legal thing to push us out of the drug business.

About 10 percent of the dealers and smugglers we observed began tapering off their drug world involvement gradually, transferring their time and money into a selected legitimate endeavor. They did not try to quit drug trafficking altogether until they felt confident that their legitimate business could support them. Like spontaneous phase-outs, many of these planned withdrawals into legitimate endeavors failed to generate enough money to keep individuals from being lured into the drug world.

In addition to voluntary phase-outs caused by burnout, about 40 percent of the Southwest County dealers and smugglers we observed experienced a "bustout" at some point in their careers.[7] Forced withdrawals from dealing or smuggling were usually sudden and motivated by external factors, either financial, legal, or reputational. Financial bustouts generally occurred when dealers or smugglers were either "burned" or "ripped-off" by others, leaving them in too much debt to rebuild their base of operation. Legal bustouts followed arrest and possibly incarceration: arrested individuals were so "hot" that few of their former associates would deal with them. Reputational bustouts occurred when individuals "burned" or "ripped-off" others (regardless of whether they intended to do so) and were banned from business by their former circle of associates. One smuggler gave his opinion on the pervasive nature of forced phase-outs:

> Some people are smart enough to get out of it because they realize, physically, they have to. Others realize, monetarily, that they want to get out of this world before this world gets them. Those are the lucky ones. Then there are the ones who have to get out because they're hot or someone else close to them is so hot that they'd better get out. But in the end when you get out of it, nobody gets out of it out of free choice; you do it because you have to.

Death, of course, was the ultimate bustout. Some pilots met this fate because of the dangerous routes they navigated (hugging mountains, treetops, other aircrafts) and the sometimes ill-maintained and overloaded planes they flew. However, despite much talk of violence, few Southwest County drug traffickers died at the hands of fellow dealers.

Re-entry

Phasing-out of the drug world was more often than not temporary. For many dealers and smugglers, it represented but another stage of their drug careers (although this may not have been their original intention), to be followed by a period of reinvolvement. Depending on the individual's perspective, re-entry into the drug world could be viewed as either a comeback (from a forced withdrawal) or a relapse (from a voluntary withdrawal).

Most people forced out of drug trafficking were anxious to return. The decision to phase-out was never theirs, and the desire to get back into dealing or smuggling was based on many of the same reasons which drew them into the field originally. Coming back from financial, legal, and reputational bustouts was possible but difficult and was not always successfully accomplished. They had to re-establish contacts, rebuild their organization and fronting arrangements, and raise the operating capital to resume dealing. More difficult was the problem of overcoming the circumstances surrounding their departure. Once smugglers and dealers resumed operating, they often found their former colleagues suspicious of them. One frustrated dealer described the effects of his prison experience:

> When I first got out of the joint [jail], none of my old friends would have anything to do with me. Finally, one guy who had been my partner told me it was because everyone was suspicious of my getting out early and thought I made a deal [with police to inform on his colleagues].

Dealers and smugglers who returned from bustouts were thus informally subjected to a trial period in which they had to re-establish their trustworthiness and reliability before they could once again move in the drug world with ease.

Re-entry from voluntary withdrawal involved a more difficult decision-making process, but was easier to implement. The factors enticing individuals to re-enter the drug world were not the same as those which motivated their original entry. As we noted above, experienced dealers and smugglers often privately weighed their reasons for wanting to quit and wanting to stay in. Once they left, their images of and hopes for the straight world failed to materialize. They could not make the shift to the norms, values, and lifestyle of the straight society and could not earn a living within it. Thus, dealers and smugglers decided to re-enter the drug business for basic reasons: the material perquisites, the hedonistic gratifications, the social ties, and the fact that they had nowhere else to go.

Once this decision was made, the actual process of re-entry was relatively easy. One dealer described how the door back into dealing remained open for those who left voluntarily:

> I still see my dealer friends, I can still buy grams from them when I want to. It's the respect they have for me because I stepped out of it without being busted or burning someone. I'm coming out with a good reputation, and even though the scene is a whirlwind—people moving up, moving down, in, out—if I didn't see anybody for a year I could call them up and get right back in that day.

People who relapsed thus had little problem obtaining fronts, re-establishing their reputations, or readjusting to the scene.

Career Shifts

Dealers and smugglers who re-entered the drug world, whether from a voluntary or forced phase-out, did not always return to the same level of transacting or commodity which characterized their previous style of operation. Many individuals underwent a "career shift" (Luckenbill and Best, 1981) and became involved in some new segment of the drug world. These shifts were sometimes lateral, as when a member of a smuggling crew took on a new specialization, switching from piloting to operating a stash house, for example. One dealer described how he utilized friendship networks upon his re-entry to shift from cocaine to marijuana trafficking:

> Before, when I was dealing cocaine, I was too caught up in using the drug and people around me were starting to go under from getting into "base" [another form of cocaine]. That's why I got out. But now I think I've got myself together and even though I'm dealing again I'm staying away from coke. I've switched over to dealing grass. It's a whole different circle of people. I got into it through a close friend I used to know before, but I never did business with him because he did grass and I did coke.

Vertical shifts moved operators to different levels. For example, one former smuggler returned and began dealing; another top-level marijuana dealer came back to find that the smugglers he knew had disappeared and he was forced to buy in smaller quantities from other dealers.

Another type of shift relocated drug traffickers in different styles of operation. One dealer described how, after being arrested, he tightened his security measures:

> I just had to cut back after I went through those changes. Hell, I'm not getting any younger and the idea of going to prison bothers me a lot more than it did 10 years ago. The risks are no longer worth it when I can have a comfortable income with less risk. So I only sell to four people now. I don't care if they buy a pound or a gram.

A former smuggler who sold his operation and lost all his money during phase-out returned as a consultant to the industry, selling his expertise to those with new money and fresh manpower:

> What I've been doing lately is setting up deals for people. I've got fool-proof plans for smuggling cocaine up here from Colombia; I tell them how to modify their airplanes to add on extra fuel tanks and to fit in more week, coke, or whatever they bring up. Then I set them up with refueling points all up and down Central America, tell them how to bring it up here, what points to come in at, and what kind of receiving unit to use. Then they do it all and I get 10 percent of what they make.

Re-entry did not always involve a shift to a new niche, however. Some dealers and smugglers returned to the same circle of associates, trafficking activity, and commodity they worked with prior to their departure. Thus, drug dealers' careers often peaked early and then displayed a variety of shifts, from lateral mobility, to decline, to holding fairly steady.

A final alternative involved neither completely leaving nor remaining within the deviant world. Many individuals straddled the deviant and respectable worlds forever by continuing to dabble in drug trafficking. As a result of their experiences in the drug world they developed a deviant self-identity and a deviant *modus operandi*. They might not have wanted to bear the social and legal burden of full-time deviant work but neither were they willing to assume the perceived confines and limitations of the straight world. They therefore moved into the entrepreneurial realm, where their daily activities involved some kind of hustling or "wheeling and dealing" in an assortment of legitimate, quasi-legitimate, and deviant ventures, and where they could be their own boss. This enabled them to retain certain elements of the deviant lifestyle, and to socialize on the fringes of the drug community. For these individuals, drug dealing shifted from a primary occupation to a sideline, though they never abandoned it altogether.

LEAVING DRUG TRAFFICKING

This career pattern of oscillation into and out of active drug trafficking makes it difficult to speak of leaving drug trafficking in the sense of final retirement. Clearly, some people succeeded in voluntarily retiring. Of these, a few managed to prepare a post-deviant career for themselves by transferring their drug money into a legitimate enterprise. A larger group was forced out of dealing and either didn't or couldn't return; the bustouts were sufficiently damaging that they never attempted re-entry, or they abandoned efforts after a series of unsuccessful attempts. But there was no way of structurally determining in advance whether an exit from the business would be temporary or permanent. The vacillations in dealers' intentions were compounded by the complexity of operating successfully in the drug world. For many, then, no phase-out could ever be definitely assessed as permanent. As long as individuals had skills, knowledge, and connections to deal they retained the potential to re-enter the occupation at any time. Leaving drug trafficking may thus be a relative phenomenon, characterized by a trailing-off process where spurts of involvement appear with decreasing frequency and intensity.

SUMMARY

Drug dealing and smuggling careers are temporary and fraught with multiple attempts at retirement. Veteran drug traffickers quit their occupation because of the ambivalent feelings they develop toward their deviant life. As they age in the career their experience changes, shifting from a work life that is exhilarating and free to one that becomes increasingly dangerous and confining. But just as their deviant careers are temporary, so too are their retirements. Potential recruits are lured into the drug business by materialism, hedonism, glamor, and excitement. Established dealers are lured away from the deviant life and back into the mainstream by the attractions of security and social ease. Retired dealers and smugglers are lured back in by their expertise, and by their ability to make money quickly and easily. People who have been exposed to the upper levels of drug trafficking therefore find it extremely difficult to quit their deviant occupation permanently. This stems, in part, from their difficulty in moving from the illegitimate to the legitimate business sector. Even more significant is the affinity they form for their deviant values and lifestyle. Thus few, if any, of our subjects were successful in leaving deviance entirely. What dealers and smugglers intend, at the time, to be a permanent withdrawal from drug trafficking can be seen in retrospect as a pervasive occupational pattern of mid-career shifts and oscillations. More research is needed into the complex process of how people get out of deviance and enter the world of legitimate work.

NOTES

1.   The term "soft" drugs generally refers to marijuana, cocaine and such psychedelics as LSD and mescaline (Carey, 1968). In this paper we do not address trafficking in psychedelics because, since they are manufactured in the United States, they are neither imported nor distributed by the group we studied.

2.   This is an idealized figure representing the profit a dealer or smuggler could potentially earn and does not include deductions for such miscellaneous and hard-to-

calculate costs as: time or money spent in arranging deals (some of which never materialize); lost, stolen, or unrepaid money or drugs; and the personal drug consumption of a drug trafficker and his or her entourage. Of these, the single largest expense is the last one, accounting for the bulk of most Southwest County dealers' and smugglers' earnings.

3.   We continued to conduct follow-up interviews with key informants through 1983.

4. While other studies of drug dealing have also noted that participants did not maintain an uninterrupted stream of career involvement (Blum et al., 1972; Carey, 1968; Lieb and Olson, 1976; Waldorf et al., 1977), none have isolated or described the oscillating nature of this pattern.

5. In the dealers' vernacular, this term is not used in the clinical sense of an individual psychopathology rooted in early childhood traumas. Instead, it resembles Lemert's (1962) more sociological definition which focuses on such behavioral dynamics as suspicion, hostility, aggressiveness, and even delusion. Not only Lemert, but also Waldorf et al. (1977) and Wedow (1979) assert that feelings of paranoia can have a sound basis in reality, and are therefore readily comprehended and even empathized with others.

6. At this point, a limitation to our data must be noted. Many of the dealers and smugglers we observed simply "disappeared" from the scene and were never heard from again. We therefore have no way of knowing if they phased-out (voluntarily or involuntarily), shifted to another scene, or were killed in some remote place. We cannot, therefore, estimate the numbers of people who left the Southwest County drug scene via each of the routes discussed here.

7. It is impossible to determine the exact percentage of people falling into the different phase-out categories: due to oscillation, people could experience several types and thus appear in multiple categories.

REFERENCES

Adler, Patricia A., and Peter Adler. 1982. "Criminal commitment among drug dealers." *Deviant Behavior* 3: 117–135.

Anonymous. 1969. "On selling marijuana." In Erich Goode (ed.), *Marijuana* (pp. 92–102). New York: Atherton.

Atkyns, Robert L., and Gerhard J. Hanneman. 1974. "Illicit drug distribution and dealer communication behavior." *Journal of Health and Social Behavior* 15(March): 36–43.

Biernacki, Patrick, and Dan Waldorf. 1981. "Snowball sampling." *Sociological Methods and Research* 10(2): 141–163.

Blum, Richard H., and Associates. 1972. *The Dream Sellers*. San Francisco: Jossey-Bass.

Carey, James T. 1968. *The College Drug Scene*. Englewood Cliffs, NJ: Prentice-Hall.

Douglas, Jack D. 1976. *Investigative Social Research*. Beverly Hills, CA: Sage.

Goode, Erich. 1970. *The Marijuana Smokers*. New York: Basic.

Langer, John. 1977. "Drug entrepreneurs and dealing culture." *Social Problems* 24(3): 377–385.

Lemert, Edwin. 1962. "Paranoia and the dynamics of exclusion." *Sociometry* 25(March): 2–20.

Lieb, John, and Sheldon Olson. 1976. "Prestige, paranoia, and profit: On becoming a dealer of illicit drugs in a university community." *Journal of Drug Issues* 6(Fall): 356–369.

Luckenbill, David F., and Joel Best. 1981. "Careers in deviance and respectability: The analogy's limitations." *Social Problems* 29(2): 197–206.

Mouledoux, James. 1972. "Ideological aspects of drug dealership." In Ken Westhues (ed.), *Society's Shadow: Studies in the Sociology of Countercultures* (pp. 110–122). Toronto: McGraw-Hill, Ryerson.

Riemer, Jeffrey W. 1977. "Varieties of opportunistic research." *Urban Life* 5(4): 467–477.

Waldorf, Dan, Sheigla Murphy, Craig Reinarman, and Bridget Joyce. 1977. *Doing Coke: An Ethnography of Cocaine Users and Sellers*. Washington, DC: Drug Abuse Council.

Wedow, Suzanne. 1979. "Feeling paranoid: The organization of an ideology." *Urban Life* 8(1): 72–93.

41

Getting Out of the Life

Crime Desistance by Female Street Offenders

IRA SOMMERS,
DEBORAH R. BASKIN,
AND JEFFREY FAGAN

S tudies over the past decade have provided a great deal of information about the criminal careers of male offenders. (See Blumstein et al. 1986 and Weiner and Wolfgang 1989 for reviews.) Unfortunately, much less is known about the initiation, escalation, and termination of criminal careers by female offenders. The general tendency to exclude female offenders from research on crime and delinquency may be due, at least in part, to the lower frequency and comparatively less serious nature of offending among women. Recent trends and studies, however, suggest that the omission of women may seriously bias both research and theory on crime.

Although a growing body of work on female crime has emerged within the last few years, much of this research continues to focus on what Daly and Chesney-Lind (1988) called generalizability and gender-ratio problems. The former concerns the degree to which traditional (i.e., male) theories of deviance and crime apply to women, and the latter focuses on what explains gender differences in rates and types of criminal activity. Although this article also examines women in crime, questions of inter- and intragender variability in crime are not specifically addressed. Instead, the aim of the paper is to describe the pathways out of deviance for a sample of women who have significantly invested themselves in criminal social worlds. To what extent are the social and psychological processes of stopping criminal behavior similar for men and women? Do the behavioral antecedents of such processes vary by gender? These questions remained unexplored.

Specifically, two main issues are addressed in this paper: (1) the role of life events in triggering the cessation process, and (2) the relationship between cognitive and life situation changes in the desistance process. First, the crime desistance literature is reviewed briefly. Second, the broader deviance literature is drawn upon to construct a social-psychological model of cessation. Then the model is evaluated using life history data from a sample of female offenders convicted of serious street crimes.

From "Getting Out of the Life: Crime Desistance by Female Street Offenders," Ira Sommers et al., *Deviant Behavior*, V. 15, No. 2, Taylor & Francis, Inc., Washington, D.C., pp. 125–149, 1994. Reprinted with permission. All rights reserved.

THE DESISTANCE PROCESS

The common themes in the literature on exiting deviant careers offer useful perspectives for developing a theory of cessation. The decision to stop deviant behavior appears to be preceded by a variety of factors, most of which are negative social sanctions or consequences. Health problems, difficulties with the law or with maintaining a current lifestyle, threats of other social sanctions from family or close relations, and a general rejection of the social world in which the behaviors thrive are often antecedents of the decision to quit. For some, religious conversions or immersion into alternative sociocultural settings with powerful norms (e.g., treatment ideology) provide paths for cessation (Mulvey and LaRosa 1986; Stall and Biernacki 1986).

. . . A model for understanding desistance from crime is presented below. Three stages characterize the cessation process: building resolve or discovering motivation to stop (i.e., socially disjunctive experiences), making and publicly disclosing the decision to stop, and maintaining the new behaviors and integrating into new social networks (Stall and Biernacki 1986; Mulvey and Aber 1988). These phases . . . describe three ideal-typical phases of desistance: "turning points" where offenders begin consciously to experience negative effects (socially disjunctive experiences); "active quitting" where they take steps to exit crime (public pronouncement); and "maintaining cessation" (identity transformation):

Stage 1   Catalysts for change

Socially disjunctive experiences
- Hitting rock bottom
- Fear of death
- Tiredness
- Illness

Delayed deterrence
- Increased probability of punishment
- Increased difficulty in doing time
- Increased severity of sanctions
- Increasing fear

Assessment
- Reappraisal of life and goals
- Psychic change

Decision
- Decision to quit and/or initial attempts at desistance
- Continuing possibility of criminal participation

Stage 2   Discontinuance
- Public pronouncement of decision to end criminal participation
- Claim to a new social identity

Stage 3 **Maintenance of the decision to stop**
- Ability to successfully renegotiate identity
- Support of significant others
- Integration into new social networks
- Ties to conventional roles
- Stabilization of new social identity

Stage 1: Catalysts for Change

When external conditions change and reduce the rewards of deviant behavior, motivation may build to end criminal involvement. That process, and the resulting decision, seem to be associated with two related conditions: a series of negative, aversive, unpleasant experiences with criminal behavior, or corollary situations where the positive rewards, status, or gratification from crime are reduced. Shover and Thompson's (1992) research suggests that the probability of desistance from criminal participation increases as expectations for achieving rewards (e.g., friends, money, autonomy) via crime decrease and that changes in expectations are age-related. Shover (1983) contended that the daily routines of managing criminal involvement become tiring and burdensome to aging offenders. Consequently, the allure of crime diminishes as offenders get older. Aging may also increase the perceived formal risk of criminal participation. Cusson and Pinsonneault (1986, p. 76) posited that "with age, criminals raise their estimates of the certainty of punishment." Fear of reimprisonment, fear of longer sentences, and the increasing difficulty of "doing time" have often been reported by investigators who have explored desistance.

Stage 2: Discontinuance

The second stage of the model begins with the public announcement that the offender has decided to end her criminal participation. Such an announcement forces the start of a process of renegotiation of the offender's social identity (Stall and Biernacki 1986). After this announcement, the offender must not only cope with the instrumental aspects (e.g., financial) of her life but must also begin to redefine important emotional and social relationships that are influenced by or predicated upon criminal behavior.

Leaving a deviant subculture is difficult. Biernacki (1986) noted the exclusiveness of the social involvements maintained by former addicts during initial stages of abstinence. With social embedment comes the gratification of social acceptance and identity. The decision to end a behavior that is socially determined and supported implies withdrawal of the social gratification it brings. Thus, the more deeply embedded in a criminal social context, the more dependent the offender is on that social world for her primary sources of approval and social definition.

The responses by social control agents, family members, and peer supporters to further criminal participation are critical to shaping the outcome of discontinuance. New social and emotional worlds to replace the old ones may strengthen the decision to stop. Adler (1992) found that outside associations and involvements provide a critical bridge back into society for dealers who have decided to leave the drug subculture. With discontinuance comes the difficult work of identity transformations (Biernacki 1986) and establishing new social definitions of behavior and relationships to reinforce them.

Stage 3: Maintenance

Following the initial stages of discontinuance, strategies to avoid a return to crime build on the strategies first used to break from a lengthy pattern of criminal participation: further integration into a noncriminal identity and social world and maintenance of this new identity. Maintenance depends in part on replacing deviant networks of peers and associates with supports that both censure criminal participation and approve of new nondeviant beliefs. Treatment interventions (e.g., drug treatment, social service programs) are important sources of alternative social supports to maintain a noncriminal lifestyle. In other words, maintenance depends on immersion into a social world where criminal behavior meets immediately with strong formal and informal sanctions.

Despite efforts to maintain noncriminal involvement, desistance is likely to be episodic, with occasional relapses interspersed with lengthening of lulls in criminal activity. Le Blanc and Frechette (1989) proposed the possibility that criminal activity slows down before coming to an end and that this slowing down process becomes apparent in three ways: deceleration, specialization, and reaching a ceiling. Thus, before stopping criminal activity, the offender gradually acts out less frequently, limits the variety of crimes more and more, and ceases increasing the seriousness of criminal involvement.

Age is a critical variable in desistance research, regardless of whether it is associated with maturation or similar developmental concepts. Cessation is part of a social-psychological transformation for the offender. A strategy to stabilize the transition to a noncriminal lifestyle requires active use of supports to maintain the norms that have been substituted for the forces that supported criminal behavior in the past.

METHODS

Constructing a sample of women who have desisted from crime poses several challenges. . . . Through the use of snowball sampling, 30 women were recruited and interviewed from January to May 1992. An initial pool of 16 women was recruited through various offender and drug treatment programs in New York City. Fourteen additional women were recruited through chain referrals. To be included in the study, the women had to have at least one official arrest for a violent street crime (robbery, assault, burglary, weapons possession, arson, kidnapping) and to have desisted from all criminal involvement for at least 2 years prior to the interview.[1] Eligibility was verified through official arrest records and through contact with program staff for those women participating in treatment programs (87%).

The life history interviews . . . were open-ended, in-depth, and audiotaped. The open-ended technique created a context in which respondents could speak freely and in their own words. Furthermore, it facilitated the pursuit of issues raised by respondents during the interview but not recognized previously by the researcher. Use of this approach enabled us to probe for information about specific events and provided an opportunity for respondents to reflect on those events. As a result, we were able to gain insight into their attitudes, feelings, and other subjective orientations to their experiences. Finally, tape-recording the interviews allowed the interviewer to adopt a more conversational style, devote complete attention to the respondent, and concentrate on the discussion. By not being preoccupied with notetaking, interviewers were able to create a more relaxed atmosphere for the participants.

All interviews were conducted by the first two authors in a private university office. They were typically held in the morning or early afternoon and lasted approximately 2 hours. A stipend of $50 was paid to each respondent. In total, 30 interviews were conducted, comprising approximately 60 hours of talk.

With regard to self-reported drug and criminal histories, the data indicate that the study respondents had engaged in a wide range of criminal and deviant activities. All of the respondents were experienced drug users. Eighty-seven percent were addicted to crack, 70% used cocaine regularly, and 10% were addicted to heroin. Of the 30 women interviewed, 19 (63%) reported involvement in robbery, 60% reported involvement in burglary, 94% were involved in selling drugs, and 47% were at some time involved in prostitution. A great deal of variation was found with regard to incarceration history. The median number of incarcerations was 3 (the mean was 4.29); however, 17% of the women ($n = 5$) had never been incarcerated.

Patterns of offending for 17 of the 30 women (57%) resemble those of a sample of addicts studied by Anglin and Speckhart (1988, p. 223) in that "while some criminality precedes the addiction career, the great majority is found during the addiction career." It seems that for these women, the cost of their increasing substance use was a major influence to engage in offending, especially robbery and selling drugs. For the remaining 13 women (43%), drug abuse appears not to be as etiologically important to violent behavior, in spite of the high correlations between these two behaviors. Instead, addiction seems to be part of a more generalized lifestyle (Peterson and Braiker 1980; Collins et al. 1985) in which involvement in violent criminal careers precedes, yet may be amplified by, addiction.

RESULTS

Resolving to Stop

Despite its initial excitement and allure, the life of a street criminal is a hard one. A host of severe personal problems plague most street offenders and normally become progressively worse as their careers continue. In the present study, the women's lives were dominated by a powerful, often incapacitating, need for drugs. Consequently, economic problems were the most frequent complaint voiced by the respondents. Savings were quickly exhausted, and the culture of addiction justified virtually any means to get money to support their habits. For the majority of the women, the problem of maintaining an addiction took precedence over other interests and participation in other social worlds.

People the respondents associated with, their primary reference group, were involved in illicit behaviors. Over time, the women in the study became further enmeshed in deviance and further alienated, both socially and psychologically, from conventional life. The women's lives became bereft of conventional involvements, obligations, and responsibilities. The excitement at the lifestyle that may have characterized their early criminal career phase gave way to a much more grave daily existence.

Thus, the women in our study could not and did not simply cease their deviant acts by "drifting" (Matza 1964) back toward conventional norms, values, and lifestyles. Unlike many of Waldorf's (1983) heroin addicts who drifted away from heroin without conscious effort, all of the women in our study made a conscious decision to stop. In short, Matza's concept of drift did not provide a useful framework for understanding our respondents' exit from crime.

The following accounts illustrate the uncertainty and vulnerability of street life for the women in our sample. Denise, a 33-year-old black woman, has participated in a wide range of street crimes including burglary, robbery, assault, and drug dealing. She began dealing drugs when she was 14 and was herself using cocaine on a regular basis by age 19.

> I was in a lot of fights: So I had fights over, uh, drugs, or, you know, just manipulation. There's a lot of manipulation in that life. Everybody's tryin' to get over. Everybody will stab you in your back, you know. Nobody gives a fuck about the next person, you know. It's just when you want it, you want it. You know, when you want that drug, you know, you want that drug. There's a lot of lyin', a lot of manipulation. It's, it's, it's crazy!

Gazella, a 38-year-old Hispanic woman, had been involved in crime for 22 years when we interviewed her.

> I'm 38 years old. I ain't no young woman no more, man. Drugs have changed, lifestyles have changed. Kids are killing you now for turf. Yeah, turf, and I was destroyin' myself. I was miserable. I was. . . I was gettin' high all the time to stay up to keep the business going, and it was really nobody I could trust.

Additional illustrations of the exigencies of street life are provided by April and Stephanie. April is a 25-year-old black woman who had been involved in crime since she was 11.

> I wasn't eating. Sometimes I wouldn't eat for two or three days. And I would. . . a lot of times I wouldn't have the time, or I wouldn't want to spend the money to eat—I've got to use it to get high.

Stephanie, a 27-year-old black woman, had used and sold crack for 5 years when we interviewed her.

> I knew that, uh, I was gonna get killed out here. I wasn't havin' no respect for myself. No one else was respecting me. Every relationship I got into, as long as I did drugs, it was gonna be constant disrespect involved, and it come. . . to the point of me gettin' killed.

When the spiral down finally reached its lowest point, the women were overwhelmed by a sense of personal despair. In reporting the early stages of this period of despair, the respondents consistently voiced two themes: the futility of their lives and their isolation.

Barbara, a 31-year-old black woman, began using crack when she was 23. By age 25, Barbara had lost her job at the Board of Education and was involved in burglary and robbery. Her account is typical of the despair the women in our sample eventually experienced.

> . . . the fact that my family didn't trust me anymore, and the way that my daughter was looking at me, and, uh, my mother wouldn't let me in her house any more, and I was sleepin' on the trains. And I was sleepin' on the beaches in the summertime. And I was really frightened. I was real scared of the fact that I had to sleep on the train. And, uh, I had to wash up in the Port Authority.

The spiral down for Gazella also resulted in her living on the streets.

> I didn't have a place to live. My kids had been taken away from me. You know, constantly being harassed like 3 days out of the week by the Tactical Narcotics Team (police). I didn't want to be bothered with people. I was gettin' tired of the lyin', schemin', you know, stayin' in abandoned buildings.

Alicia, a 29-year-old Hispanic woman, became involved in street violence at age 12. She commented on the personal isolation that was a consequence of her involvement in crime:

> When I started getting involved in crime, you know, and drugs, the friends that I had, even my family, I stayed away from them, you know. You know how you look bad and you feel bad, and you just don't want those people to see you like you are. So I avoided seeing them.

For some, the emotional depth of the rock bottom crisis was felt as a sense of mortification. The women felt as if they had nowhere to turn to salvage a sense of well-being or self-worth. Suicide was considered a better alternative than remaining in such an undesirable social and psychological state. Denise is one example:

> I ran into a girl who I went to school with that works on Wall Street. And I compared her life to mine and it was like miserable. And I just wanted out. I wanted a new life. I was tired, I was run down, looking

bad. I got out by smashing myself through a sixth-floor window. Then I went to the psychiatric ward and I met this real nice doctor, and we talked every day. She fought to keep me in the hospital because she felt I wouldn't survive. She believed in me. And she talked me into going into a drug program.

Marginalization from family, friends, children, and work—in short, the loss of traditional life structures—left the women vulnerable to chaotic street conditions. After initially being overwhelmed by despair, the women began to question and reevaluate basic assumptions about their identities and their social construction of the world. Like Shover's (1983) male property offenders, the women also began to view the criminal justice system as "an imposing accumulation of aggravations and deprivations" (p. 212). They grew tired of the street experiences and the problems and consequences of criminal involvement.

Many of the women acknowledged that, with age, it is more difficult to do time and that the fear of incurring a long prison sentence the next time influenced their decision to stop. Cusson and Pinsonneault (1986, p. 76) made the same observation with male robbers. Gazella, April, and Denise, quoted earlier, recall:

Gazella: First of all, when I was in prison I was like, I was so humiliated. At my age [38] I was really kind of embarrassed, but I knew that was the lifestyle that I was leadin'. And people I used to talk to would tell me, well, you could do this, and you don't have to get busted. But then I started thinking why are all these people here. So it doesn't, you know, really work. So I came home, and I did go back to selling again, but you know I knew I was on probation. And I didn't want to do no more time.

April: Jail, being in jail. The environment, having my freedom taken away. I saw myself keep repeating the same pattern, and I didn't want to do that. Uh, I had missed my daughter. See, being in jail that long period of time, I was able to detox. And when I detoxed, I kind of like had a clear sense of thinking, and that's when I came to the realization that, uh, this is not working for me.

Denise: I saw the person that I was dealing with—my partner—I saw her go upstate to Bedford for 2 to 4 years. I didn't want to deal with it. I didn't want to go. Bedford is a prison, women's prison. And I couldn't see myself givin' up 2 years of my life for something that I knew I could change in another way.

As can be seen from the above, the influence of punishment on these women was due to their belief that if they continued to be involved in crime, they would be apprehended, convicted, and incarcerated.

For many of the women, it was the stresses of street life and the fear of dying on the streets that motivated their decision to quit the criminal life. Darlene, a 25-year-old black woman, recalled the stress associated with the latter stage of her career selling drugs:

The simple fact is that I really, I thought that I would die out there. I thought that someone would kill me out there and I would be killed; I had a fear of being on the front page one day and being in the newspaper dying. I wanted to live, and I didn't just want to exist.

Sonya, a 27-year-old Hispanic woman, provided an account of what daily life was like on the streets:

You get tired of bein' tired, you know. I got tired of hustlin', you know. I got tired of livin' the way I was livin', you know. Due to your body, your body, mentally, emotionally, you know. Everybody's tryin' to get over. Everybody will stab you in your back. Nobody gives a fuck about the next person. And I used to have people talkin' to me, "You know, you're not a bad lookin' girl. You know, why you don't get yourself together."

Perhaps even more important, the women felt that they had wasted time. They became acutely aware of time as a diminishing resource (Shover 1983). They reported that they saw themselves going nowhere. They had arrived at a point where crime seemed senseless, and their lives had reached a dead end. Implicit in this assessment was the belief that gaining a longer-range perspective on one's life was a first step in changing. Such deliberations develop as a result of "socially disjunctive experiences" that cause the offender to experience social stress, feelings of alienation, and dissatisfaction with her present identity (Ray 1961).

Breaking Away from the Life

Forming a commitment to change is only the first step toward the termination of a criminal career. The offender enters a period that has been characterized as a "running struggle" with problems of social identity (Ray 1961, p. 136). Successful desisters must work to clarify and strengthen their nondeviant identity and redefine their street experience in terms more compatible with a conventional lifestyle. The second stage of the desistance process begins with the public announcement or "certification" (Meisenhelder 1977, p. 329) that the offender has decided to end her deviant behavior. After this announcement, the offender must begin to redefine economic, social, and emotional relationships that were based on a deviant street subculture.

The time following the announcement was generally a period of ambivalence and crisis for the study participants, because so much of their lives revolved around street life and because they had, at best, weak associations with the conventional world. Many of the women remembered the uncertainty they felt and the social dilemmas they faced after they decided to stop their involvement in crime.

Denise: I went and looked up my friends and to see what was doing, and my girlfriend Mia was like, she was gettin' paid. And I was livin' on a $60 stipend. And I wasn't with it. Mia was good to me, she always kept money in my pocket when I came home. I would walk into her closet and change into clothes that I'm more accustomed to. She started calling me Pen again. She stopped calling me Denise. And I would ride with her knowing that she had a gun or a package in the car. But I wouldn't touch nothin'. But that was my rationale. As long as I don't fuck with nothin. Yeah, she was like I can give you a grand and get you started. I said I know you can, but I can't. She said I can give you a grand, and she kept telling me that over and over; and I wasn't that far from taking the grand and getting started again.

Barbara: After I decided to change, I went to a party with my friend. And people was around me and they was drinkin' and stuff, and I didn't want to drink. I don't have the urge of drinking. If anything, it would be smokin' crack. And when I left the party, I felt like I was missing something—like something was missing. And it was the fact that I wasn't gettin' high. But I know the consequences of it. If I take a drink, I'm gonna smoke crack. If I, uh, sniff some blow, I'm gonna smoke crack. I might do some things like rob a store or something stupid and go to jail. So I don't want to put myself in that position.

At this stage of their transition, the women had to decide how to establish and maintain conventional relationships and what to do with themselves and their lives. Few of the women had maintained good relationships with people who were not involved in crime and drugs. Given this situation, the women had to seek alternatives to their present situation.

The large majority of study participants were aided in their social reintegration by outside help. These respondents sought formal treatment of some kind, typically residential drug treatment, to provide structure, social support, and a pathway to behavioral change. The women perceived clearly he need to remove themselves from the "scene," to meet new friends, and to begin the process of identity reformation. The following account by Alicia typifies the importance of a "geographic" cure:

I love to get high, you know, and I love the way crack makes me feel. I knew that I needed long-term, I knew that I needed to go somewhere. All away from everything, and I just needed to away from everything. And I couldn't deal with responsibility at all. And, uh, I was just so ashamed of the way that I had, you know, became and the person that I became that I just wanted to start over again.

Social avoidance strategies were common to all attempts at stopping. When the women removed themselves from their old world and old locations, involvement in crime and drugs was more difficult.

April: Yeah, I go home, but I don't, I don't socialize with the people. I don't even speak to anybody really. I go and I come. I don't go to the areas that I used to be in. I don't go there anymore. I don't walk down the same blocks I used to walk down. I always take different locations.

Denise: I miss the fast money; otherwise, I don't miss my old life. I get support from my positive friends, and in the program. I talk about how I felt being around my old associates, seeing them, you know, going back to my old neighborhood. It's hard to deal with, I have to push away.

Maintaining a Conventional Life

Desisters have little chance of staying out of the life for an extended period of time if they stay in the social world of crime and addiction. They must rebuild and maintain a network of primary relations who accept and support their nondeviant identity if they are to be successful (the third stage of this model). This is no easy task, since in most cases the desisters have alienated their old nondeviant primary relations.

To a great extent, the women in this study most resemble religious converts in their attempts to establish and maintain support networks that validate their new sense of self. Treatment programs not only provide a ready-made primary group for desisters, but also a well-established pervasive identity (Travisano 1970), that of ex-con and/or ex-addict, that informs the women's view of themselves in a variety of interactions. Reminders of "spoiled identities" (Goffman 1963) such as criminal, "con," and "junkie" serve as a constant reference point for new experiences and keep salient the ideology of conventional living (Faupel 1991). Perhaps most important, these programs provide the women with an alternative basis for life structure—one that is devoid of crime, drugs, and other subcultural elements.

The successful treatment program, however, is one that ultimately facilitates dissociation from the program and promotes independent living. Dissociation from programs to participate in conventional living requires association, or reintegration, with conventional society. Friends and educational and occupational roles helped study participants reaffirm their noncriminal identities and bond themselves to conventional lifestyles. Barbara described the assistance she receives from friends and treatment groups:

> . . . a bunch of friends that always confronts me on what I'm doing' and where I'm goin', and they just want the best for me. And none of them use drugs. I go to a lot like outside support groups, you know. They help me have more confidence in myself. I have new friends now. Some of them are in treatment. Some have always been straight. They know. You know, they glad, you know, when I see them.

In the course of experiencing relationships with conventional others and participating in conventional roles, the women developed a strong social-psychological commitment not to return to crime and drug use. These commitments most often revolved around renewed affiliations with their children, relationships with new friends, and the acquisition of educational and vocational skills. The social relationships, interests, and investments that develop in the course of desistance reflect the gradual emergence of new identities. Such stakes in conventional identity form the social-psychological context within which control and desistance are possible (Waldorf et al. 1991).

In short, the women in the study developed a stake in their new lives that was incompatible with street life. This new stake served as a wedge to help maintain the separation of the women from the world of the streets (Biernacki 1986). The desire to maintain one's sense of self was an important incentive for avoiding return to crime.

> **Alicia:** I like the fact that I have my respect back. I like the fact that, uh, my daughter trusts me again. And my mother don't mind leavin' me in the house, and she don't have to worry that when she come in her TV might be gone.

> **Barbara:** I have new friends. I have my children back in my life. I have my education. It keeps me straight. I can't forget where I came from because I get scared to go back. I don't want to hurt nobody. I just want to live a normal life.

Janelle, a 22-year-old black woman, started dealing drugs and carrying a .38-caliber gun when she was 15. She described the ongoing tension between staying straight and returning to her old social world:

> It's hard, it's hard stayin' on the right track. But letting myself know that I'm worth more. I don't have to go in a store today and steal anything. I don't deserve that. I don't deserve to make myself feel really bad. Then once again I would be steppin' back and feel that this is all I can do.

Overall, the success of identity transformations hinges on the women's abilities to establish and maintain commitments and involvements in conventional aspects of life. As the women began to feel accepted and trusted within some conventional social circles, their determination to exit from crime was strengthened, as were their social and personal identities as noncriminals.

DISCUSSION

The primary purpose of this study was to describe—from the offenders' perspective—how women embedded in criminal street subcultures could end their deviance. Desistance appears to be a process as complex and lengthy as that of initial involvement. It was interesting to find that some of the key concepts in initiation of deviance—social bond, differential association, deterrence, age—were important in our analysis. We saw the aging offender take the threat of punishment seriously, reestablish links with conventional society, and sever association with subcultural street elements.

Our research supports Adler's (1992) finding that shame plays a limited role in the decision to return to conventional life for individuals who are entrenched in deviant subcultures. Rather, they exit deviance because they have evolved through the typical phases of their deviant careers.

In the present study, we found that the decision to give up crime was triggered by a shock of some sort (i.e., a socially disjunctive experience), by a delayed deterrence process, or both. The women then entered a period of crisis. Anxious and dissatisfied, they took stock of their lives and criminal activity. They arrived at a point where their way of life seemed senseless. Having made this assessment, the women then worked to clarify and strengthen their nondeviant identities. This phase began with the reevaluation of life goals and the public announcement of their decision to end involvement in crime. Once the decision to quit was made, the women turned to relationships that had not been ruined by their deviance, or they created new relationships. The final stage, maintaining cessation, involved integration into a nondeviant lifestyle. This meant restructuring the entire pattern of their lives (i.e., primary relationships, daily routines, social situations). For most women, treatment groups provided the continuing support needed to maintain a nondeviant status.

The change processes and turning points described by the women in the present research were quite similar to those reported by men in previous studies (Shover 1983, 1985; Cusson and Pinsonneault 1986). Collectively, these findings suggest that desistance is a pragmatically constructed project of action created by the individual within a given social context. Turning points occur as a "part of a process over time and not as a dramatic lasting change that takes place at any one time" (Pickles and Rutter 1991, p. 134). Thus, the return to conventional life occurs more because of "push" than "pull" factors (Adler 1992), because the career of involvement in crime moves offenders beyond the point at which they find it enjoyable to the point at which it is debilitating and anxiety-provoking.

Considering the narrow confines of our empirical data, it is hardly necessary to point out the limits of generalizability. Our analysis refers to the woman deeply involved in crime and immersed in a street subculture who finds the strength and resources to change her way of life. The fact that all the women in this study experienced a long period of personal deterioration and a "rock bottom" experience before they were able to exit crime does not justify a conclusion that this process occurs with all offenders. Undoubtedly, there are other scenarios (e.g., the occasional offender who drifts in and out of crime, the offender who stops when criminal involvement conflicts with commitments to conventional life, the battered woman who kills) in which the question of desistance does not arise. Hence, there is a need to conceptualize and measure the objective and subjective elements of change among various male and female offender subgroups.

Furthermore, the evidence presented here does not warrant the conclusion that none of the women ever renewed their involvement in crime. Because the study materials consist of retrospective information, with all its attendant problems, we cannot state with certainty whether desistance from crime is permanent. Still, it is also clear that these women broke their pattern of involvement in crime for substantial lengths of time and have substantially changed their lives.

NOTE

1. A problematic aspect of the definition of desistance is its permanence. Termination that is followed by criminal involvement might be considered false (Blumstein et al. 1985). Elliott et al. (1989) have avoided the variable "termination" by using the variable "suspension," a temporary or permanent cessation of criminal activity during a particular period of time.

Clearly, we cannot know if the women in this study have demonstrated true desistance. However, a 2-year hiatus from crime certainly indicates temporary cessation and, more importantly, it is a long enough period of time to demonstrate the processes that initiate and sustain desistance.

REFERENCES

Adler, Patricia. 1992. "The 'Post' Phase of Deviant Careers: Reintegrating Drug Traffickers." *Deviant Behavior* 13: 103–126.

Anglin, Douglas, and George Speckhart. 1988. "Narcotics Use and Crime: A Multisample, Multimethod Analysis." *Criminology* 26: 197–234.

Biernacki, Patrick A. 1986. *Pathways from Heroin Addiction: Recovery Without Treatment*. Philadelphia: Temple University Press.

Biernacki, Patrick A., and Dan Waldorf. 1981. "Snowball Sampling: Problems and Techniques of Chain Referral Sampling." *Sociological Methods and Research* 10: 141–163.

Blumstein, Alfred, Jacqueline Cohen, Jeffrey A. Roth, and Christy A. Visher. 1986. *Criminal Careers and Career Criminals*. Washington, DC: National Academy Press.

Collins, J., R. Hubbard, and J. V. Rachal. 1985. "Expensive Drug Use and Illegal Income: A Test of Explanatory Hypotheses." *Criminology* 23: 743–764.

Cusson, Maurice, and Pierre Pinsonneault. 1986. "The Decision to Give Up Crime." In *The Reasoning Criminal: Rational Choice Perspectives on Offending*, edited by Derek Cornish and Ronald Clarke. New York: Springer-Verlag.

Daly, Kathy, and Meda Chesney-Lind. 1988. "Feminism and Criminology." *Justice Quarterly* 5: 101–143.

Faupel, Charles. 1991. *Shooting Dope: Career Patterns of Hard-Core Heroin Users*. Gainesville: University of Florida Press.

Goffman, Erving. 1963. *Stigma: Notes on the Management of Spoiled Identity*. Englewood Cliffs, NJ: Prentice-Hall.

Hirschi, Travis, and H. C. Selvin. 1967. *Delinquency Research: An Appraisal of Analytic Methods*. New York: Free Press.

Le Blanc, Marc, and M. Frechette. 1989. *Male Criminal Activity from Childhood Through Youth: Multilevel and Developmental Perspective*. New York: Springer-Verlag.

Matza, David. 1964. *Delinquency and Drift*. New York: Wiley.

Meisenhelder, Thomas. 1977. "An Exploratory Study of Exiting from Criminal Careers." *Criminology* 15: 319–334.

Mulvey, Edward P., and John F. LaRosa. 1986. "Delinquency Cessation and Adolescent Development: Preliminary Data." *American Journal of Orthopsychiatry* 56: 212–224.

Petersilia, Joan, Peter Greenwood, and Marvin Lavin. 1978. *Criminal Careers of Habitual Felons*. Washington, DC: Law Enforcement Assistance Administration, U.S. Department of Justice.

Peterson, M., and H. Braiker. 1980. *Doing Crime: A Survey of California Prison Inmates*. Santa Monica, CA: Rand.

Pickles, Andrew, and Michael Rutter. 1991. "Statistical and Conceptual Models of 'Turning Points' in Developmental Processes." In *Problems and Methods in Longitudinal Research: Stability and Change,* edited by D. Magnusson, L. Bergman, G. Rudinger, and B. Torestad (pp. 110–136). New York: Cambridge University Press.

Ray, Marsh. 1961. "The Cycle of Abstinence and Relapse Among Heroin Addicts." *Social Problems* 9: 132–140.

Shover, Neil. 1983. "The Latter Stages of Ordinary Property Offenders' Careers." *Social Problems* 31: 208–218.

———. 1985. *Aging Criminals.* Newbury Park, CA: Sage.

Shover, Neil, and Carol Thompson. 1992. "Age, Differential Expectations, and Crime Desistance." *Criminology* 30: 89–104.

Stall, Ron, and Patrick Biernacki. 1986. "Spontaneous Remission from the Problematic Use of Substances: An Inductive Model Derived from a Comparative Analysis of the Alcohol, Opiate, Tobacco, and Food/Obesity Literatures." *International Journal of the Addictions* 2: 1–23.

Travisano, R. 1970. "Alteration and Conversion as Qualitatively Different Transformations." In *Social Psychology Through Symbolic Interaction,* edited by G. Stone and H. Farberman (pp. 594–605). Boston: Ginn-Blaisdell.

Waldorf, Dan. 1983. "Natural Recovery from Opiate Addiction: Some Social-Psychological Processes of Untreated Recovery." *Journal of Drug Issues* 13: 237–280.

Waldorf, Dan, Craig Reinerman, and Sheila Murphy. 1991. *Cocaine Changes.* Philadelphia: Temple University Press.

Weiner, Neil, and Marvin E. Wolfgang. 1989. *Violent Crime, Violent Criminals.* Newbury Park, CA: Sage.

Weis, Joseph G. 1989. "Family Violence Research Methodology and Design." In *Family Violence,* edited by Lloyd Ohlin and Michael Tonry (pp. 117–162). Chicago: University of Chicago Press.

References for the General and Part Introductions

Adler, Patricia A., and Peter Adler. 1987. *Membership Roles in Field Research.* Newbury Park, CA: Sage.

———. 1995. "The Demography of Ethnography." *Journal of Contemporary Ethnography* 24: 3–29.

Becker, Howard S. 1963. *Outsiders: Studies in the Sociology of Deviance.* New York: Free Press.

———. 1973. "Labeling Theory Reconsidered." In *Outsiders* (pp. 177–212). New York: Free Press.

Best, Joel, and David F. Luckenbill. 1980. "The Social Organization of Deviants." *Social Problems* 28(1): 14–31.

———. 1981. "The Social Organization of Deviance." *Deviant Behavior* 2: 231–59.

Cloward, Richard, and Lloyd Ohlin. 1960. *Delinquency and Opportunity.* Glencoe, IL: Free Press.

Cohen, Albert. 1955. *Delinquent Boys.* Glencoe, IL: Free Press.

Davis, Fred. 1961. "Deviance Disavowal: The Management of Strained Interaction by the Visibly Handicapped." *Social Problems* 9: 120–32.

Erikson, Kai T. 1966. *Wayward Puritans.* New York: Wiley.

Galliher, John. 1995. "Chicago's Two Worlds of Deviance Research: Whose Side Are They On?" In *A Second Chicago School?,* edited by Gary Alan Fine (pp. 164–87). Chicago: University of Chicago Press.

Goffman, Erving. 1961. *Asylums.* New York: Doubleday.

———. 1963. *Stigma.* Englewood Cliffs, NJ: Prentice-Hall.

Gusfield, Joseph. 1967. "Moral Passage: The Symbolic Process in Public Designations of Deviance." *Social Problems* 15: 175–88.

Henry, Jules. 1964. *Jungle People.* New York: Vintage.

Hughes, Everett. 1945. "Dilemmas and Contradictions of Status." *American Journal of Sociology* (March): 353–59.

Kitsuse, John. 1962. "Societal Reactions to Deviant Behavior: Problems of

Theory and Method." *Social Problems* 9: 247–56.

———. 1980. "Coming Out All Over: Deviants and the Politics of Social Problems." *Social Problems* 28: 1–13.

Krauthammer, Charles. 1993. "Defining Deviancy Up." *The New Republic* November 22: 20–25.

Lemert, Edwin. 1951. *Social Pathology.* New York: McGraw-Hill.

———. 1967. *Human Deviance, Social Problems, and Social Control.* New York: Prentice-Hall.

Luckenbill, David F., and Joel Best. 1981. "Careers in Deviance and Respectability: The Analogy's Limitations." *Social Problems* 29(2): 197–206.

Lyman, Stanford M. 1970. *The Asian in the West.* Reno/Las Vegas, Nevada: Western Studies Center, Desert Research Institute.

Matza, David. 1964. *Delinquency and Drift.* New York: Wiley.

Merton, Robert. 1938. "Social Structure and Anomie." *American Sociological Review* 3 (October): 672–82.

Miller, Walter. 1958. "Lower Class Culture as a Generating Milieu of Gang Delinquency." *Journal of Social Issues* 14(3): 5–19.

Moynihan, Daniel Patrick. 1993. "Defining Deviancy Down." *The American Scholar* 62(1): 17–30.

Polsky, N. 1967. *Hustlers, Beats, and Others.* Chicago: Aldine.

Schrager, Laura Shill, and James F. Short, Jr. 1978. "Toward a Sociology of Organizational Crime." *Social Problems* 25(4): 407–19.

Schur, Edwin. 1979. *Interpreting Deviance.* New York: Harper and Row.

Scott, Marvin, and Stanford Lyman. 1968. "Accounts." *American Sociological Review* 33(1): 46–62.

Sellin, Thorsten. 1938. "Culture Conflict and Crime." A Report of the Subcommittee on Delinquency of the Committee on Personality and Culture, *Social Science Research Council Bulletin* 41. New York.

Sumner, Colin. 1994. *The Sociology of Deviance: An Obituary.* New York: Continuum.

Sumner, William. 1906. *Folkways.* New York: Vintage.

Sutherland, Edwin. 1934. *Principles of Criminology.* Philadelphia: J.B. Lippincott.

Sykes, Gresham, and David Matza. 1957. "Techniques of Neutralization: A Theory of Delinquency." *American Sociological Review* 22: 664–70.

Tannenbaum, Frank. 1938. *Crime and the Community.* Boston: Ginn.

Turner, Ralph H. 1972. "Deviance Avowal as Neutralization of Commitment." *Social Problems* 19 (Winter): 308–21.

Weatherford, Jack. 1986. *Porn Row.* New York: Arbor House.